Two Hundred Years of Magma Transport and Storage at Kīlauea Volcano, Hawai‘i, 1790–2008

By Thomas L. Wright and Fred W. Klein

Professional Paper 1806

U.S. Department of the Interior
U.S. Geological Survey

U.S. Department of the Interior
SALLY JEWELL, Secretary

U.S. Geological Survey
Suzette M. Kimball, Acting Director

U.S. Geological Survey, Reston, Virginia: 2014

For more information on the USGS—the Federal source for science about the Earth, its natural and living resources, natural hazards, and the environment—visit http://www.usgs.gov or call 1–888–ASK–USGS

For an overview of USGS information products, including maps, imagery, and publications, visit http://www.usgs.gov/pubprod

To order this and other USGS information products, visit http://store.usgs.gov

Suggested citation:
Wright, T.L., and Klein, F.W., 2014, Two hundred years of magma transport and storage at Kīlauea Volcano, Hawai‘i, 1790–2008: U.S. Geological Survey Professional Paper 1806, 240 p., 9 appendixes, http://dx.doi.org/10.3133/pp1806.

ISSN 1044-9612 (print)
ISSN 2330-7102 (online)

Acknowledgments

We are grateful for a broad range of support during the preparation of this long and difficult manuscript. A three-year (2007–2009) U.S. Geological Survey (USGS) Bradley scholarship to the senior author funded completion of the research and intitial writing. Our ideas were refined by dialogues developed over many years with present and former staff members of the Hawaiian Volcano Observatory (HVO). Many concepts about how Kīlauea works were developed in the authors' brains during staff discussions while working at HVO. Both authors appreciate in particular discussions over many years with Bob Koyanagi, who impressed on us the importance of earthquake sequences in interpreting Kīlauea's history. During preparation of the manuscript, Asta Miklius has been particularly helpful in providing information about HVO's computer database (VALVE) and in clarifying for us information obtained from VALVE. Jennifer Nakata helped develop our earthquake classification and has facilitated acquisition of the seismic data from the HVO catalog. Jerry Eaton, Don Swanson, and Dick Fiske (Smithsonian Natural History Museum Mineral Sciences Department) have been sounding boards for discussion of many of our ideas in the formative stages of producing this manuscript. Among non-USGS sources, Emily Montgomery-Brown (Stanford University, University of Wisconsin, and presently Mendenhall fellow at the USGS in Menlo Park) has been particularly helpful in providing and discussing data that clarified our association of suspected deep intrusions with silent earthquakes.The ideas not specifically attributed to published and unpublished sources are our own syntheses of events described in those sources. We greatly appreciate reviews from Dan Dzurisin, John Power, and an exceptionally thorough and incisive review by Jim Kauahikaua, all of which resulted in significant revisions of many parts of the manuscript. Jim Kauahikaua also checked all Hawaiian place names in the manuscript and ensured that their spelling was in conformance with the Geographic Names Information System (GNIS) as of 2010. Finally, we greatly benefitted from Peter Stauffer's thorough editing and appreciate Jeanne DiLeo's design of the book.

Contents

Figures

Tables

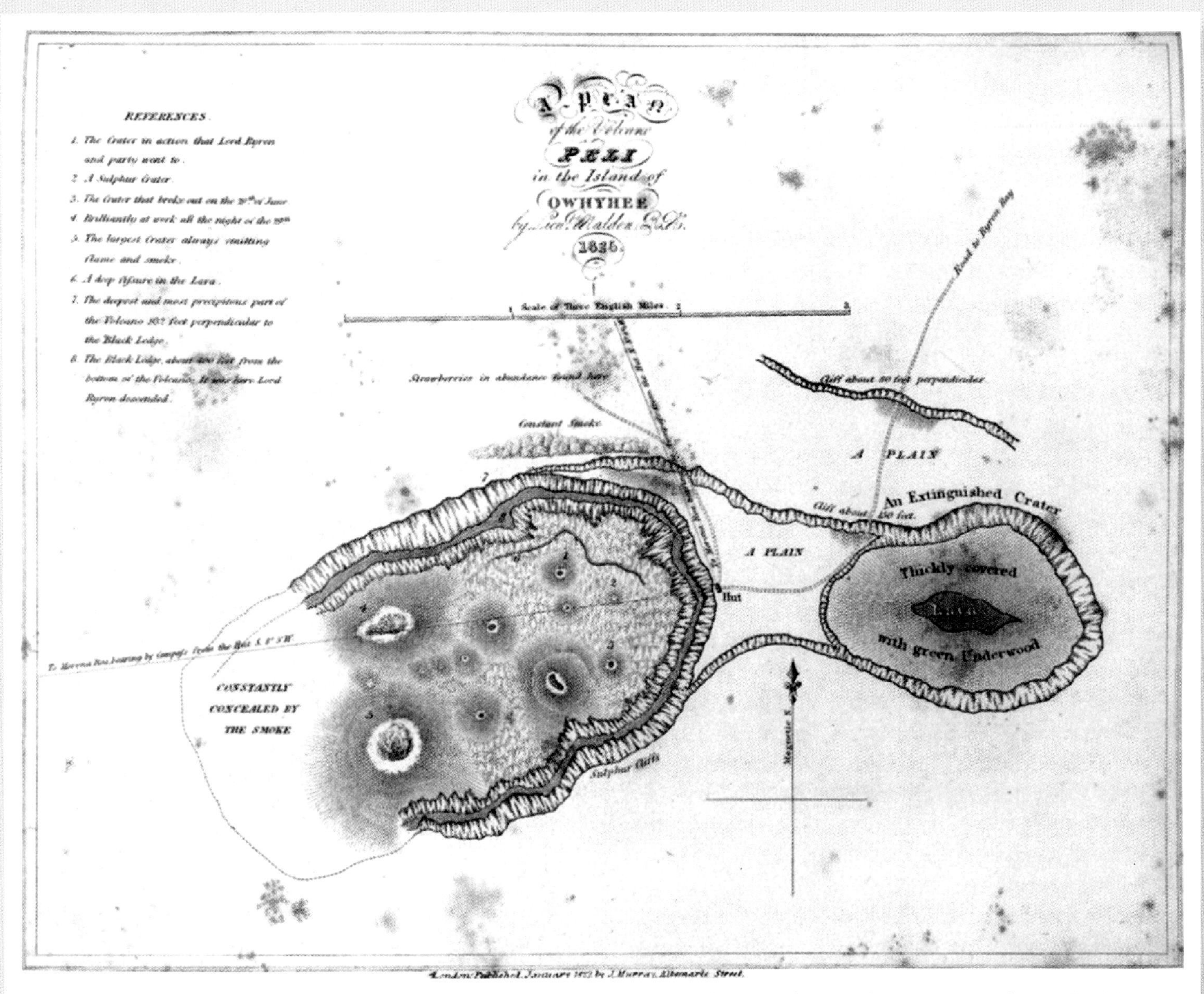

The first map of Kīlauea Caldera, as made by Lieutenant Malden during the visit of Lord Byron in 1825 (Fitzpatrick, G.L., 1986, The early mapping of Hawaii: Honolulu, Editions Limited, v. 1., p. 89).

Two Hundred Years of Magma Transport and Storage at Kīlauea Volcano, Hawai'i, 1790–2008

By Thomas L. Wright[1] and Fred W. Klein[2]

Abstract

This publication summarizes the evolution of the internal plumbing of Kīlauea Volcano on the Island of Hawai'i from the first documented eruption in 1790 to the explosive eruption of March 2008 in Halema'uma'u Crater. For the period before the founding of the Hawaiian Volcano Observatory in 1912, we rely on written observations of eruptive activity, earthquake swarms, and periodic draining of magma from the lava lake present in Kīlauea Caldera. After 1912 the written observations are supplemented by continuous measurement of tilting of the ground at Kīlauea's summit and by a continuous instrumental record of earthquakes, both measurements made during 1912–56 by a single pendulum seismometer housed on the northeast edge of Kīlauea's summit. Interpretations become more robust following the installation of seismic and deformation networks in the 1960s. A major advance in the 1990s was the ability to continuously record and telemeter ground deformation to allow its precise correlation with seismic activity before and after eruptions, intrusions, and large earthquakes.

We interpret specific events in Kīlauea's 200-year written history as steps in a broad transition from summit lava-lake activity in Kīlauea Caldera to shield building on the east rift zone. The ability of the magmatic plumbing to deliver magma to eruption is critical to the history of eruption and intrusion. When the rate of magma supply equals the rate of eruption, there is little ground deformation or intrusion. When the magma supply rate is greater than the rate of eruption, then the edifice responds through any or all of summit inflation, intrusion, increased spreading rate, and large flank earthquakes.

In Kīlauea's 200-year history we identify three regions of the volcano in which magma is stored and supplied from below. Source 1 is at 1-km depth or less beneath Kīlauea's summit and fed Kīlauea's summit lava lakes throughout most of the 19th century and again from 1907 to 1924. Source 1 was used up in the series of small Halema'uma'u eruptions following the end of lava-lake activity in the summit collapse of 1924. Source 2 is the magma reservoir at a depth of 2–6 km beneath Kīlauea's summit that has been imaged by seismic and deformation measurements beginning in the 1960s. This source was first identified in the summit collapses of 1922 and 1924. Source 3 is a diffuse volume of magma-permeated rock between 5 and 11 km depth beneath the east rift zone and above the near-horizontal decollement at the base of the Kīlauea edifice.

Magma distribution within source 2 has been derived by combining petrologic study of the three chemically uniform summit eruptions of 1952, 1961, and 1967–68 and the east rift eruptions within this interval with both observation of migrating centers of inflation determined from leveling surveys conducted before the 1967–68 eruption and with published models of expected deformation from different source geometries. We adopt a model of concatenated magmatic plugs with nodes beneath the inflation centers. Addition of erupted and intruded volumes of the three summit magma batches yields a liquid magma volume of about 0.2 km^3, with dimensions of ~1 km by 1 km by 200 m centered at about 3-km depth within source 2. Following the Halema'uma'u eruption of 1967–68, the chemistry of magma coming into Kīlauea's summit reservoir has changed frequently, and during the eruption that began in 1983, chemical changes have been subtle and continuous. In this period we interpret changes in chemistry as related to an increase in magma supply resulting from increased partial melting in an expanding mantle source volume.

We know from instrumental recording of eruptions since the long Halema'uma'u eruption in 1952 that stress in the edifice accumulates as magma is added underground and is relieved by eruption and by dilation of the rift zones associated with seaward movement (spreading) of Kīlauea's south flank. During

[1]U.S. Geological Survey, Johns Hopkins University, Department of Earth and Planetary Sciences, Baltimore MD

[2]U.S. Geological Survey, 345 Middlefield Rd., Menlo Park, CA

and after the last half of the 20th century, magma transfer to the rift zone has dominantly occurred from source 2. High rates of flank motion have been correlated with high rates of endogenous growth; alternatively, lower rates of motion have characterized periods when the underground magmatic plumbing was being refilled following lateral removal of magma, as well as periods when a more open magmatic plumbing favored continuous eruption.

Since at least 1952, source 3 has not drained during deflations, which was apparently not the case before 1924. Triangulation and leveling conducted in 1912, 1921, and 1926, combined with post-1912 tilt measurements, identified a broad regional uplift in 1918–19 and an equally broad collapse in 1924, neither of which has been seen since. We associate these elevation changes with addition or subtraction of magma from all three magma sources, dominantly source 3. We interpret the intrusion beneath the east rift zone during the 1924 collapse to have stabilized the rift zone-south flank relationship, preventing loss of magma from source 3 in subsequent collapses. Rates of seaward spreading were low until 1952, when earthquakes in 1950 and 1951 associated with surges of magma from the hotspot triggered a large offshore south flank earthquake swarm that unlocked the south flank and enabled a greatly increased rate of seaward spreading.

Magma supply rates have been derived for the entire period of study. Between 1823 and 1840, magma was supplied from source 1 at a very high rate of more than 0.2 km^3/yr, which we interpret as recovery from a substantial draining of magma from beneath Kīlauea in 1790. Inferred magma supply rates diminished to one-tenth of that value after 1840, in part because of increase in the activity of Mauna Loa beginning in 1843. Magma supply rates between 1918 and 1924 were about 0.024 km^3/yr, matching that of the period from 1840 to 1894. During 1950–52 the magma supply rate increased to about 0.06 km^3/yr, in part because of the great reduction in Mauna Loa activity following its large eruption in June 1950. Following the summit eruption of 1967–68, magma supply increased further to ~0.1 km^3/yr, and further increases to more than 0.2 km^3/yr occurred during the east rift eruption that began in 1983.

Eruption at Kīlauea's summit took place in 1952, and eruptive activity steadily increased as increased magma supply also drove increased spreading rates. The inability of magma supply to be accommodated by a combination of eruption and spreading during the 1969–74 Mauna Ulu period stressed Kīlauea's south flank. The stress was relieved in part by the *M*7.2 earthquake of 29 November 1975. That earthquake, in turn, dilated Kīlauea's east rift zone as the south flank moved seaward, producing a favorable condition for continuous east rift eruption, which began in 1983. The 1975 earthquake also resulted in the ability of the south flank to move independently under the influence of gravity, effectively decoupling the spreading rate from changes in the magma supply rate. The continuing increase in magma supply after 1983 was instead manifested in rift dilation, increased intrusion, and ultimately in the launching of a second eruption in Halema'uma'u in March 2008, the first instance in Kīlauea's recorded history of simultaneous eruption at the summit and on the east rift zone.

Kīlauea's history can be considered in cycles of equilibrium, crisis, and recovery. The approach of a crisis is driven by a magma supply rate that greatly exceeds the capacity of the plumbing to deliver magma to the surface. Crises can be anticipated by inflation measured at Kīlauea's summit coupled with an increase in overall seismicity, particularly manifest by intrusion and eruption in the southwest sector of the volcano. Unfortunately the nature of the crisis—for example, large earthquake, new eruption, or edifice-changing intrusion—cannot be specified ahead of time. We conclude that Kīlauea's cycles are controlled by nonlinear dynamics, which underscores the difficulty in predicting eruptions and earthquakes.

Halema'uma'u lava lake in 1894. Photograph by D.H. Hitchcock. Image courtesy of Hawaii National Park.

Chapter 1

Introduction

Introduces Kīlauea Volcano and its history by means of index maps, a summary of previous work, a three-fold classification of eruptions and intrusions, explication of our methodology for presentation of seismic and tilt data, and a discussion of our approach to studying the historical data.

Appendix A at the end of the printed book supplements this chapter. Appendixes B–I are available only in the digital versions of this work—on the DVD accompanying the book and in the online version. Appendixes B–H provide for chapters 2–8 expanded views of time series and map plots of all earthquake swarms listed in event tables. Finally, appendix I outlines the Mogi method for determining inflation and deflation centers and presents many examples of its application.

Our present understanding of Kīlauea Volcano has been largely derived from 100 years of instrumental monitoring and detailed observations conducted at the Hawaiian Volcano Observatory (HVO), presently and through much of its history as part of the U.S. Geological Survey. Descriptions of Kīlauea, however, extend back more than 200 years, beginning with native Hawaiian accounts of the explosive eruption of 1790. One purpose of this paper is to join our modern understanding with these historical descriptions to determine magmatic, tectonic and seismic processes that have acted within Kīlauea subsequent to 1790.

Descriptions dating from the 19th century are qualitative, becoming more quantitative with the founding of HVO in 1912. Before 1912, we know the time and place of most eruptions, times of deflation of Kīlauea's summit, as evidenced by draining of the active lava lake, and the times of occurrence of large, felt earthquakes, including some earthquake swarms associated with magma withdrawal and intrusion beneath Kīlauea's rift zones. While recognizing the limitations imposed by the nature of these descriptions, the thesis of this paper is that the volcanic and seismic events, beginning with an explosive eruption in 1790 and ending with the east rift zone and summit eruption continuing today, can be connected by causes and effects inferred at Kīlauea using modern instrumental observations. Our interpretations are in large part original, being a synthesis of information from a range of time periods and not confined to a restatement of already published conclusions. Our interpretations are presented as hypotheses that, while not rigorously testable with additional data, can challenge present and future workers to think more broadly and synoptically about volcanic process at Kīlauea than can be done using only the more quantitative measurements being made today.

Previous Work on Kīlauea Volcano

Our study builds on a body of previously published work. For the period before the founding of HVO in 1912, we rely on original published accounts of volcanic activity compiled and annotated as part of the Hawaiian bibliographic database (Wright and Takahashi, 1998)[3]. This database includes information found in early newspapers and magazines, as well as references to missionary diaries and other unpublished accounts of volcanic and seismic activity in Hawai'i. Early observations of Kīlauea and Mauna Loa were summarized in books by Hitchcock and Brigham (Brigham, 1909; Hitchcock, 1909). Especially important to us in interpreting the volcanic and seismic history are the series of papers by J.D. Dana published in the American Journal of Science (Dana, 1887a; Dana, 1887b; Dana, 1888).

The early work of HVO has been compiled in two publications (Bevens and others, 1988; Fiske and others, 1987). These works were compiled from summaries of volcanic activity issued weekly or monthly between 1912 and 1955. Beginning in 1950 yearly summaries were published covering all activity of the Hawaiian volcanoes (Finch and Macdonald, 1953; Macdonald, 1954; Macdonald, 1955; Macdonald, 1959; Macdonald and Eaton, 1955; Macdonald and Eaton, 1957; Macdonald and Eaton, 1964; Macdonald and Wentworth, 1954). Seismic summaries were published as administrative reports from 1956 to 2006 and later; yearly summaries date from 1974 and deformation summaries were appended for 1988–90. The summaries are available as open-file reports that can be downloaded from the U.S. Geological Survey publication warehouse at: http://pubs.er.usgs.gov/.

HVO's monthly or bimonthly unpublished reports summarizing contemporary eruptive and intrusive activity began with a description of the early stages of the 1967–68 summit eruption. The reports have continued to the present and are a valuable unpublished archive of the results of instrumental monitoring and eruption observation, and we have used the monthly reports as an essential resource during preparation of this paper. The citations are formatted as personal communications.

In many instances our observations may include data overlooked in the monthly reports, and our interpretations may differ from those in the monthly reports. In writing about events close in time to the preparation of the report, the HVO staff at any given time (including the authors of this paper) cannot know what lies ahead and therefore necessarily focus on only the most obvious changes in seismic activity and ground deformation. Our synoptic view looks at the same periods of time with the added benefit of being able to interpret the same events within a context that recognizes both what happened after as well as what happened before. In cases where we can identify the differences between our summary and what is published in the monthly reports the differences reflect errors of omission, and our new interpretations build on rather than contradict anything published in the monthly reports.

The history of Kīlauea's seismic network has been published (Klein and Koyanagi, 1980). Hawai'i earthquake data for the period prior to 1959 have been compiled and published by the present authors (Klein and Wright, 2000). Hawai'i earthquake data obtained by HVO since 1959 are available from the composite ANSS catalog: http://www.ncedc.org/anss/; http://www.ncedc.org/anss/anss-caveats.html.

[3]The database is available for download from ftp://ftpext.usgs.gov/pub/wr/=hi/Kīlauea/Hawai'ibibliography.public. It is updated every four months and may be opened using the EndNote bibliographic database program. Instructions are given in the downloaded ReadMe and info files.

Kīlauea's seismicity and tectonics have been interpreted in two landmark papers (Klein and Koyanagi, 1989; Klein and others, 1987). A recent paper covering the period between the 1967–68 summit eruption and the *M*7.2 south flank earthquake (Wright and Klein, 2008) introduced the classification of eruptions and intrusions and the methodology of plotting seismic datasets used in the present study. A new methodology for earthquake classification is adopted from a paper covering deep magma transport at Kīlauea (Wright and Klein, 2006).

History of Instrumental Monitoring at Kīlauea

Ground deformation data began to be acquired at Kīlauea in the early twentieth century, with triangulation of the east part of the Island of Hawai'i completed in 1914 (Swanson and others, 1976a). In 1912, a Bosch-Omori long-period seismometer that doubled as a tiltmeter was installed in the Whitney Vault on the northeast rim of Kīlauea Caldera (Mitchell, 1930; Powers, 1946). The great subsidence of Kīlauea Caldera in 1924 produced large ground motions and proved the utility of both tilt and triangulation to monitor such changes (Powers, 1946; Wilson, 1935).

In the modern era, surface deformation measurements are critical to our understanding of Kīlauea's magmatic plumbing. Measurement of vertical uplift or subsidence, horizontal extension or contraction, and outward- or inward-directed ground tilt made during cycles of inflation and deflation define the depth(s) and location(s) of magma storage beneath Kīlauea's summit. The same measurements can be used to infer whether Kīlauea's magmatic system has changed over time. Finally, we use deformation data to quantify rates of magma supply and seaward spreading.

Seismology in Hawai'i began with installation of a seismometer at Honolulu in 1903 (Klein and Wright, 2000), followed within a decade by installation of a single seismometer of Bosch-Omori design housed in the Whitney Vault (Wood, 1912) coincident with the founding of HVO (Jaggar, 1912). Up until the mid-1950s the HVO network consisted of no more than five seismometers on the island, only three of which—Whitney Vault, Hilo, and Mauna Loa—were close enough to be useful in estimating locations of Kīlauea earthquakes. Earthquake locations and magnitudes were recorded in publications listed under "Previous work."

A modern seismic network was established at Kīlauea in the mid-1950s (Eaton, 1986a; Eaton, 1986b; Klein and Koyanagi, 1980) Over the following decades the seismic network was expanded (Klein and Koyanagi, 1980). In the early 1960s, smoked paper recording of a handful of stations connected to the observatory by overland cable was the standard. In 1967, the expanding network with radio telemetry recorded data on 16-mm film with a Develocorder. In 1979, an Eclipse mini-computer capable of digitizing the 1-inch analog magnetic tape used to record most of the 44-station network was introduced. This was the time when picking seismic traces on a computer screen improved the earthquake location quality. The Caltech-USGS Seismic Processing (CUSP) real-time digitizing and timing system came online in 1985, and the Earthworm system replaced CUSP in the mid 2000s for greater catalog completeness and precision compared to the predigital years. Maps of the current seismic network are given in numerous recent papers (for example, Got and others, 2008).

A modern geodetic network was established in the late 1950s (Eaton, 1986a; Eaton, 1986b). A network of water-tube tiltmeters was installed to record ground deformation (Eaton, 1959), including one short-base water-tube tiltmeter installed in the Uwēkahuna Vault on the northwest rim of Kīlauea Caldera that was read daily beginning in 1956. A similar tiltmeter was installed in the Whitney Vault, where it was shown to satisfactorily track both the daily tilt read from the Bosch-Omori seismometer for the five years preceding abandonment of the Whitney Vault in 1962 as well as the daily tilt read from an identical tiltmeter installed in Uwēkahuna Vault (Eaton, 1986a, p. 13–14). This 5-year period of redundancy is important, because it allows us to more confidently interpret the nearly 40 years of seismometric tilt measured when the Whitney Vault was the only installation (see discussion of data reduction in appendix A).

The geodetic network was also steadily expanded after the 1950s (Decker and others, 2008). Vertical (leveling networks) and horizontal electronic distance measurements (EDM) were made periodically to augment the tilt measurements. From the mid-1950s, it has been possible to specify rather closely the seismic and geodetic response of Kīlauea across cycles of eruptions, intrusions, and large earthquakes. An important limitation of the earlier deformation studies was the lack of continuous measurement. Precise data could be gathered soon after an important eruption or earthquake, but one had to be lucky to make measurements just before such an event. Global Positioning System (GPS) geodesy, which measures both vertical and horizontal changes, was begun in the late 1980's (Decker and others, 2008; Dvorak and others, 1989). When continuously recording GPS networks were installed in the 1990s (Decker and others, 2008) it was possible for the first time to view correlated seismic and geodetic activity through an entire eruption or earthquake cycle.

A Tectonic Context for Interpreting Ground Deformation and Seismicity

Figure 1.1 summarizes the elements important to an understanding of the evolution of Kīlauea. In plan view Kīlauea is built on the south flank of Mauna Loa (fig. 1.1*A*), and its history is therefore intimately tied to the presence and activity of Mauna Loa. Additional tectonic features found offshore (fig. 1.1*B*) are two parts of a raised platform ("midslope and outer benches") associated with an early landslide called the Hilina slump (Lipman and others, 2001; Morgan and others, 2003) and a linear topographic feature called the Pāpa'u Seamount. Recent bathymetry around the Island of Hawai'i shows additional offshore features (fig. 1.1*C*), including the continuation of Kīlauea's east rift zone, which extends more than 50 km beyond the shoreline, and Hawai'i's newest active volcano, "Lō'ihi Seamount." Kīlauea's long east rift zone contrasts with the much less active southwest rift zone, which extends less than 1 km beyond the shoreline. Mauna Loa, in contrast, has a stunted northeast rift zone, its active part extending only about 10 km from the summit, and a southwest rift zone that extends for about 40 km beyond the shoreline.

Important tectonic elements of Kīlauea are (1) its summit caldera containing a nested pit crater, Halema'uma'u, (2) rift zones extending to the east and southwest connected by the Koa'e Fault Zone, (3) an immobile north flank buttressed by Mauna Loa, and (4) a mobile south flank, described in the next section, extending seaward of the Koa'e Fault Zone and rift zones. The traditionally defined southwest rift zone (SWR) is defined by eruption centers, pit craters, and open cracks that represent eruptive activity over the last hundreds or thousands of years. An additional important tectonic element is the "seismic southwest rift zone" (SSWR), which geographically forms a mirror image of the east rift zone. The SSWR is defined by earthquake swarms that trace the path of numerous recent intrusions on the southwest side of the volcano (Klein and others, 1987; Wright and Klein, 2006), and extends south from Kīlauea Caldera, then trends southwest to join the traditional southwest rift zone in the vicinity of the Kamakai'a Hills (fig. 1.1*A*).

In an oblique cutaway view (fig. 1.1*D*, modified many times from plate 1 of Ryan, 1988) we show a shallow magma reservoir beneath Kīlauea's summit, a magma conduit connecting the site of magma generation within the Hawaiian mantle source region to the shallow magma chamber, a subhorizontal decollement (see, for comparison, Ando, 1979) at the base of the volcanic pile (10–12 km depth) along which the mobile south flank is spreading seaward, and a magma system that extends beneath the rift zones to the decollement (Delaney and others, 1990). Recent offshore geophysical studies have shown that the distal portion of the midslope bench ("outer bench" in fig. 1.1*C*) acts as a barrier to free movement of Kīlauea's south flank (Morgan and others, 2000, 2003; Phillips and others, 2008).

Kīlauea's shallow magma reservoir is located at the point where the density of the rising magma matches the density of the surrounding rocks (Ryan, 1987). Both seismic and deformation studies infer a magma storage zone 2–6 km beneath Kīlauea's summit with a diameter of about 2–4 km (Koyanagi and others, 1974; Tilling and Dvorak, 1993). More recent work has demonstrated the existence of magma pockets at depths of less than 1 km as part of the feeding system for the east rift eruption that began in 1983 (Almendros and others, 2002a).

Figure 1.1. Index maps and section of Kīlauea Volcano and its setting on the Island of Hawai'i. ***A***, Shaded relief map of Kīlauea and part of Mauna Loa (adapted from Moore and Mark, 1992) showing tectonic features of Kīlauea Volcano and names of features mentioned in the text. Rift zones (red) marked by open fractures, cinder cones, and pit craters extend to the east and southwest. The east rift has been far more active than the southwest rift within historical time. The seismic southwest rift zone (seismic SWRZ) is the locus of intrusions and is a mirror image of the active east rift zone. The Koa'e Fault Zone (orange) is a zone of tensional faulting connecting the active rift zones. The small double arrow shows dilation within the Koa'e. Kīlauea's flank, south of the Koa'e and the two rift zones, is spreading seaward (single arrows) under the pressure of continuing magma supply aided by gravity. Movement occurs principally along the decollement at 12-km depth (see part *D*). The south flank's Hilina Pali fault system consists of normal faults formed as a gravitational response to spreading. ***B***, Shaded relief map showing onshore terrain, with added submarine features mentioned in the text. Notable among these is the Pāpa'u Seamount and the Hilina slump, an older submarine landslide from Kīlauea. Within the slump are a midslope bench and an outer bench, the latter acting as a barrier to south flank spreading. ***C***, Shaded relief map (taken from Eakins and others, 2003) with additional colored bathymetry and tectonic features near the distal ends of Kīlauea Volcano. Hawai'i's newest volcano, Lō'ihi submarine volcano, is also shown. ***D***, Oblique cut-away view of the Kīlauea plumbing system (adapted from Tilling and others, 1987). A near-horizontal landward-dipping basal decollement is the principal surface along which seaward spreading occurs away from the Mauna Loa buttress. Spreading is associated with dilation of the two rift zones creating room to house a deep magma system within the Kīlauea edifice. Magma is supplied from the mantle at depths greater than 60 km through a subhorizontal deep seismic zone at depths of about 25–35 km connected to a vertical conduit extending to 20 km beneath Kīlauea's summit to feed a magma chamber at 2–6 km depth that contains a liquid core.

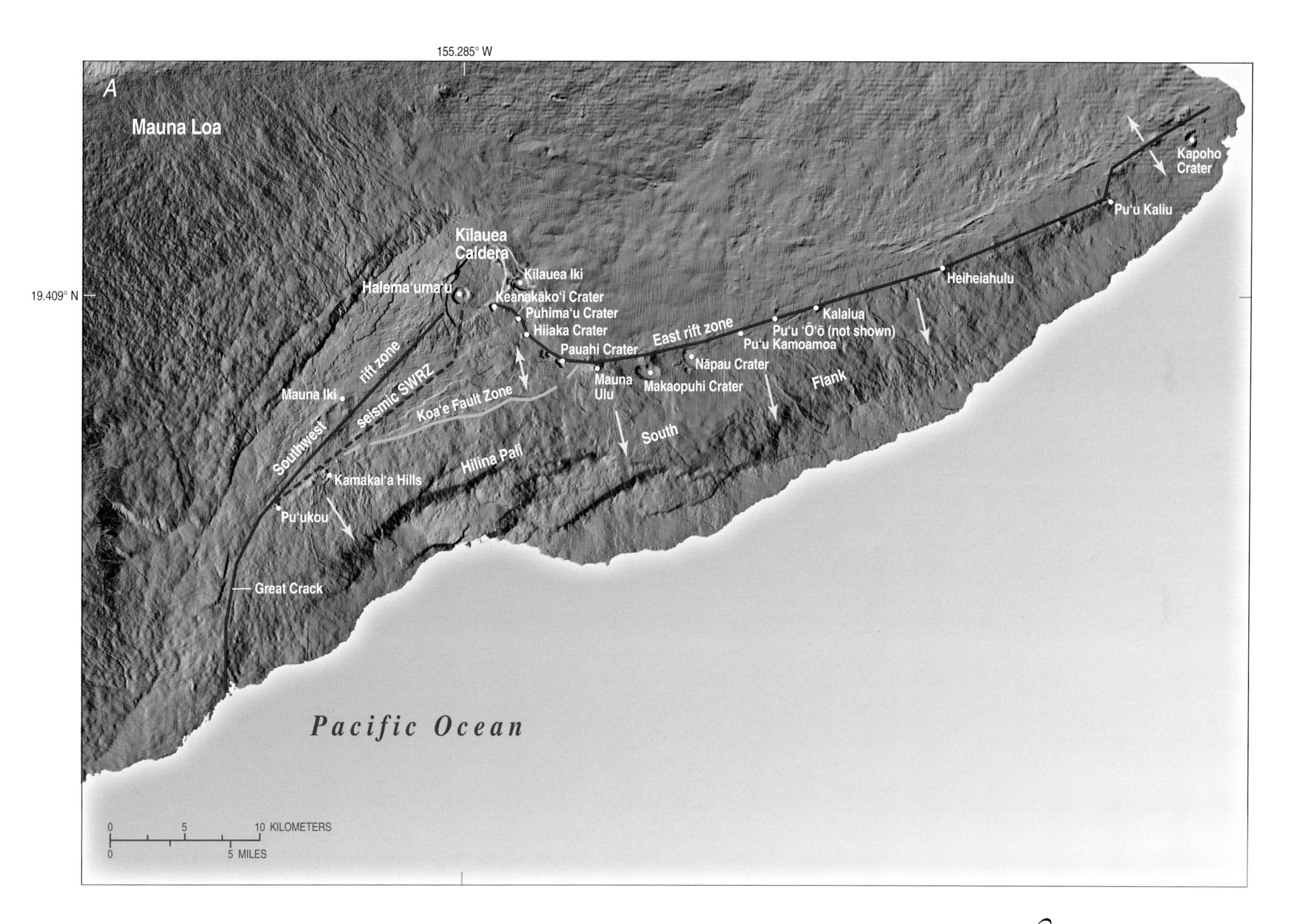

155.285° W
19.409° N
A
Mauna Loa
Kīlauea
Caldera
Halema'uma'u
Kīlauea Iki
Keanakāko'i Crater
Puhima'u Crater
Hiiaka Crater
Pauahi Crater
East rift zone
Mauna
Ulu
Makaopuhi Crater
Nāpau Crater
Pu'u Kamoamoa
Pu'u 'Ō'ō (not shown)
Kalalua
Heiheiahulu
Pu'u Kaliu
Kapoho
Crater
Flank
South
Koa'e Fault Zone
seismic SWRZ
rift zone
Southwest
Mauna Iki
Hilina Pali
Kamakai'a Hills
Pu'ukou
Great Crack
Pacific Ocean
0
5
10 KILOMETERS
0
5 MILES

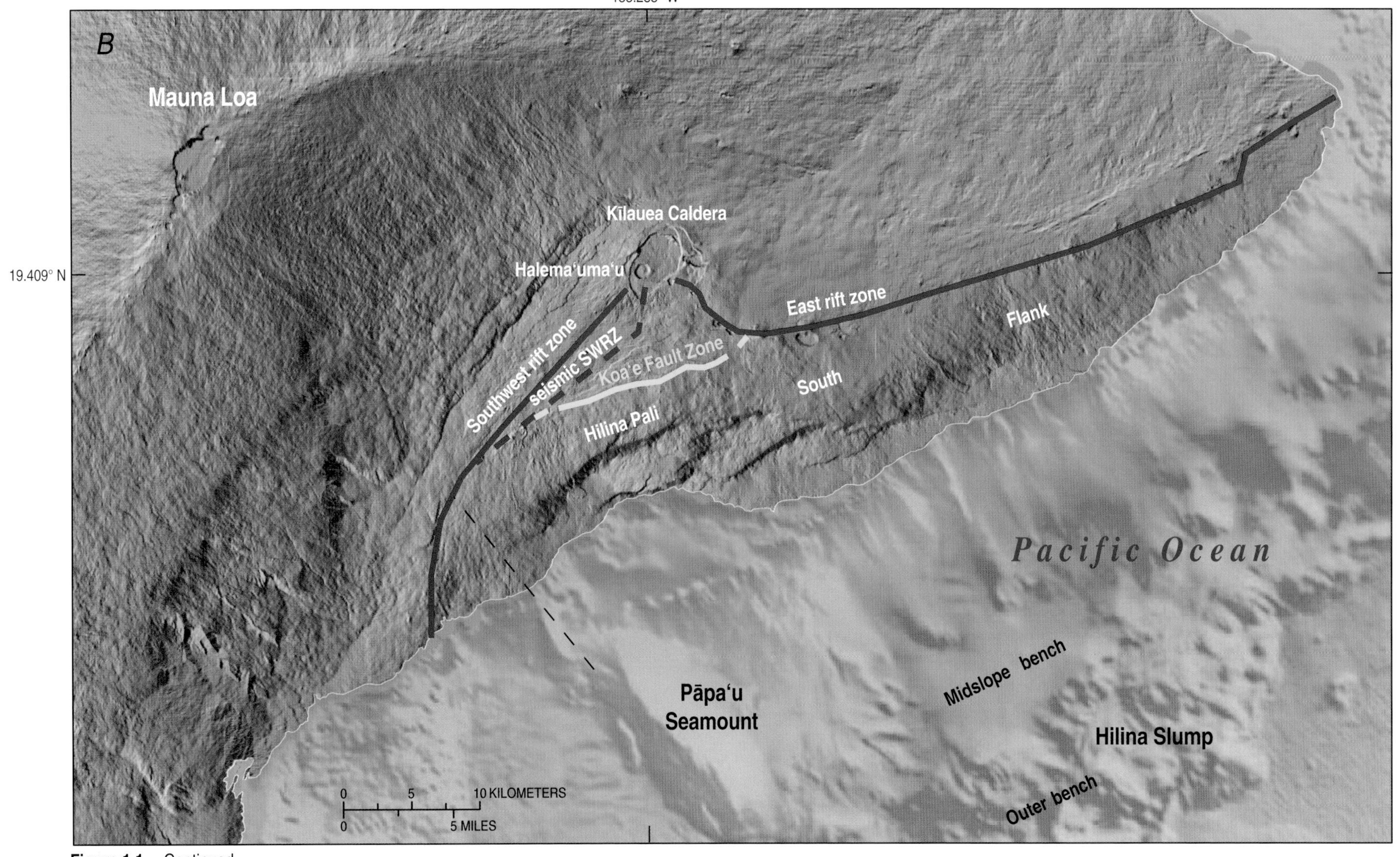

Figure 1.1.—Continued

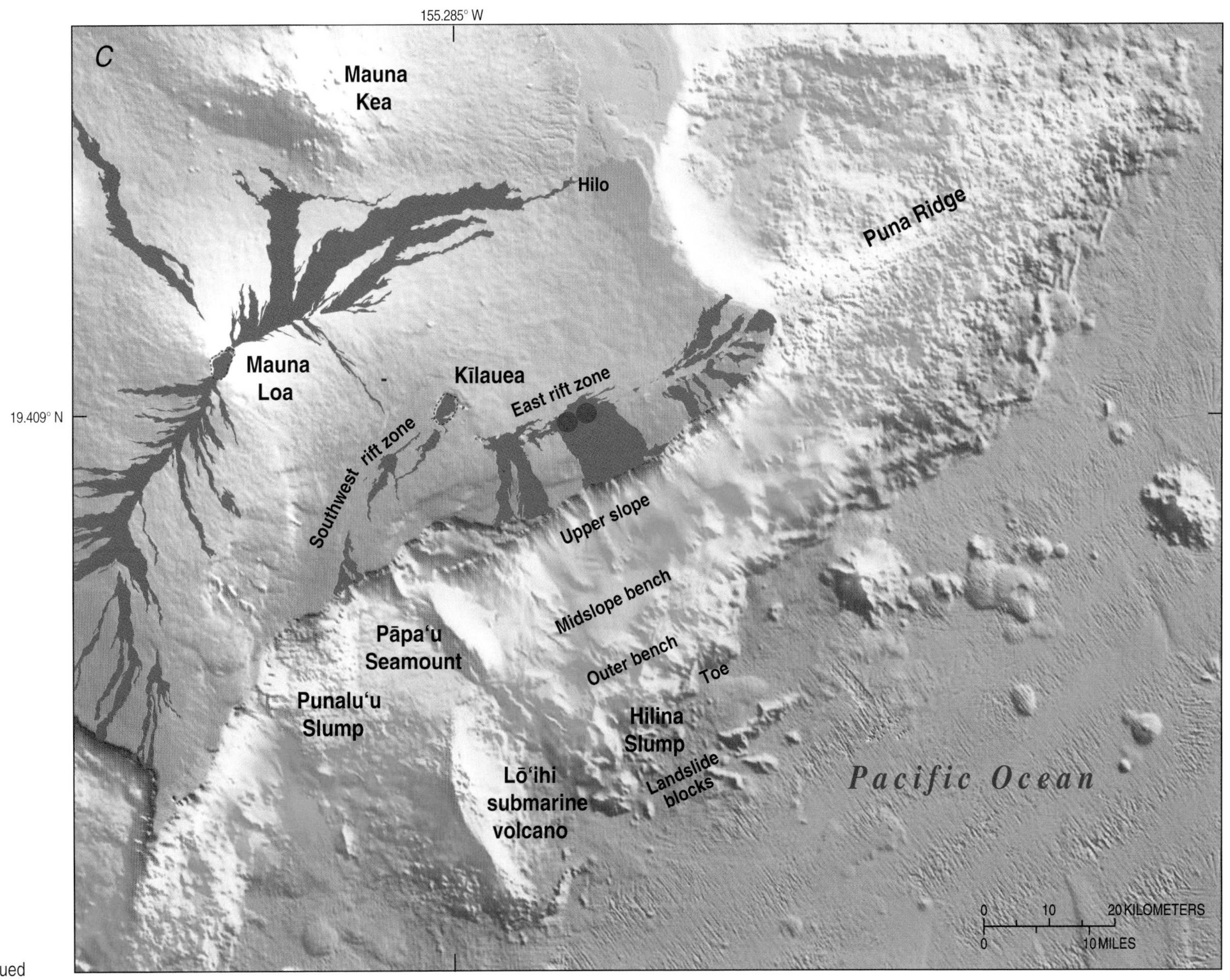

Figure 1.1.—Continued

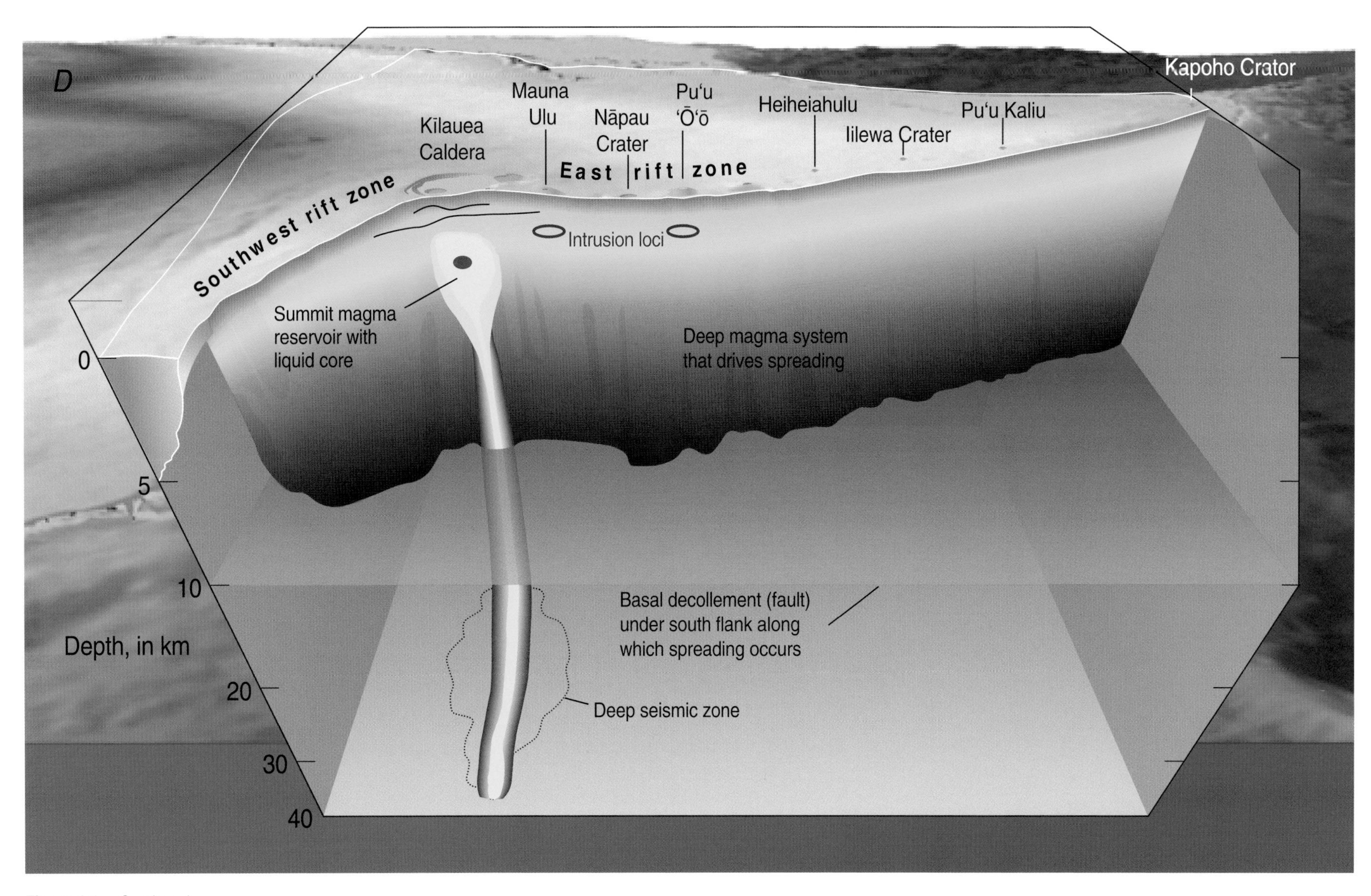

Figure 1.1.—Continued

Active Seaward Spreading of Kīlauea's South Flank

Seaward spreading of Kīlauea's south flank was recognized in the early 1970s with data obtained from an expanded geodetic monitoring network (Swanson and others, 1976a). A surface along which spreading took place was located at the base of the Kīlauea edifice from analysis of the focal mechanism of the *M*7.2 south flank earthquake that occurred on November 29, 1975 (Ando, 1979). Subsequent studies have further quantified spreading rates and give support to the idea originally expressed by Swanson and others (1976a) that spreading was distributed among several subhorizontal planes over a depth range of several kilometers (see, for example, Arnadottir and others, 1991; Morgan and McGovern, 2005b; Morgan and others, 2003). Theoretical models of volcano spreading have also been applied to Hawai'i (Borgia, 1994; Borgia and Treves, 1992). We address whether seaward-directed south flank spreading has occurred throughout Kīlauea's history and, during the time period(s) during which spreading has occurred, we attempt to quantify the spreading rate.

Gravity

A regional gravity study (fig. 1.2*A*; Kauahikaua, 1993, figure 1.3) depicts a Bouguer gravity high centered on Kīlauea's Caldera, indicating that Kīlauea's summit magma system has remained in place over the lifetime of the volcano (Kauahikaua and others, 2000). Linear highs on the Bouguer gravity map extend to the east and southwest. To the east, there are gravity lobes both for the present and for a "fossil" rift to the north (Swanson and others, 1976a). The dominant eastern gravity lobe is curved near Kīlauea's summit, becoming straight east of the bend in the east rift zone. It lies north of the east rift zone (fig. 1.2*A*), merging with the rift zone within a few kilometers of where the rift zone disappears offshore. The curvature of Kīlauea's east rift zone, as well as the distance between the gravity high and the present position of the east rift zone, is a measure of accumulated southward dilation and translation of the rift zone during spreading of Kīlauea's south flank. The coincidence of the eastern gravity high with the lower east rift zone reflects the disappearance of Mauna Loa as a buttress and is consistent with seismicity and dilation being observed both north and south of the rift zone (see also Zablocki and Koyanagi, 1979).

The relation between the southwestern gravity high and the southwest rift zone tells a different story. The gravity high lies mostly west of the seismic southwest rift zone, and ends at the landward extension of the Pāpa'u lineament. The Kamakai'a Hills and Yellow cone (Pu'ukou) lie towards Kīlauea's summit from the end of the gravity anomaly and are composed in whole or in part of highly fractionated Kīlauea lava (Wright and Fiske, 1971). The coincidence of the end of the gravity high and the disappearance of fractionated lava to the southwest (Wright, 1971) defines the southwestern extent of Kīlauea's magma system, estimated from intrusions to lie at depths between 2 and 5 km. The southwest continuation of the rift zone, known as the "Great Crack," is a shallow feature and suggests that Kīlauea might originally have formed within an asymmetric landslide scar—steep on the west side, flattening toward the east—associated with the larger Mauna Loa Volcano. The two rift zones are thus a bit different: the eastern gravity high marks the fossil rift north of the present volcanic and seismic rift, and the southwest gravity high in the upper rift coincides with the seismic southwest rift, which is displaced south of the volcanic rift (fig. 1.2*A*).

Seismicity as an Index of Stress Release

Kīlauea's early seismic history is known to a completeness level of *M*4–*M*5.2. Many *M*4 earthquakes are reported as far back as the early 19th century (Klein and Wright, 2000, 2008). Earthquake swarms and earthquakes of magnitude smaller than 4 were often reported and are also noted in the catalog (Klein and Wright, 2000, 2008). Following the start of the Lyman earthquake diaries (Wyss and others, 1992b) in 1833, all *M*5.2 and larger earthquakes are probably known. *M*5.2 corresponds to Mercalli shaking intensity V (at which sleepers are awakened). Magnitudes are estimated from felt intensity information (Klein and Wright, 2008). The modern interpretive seismic history (Klein and others, 1987) identifies shallow earthquake swarms associated with eruptions and intrusions, deep earthquake swarms associated with the magma supply path (Klein and others, 1987; Wright and Klein, 2006) and swarms beneath Kīlauea's south flank caused by the pressure of incoming magma (Dvorak and others, 1986; Klein and others, 1987; Wright and Klein, 2008). Throughout this paper we use seismicity as an indicator of stress release within the Kīlauea edifice in response to magma pressure applied at different depths (compare with Cayol and others, 2000). Rift seismicity is a direct response to magma on its way to eruption or intrusion, but flank seismicity is a nonlinear function of applied stress rate (Dieterich and others, 2000). As applied specifically to rift zone intrusions, seismicity is a useful but imperfect measure of the stress applied to the flank, because some magma movement may be aseismic and magma movement at different depths may generate a different stress response.

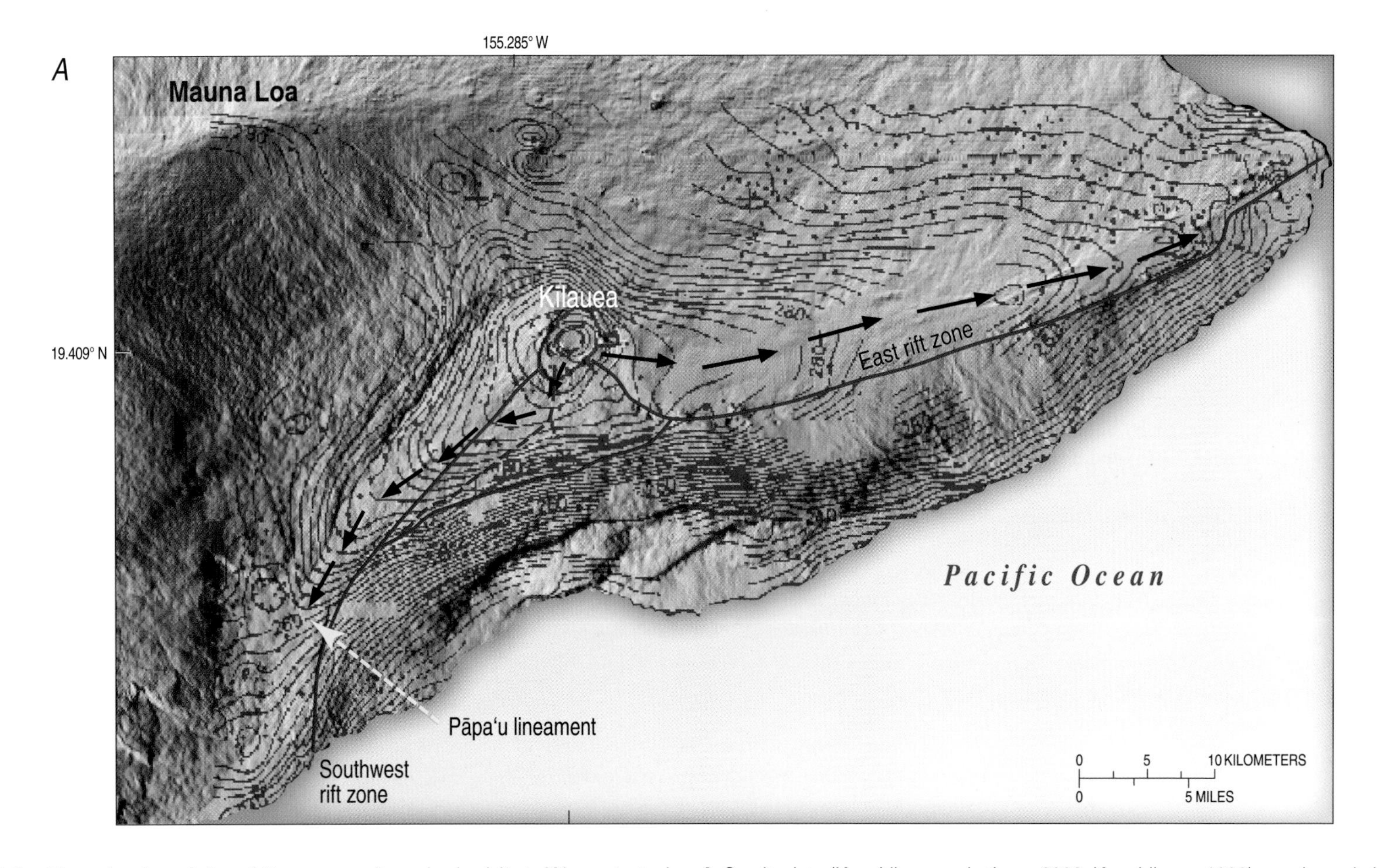

Figure 1.2. Maps showing relation of Bouguer gravity and seismicity to Kīlauea tectonics. ***A***, Gravity data (Kauahikaua and others, 2000; Kauahikaua, 1993) are shown in blue on top of Kīlauea topography. A gravity high is centered in the northeastern part of Kīlauea Caldera. Gravity contours to the southwest initially follow the seismic southwest rift zone (fig. 1.1*A, B*) and then lie west of the seismic southwest rift, terminating west of the southwest rift zone. Gravity contours to the east show a maximum to the north of Kīlauea's east rift zone, joining the rift zone at its eastern end. ***B***, Epicenter maps for short-period (brittle failure) Kīlauea earthquakes in four depth ranges. The plotted earthquakes cover the period from 1970 through 2007. The 2.0 minimum magnitude for plotting means that small, poorly located events and those too small to have magnitude information are not plotted. The locations are standard hypoinverse locations from the HVO catalog. Earthquakes tagged as LP (long period) are excluded from this plot: these LP events often have emergent arrivals, are difficult to locate, and therefore would blur volcanic features if included in these maps. The four depth ranges are (1) 0–5 km: this includes the active and shallow rift zone magma conduits, shallow earthquakes above the magma reservoir under Kīlauea Caldera, and the shallowest events in the south flank and Ka'ōiki zones; (2) 5–15 km: the Kīlauea south flank, Ka'ōiki, and Hīlea zones that are active with large (up to *M7*) earthquakes are in this depth range, where the decollement and flank spreading are located; (3) 15–20 km, this zone is a minimum of seismicity except for the vertical magma conduit under Kīlauea Caldera; (4) 20–60 km; the upper-mantle magma conduit is visible as a cloud of earthquakes displaced just south of the caldera and centered near 30-km depth, plus deeper earthquakes of the conduit system in the lower left of the plot frame. Many of these deep earthquakes are caused by bending stresses in the lithosphere under Hawai'i and are localized near the deep magma conduits.

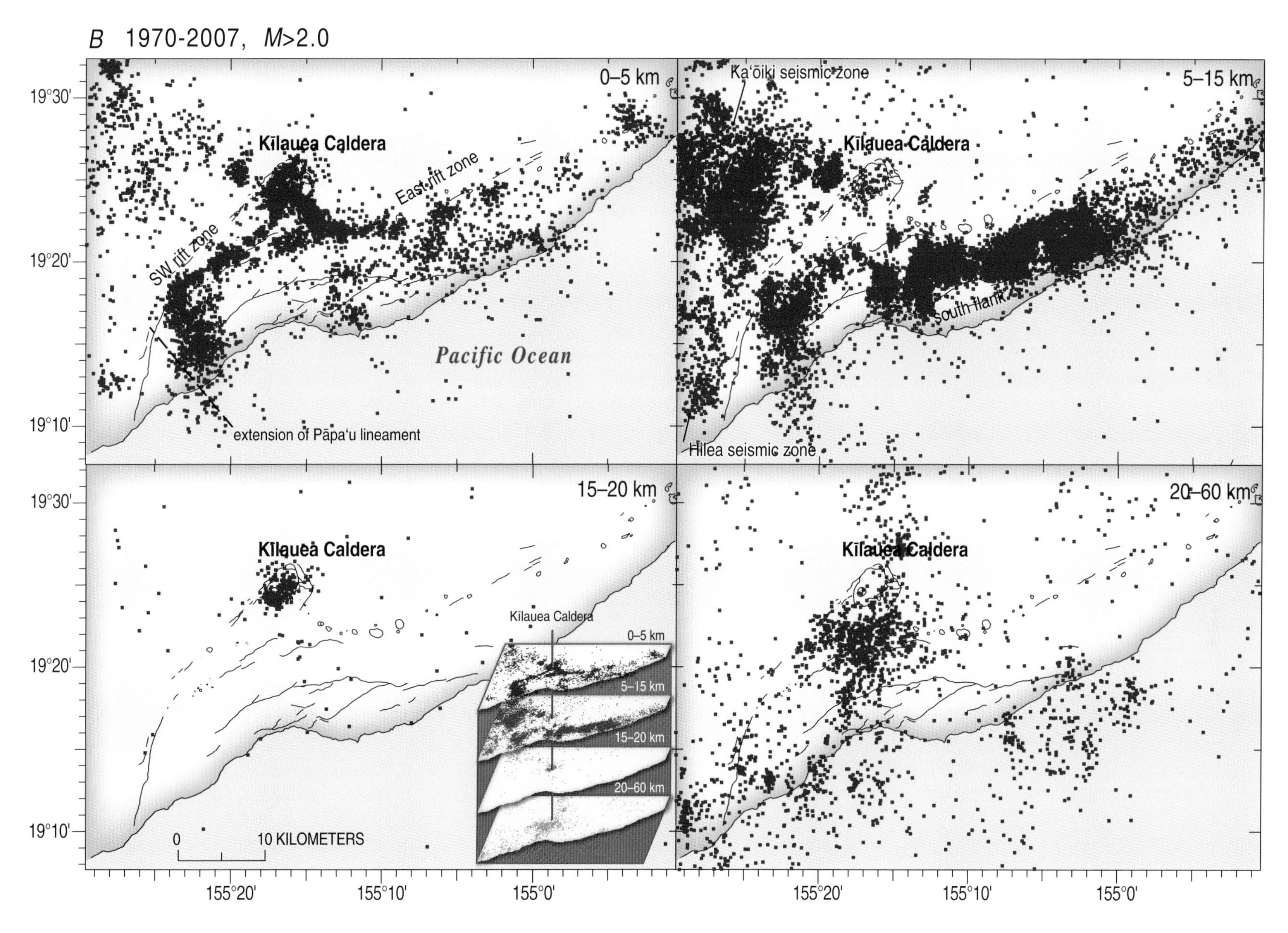

Figure 1.2.—Continued

The general distribution of earthquakes beneath Kīlauea is shown for four depth ranges (fig. 1.2*B*). The primary seismic features are (1) eruption- and intrusion-related earthquakes between 2 and 4 km depth below rift zone magma conduits, (2) active flank and decollement earthquakes between the surface and the decollement, (3) earthquakes associated with a vertical magma conduit under Kīlauea Caldera between 7 and 20 km deep, and (4) earthquakes associated with a deeper and wider conduit plus flexure earthquakes at depths below 20 km. The earthquake distribution is shown for the very active period from 1970 through 2007 (fig. 1.2*B*). Swarms associated with eruptions and intrusions (fig. 1.2*B*, 0–5 km), and along the magma supply path (fig. 1.2*B*, 5–20 km beneath Kīlauea Caldera and extending south-southwest at depths greater than 20 km) are shown for the same time period. Flank swarms are mainly concentrated between depths of 5 and 15 km and occur in response to stress generated by upward and lateral movement between the magma source and sites of storage, eruption, and intrusion. Identical patterns characterize periods when no swarm activity occurs, suggesting that the general stress distribution within Kīlauea is nearly constant over time.

Seismic moment is calculated according to the moment-magnitude formulation for Hawai'i (Zuniga and others, 1988). Seismic moment is accumulated for the different regions and plotted as time series. Changes in the rate of moment accumulation reflect the level of seismic activity within the different tectonic regions. In our interpretations, particular attention is paid to changes in moment rate for periods not affected by earthquakes of magnitude greater than 5.

Eruptions and Intrusions

Translating the historical and geologic record into an understanding of magmatic history and volcano tectonics is based on the recognition that Kīlauea has been built by two different processes, both of which involve solidification of magma. "Exogenous growth" occurs by addition of lava flows to Kīlauea's surface. "Endogenous growth" occurs through subsurface magmatic intrusion. We presume that exogenous growth was dominant during Kīlauea's very earliest history, before the development of rift zones and a shallow magma storage reservoir beneath the summit. Once these features formed, affording the means to store magma in the edifice, endogenous growth became increasingly important and in the modern history is often the most frequent endpoint of magma transport to the rift zones.

During historical time, eruptions have occurred at the summit and along either rift zone. Intrusions may accompany eruptions or occur independently without associated eruption. During transport away from the summit reservoir, small pockets of magma can be isolated beneath the rift zones, where they cool and fractionate (Wright and Klein, 2006, figure 1.3; Wright and others, 1975). Spreading dilates the rift zones, and an additional volume of magma goes to fill the room created during spreading (see, for example, Cayol and others, 2000).

We recognize three different styles of eruption, "traditional," "episodic," and "sustained" (Wright and Klein, 2008).

Traditional eruptions are short-lived, usually lasting a few days to weeks, and are characterized by rapid transfer of magma causing a pronounced downward tilting (deflation) at Kīlauea's summit. Succeeding eruptions occur at a different location. A series of six traditional eruptions occurred between September 1961 and December 1965.

Episodic eruptions return to the same vent location many times at intervals of days to weeks; individual episodes are short, lasting hours to days, and are also accompanied by sharp deflation of Kīlauea's summit. The first 8 months of the Mauna Ulu eruption in 1969 is one example.

Sustained eruptions occupy one vent for a continuous period lasting for months or years, generally feeding a lava lake bounded by either a preexisting crater or by its own levees. Continuous eruption at Mauna Ulu between January 1970 and June 1971 is one example.

We recognize three types of intrusion, "traditional," "inflationary" and "suspected deep" (Klein and others, 1987; Wright and Klein, 2008).

Traditional intrusions are marked by rapid tilt change, either inflation (near-summit intrusion) or deflation (rift intrusion), accompanied by shallow earthquake swarms near the site of intrusion. Traditional intrusions frequently accompany eruptions and indicate transfer of magma to a rift eruption site at a rate greater than the rate at which magma can be erupted. In some cases, accompanying earthquakes can be shown to migrate uprift, downrift, or to shallower depth from their starting point (Klein and others, 1987).

Inflationary intrusions, as previously defined (Klein and others, 1987, p. 1055–1056, table 43.1), are accompanied by intense shallow seismic activity but little summit tilt change or obvious rift deformation. We visualize magma as "leaking" into the uppermost parts of the east and seismic southwest rift zones at times when the summit reservoir is unable to further inflate in response to a steady magma supply from depth. The paired inflationary intrusion of 4 to 8 April 1970 is one example (compare this report, chap. 5; Wright and Klein, 2008, figure 4). It is possible that some of the inflationary intrusions are fed all or in part by magma left in the upper rift zone following traditional intrusions. Earthquakes often migrate downrift during inflationary intrusions, just as they typically do during some traditional intrusions. When relocated, the earthquakes may be aligned to reflect a left-lateral tectonic adjustment to seaward spreading of the south flank (Rubin and Gillard, 1998).

Suspected deep intrusions, previously defined as "slow intrusions" (Klein and others, 1987, p. 1055–1056, table 43.1), are defined by concentrations of mostly small-magnitude earthquakes occurring beneath Kīlauea's south flank at an elevated but slow rate for several days. Suspected deep intrusions are accompanied by little or no rift seismicity and small summit tilt change. The intrusion of 7 to 10 October 1969 is one example (compare this report, chap. 5; Wright and Klein, 2008, fig. 4). We interpret suspected deep intrusions as being related to magma supplied to the deeper (5–10 km) magma system beneath the rift zones and, by extension, to incremental increases in rates of rift dilation and seaward spreading (Wright and Klein, 2008). The relationship between our "suspected deep intrusions" and recently defined "slow" or "silent" earthquakes (compare Brooks and others, 2006; Cervelli and others, 2002b; Segall and others, 2006), marked by a discrete step in the seaward spreading of Kīlauea's south flank, is discussed in chapter 8 of this report.

Role of Volatiles

Carbon dioxide (CO_2), sulfur dioxide (SO_2), and water (H_2O), listed in order of increasing solubility, are the principal volatile species contained in Kīlauean magma. Degassing of CO_2 is an important process that occurs as magma moves upward to shallow storage (Gerlach and Graeber, 1985; Gerlach and others, 2002; Greenland and others, 1985). Most of the CO_2 is lost before magma moves into the rift zone. Analysis of CO_2 in rift and summit lavas, combined with experiments indicates that escape of CO_2 occurs in the summit reservoir at pressures of about 1.1 kbar corresponding to depths of 4–5 km (Dixon and others, 1995; Gerlach and others, 2002).

SO_2 is degassed both at Kīlauea's summit and rift zones and has been used as an indicator of eruption rate (Sutton and others, 2003). H_2O is released during summit storage and adds to the pressure exerted on the edifice that eventually results in eruption or intrusion.

Presentation of Seismic and Geodetic Data

The catalogs of seismicity and the record of tilting at Kīlauea's summit are the only continuously monitored data extending back to the founding of HVO. As such they are critical to an interpretation of eruptive and intrusive activity. Seismic expression of volcanic activity at Kīlauea takes several forms: (1) seismic swarms associated with magma moving from summit storage into the rift zones, or magma moving within the rift zones as both precursors and response to eruptions and intrusions (Klein and others, 1987); (2) earthquakes and earthquake swarms along the magma supply path leading to the summit reservoir (Wright and Klein, 2006; Wright and Klein, 2008); (3) release of stress built up beneath Kīlauea's south flank by the pressure of incoming magma, including seismic swarms related to flank spreading (Wright and Klein, 2008); and (4) mantle earthquakes related to flexure of the Hawaiian lithosphere in response to the island load (Klein, 2007; Klein and others, 1987).

The early, pre-1959, Hawai'i earthquake catalog (Klein and Wright, 2000) classifies known and possible Kīlauea earthquakes as "kl" [k(i) l(auea) s(outh) f(lank)]; rift zone ["uer" u(pper) e(ast) r(ift), "mer" m(iddle) e(ast) r(ift), "ler" l(ower) e(ast) r(ift), "sswr" s(eismic) s(outh) w(est) r(ift), and "swr" s(outh) w(est) r(ift)]; and summit ["kc" k(ilauea) c(aldera)], including a depth designation. "South Hawai'i," and "Kīlauea" are used when information is insufficient to make a more specific classification. A modern classification is presented in the section on "Earthquake swarms" in appendix A, which appears both at the end of the printed book and in the digital versions of this report. Additional supplementary material is in appendixes B–I, which are only available in the digital versions of this work—in the DVD that accompanies the printed volume and as a separate file accompanying this volume on the Web at http://pubs.usgs.gov/pp/1806/. We recognize that it is not possible to equate the terminology for the older periods with the more precise terminology used for the period beginning about 1967 when an adequate seismic network and detailed earthquake classification was possible. Differences in terminology for the older time periods are clarified in the figure captions. Classification of earthquakes from the modern catalog and a definition of earthquake swarms are given in appendix A (table A3; figure A4).

In this report we emphasize the only continuous record of geodetic change, which is the daily tilt records measured in the two vaults located near the edge of Kīlauea Caldera. In the text we report tilt as an azimuth (in degrees) clockwise from north and a magnitude (in microradians) (µrad) in conformity with current practice. In tables and figures for time periods before 1952 (chaps. 2 and 3) we report tilt magnitude in arc-seconds because the early seismic tilt was reported and plotted in arc-seconds The conversion factors are: 1 arc-second = 4.85 µrad, or 1 µrad = 0.21 arc-seconds. Errors in the tilt records from the seismometer in the Whitney Vault are estimated to be ±1 arc-second or 5 µrad, and from the short-base water-tube tiltmeter in the Uwēkahuna Vault are estimated to be ±0.2 arc-second or 1 µrad. Geologically significant tilt changes are ~10 µrad at Whitney and 2–3 µrad at Uwēkahuna. A more detailed discussion is given in appendix A.

Philosophical Approach to Interpretation of Disparate Datasets

Interpretations Based on Imprecise and Incomplete Historical Datasets

Data analyzed in this report, such as caldera filling rates in the 19th century, daily reading of tilt records at Kīlauea's summit, and early earthquake records from a very limited network of seismometers, are impossible to evaluate with statistical certainty. Lava levels in Kīlauea Caldera are visual estimates made by different persons. Daily tilt was not always measured and recorded by the same person. When we interpret the tilt or filling records, coincidence with other events is the critical factor that tells us whether changes are associated with known correlative events, such as eruptions, draining episodes, or earthquake swarms. We can formulate null hypotheses, such as the statement "Filling rates were constant before and after 1840." For example, within the ranges of actual filling estimates the null hypothesis is true and the observed variations shown before and after 1840 in chapter 2, figure 2.1 are the measure of uncertainty and cannot be accepted as precise numbers. However the difference in the observations before and after 1840 are real, as noted by many persons who have looked at the data.

Intrusions before the founding of HVO are identified by concurrent draining of Kīlauea Caldera and an earthquake swarm with some earthquakes large enough to be felt. Earthquakes felt at Kapāpala Ranch are evidence of intrusion on Kīlauea's southwest rift zone, and those felt at Hilo are evidence of intrusion beneath Kīlauea's east rift zone. A Mauna Loa origin for the earthquake swarms is rejected on the basis of observed draining at Kīlauea.

Interpretations Based on Data from Established Seismic and Geodetic Networks

Typical background levels of summit tilt data and earthquakes located with a seismic network can be calculated in order to interpret small tilt changes and elevated numbers of earthquakes that are not part of a swarm. However, when tilt changes or small numbers of earthquakes are associated with a documented volcanological event the criteria are loosened. For example, the tilt magnitude drop between 11 and 24 November 1959 is 5.41 µrad at an unusual azimuth. Although the magnitude is less than twice the average standard deviation calculated for quiet periods (appendix A, table A2, column 4), the azimuth points toward the shallow source path of magma supplied to the 1959 eruption in Kīlauea Iki Crater. The error in the tilt measurement may be high but is accepted, because it is consistent with the existence of a syneruption collapse of Kīlauea Iki Crater identified in later drilling. The timing, following the long first episode of the 1959 eruption, increases the credibility of this interpretation (see discussion of the 1959 eruption in chapter 4).

The number of earthquakes per day for any given region in the modern catalog is sufficiently variable to render calculation of an average background value meaningless. We use the occurrence of earthquake swarms beneath Kīlauea's east rift zone to identify periods of intrusion. There are many weeks in which no earthquakes are located beneath the rift zone. If the seismicity increases in the few days preceding a swarm, we interpret that not as background but as the first indication of an approaching intrusion defined by the swarm. Similarly, if lower seismicity in the few days following a swarm is followed by a week of no earthquakes, we interpret the lowered seismicity not as a return to background but as a waning of the pressure of intrusion.

Interpretations in chapters 5–7 cite previously published conclusions to the extent possible. Our contribution is to identify longer term comparisons and contrasts among the various modern periods or between the modern studies and earlier work with the aim of presenting the evolution of the Kīlauea magma plumbing over time.

Volcano Observation: the Historical Record, 1790–Present

We have divided this 200-year history into six periods (chaps. 2–7), each separated by a major eruption or earthquake: 1790 to 1924 (chap. 2), 1924 to 1952 (3), 1952 to 1967 (4), 1967 to 1976 (5), 1976 to 1983 (6), and 1983 to 2008 (7). Master tables of all identified events are given as the first table in each chapter. Additional earthquakes of *M*5–6 are summarized in appendixes B and C for earlier periods when locations were uncertain. Map plots for all tabulated eruptions and intrusions and time-series figures for periods of one year or longer are given in appendices B–G, corresponding to chapters 2–7. A small subset of representative events are included in each chapter as reduced text tables and figures.

Calculations of magma supply rates and eruption efficiency are discussed in appendix A. Appendixes D–G includes tables summarizing the calculation of magma supply and eruption efficiency for intervals of time covered in chapters 4–7. Results are discussed and plotted in chapter 8.

Our presentation of seismic data emphasizes swarm activity accompanying eruptions and intrusions; examples of both time-series and map plots of representative events are shown in chapters 4–7. Time-series and map plots covering all events in the time periods covered by chapters 4–7 are included in appendixes D–G. The presentation of seismic data in chapters 2 and 3 is modified to reflect the lack of a robust seismic network and the smaller number of accurately located events. Taken together, the seismic and tilt plots in chapters 2 and 3 and appendixes D–G present a synoptic view of monitored changes in Kīlauea volcanism over time.

The numbering of figures and tables is made as consistent as possible, with a chapter number as the leading number and a minimum number of subheadings. For example, the first text table and figure in chapter 3 are labeled 3.1. Where multiple closely related illustrations (as in illustrating different aspects of a single eruption) are combined as parts in one, the figure and table number are followed by letters identifying the parts, for example figure 3.1*B*. The numbering in the printed appendix A is independent, beginning with A1. Tables and figures in appendixes B–G follow a similar numbering convention—the appendix letter followed by a figure or table number beginning with 1. For example, the first appendix D table is table D1.

In March 2008, explosive activity began in Halema‘uma‘u with a lava lake in the bottom of Halema‘uma‘u occasionally in view. This activity continues as of this writing (2013), as has the eruption at Pu‘u ‘Ō‘ō that began in 1983. This is the first time in Kīlauea's recorded history that activity has occurred simultaneously at the rift and summit. We take the beginning of the 2008 activity in Halema‘uma‘u as the end of our study period.

Supplementary material for this chapter appears in appendix A, which is printed at the end of this report. Appendix A contains the following figures and tables:

Tables A1 contains volume factors (km^3/µrad) for Uwēkahuna and Whitney tiltmeters as a function of distance to various summit Mogi sources.

Table A2 contains Whitney tilt rainfall correlation and error magnitudes.

Table A3 contains our earthquake classification by tectonic region, and our working definition of earthquake swarms

Figure A1 shows a comparison of tilt records from (1) the Bosch-Omori seismometer in the Whitney Vault, (2) the daily short–base water tube tiltmeter in the Whitney Vault and (3) the daily short–base water tube tiltmeter in the Uwēkahuna Vault over the time period when all three tiltmeters were in operation.

Figure A2 shows the location of centers of inflation recorded by leveling surveys conducted before the 1967–68 eruption in Halema‘uma‘u.

Figure A3 compares distances to deformation centers from the Uwēkahuna and Whitney Vaults.

Figure A4 shows the map location of the regions defined in the earthquake classification of table A1.

Figure A5 illustrates our methodology for calculating magma supply.

Thomas A. Jaggar, founder of the Hawaiian Volcano Observatory, in his laboratory in the Whitney Vault, located on the northeast edge of Kilauea Caldera. Image courtesy of Bishop Museum.

Halema‘uma‘u lava lake in 1894. Painting by D.H. Hitchcock. Image courtesy of Hawaii National Park.

Chapter 2

Eruptive and Intrusive Activity, 1790–1924

Documents the changes from a rapidly filling caldera at the time when missionaries arrived in 1823 through a period of quiescence to refilling of Halema‘uma‘u Crater and establishment of a new lava lake. The 1924 deflation marks the last draining of Kīlauea's entire magma system, evidenced by deformation far beyond the bounds of the caldera. The lower east rift intrusion in 1924 stabilized the east rift and south flank and allowed seaward spreading documented in the post-1952 period.

The historical period of direct observation in Hawai'i begins with the European discovery of the islands in 1778–79 by Capt. James Cook (Wright and others, 1992). The written record of observations at Kīlauea Volcano begins in 1823 with the arrival of the missionary William Ellis (Ellis, 1825). The history can be extrapolated back to 1790 from oral reports that recount a great explosive eruption (Dibble, 1843; Swanson and Christiansen, 1973).

We subdivide this period into five segments, starting in 1790, 1840, 1868, 1895, and 1918, and ending in 1924. During the first three periods a lava lake was present within a broad region of Kīlauea Caldera, and filling rates diminished after 1840. Between 1840 and 1894 the lava lake was gradually reduced to a single vent at the position of the present-day Halema'uma'u. Within this period the great Ka'ū earthquake of April 1868 (Wyss and Koyanagi, 1992, and references therein) was manifested at Kīlauea by intense shaking accompanied by loss of parts of the caldera wall, rapid withdrawal of magma from Kīlauea Caldera, eruptions in Kīlauea Iki and the southwest rift zone and by the occurrence of a possible strong aftershock beneath Kīlauea's south flank.

The disappearance of lava from Halema'uma'u from July through December 1894 marked the end of continuous activity in Kīlauea Caldera. During the period from 1895 to 1918 Halema'uma'u was gradually refilled to initiate a new era of lava lake activity. The lava lake ended with phreatic eruptions in May 1924 that accompanied emptying and enlargement of Halema'uma'u. A regional uplift in 1918–19 and a regional subsidence in 1924 bracketed several eruptions and intrusions both within and outside of the caldera.

1790–1840

Events of this period, including earthquakes with magnitudes ≥6.0, are summarized in table 2.1.

The earliest observations of Kīlauea in 1823 document the existence of a lava lake active over much of the summit caldera more than 400 m below the caldera rim. The subsequent history is marked by alternating episodes of filling and withdrawal, the latter sometimes accompanied by eruptions outside of Kīlauea Caldera. Withdrawal of magma left a ring of lava, termed a "black ledge", and subsequent filling occurred both within and over the previously existing black ledge. The active lava surface was often described as a broad dome with relief of as much as 90 m. We use areas within an inner and outer black ledge as well as the entire crater floor (Mastin, 1997; table A1) to calculate lava volumes that correspond to lava thicknesses added to the caldera over time, as cited in the various references. The thickness values are estimates, the most reliable coming from persons who made frequent visits to the crater, such as the missionary Titus Coan.

The occurrence of large earthquakes and earthquake swarms have been tabulated and their sources tentatively identified (Klein and Wright, 2000). In some cases earthquake locations are made more certain by viewing them within the context of documented filling and draining.

In this early period we estimate magma supply rates using data on caldera filling rates as shown in appendix A, "Calculation of magma supply" section. Filling rates within Kīlauea Caldera have been estimated from early drawings, crude maps, and measurement of the depth to the top of the lake surface (fig. 2.1; Mastin, 1997[4]). Filling is calibrated against maps made in 1825 (Byron, 1826) and 1840 (Wilkes, 1845). The pre-1840 eruption rate was very high; filling rates, corrected for vesicles, begin at values of 0.26 km^3/year, significantly higher than the 0.11–0.18 km^3/year vesicle-free eruption rate estimated at the start of the 1983 and ongoing eruption of Kīlauea (Dvorak and Dzurisin, 1993; Heliker and Mattox, 2003)[5]. The rates diminish to 1832, then increase as a result of refilling the very large volume lost in the 1832 draining, then diminish again after 1840. There are inconsistencies in reports of different persons on the same expedition, but any adjustments cannot deny the very high rates of filling before 1840 contrasting with a fivefold decrease in filling rate after 1840 (fig. 2.1)[6].

Two early eruptions from the southwest rift zone (1823) and east rift zone (1840) also had rates that were much higher than those measured in any modern eruption. Both eruptions were accompanied by rapid and temporary draining of Kīlauea's summit lava lake. Lava flows of the 1823 eruption were mapped and described by Harold Stearns, who noted that, although the lava flow was very thin, splash from the flow was found 10 m above the flow surface on cinder cones adjacent to the flow (Stearns, 1926). A flow velocity of 12–15 m/sec has been estimated by Guest and others (1995). This rate exceeds by about an order of magnitude the median velocity of 1–2 m/sec for flow in tubes feeding the current Pu'u 'Ō'ō-Kupaianaha eruption (Hon and others,

[4]Mastin's calculations, summarized in his appendix table A3, are uncorrected for vesicularity. We use his volumes corrected to magma volume.

[5]We made new calculations of filling rate, based on reconciling the early reports of dimensions of the intra-caldera lava lake and the depth to active magma.

[6]Our estimates of filling rate are higher than those made by earlier workers (Finch, 1941; Macdonald, 1955) , but we are in full agreement with their conclusion that filling rates prior to 1840 were greater than those following 1840.

1994, p. 361). More recent work (Soule and others, 2004) suggests a lower emplacement rate closer to that of other channeled ʻaʻā flows.

The 1840 lava field was visited several months after the eruption by Titus Coan (1841) and James Dana (1849). Eruption took place at several points along the rift zone, extending from ʻAlae Crater, several kilometers from Kīlauea's summit, to the lower east rift zone about 12 km west-southwest of Kapoho Crater. Coan also reported steaming open fractures between Kīlauea's summit and the uppermost point where lava reached the surface[7]. The vents farthest from the summit erupted picritic lava that traveled rapidly to the ocean over a three-week period, producing a series of littoral cones that are still extant. Lava entering the ocean was dispersed downslope and has not been separately identified in recent bathymetry obtained adjacent to Kīlauea's south coast (for example, Smith and others, 1999)

The 1840 eruption also had unusually high flow rates for a channeled ʻaʻā flow at Kīlauea. Coan (1882) indicates that the final flow to the sea took place on 3 June at rates of 0.22–2.2 m/sec (0.5 to 5 miles per hour), consistent with the flow having reached the ocean in the same day from vents 9 to 15 km upslope. The areal extent of lava erupted from the lower vents is about 21 km^2, estimated from recent geologic mapping, and the flow thickness on land ranges from 1.2 to 12.2 m (Trusdell, 1991, fig. 5 and p. 18). Macdonald (1955) estimated the on-land volume to be 0.0618 km^3, which corresponds to an average flow thickness of 3 m. If we conservatively estimate that one-quarter of the total

[7]In 2010 D.A. Swanson discovered a previously unrecognized small lava patch in the western Koaʻe Fault Zone that has a chemistry consistent with other analyses of the 1840 eruption.

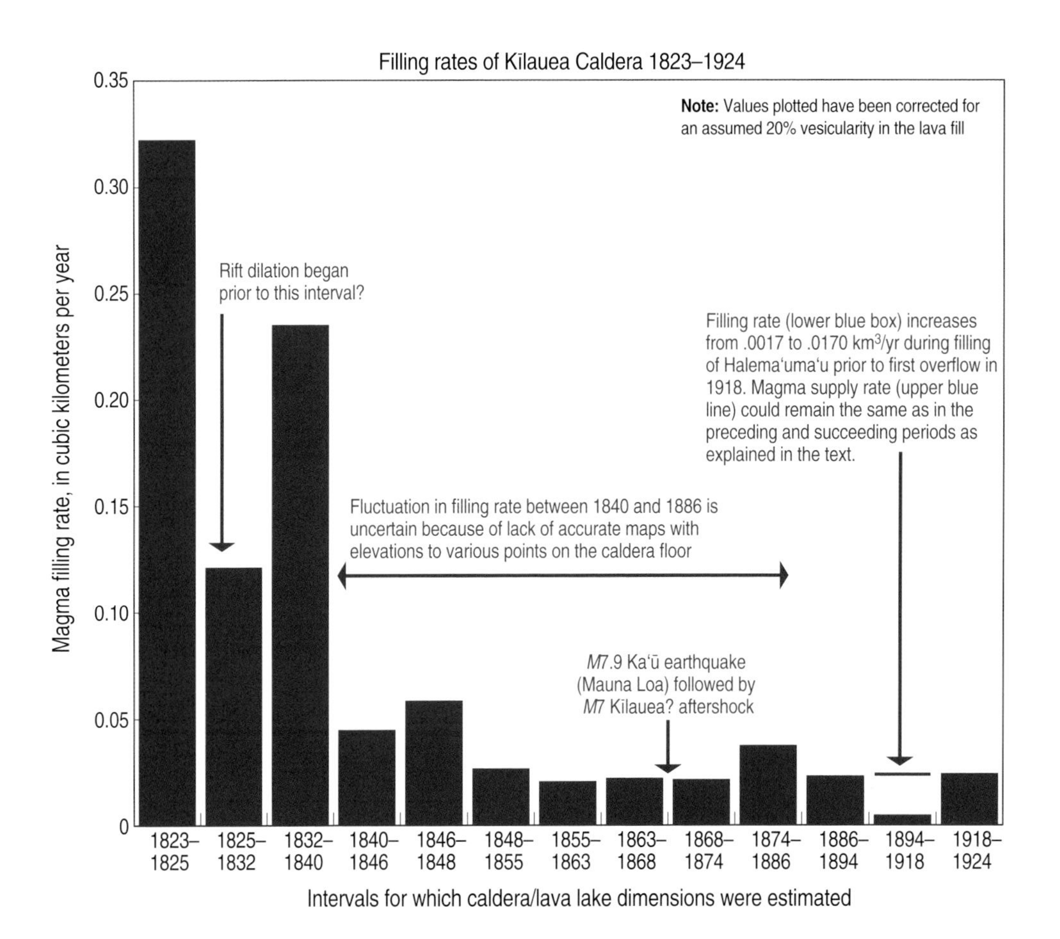

Figure 2.1. Graph of filling rates within Kīlauea Caldera 1823–1924. Magma filling rates calculated as cubic kilometers per year (within time periods of varying length) according to procedures discussed in the text. Lava volumes are corrected for 20-percent vesicularity. Uncertainties in the volume estimates are sufficiently large that we cannot deny the null hypothesis that rates could have been constant at different levels, both before and after 1840. However, the difference in rate of nearly an order of magnitude before and after 1840 is supported. The history after 1840 is critically dependent on a missing piece of data, the lack of an accurate figure for depth to the caldera floor before and after the 1868 earthquake. We use the estimate provided by James Dana (1888, p. 18). An assumption of an even greater amount of filling before 1868 would be more consistent with continuing decrease of the overall rate prior to disappearance of the caldera lava lake in 1894.

Table 2.1. Kīlauea eruptions, intrusions, and large earthquakes, 1790–1894.

[In rows with multiple entries text applies down to the next entry; data for eruptions and traditional intrusions are emphasized by grey shading; dates in m/d/yyyy format]

Start date	End date	Location[1]	Event-Type[2]	Comment	References
1790	1790	kc	E EQ	Last in a 300-year series of explosive eruptions at Kīlauea's summit *M*~6.4	Dibble, 1843; Mastin, 1997; Swanson and others, 1999
1790+	1894	kc	E	Continuous filling of Kīlauea Caldera	Brigham, 1909; Dana, 1887a; Dana, 1887b; Dana, 1888; Hitchcock, 1909
6/01?/1823		sf	EQ	Strong earthquake (*M*~7.0), reported to Ellis as preceding his visit by two months; probably located on the southwestern part of Kīlauea's south flank	Ellis, 1825; Klein and Wright, 2000
06?/1823	07?/1823	swr	E	Eruption still hot when seen by Ellis	Ellis, 1825
1/10/1832		sf	EQ	*M*5.6, 6.2—two damaging earthquakes, with aftershocks and many ground cracks	Dibble, 1843; Klein and Wright, 2000
1/10/1832	1/14/1832	kc erz?	E/C I?	Eruption on Byron's Ledge, Kīlauea's summit, strong earthquakes (see above) and a major collapse, magma withdrawing to the level first seen by Ellis	Dibble, 1839; Goodrich, 1833
11/5/1838	11/17/1838	erz/sf	EQS I	A series of 50–100 shocks felt in Hilo with magnitude range <4.5 to 5.9—probably an intrusion on the east rift zone with a south flank response	Klein and Wright, 2000
12/10/1838		Sf, erz	EQ, I	*M*6.1—preceded by elevated seismicity from 4 December; possible renewal of intrusion beneath the east rift zone	Klein and Wright, 2000
3/18/1839		kc??	EQ	*M*6.1—classified as "Hawai'i", possibly deep Kīlauea Caldera	Klein and Wright, 2000
2/1/1840		sf?	EQ	*M*6.1—classified as "south Hawai'i"	Klein and Wright, 2000
5/30/1840	6/7/1840	kc erz erz erz	C E EQS I	Eruption/intrusion migrated down the east rift zone, first appearing at 'Alae Crater and ending in Puna—accompanied by seismicity felt only near the eruption site and collapse and withdrawal of magma from Kīlauea's summit	Coan, 1841; Jarves, 1840; Jarves, 1844; Wilkes, 1845
4/5/1841	4/8/1841	sf?	EQ	*M*6.1 on 7 April preceded by *M*5.3 foreshock on 5 April and followed by several aftershocks	Klein and Wright, 2000
2/18/1844		sf?	EQ	*M*6.1	Klein and Wright, 2000
7/18/1860		sf??	EQ	*M*6.1	Klein and Wright, 2000
4/2/1868		ml	EQ	*M*7.9 Great Ka'ū earthquake. The source volume covered the entire south half of Island of Hawai'i, and the magnitude distribution of earthquakes occurring at Kīlauea and Mauna Loa up to the present are still part of the aftershock sequence following this earthquake (Klein and Wright, 2008). Alexander's map shows fractures generated by the 1868 shaking extending through the southwest flank	Anonymous, 1868a; Coan, 1868; Coan, 1869; Fornander, 1868; Hillebrand, 1868; Klein and Wright, 2000; Macdonald, 1952; Williamson, 1868b; Wyss and Koyanagi, 1992 Alexander, 1886

Table 2.1. Kīlauea eruptions, intrusions, and large earthquakes, 1790–1894.—Continued

[In rows with multiple entries text applies down to the next entry; data for eruptions and traditional intrusions are emphasized by grey shading; dates in m/d/yyyy format]

Start date	End date	Location[1]	Event-Type[2]	Comment	References
4/4/1868		sf?	EQ	*M*7? D. Cox (oral commun., 1992) notes that excessive damage on Kīlauea's south flank and an anomalous arrival of a 2nd tsunami in Hilo suggest that one of the strong aftershocks was beneath Kīlauea	Cox, 1980
4/8/1868		kc	C/E	Major collapse of Kīlauea's summit followed by eruption in Kīlauea Iki Crater	Coan, 1868; Coan, 1869; Fornander, 1868; Hillebrand, 1868
4/8/1868		erz?	I?	Report of eastern craters opening and smoke in the direction of Puna	Anonymous, 1868d
4/8/1868		swr	E/I?	Eruption (and intrusion?) on southwest rift zone	Anonymous, 1868b; 1868c; Coan, 1868
5/24/1868	5/29/1868	sf?	EQ	*M*6.2 mainshock with 8 aftershocks *M*3.5-5.9	Klein and Wright, 2000
3/21/1870 4/22/1872		sf??	EQ	*M*6.2; 6.1 aftershocks of the 1868 earthquake classified as "South Hawai'i"	Klein and Wright, 2000
5/4/1877	5/4/1877	kc	E/C	Draining of magma from Kīlauea Caldera coincident with eruption in Keanakāko'i	Anonymous, 1877a
5/5/1877	5/6/1877	sf?	I?	Possible south flank earthquakes associated with east rift intrusion?	Klein and Wright, 2000
5/31/1877		kc	EQ	*M*6.3 deep beneath Kīlauea Caldera—2 aftershocks	Klein and Wright, 2000
4/21/1879	4/21/1879	hm	C/I?	Collapse of several hundred feet—lava gone, no earthquakes	Anonymous, 1879; Coan, 1879; Wood, 1917
9/25/1880 ?		sf? ?	EQ I?	*M*6.0 with aftershocks Report of lava draining from Halema'uma'u before the Mauna Loa eruption	Klein and Wright, 2000 Maby, 1886; Wood, 1917
3/6/1886	3/6/1886	kc	C	Draining of Halema'uma'u lava lake	Klein and Wright, 2000; Maby, 1886
3/6/1886	3/7/1886	swr/sfswr	EQS/I	43 earthquakes, some felt at Kapāpala and Hilo map of Kīlauea Caldera made after the collapse	Emerson, 1887
8/6/1890	8/7/1890	sf?	EQ	*M*6.5 with at least 9 aftershocks, one *M*5.9	Klein and Wright, 2000
3/6/1891	3/6/1891	kc	C; I	Draining of Halema'uma'u Crater	Anonymous, 1891; Klein and Wright, 2000; Maby, 1891
3/6/1891	3/8/1891	swr/sfswr	EQS	Earthquake swarm of over 100 events, most felt at Kapāpala	Klein and Wright, 2000
7/12/1894	7/19/1894	kc	C	Halema'uma'u draining	Anonymous, 1894a; Thurston, 1894
7/13/1894		erz	EQS/I	At least 12 events; earthquakes felt at Hilo	
12/4/1894	12/6/1894	kc swr/sfswr	C EQS/I	Final breakdown of Halema'uma'u—lava disappears Earthquakes felt in Ka'ū	 Anonymous, 1894d

[1]Location abbreviations correspond to regions shown on chapter 1, figure 1.1*A*: kc, Kīlauea Caldera; hm, Halema'uma'u Crater; erz, east rift zone; swr, southwest rift zone; sf, south flank.

[2]Event abbreviations: E, Eruption; I, intrusion; EQ, Earthquake ≥*M*5; EQS, Earthquake swarm; fs, foreshock; C, Collapse of Kīlauea's summit or sharp summit tilt drop indicating transfer of magma to rift zone.

was emplaced in 18 hours on the first day, then the volumetric rate into the ocean is 0.02 km^3 per day (20 million m^3 per day). Another calculation, taking the cross-sectional area of the distributary tubes estimated by Trusdell at 360 m^2, and the median flow velocity seen in recent eruptions (1 m/sec), yields a volumetric flow rate of 0.031 km^3 per day. Even given the large uncertainties in these estimates imposed by required assumptions, we can say with confidence that the 1840 eruption exceeds by a considerable factor the discharge rate of other Kīlauea eruptions.

In addition to its high rate, the 1840 eruption was unlike any more recent Kīlauea eruptions in the following respects:

1. The distance that magma migrated along the east rift zone. The 1840 eruption propagated from Kīlauea's summit a total distance of 50 km. Only large east rift zone eruptions since 1840 have propagated more than 10 km within the rift zone and none has shown surface faulting propagating downrift beginning at Kīlauea's summit.
2. The rate of propagation down the rift. The total distance of 50 km was traversed in 4 days, or 12.5 km per day. Magma resupply from Kīlauea's summit to Puna during 1955 took 13–18 days to go approximately the same distance (Wright and Fiske, 1971). The 1840 rate may also be compared to rates of sustained downrift earthquake propagation over shorter (1 to 10 km) sections of rift accompanying emplacement of dikes at shallow depths. Well-defined rates range from 0.06 to 3.8 km/hr (1.4–91 km/day), the rate decreasing logarithmically with increasing time of propagation (Klein and others, 1987, table 43.1, p. 1056). The highest rates were sustained for times on the order of 1 hour over distances of 4 km or less; the lowest rates were sustained for as long as 48 hours over about the same distance. One of the best defined swarms, one that marked the beginning of the 1983 Puʻu ʻŌʻō eruption, propagated 8 km in 12 hours, a rate of 16 km per day. This is the only swarm to have a propagation rate close to that inferred for 1840, but the propagation distance was far less.
3. The absence of significant seismic activity during or preceding the eruption. This is explained by looking at the earthquake history prior to 1840 (Klein and Wright, 2000). J.J. Jarves, editor of the Polynesian Magazine, visited the site of the 1840 eruption a few months after the eruption ended. Jarves concludes his narrative with comments on the eruptive process as it relates to earthquakes:

> It is singular that an eruption of this magnitude should occur without the slightest shock of an earthquake, at least none was noticed if any happened, which proves that this was the effect of no sudden, violent action, but one of long and gradual preparation. . . . . .
>
> Three years since, smoke and steam were seen issuing from near where the present eruption commenced, and two years ago a great rent was made in the ground, and all the springs in the vicinity dried up.

The latter statement is supported by the documentation of an earthquake swarm of hundreds of events, between 5 and 17 November 1838, culminating in a *M*5.9 earthquake beneath Kīlauea's south flank, and again in the first week of December, culminating in a *M*6.1 earthquake beneath Kīlauea's south flank (Klein and Wright, 2000, and references cited therein). It appears that a relatively slow intrusion into the east rift zone culminated in two large south flank earthquakes. The dilation of the rift that can be inferred from such an event made it possible for the eruption of 1840, two years later, to occur without an accompanying earthquake swarm.

A small eruption at Kīlauea's summit occurred in January 1832, between the two high-volume eruptions of 1823 and 1840. The eruption was located on Byron Ledge, which separates Kīlauea Caldera from Kīlauea Iki Crater, and was accompanied by a series of earthquakes felt in Hilo, culminating in an earthquake of *M*6.2 beneath Kīlauea's south flank (Klein and Wright, 2000, and references therein). This earthquake was strong enough to cause draining of lava from the caldera floor, subsidence adjacent to Kīlauea's summit, and fault opening over a wide region of the volcano (Dibble, 1839; Dibble, 1843). We interpret the 1832 event as an east rift intrusion triggering a broad draining of Kīlauea Caldera's lava lake and a possible lowering of the caldera floor (Dana, 1891).

1840–1868

Of inestimable help in our interpretations of the period between 1840 and 1887 are the excellent summaries of James Dana, who was a scientist on the U.S. Exploring Expedition that arrived at Kīlauea just after the 1840 eruption (Dana, 1887a,b, 1888).

The filling of Kīlauea caldera is imperfectly documented by (1) many anecdotal observations by visitors to the active lava lake, (2) more authoritative observations of scientists and missionaries who visited the caldera at several different times, and (3) the occasional production of maps and cross sections that define the depth to the caldera floor and the size of the active areas. From these observations both short- and long-term filling rates have been calculated (figs. 2.1, 2.2). Emerson's 1887 map (Emerson, 1887) is the only map on which elevations to various points on the caldera floor were accurately surveyed. The largest uncertainty in the caldera filling rates is before and after the events of 1868. W.T. Brigham made maps both before and following the 1868 draining, but he gives no elevations to indicate how much the caldera adjacent to its walls had filled since 1840. His one citation of vertical distance to the crater floor below the Volcano House differs only slightly from a similar estimate made in 1840, an observation at variance with many other reports. Dana also questioned Brigham's observation (Dana, 1888, p. 18) and assumed a filling rate of ~0.03 km^3/yr between 1840 and 1868, which we accept, even though it has no quantitative basis.

The filling process is inferred from the early descriptions and verified by observation of more recent lava lakes. It involves four distinct processes:

1. Filling of an empty crater. An active lava lake of some dimension is enclosed within a levee of previously cooled lava, termed "black ledge" in the early literature. The filling is calculated as the depth of fill multiplied by the area of the upper and lower bounds of the cone or cylinder assumed for the shape of the crater being filled within its black ledge.

2. Overflow beyond the confines of a crater being filled. In this instance the active lava spreads beyond the bounds of the initial crater to form a new black ledge. In some cases the overflow extends to the boundaries of the caldera. Fill is calculated as depth multiplied by the area covered. Reference areas (Mastin, 1997; appendix A, table A1) are given for the area within the inner black ledge before the 1840 draining and eruption (3.42 km^2), the area within the outer black ledge after the 1840 draining (8.13 km^2) and the area within the entire caldera (9.85 km^2).

3. Endogenous filling and dome growth. There are numerous early accounts of uplift in which the central active zone assumes the form of a dome whose height may be as much as 100 m. (for example, Brigham, 1887; Coan, 1851, 1854) The additional volume is calculated as half of a prolate spheroid. There are also numerous accounts of uplift of the central caldera floor with little or no evidence of active overflow (for example, Clarke, 1886; Coan, 1870; Lyman, 1851). This process was identified more recently during the partial filling of Halema'uma'u in 1968 (Kinoshita and others, 1969, p. 62). There is no direct account of uplift at the margins of the caldera, so we have assumed that increase in elevation of the caldera floor at its edges is by overflow only.

4. The volumes of endogenous uplift are difficult to quantify because the dimensions of the uplifted area are generally not given. We have used the area within the pre-1840 black ledge in our calculations.

5. Draining, leaving an empty crater. For all such events an estimate of the drained volume has been made (table 2.2). Most of these are associated with earthquake swarms that include felt events far from the caldera, indicating intrusion beneath one or both rift zones. When earthquakes are felt in Hilo the intrusion is presumed to be beneath the east rift zone. When earthquakes are felt in Ka'ū, the intrusion is assumed to be beneath the southwest rift zone. In some instances reports of felt earthquakes suggest intrusion beneath both rift zones and (or) large earthquakes beneath the south flank.

Some of the best values for short-term filling rates come from descriptions of lava levels and crater dimensions following a draining. For the long-term calculations the draining volumes are added to increases in the caldera level; that is, it is assumed that new magma first fills the drained volume, then may flow into the broader caldera. The long-term estimates of filling of the caldera integrate the combination of exogenous and endogenous growth.

The Draining of 1868

An active lava lake returned to Kīlauea Caldera after the 1840 eruption, and caldera filling continued at a greatly diminished rate (fig. 2.1). The next major draining took place following the great Ka'ū earthquake of 2 April 1868. The magnitude 7.9 earthquake occurred beneath the Hīlea zone on Mauna Loa's south flank (Wyss and Koyanagi, 1992). Eyewitness reports of what happened at Kīlauea are quoted by Brigham (1868) and by Dana (1887a). From 20 January the lava lakes at Kīlauea were increasingly active, becoming even more markedly active following a felt foreshock of the Ka'ū earthquake on 27 March. On 28 March an earthquake sequence began beneath the Hīlea district of Mauna Loa. On 29 March the lava lakes overflowed and began to cover the caldera floor. The great earthquake occurred at 4 p.m. on 2 April. Recession of the lava began as parts of the caldera wall collapsed. By 6 p.m., lava was 30 m down in the main lake, the caldera floor was rent by gaping fractures, and a short eruption had taken place in Kīlauea Iki Crater. Over the next 3 days, lava continued to drain away; by 5 April, all active lava had left the caldera, leaving a double collapse—a central, empty pit about 150 m deep, surrounded by a broad sag with walls from 30 to 90 m high. There was an unconfirmed report of eruption on the east rift zone and a confirmed sighting of a glow emanating from the southwest rift zone. Later, fresh lava flows on the southwest rift zone were located and carefully described by Titus Coan (1869). The timing of the southwest rift eruption is not clear, although it undoubtedly took place within the 3-day period during which the caldera was draining. Following the *M*7.9 mainshock, Kīlauea's southwest rift zone and adjacent south flank were cracked, as documented by fault-related fractures shown on an early map of the island (Alexander, 1886) and the southern coast of Hawai'i subsided 1–2 m (Coan, 1869). It is not known whether the east rift zone and eastern south flank were also affected.

The events of 1868 have been interpreted in the light of more recent knowledge by Wyss and Koyanagi (1992) and by Clague and Denlinger (1993). Both sets of authors place the epicenter of the foreshock on Mauna Loa's southwest rift zone. Wyss and Koyanagi placed the mainshock on the south flank of Mauna Loa and concluded that the mainshock rupture extended from the Hīlea area on Mauna Loa's south flank across Kīlauea's southwest rift zone and also ruptured Kīlauea's south flank. Clague and Denlinger place the mainshock near the lower end of Kīlauea's southwest rift zone. They interpret the ground cracking to indicate that a large landslide comprising the entire south flank of Kīlauea moved seaward during the *M*7.9 event. In our view the effects of the earthquake were not simple. The observation that propagation of magma down the full length of the east rift zone has not been seen at any time after 1868 suggest that the east rift zone was activated but later sealed as a consequence of the earthquake, but the mechanism is a matter of speculation.

One additional piece of modern analysis makes the events of 1868 more easily understandable. Doak Cox (1980 and oral commun., 2000) pointed to the timing of arrival in Hilo of a tsunami triggered by the great Ka'ū earthquake as indicating the existence of a second tsunami source closer to Hilo. This is supported by comparing damage reports from (1) the south coast of Mauna Loa near the epicenter, (2) the city of Hilo, and (3) the summit of Kīlauea. It appears that Kīlauea had more damage than one would expect from damage at the other two localities. Wyss and Koyanagi (1992) concluded that the mainshock rupture extended from the Hīlea area on Mauna Loa's south flank across Kīlauea's southwest rift zone and ruptured Kīlauea's south flank also. Cox concluded that the 1868 earthquake aftershock sequence included a second earthquake of high magnitude (~*M*7) beneath Kīlauea's south flank, resulting in the generation of a second tsunami. The delay of at least one day in the major draining of Kīlauea Caldera and in the initiation of a southwest rift intrusion and eruption is more consistent with events local to Kīlauea than to the effects of a Mauna Loa earthquake, however large. The source volume estimated for the mainshock (Wyss and Koyanagi, 1992) includes all of Kīlauea and is consistent with the possibility of a large aftershock beneath Kīlauea.

1868–1895

During the remainder of the 19th century, lava was added to Kīlauea Caldera at a low and possibly decreasing rate (fig. 2.1) and the center of eruptive activity became more localized near the present site of Halema'uma'u Crater. Major, well documented draining of Kīlauea summit lava lake occurred in 1886, 1891, and 1894, each accompanied by felt earthquake sequences suggesting rift intrusion (table 2.1). Minor draining with less compelling evidence for intrusion occurred in 1871, 1877 (associated with an eruption in Keanakāko'i Crater—a minor repeat of the 1868 sequence in Kīlauea Iki?), 1879, and 1880. The 1894 draining occurred in two stages, July and December, with increments of additional draining occurring between the two main events. Following the December 1894 earthquake swarm, lava disappeared from the crater, ending the 19th century period of near-continuous eruptive activity. Halema'uma'u was left as a circular pit of radius 180 m at the top and unspecified radius at its bottom.

1895–1918

Following the 1894 draining, the lava stored at shallow depth beneath the caldera remained relatively inactive, erupting sporadically deep within Halema'uma'u over the next decade (fig. 2.2). Halema'uma'u Crater also continued to deepen, reaching a maximum depth of 300 m in March 1899 (Wood, 1917). From 1899 on, we calculate the volume of fill as a frustum of a cone (the volume of part of a cone bounded by two horizontal planes) whose radius varies between 180 m at the surface to 120 m at 300-m depth. The altitude at the surface, used as a datum in figure 2.2, is 1,130 m (3,700 feet). Beginning at the end of 1906, filling became almost continuous and Halema'uma'u was nearly full toward the end of 1908 (fig. 2.2). However, reminiscent of the 19th century, the filling was punctuated by drainings, many of which occurred close in time to large Kīlauea earthquakes or earthquake swarms noted in the catalog (tables 2.3, 2.4; Klein and Wright, 2000). At the time of the founding of the Hawaiian Volcano Observatory in 1912 the lake was still nearly full (fig. 2.2).

Between 1912 and the first overflow in 1918, the level of the lake continued to fluctuate and most of the periods of draining were accompanied by earthquake swarms implying an intrusion (fig. 2.3*A*). In 1913, during a period when the lava lake was described as "dormant," there was a swarm of earthquakes beneath the south flank, possibly signifying a suspected deep intrusion (Klein and others, 1987; Wright and Klein, 2008). Filling rates (table 2.5, fig. 2.1) were much lower than in the 19th century, increasing from 0.0017 km^3/yr to 0.017 km^3/yr before overflow in 1918[8].

[8]When intrusions inferred during the refilling of Halema'uma'u (fig. 2.2) are taken into account, the magma supply rate could be close to that of periods preceding and following the refilling, as discussed in chapter 8, section on "Relation between eruption and intrusion expressed as eruption efficiency."

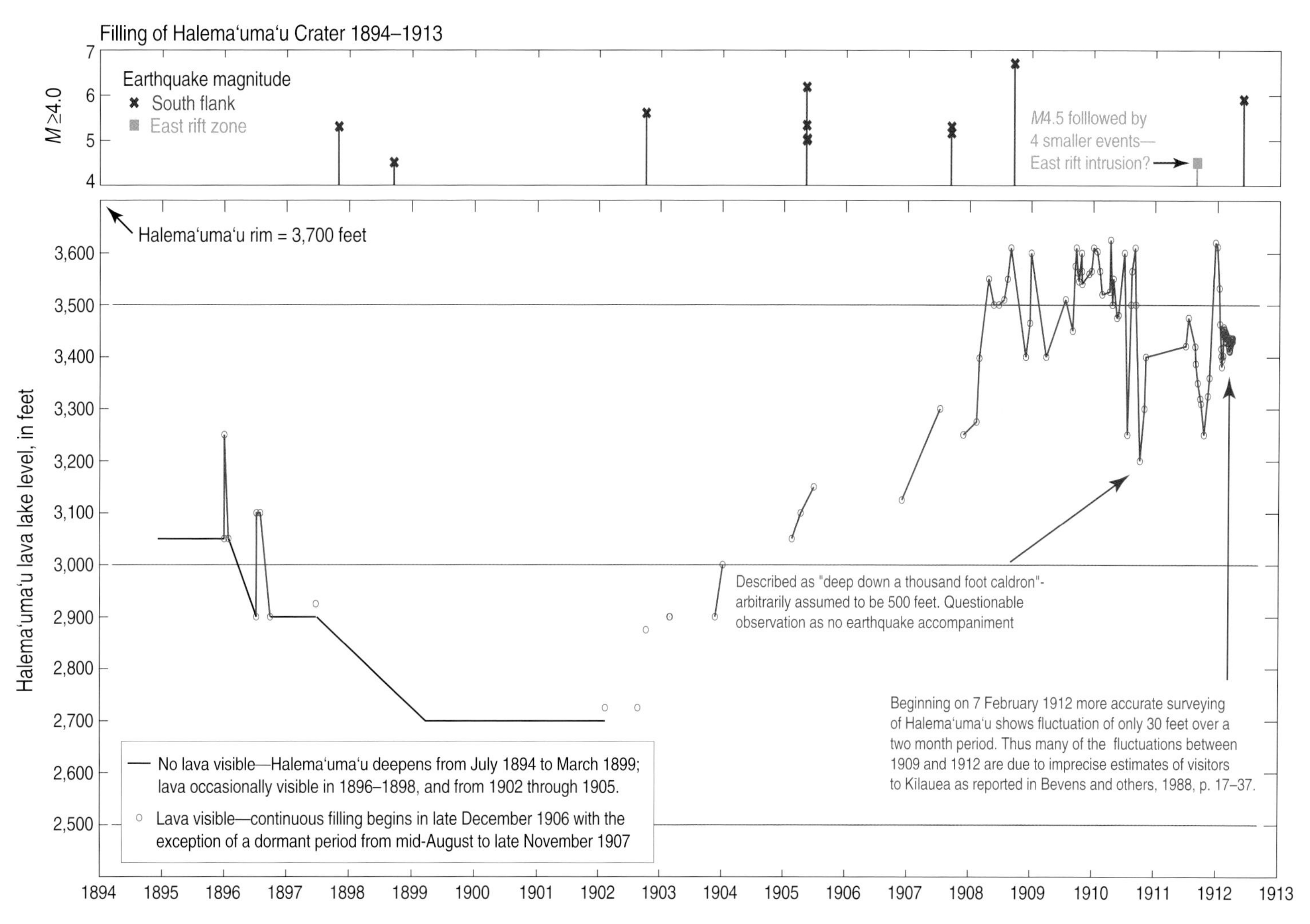

Figure 2.2. Graph showing refilling of Halemaʻumaʻu, 1895–1912. Dates and tick marks on the time (x) axis are centered at the beginning of the year. Elevation (above sea level) of the lava surface visible in Halemaʻumaʻu Crater, following disappearance of the caldera lava lake in December 1894 is plotted at times for which there exist published observations. The record is shown up to the founding of the Hawaiian Volcano Observatory in 1912. The pattern is one of gradual rise punctuated by periods of fall. The latter often correspond in time to large-magnitude earthquakes with aftershocks, tentatively assigned to Kīlauea's south flank. There is no evidence of preceding swarm activity that might suggest intrusion, and any possible swarm earthquakes would have to be large enough to be felt in this preinstrumental era. A single earthquake assigned to the east rift zone may be associated with intrusion. Estimates before 1912 have high errors as they are visual estimates made by visitors to the volcano without benefit of instruments. Instrumental measurements made after 1912 vary less than ±30 feet and probably represent the true variability between 1907 and 1912. Data from Bevens and others, 1988, v. 1, including Kīlauea activity from 1865–1011 (Wood, 1917), Kīlauea activity in 1909–1912, and F. Perret's observations in 1911, included in the first report of the Hawaiian Volcano Observatory.

Table 2.2. Kīlauea Caldera magma draining 1840–1900.

[In rows with multiple entries text applies down to the next entry; dates in m/d/yyyy format]

Start date	End date[1]	Return of lava[2]	Eqs yes/no	Volume (km^3)[3]	Comment[4,5]	References
5/30/1840	?	?	yes	0.22	Earthquake swarm on lower east rift zone. Eruption/intrusion?	Coan, 1841; Mastin, 1997
10/5/1855	10/6/1855			0.0371[6]		Dana, 1887a,b
4/2/1868	4/5/1868		yes	KC: 0.2658 Hm: 0.0537 Total: 0.3195[7]	Great Ka'ū earthquake of 04/02/1868 preceded draining of lava from Kīlauea Caldera by more than 24 hours; many aftershocks, including one of *M*7+ located beneath Kīlauea(?)	Coan, 1869; Dana, 1887b
5/25/1871	5/25/1871	08/10/1871	yes	0.0267	Earthquakes felt in Hilo two days before lava disappeared	Anonymous, 1868, 1871a, 1871b; Wood, 1917
5/4/1877	5/4/1877	05/14/1877	yes	0.0278	Eruption in Keanakāko'i accom. by intrusion beneath east rift zone.	Anonymous, 1877a, b; Wood, 1917
					Additional intrusion beneath southwest rift zone?	Anonymous, 1877b
4/21/1879	4/22/1879	04/23/1879	no	0.0500	Lava (assumed 300) feet down; probable intrusion	Coan, 1879; Dana, 1887a, b
9/1880	9/1880?	10/01/1880	yes?	0.0250	Lava draining from Halema'uma'u before the Mauna Loa eruption; associated with *M*6.1 south flank earthquake on 09/25/180?	Maby, 1886; Wood, 1917
3/6/1886	3/8/1886	03/20/1886	yes	0.056	Combine outer and inner collapses from elevations given on Emerson's map (plate)	Emerson, 1887
						Klein and Wright, 2000
			yes	0.109	Intrusion beneath upper east and sw rift	Anonymous, 1907; Klein and Wright, 2000
					Intrusion beneath southwest rift zone	
7/10/1894	7/15/1894	01/03/1896		0.0079	Initial collapse to depth of 275 feet; probable intrusion (east rift zone)	Armstrong, 1894; Anonymous, 1894b
	8/28/1894			0.0154	Further collapse to 600 feet	Anonymous, 1894e
12/3/1894	12/6/1894				All lava gone; probable intrusion (southwest rift zone)	Anonymous, 1894c, d
	3/24/1899			0.0225	Final collapse to 1,000 feet below rim	Anonymous, 1902; Wood, 1917

[1]Lava gone.

[2]Lava reappears.

[3]Volumes given as reported, estimated from cross-sections or description in reference(s). These are converted to magma volumes before calculation of caldera filling rates. KC, Kīlauea; Hm, Halema'uma'u Crater.

[4]Earthquake swarms are reported in Klein and Wright, 2000, Wyss and others, 1992 (Earthquake diary kept by the Lyman family), and Wood, 1917, in addition to anonymous newspaper reports cited in the table. Earthquakes felt in Hilo are presumed to indicate intrusion beneath Kīlauea's east rift zone. Earthquakes felt in Ka'ū are presumed to indicate intrusion beneath Kīlauea's southwest rift zone.

[5]Intrusion locations estimated from felt earthquakes in earthquake swarms. If felt at Hilo, intrusion is beneath east rift zone. If felt in Ka'ū, intrusion is beneath southwest rift zone. For some events, intrusion beneath both rift zones is indicated.

[6]Assume dome with area of inner black ledge (3.42 km^2 from Mastin, 1997) and height of 100 feet. Collapse reflects loss of dome and an additional loss of 50 feet of lava in Halema'uma'u (ellipse of 400 × 250 feet from Coan, 1910).

[7]The area of collapse (4.36 km^2) obtained from Lydgate's map (Dana, 1887b, p. 94) and depth of 300 feet from Coan (1869; quoted in Dana, 1887b, p. 92). Dimensions of Halema'uma'u (cone with 3,000 feet diameter at top and 1,500 feet at bottom and depth of 500 feet) from Dana (1887b, p. 92).

Table 2.3. Kilauea eruptions, intrusions, and large earthquakes, 1895–1925.

[In rows with multiple entries text applies down to the next entry; data for eruptions and traditional intrusions are emphasized by grey shading; dates in m/d/yyyy format]

Start date	End date	Location[1]	Event type[2]	Comment	References[3]
1/3/1896	1/26/1896	hm	E	Lava returned briefly to bottom of Halema'uma'u	Wood, 1917
1/29/1896	1/31/1896	swr/sf	EQS/I?	Several earthquakes felt at Kapāpala	Anonymous, 1896
7/11/1896	9/30/1897	hm	E	Lava returned briefly to bottom of Halema'uma'u	Wood, 1917
6/24/1897	6/27/1897	hm	E	Lava returned briefly to bottom of Halema'uma'u	Wood, 1917
3/24/1899		hm	C	Halema'uma'u drained to 1,000 feet depth	KW; Wood, 1917
2/14/1902	2/15/1902	hm	E	Lava returned briefly to bottom of Halema'uma'u	KW; Wood, 1917
3/1/1903	3/5/1903	hm	E	Lava returned briefly to bottom of Halema'uma'u	KW
8/25/1903 10/13/1903 2/22/1905	9/12/1903 1/10/1904 9/15?/1905	hm	E	Intermittent addition of lava to the bottom of Halema'uma'u Crater	Wood, 1917
5/3/1905	5/7/1905	sf?	EQ	*M*6.18 preceded by M5.3 foreshock and followed by many aftershocks, some as strong as $M \geq 5$—classified as "Kīlauea south flank?"	KW
11/1905	5/1/1906	hm	E	Intermittent activity	Wood, 1917
12/2/1906 5/12/1907	4/15/1907 8/15/1907	hm	E	Intermittent but increasing activity from this date	Wood, 1917
11/30/1907	2/23/1918	hm	E	Continuous filling with occasional drawdown building to first overflow	Wood, 1917
9/2/1908	9/6/1908	kc?	EQS/C/I	Earthquakes classified as 5–10 km beneath Kīlauea Caldera—most likely east rift and south flank	KW
9/20/1908	9/30/1908	sf	EQ	*M*6.7 Kīlauea south flank w many aftershocks	KW
4/19/1910		kc? or sf?		*M*5.3 classified as Kīlauea?	KW
4/26/1910		kc?	EQ	*M*? classified as Hawai'i—felt over entire island, might be deep beneath Kīlauea Caldera	KW
2/9/1911		kc?	EQ	do	KW
7/24/1911	8/7/1911	kc	EQS?	Kīlauea Caldera 0–5 km? Halema'uma'u lava lake falling between July 17 and August 7	ESPHVO, v. 1, p. 40; KW
8/25/1911	8/26/1911	erz/kc?	EQS	Classified as east rift—Ass. with modest subsidence of Halema'uma'u lava lake	ESPHVO, v. 1, p. 45; KW
12/26/1911	12/27/1912	kc	EQS	Kīlauea Caldera 0–5 km—associated with perturbations in Halema'uma'u lava lake	ESPHVO, v. 1, p. 57; KW
9/1/1912				Systematic earthquake reports from the Whitney Laboratory of Seismology begin	KW
9/2/1912	9/14/1912	kc erz/sf?	EQS I?	Kīlauea Caldera 0-5 km?—apparently preceded draining of magma in Halema'uma'u Possible east rift intrusion with south flank response on 09/15, 16, and 20	Jaggar, 1947, p. 37-40; KW

Table 2.3. Kīlauea eruptions, intrusions, and large earthquakes, 1895–1925.—Continued

[In rows with multiple entries text applies down to the next entry; data for eruptions and traditional intrusions are emphasized by grey shading; dates in m/d/yyyy format]

Start date	End date	Location[1]	Event type[2]	Comment	References[3]
9/12/1912	9/14/1912	kc erz/sf?	EQS I?	Kīlauea Caldera 0–5 km?—apparently preceded draining of magma in Halema'uma'u Possible east rift intrusion with south flank response on 09/15, 16, and 20.	Jaggar, 1947, p. 37-40; KW
3/25/1913	3/26/1913	erz?	EQS	Classified as "Lower east rift zone and Kīlauea south flank?"—probably related to subsidence of Halema'uma'u lava lake between 03/19 and 05/5, 1913	Jaggar, 1947, p. 79-81; KW
5/18/1913		sf?	EQ	Classified as "Kīlauea south flank"—Halema'uma'u still empty—revival near 10/01/1913	Jaggar, 1947, p. 86-88; KW
6/5/1913 7/9/1913 9/20/1913	6/16/1913 7/12/1913 9/26/1913	sf?	EQS/I?	Many earthquakes classified as "Kīlauea south flank?" during period when Halema'uma'u was empty—possible "suspected deep" intrusion?	Jaggar, 1947, p. 88; KW
11/29/1913 1/25/1914 3/4/1914 3/30/1914	12/9/1913 2/23/1914 3/9/1914 4/13/1914	sf?	EQS/I?	Many earthquakes classified as "Kīlauea south flank?" during period when Halema'uma'u was still minimally active—possible series of "suspected deep" intrusions?	KW
1/11/1915	1/18/1915	kc?	EQS?	Several earthquakes associated with beginning of withdrawal of magma from Halema'uma'u	KW
4/11/1915	4/18/1915	erz? kc?	EQS/I?	Possible east rift-south flank activity associated with the end of withdrawal of magma from Halema'uma'u	KW
8/27/1915 9/5/1915	8/31/1915 9/18/1915	erz sf	EQS/I? EQS	Swarm at distance of 20–25 km (middle east rift-south flank)—some events felt at Kīlauea's summit and Hilo—"suspected deep?" intrusion preceding normal intrusion beginning on 09/25/1915	KW
9/19/1915	9/20/1915	kc	EQS	3 events at distance of 30–35 km (deep beneath Kīlauea Caldera?)	KW
9/22/1915	9/29/1915	erz, sf	EQS/C/I	Middle/upper? east rift intrusion (110 events) with south flank response (8 events) and some intercalation of events at 30–35 km.	KW
9/30/1915	10/10/1915	sf		Continued south flank response (21 events)	
6/4/1916	6/12/1916	erz	EQS/C/I	Middle/upper? east rift intrusion (281 events) with south flank response (19 events) and some intercalation of events at 30–35 km.	KW
6/12/1916	6/25/1916	sf		Continued south flank response (25 events)	
2/23/1918	3/10/1918	kc	E	Halema'uma'u overflow	KW
3/3/1918	3/5/1918	kc	EQS	5 events (0–5 km beneath Kīlauea Caldera?) associated with lowering of Halema'uma'u lake level	KW
3/26/1918	4/6/1918	hm	EQS/C	41 events (0–5 km beneath Kīlauea Caldera?) associated with lowering of Halema'uma'u lake level	KW
11/13/1918	11/17/1918	hm	EQS/C	72 events associated with lowering of Halema'uma'u lake level	KW
2/7/1919	11/28/1919	kc	E	"Postal Rift" eruption on Kīlauea Caldera floor	ESPHVO, v. 2, p. 888, 1055
6/18/1919		hm	EQS?/C	Earthquakes associated with lowering of Halema'uma'u lava lake level?	KW
8/26/1919		kc?	EQ	*M*5 deep beneath Kīlauea Caldera?	KW

Table 2.3. Kīlauea eruptions, intrusions, and large earthquakes, 1895–1925.—Continued

[In rows with multiple entries text applies down to the next entry; data for eruptions and traditional intrusions are emphasized by grey shading; dates in m/d/yyyy format]

Start date	End date	Location[1]	Event type[2]	Comment	References[3]
11/28/1919	12/4/1919	erz	EQS/C/I	> 200 events associated with draining of Halema'uma'u lava lake. Assume middle east rift intrusion with south flank response	ESPHVO, v. 2, p. 1059; KW
12/15/1919	8/15/1920	swr	E	Mauna Iki eruption on southwest rift zone	KW
12/15/1919	12/19/1919	swr	EQS/I	Earthquake swarm of 79 events—probable intrusion into seismic southwest rift zone?	
12/22/1919	2/8/1920	sf	EQS	South flank response—25 events	
3/18/1921	3/27/1921	kc	E	Overflow onto caldera floor	ESPHVO, v. 3, p. 63-76
5/17/1922	6/1/1922	kc/ erz	EQS/ C/I	Earthquake swarm of 560 events (108 at 0–10 km beneath Kīlauea Caldera; 432 beneath east rift zone) associated with 850-foot lowering of lava lake	ESPHVO, v. 3, p. 287-290; KW
6/1/1922	6/8/1922	sf	EQS	South flank response (67 events)	
5/28/1922	5/30/1922	erz	E/I/C	Eruption at Makaopuhi and Nāpau Craters. Intrusion beneath upper and middle east rift zone; draining of Halema'uma'u	ESPHVO, v. 3, p. 275, 282-285
11/21/1922	11/23/1922	sf	EQ	*M*5.5 classified as "Kīlauea south flank??" with several aftershocks	KW
12/31/1922	1/4/1923	erz	EQS/C/I	Earthquake swarm of 133 events—classified as upper east rift zone with minor south flank response	ESPHVO, v. 3, p. 375, 384-386; KW
4/1/1923	4/5/1923	kc	EQS	Earthquake swarm of 22 events 0–5 km beneath Kīlauea Caldera	ESPHVO, v. 3, p. 413; KW
8/3/1923	8/7/1923	erz	EQS	Earthquake swarm of 45 events beneath upper east rift zone?; draining of Halema'uma'u	ESPHVO, v. 3, p. 457, 461-462; KW
8/24/1923	8/27/1923	erz	EQS/C	Earthquake swarm of 92 events beneath upper east rift zone; draining of Halema'uma'u	ESPHVO, v. 3, p. 461-462; KW
8/25/1923	8/26/1923	erz	E	Eruption west of Makaopuhi Crater	ESPHVO, v. 3, p. 460
2/13/1924	2/20/1924	kc	EQS	Broken earthquake swarm of 22 events 0–5 km beneath Kīlauea Caldera?	KW
3/7/1924	4/17/1924	erz	EQS/I	Earthquake swarm migrating down east rift zone (83 events beneath east rift zone; 3 south flank) associated with Halema'uma'u draining	ESPHVO, v. 3, p. 513; KW
4/17/1924	5/1/1924	erz	EQS/I	Continuation of earthquake swarm (324 events), now located beneath the lower east rift zone associated with intense ground deformation	ESPHVO, v. 3, p. 519-528; KW
5/1/1924	5/10/1924	kc	EQS	Continuing subsidence of Halema'uma'u	ESPHVO, v. 3, p. 536-539; KW
5/10/1924	5/28/1924	kc	EQS/E	Explosive (phreatic) eruption from Halema'uma'u Crater accompanied by continuing earthquake swarm	ESPHVO, v. 3, p. 540-560; KW
5/28/1924	6/30/1924			Earthquakes continue through the end of June	
7/19/1924	7/30/1924	hm	E	Return of lava to bottom of Halema'uma'u	ESPHVO, v. 3, p. 576; KW

[1]Location abbreviations correspond to regions shown on chapter 1, figure 1.1a: kc, Kīlauea Caldera; hm, Halema'uma'u Crater; erz, East rift zone; swr, Southwest rift zone; sf, South flank.

[2]Event abbreviation: E, Eruption; I, intrusion; EQ, Earthquake ≥*M*5; EQS, Earthquake swarm; C, Collapse of Kīlauea's summit or sharp summit tilt drop indicating transfer of magma to rift zone.

[3]ESPHVO, Early Serial Publications of the Hawaiian Volcano Observatory (Bevens and others, 1988); KW, Klein and Wright (2000).

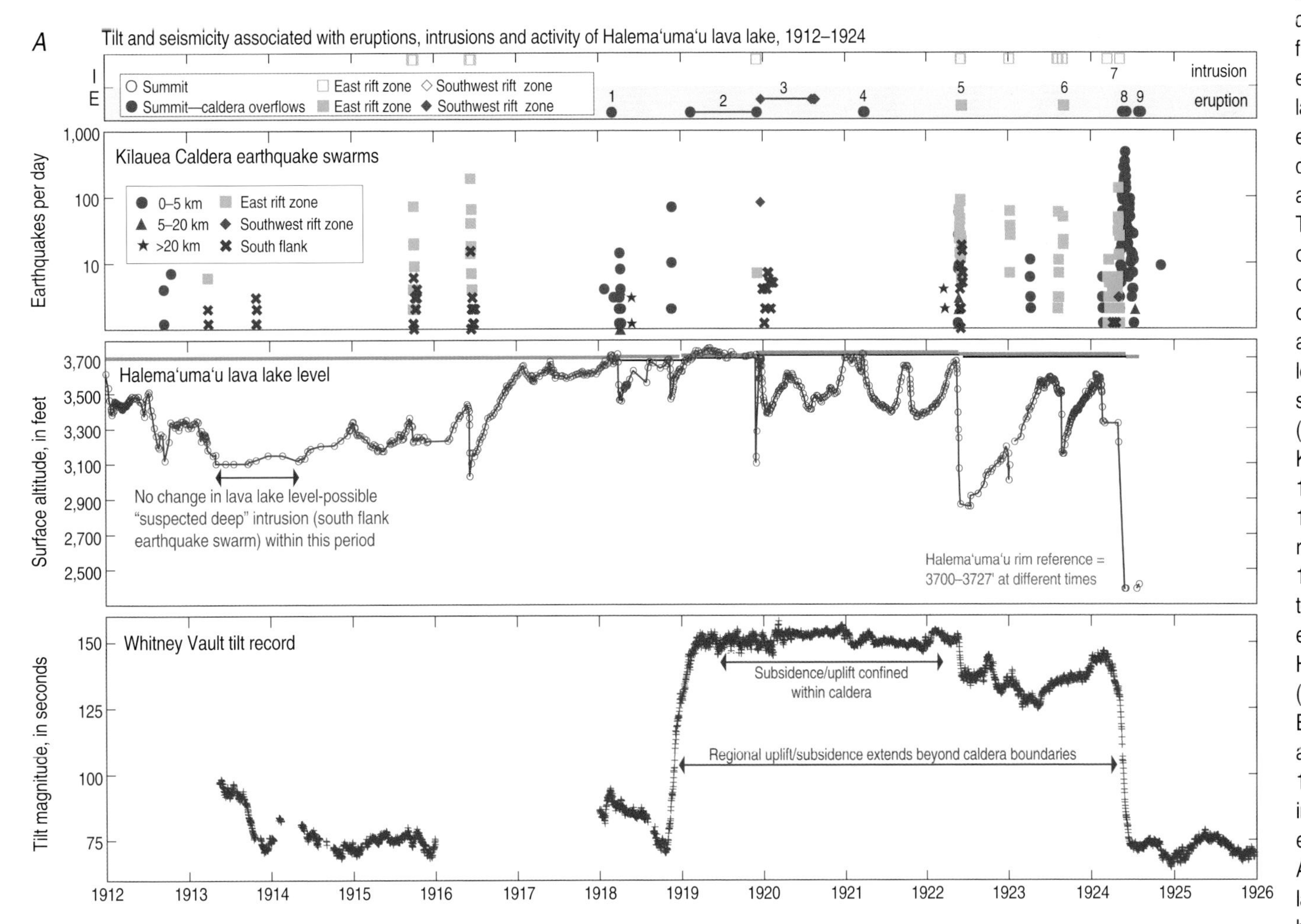

Figure 2.3. Graphs comparing Halemaʻumaʻu lava lake level, tilt magnitude, earthquake swarms, and times of eruption and intrusion. ***A***, 1912–1926. Panels show, from bottom to top: Whitney tilt magnitude, elevation of the surface of Halemaʻumaʻu lava lake, earthquake swarms, and eruptions(E)/intrusions(I), plotted against date. Dates on the time (x) axis are centered at the beginning of the year, as in figure 2.2. The rim of the lava lake is shown as a heavy orange line. The altitude varies within 20 feet or so (6 m) from overflows and subsidence over short time scales. Sites of intrusions are identified from the earthquake swarm locations. Numbers in the top panel identify selected eruptions and intrusions as follows: (1) First overflow of Halemaʻumaʻu. (2) 1919 Kīlauea Caldera "Postal rift" eruption. (3) 1919–1920 SW rift "Mauna Iki" eruption. (4) 1921 Halemaʻumaʻu overflow. (5) 1922 East rift eruption. (6) 1923 East rift eruption. (7) 1924 East rift intrusion migrates from upper to lower east rift zone. (8) 1924 explosive eruption associated with deepening of Halemaʻumaʻu and end of lava lake activity. (9) 1924 Return of lava to Halemaʻumaʻu. Eruptions on the east rift zone in May 1922 and May 1924 and an intrusion earlier in 1924 show a clear correlation of a drop in lava lake level with deflation and an earthquake swarm. The east rift eruption of August 1923 is accompanied by a drop in lava lake level and an earthquake swarm, but is accompanied by only a small tilt change at an unusual azimuth (appendix B, table B1). The large inflation at the end of 1918 occurs with no change in lava lake

level or increase in earthquakes. Both this tilt increase and the subsequent tilt decrease in 1924 are consistent with triangulation and leveling data that show changes extending well beyond the confines of Kīlauea Caldera. Data for all events are summarized in table 2.1. See text for interpretation of these events. ***B***, An expanded plot of earthquakes, tilt and lava lake level for the period between 1919 and 1924. Dates on scale are in m/d/yyyy format. Panels follow the same order as in figure 2.3*A*. Numbers in the top panel identify selected eruptions and intrusions as follows: (1) 1919 Kīlauea Caldera "Postal rift" eruption. (2) 1919 intrusion at end of the postal rift eruption. (3) 1919–1920 SW rift "Mauna Iki" eruption. (4) 1921 Halema'uma'u overflow. (5) 1922 East rift eruption/intrusion (vertical dotted line). (6) Minor East rift zone intrusion. (7) Minor East rift zone intrusion. (8) 1923 East rift eruption/intrusion (vertical dotted line). The end of the 1919 "postal rift" eruption in Kīlauea Caldera is accompanied by a major draining of the lava lake, a moderately intense earthquake swarm beneath the east rift zone, and a small deflationary tilt change. The beginning of the "Mauna Iki" eruption on the southwest rift zone is accompanied by an intense earthquake swarm, after which there is a lowering of the lava lake level and a period of deflationary tilt. The tilt signal during these two events falls within the scatter for the longer period and, taken alone, would not identify eruption or intrusion. Likewise, the 1922 and 1923 eruptions/intrusions are both accompanied by a drop in lava lake level, but only the 1922 event shows a tilt drop. See text for further explanation.

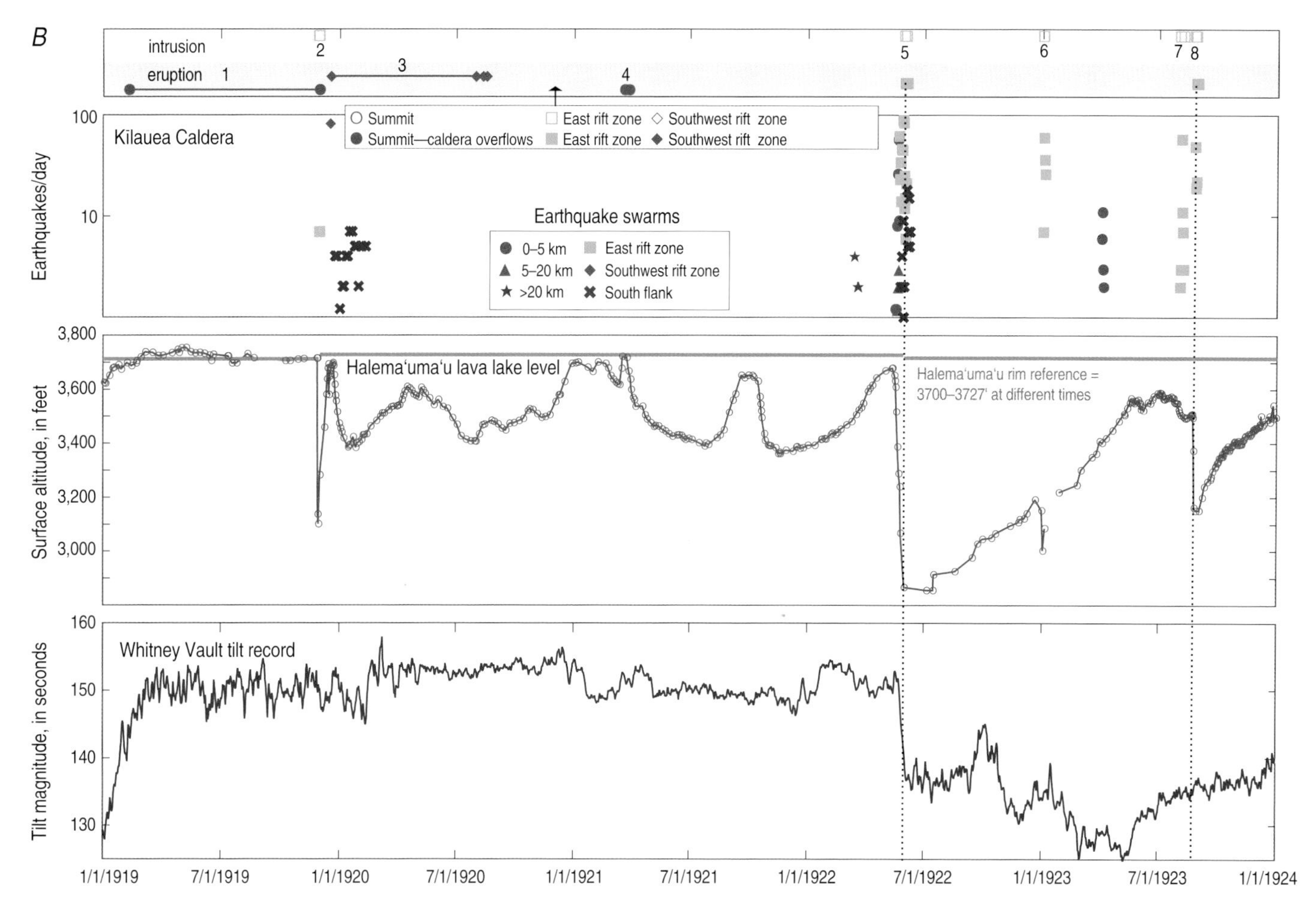

Table 2.4. Halema'uma'u Crater filling and draining 1900–1924.

[Volume figures all represent net filling; in rows with multiple entries text applies down to the next entry; dates in m/d/yyyy format; Do., same as previous entry]

Start date	End date	Return of lava	Eqs[1] yes/no	Volume (km^3)	Comment	References[2]
					Filling of Halema'uma'u Crater to first overflow on 2/23/1918	
1/10/1904	1/10/1904	2/22/1905	no	0.0091	Estimated from change of crater depth	Wood, 1917
5/1/1906?	5/1/1906?		no	0.0069	Maximum collapse calculated after 400 foot fill in 2005; amount of draining unspecified	
		12/2/1906			Initiation of continuous fill	Wood, 1917
9/7/1908	4/4/1909		yes	0.0059	Presumed intrusion beneath east rift zone; south flank slow intrusion??	KW; Wood, 1917
10/5/1909	10/15/1909		no	0.0019	Estimated from change of crater depth	Wood, 1917
1/1/1910	8/1/1910		no	0.0096	Estimated from change of crater depth	Wood, 1917
1/1/1912	2/1/1912		yes?	0.0115	Read from chart showing lava rise and fall in Halema'uma'u	ESPHVO, v. 2, p. 295
7/12/1912	8/26/1912		no	0.0080	Local earthquakes in Jan., Sept., and 10/15 (accompanying. inflation?)	KW
2/8/1913	5/5/1913		no	0.0055	Read from chart showing lava rise and fall in Halema'uma'u	ESPHVO, v. 2, p. 295
1/4/1915	5/13/1915		no	0.0038	Do.	ESPHVO, v. 2, p. 647
9/15/1915	9/28/1915		yes	0.0033	Read from chart showing lava rise and fall in Halema'uma'u	ESPHVO, v. 2, p. 647
					Major earthquake swarm on 25–30 Sept.1915 tentatively identified as beneath east rift zone followed on 4–10 Oct. by south flank response	KW
6/4/1916	6/6/1916		yes	0.0119	Do.; last collapse before overflow 2/23/1918	ESPHVO, v. 2, p. 647
					Major earthquake swarm on 6/4–11/1916 tentatively identified as beneath east rift zone followed on 10/12–24 by south flank response	KW
3/11/1917	3/15/1917		yes	0.0008	Local earthquakes	ESPHVO, v. 2, p. 578; KW
					Halema'uma'u lava lake active 2/23/1918–5/31/1924	
3/28/1918	4/15/1918		yes	0.0077	Local earthquakes 3/26–4/4/1918	ESPHVO, v. 2, p.748–751
11/5/1918	11/16/1918		yes	0.0064	Intrusion beneath east rift zone	ESPHVO, p. 844–845
11/28/1919	11/29/1919		yes	0.0154	End of postal rift eruption	ESPHVO, p. 1055–1059
12/22/1919	1/15/1920		yes	0.0098	Initiation of Mauna Iki eruption	ESPHVO, p. 1072–1092
5/14/1922	5/31/1922		yes	0.0358	East rift eruption	ESPHVO, v. 3, p. 297
8/23/1923	8/29/1923		yes	0.0266	East rift eruption	ESPHVO, v. 3, p. 461
2/9/1924	4/30/1924		yes	0.0227	Earthquake swarm and intrusion beneath lower east rift zone	ESPHVO, v. 3, p. 515–528[3]
4/30/1924	5/31/1924		yes	0.1304	Halema'uma'u phreatic eruption	ESPHVO, v. 3, p. 529–560[3]

[1]Eqs, earthquake swarm. Yes/no refers to whether or not an earthquake swarm accompanied the activity listed.

[2]ESPHVO, Early Serial Publications of the Hawaiian Volcano Observatory (Bevens and others, 1988); KW, Klein and Wright (2000).

[3]Additional published summaries of 1924 activity are: (Finch, 1924, 1925, 1947; Jaggar and Finch, 1924).

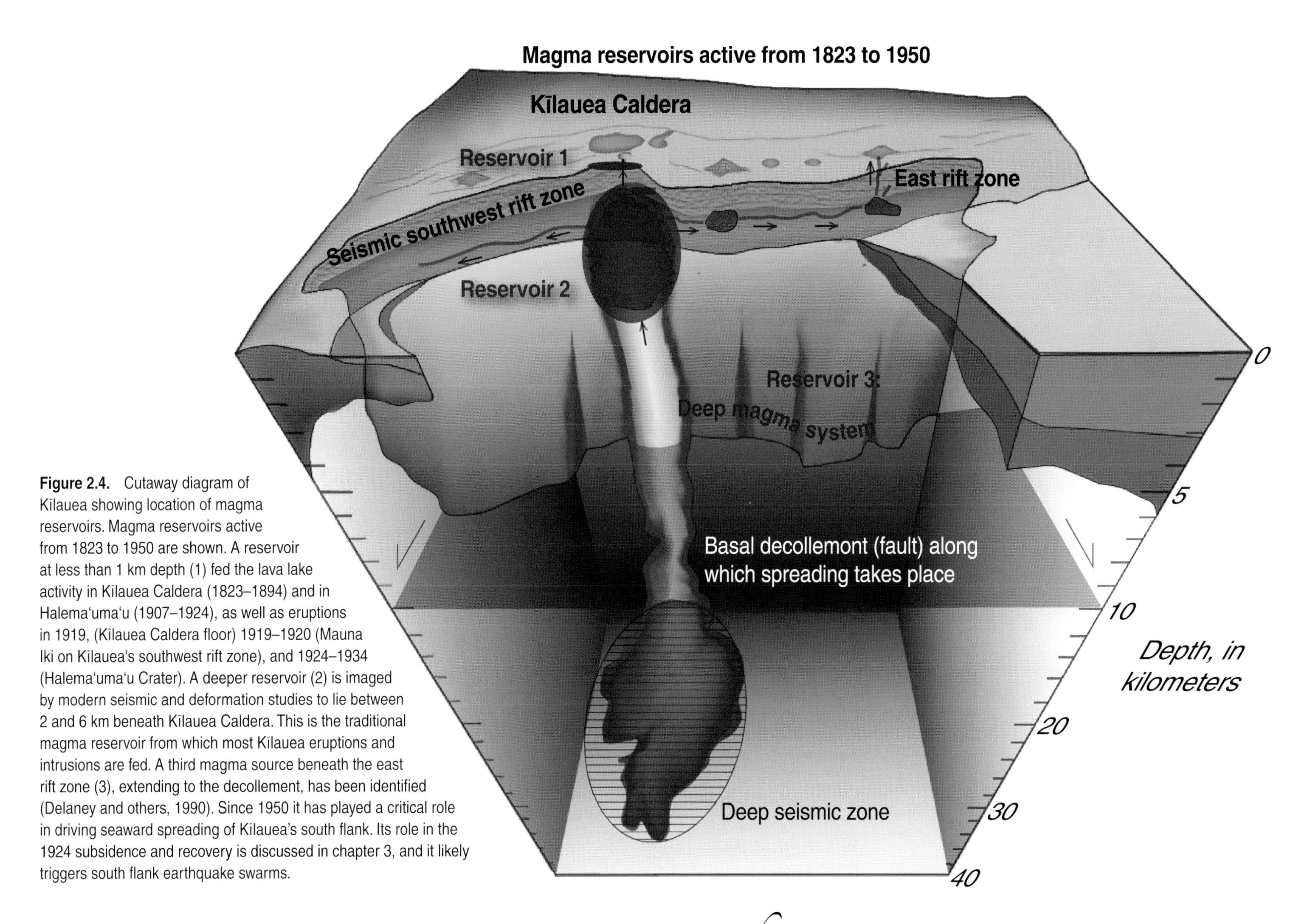

Figure 2.4. Cutaway diagram of Kīlauea showing location of magma reservoirs. Magma reservoirs active from 1823 to 1950 are shown. A reservoir at less than 1 km depth (1) fed the lava lake activity in Kīlauea Caldera (1823–1894) and in Halema‘uma‘u (1907–1924), as well as eruptions in 1919, (Kīlauea Caldera floor) 1919–1920 (Mauna Iki on Kīlauea's southwest rift zone), and 1924–1934 (Halema‘uma‘u Crater). A deeper reservoir (2) is imaged by modern seismic and deformation studies to lie between 2 and 6 km beneath Kīlauea Caldera. This is the traditional magma reservoir from which most Kīlauea eruptions and intrusions are fed. A third magma source beneath the east rift zone (3), extending to the decollement, has been identified (Delaney and others, 1990). Since 1950 it has played a critical role in driving seaward spreading of Kīlauea's south flank. Its role in the 1924 subsidence and recovery is discussed in chapter 3, and it likely triggers south flank earthquake swarms.

Table 2.5. Kīlauea Caldera filling rates 1823-1924.

[In rows with multiple entries text applies down to the next entry; dates in mm/dd/yyyy format]

Period	Start date	End date	D time (yr)	Volume (km^3)[1]	Filling rate (km^3/yr)[2]	Comment	Reference(s)
1823-1825	08/14/1823	06/30/1825	1.88	0.604	0.322	Average of Mastin volume range used	(Mastin, 1997)
1825-1832	0630/1825	01/01/1832	6.50	0.788	0.121[3]	Average of Mastin volume range used	(Mastin, 1997)
1832-1840	01/15/1832	05/31/1840	8.38	1.968	0.235	Average of Mastin volume range used	(Mastin, 1997)
1840-1846	05/31/1840	07/10/1846	6.108	0.2721	0.0446	1840 collapse volume (Mastin, 1997) added to crater fill	(Dana, 1887b; Lyman, 1851)
1846-1848	07/10/1846	08/10/1848	2.086	0.1216	0.0583	Endogenous growth between 1840 and 1848	(Coan, 1852; Dana, 1887b)
1848-1855	08/10/1848	10/05/1855	7.151	0.1886	0.0263		(Coan, 1856a, 1856b; Dana, 1887b)
1855-1863	10/5/1855	10/15/1863	8.027	0.1631	0.0203		(Coan, 1864; Dana, 1887b)
1863-1868	10/15/1863	04/01/1868	4.463	0.0973	0.0218	Central dome plus lava fill	(Coan, 1864, 1867; Dana, 1887b)
1868-1874[4]	04/01/1868	10/15/1874	6.535	0.1119	0.0212	Halema'uma'u dome + 200 feet filling of 1868 depression + 1871 collapse	(Coan, 1874; Dana, 1887b)
1874-1886[4]	10/15/1974	03/06/1886	11.389	0.4202	0.0373	Fill remainder of 1868 depression fill + 1877, 1879 and 1880 collapses + dome	(Brigham, 1868; Dana, 1887b, 1888; Emerson, 1887)
1886-1894	03/06/1886	07/10/1894	8.345	0.1899	0.0228	1886 + 1891 collapse + Halema'uma'u dome	(Anonymous, 1894a; Thurston, 1894)
1840-1894	05/31/1840	07/10/1894	54.108	1.4048	0.0307	Entire interval: fill + all collapses	
1894-1918	12/06/1894	12/02/1906	11.986	0.0043	0.0017	Halema'uma'u fill + collapses in 1904 and 1905-06; continuous filling;	(Bevens and others, 1988; Wood, 1917)
	12/02/1906	05/20/1916	9.465	0.0514	0.0054	includes 0.0042 km^3 fill and 0.0323 km^3 refill following collapses	
	05/20/1916	02/23/1918	1.763	0.0300	0.0170	Includes 0.0139 km^3 fill and 0.0082km^3 refill following collapses in 1916 and 1917	
1894-1918	12/06/1894	02/23/1918	23.214	0.1051	0.0045	Entire interval; includes 0.0214 km^3 fill and 0.0533 km^3 refill following collapses	(Bevens and others, 1988; Wood, 1917)
1918-1924	02/23/1918	02/9/1924	5.960	0.1425	0.0239	0.0617 km^3 erupted lava + 0.0808 km^3 refill after collapse[5]	(Bevens and others, 1988; Wood, 1917)

[1]Lava volume converted to magma volume by correction for 20% vesicularity. Volumes used are the sum of the net lava fill added to the volume of collapse (tables 2.3, 2.4) immediately preceding and within the filling period. Volumes of lava fill are calculated as cylinders or half spheroids to account for doming of the inner caldera (see text).

[2]Filling rate after correction of lava volume to magma volume.

[3]Mastin uses Malden's estimate that 50 feet of lava covered the black ledge in 1832. In order to support a declining fill rate from 1823 to 1840 Malden's depth must be increased by 200 feet or more or the altitude of the black ledge must be reduced by a similar amount. For example, an increase of lava depth of 250 feet between 1825 and 1832, matched by a corresponding decrease of 250 feet between 1832 and 1840 yields magma supply rates of 0.206 km^3/yr for 1825–1832 and 0.183 km^3/yr for 1832–1840.

[4]Brigham's estimate of depth to the caldera floor indicates no change since 1840. Emerson's 1886 map indicates a depth to the caldera floor yields an altitude close to the pre-1919 altitude. Accepting both maps gives a very low rate of caldera fill between 1874 and 1880 and a very high rate between 1880 and 1886. Dana (1888, v. 35, p. 18) corrects Brigham's estimate of fill by 50 feet; accepting this correction yields more reasonable values for the post-1868 filling rates.

[5]The volumes of the two east rift eruptions are very small. The volume of magma transfer is included in the collapse/refill figure.

1918–1924

The lava lake in Halema'uma'u overflowed onto the caldera floor numerous times between 1918 and 1921. A long eruption in Kīlauea Caldera, named "Postal Rift,"[9] occurred between February and December 1919, and a second eruption followed shortly thereafter to build the small Mauna Iki shield on Kīlauea's southwest rift zone. Both eruptions were fed directly from the Halema'uma'u lava lake, as evidenced by lava sighted at a few meters depth in the southwest rift preceding the beginning of surface eruption (Bevens and others, 1988; Rowland and Munro, 1993); likewise, the outlet for the Postal Rift eruption was visible during the draining of Halema'uma'u lava lake preceding the 1919–20 Mauna Iki eruption (Bevens and others, 1988). Small eruptions on Kīlauea's east rift zone near Nāpau and Makaopuhi Craters occurred in 1922 and 1923, associated with temporary draining of Halema'uma'u lava lake and large earthquake swarms (de Vis-Norton, 1922, 1923; Finch, 1926b; Jaggar, 1931). Extensive ground deformation near the eruption sites indicated significant intrusion beneath the east rift zone. These two eruptions are very similar to eruptions that occurred on the east rift zone during the 1960s. The seismicity was more intense (table 2.3), with higher average swarm magnitudes, expectable for a rift zone that had seen no activity since 1840.

Eruption rates were calculated for this period in the same way as filling rates were for the preceding period. Drainings in 1922 and 1923 were far larger than the volume of erupted lava, so the drained volumes have been used in combination with the 1919 and 1919–20 eruption volumes to calculate a magma supply rate of 0.0239 km^3/yr, similar to the rate prevailing in the latter part of the 19th century (fig. 2.1, table 2.5).

[9]Before the eruption the source was located within a hot and steaming crack that was a favorite place for tourists to singe postcards before mailing them.

Tilt Record 1913–1924

After 1913 HVO made continuous records of the tilting of the ground as measured by the Bosch-Omori seismometer installed in the Whitney Vault on the northeast rim of Kīlauea Caldera. Comparison of tilt magnitude with fluctuation in the level of Halema'uma'u lava lake is shown in figure 2.3. The correlation is not perfect, but is obvious for some major events, such as the 1922 and 1924 drainings.

The 1924 Crisis at Kīlauea

Sequence of Events

In 1924, continuous lava lake activity abruptly ended with a spectacular series of events, culminating in large phreatic explosions at Halema'uma'u. The sequence began in February with initial disappearance of lava from Halema'uma'u, followed by a series of felt earthquakes beneath the middle section of Kīlauea's east rift zone in March (Finch, 1924), and a major earthquake swarm farther east along the rift zone in April (Finch, 1925). Although both Jaggar (Bevens and others, 1988, v. 3, p. 515) and Finch (1924, 1925) report a progressive increase in epicentral distance along the rift zone, tabulated location data (Bevens and others, 1988, v. 3, p. 513) do not show an obvious time-distance progression. Rather, it appears that Jaggar and Finch were extrapolating eastward from a mild concentration of earthquakes in the vicinity of Kalalua Crater, spread over a period of about two weeks.

Three weeks after the last Kalalua earthquake, a swarm of small earthquakes felt locally began in the vicinity of Puu Kaliu, on the lower east rift zone, expanding within 3 days to an intense swarm with many widely felt earthquakes beneath the village of Kapoho. During the following 3 days the Kapoho graben subsided dramatically, accompanied by opening of fractures and uplift of the land on either side of the graben-bounding faults where they intersected the shoreline (Finch, 1925). Subsequent modeling of the triangulation done in 1933 (Wingate, 1933)[10] fits a dike emplaced beneath Kapoho (Paul Delaney, written commun., 1990). Jaggar and Finch also surmised the possibility of an undersea eruption in 1924. This is now considered unlikely, as dredging and diving on the offshore east rift zone has failed to confirm the existence of a flow of the appropriate age and chemistry.[11]

Meanwhile, at Kīlauea's summit, Halema'uma'u began draining in early February and lava disappeared by 21 February 1924 (Bevens and others, 1988, v. 3, p. 505). At the time of the peak in earthquake activity in lower Puna on 23 April, lava briefly reappeared at the bottom of Halema'uma'u, then disappeared again. On 30 April, following the Puna crisis, the summit began to collapse, accompanied by tens to hundreds of earthquakes per day. On 9 May a series of violent phreatic explosions at Halema'uma'u began, continuing to the end of the month. At this time observatory staff members and volunteers were sent to various locations to record the times of felt earthquakes, the location of falling lithic ash, and any other observations of interest. The contents of their notebooks were published

[10]Unpublished data obatined from the Hawai'i State archives in Honolulu.

[11]Submarine exploration of Kīlauea's east rift zone has to date revealed no lavas of historical age, thus suggesting that the prior drainings and east rift eruptions of 1790 and 1840 did not reach the submarine surface of the east rift zone.

much later, as one chapter in a book discussing the mechanics of crater formation (Jaggar, 1947, p. 205–259). The events were summarized in the HVO Monthly Bulletin for May 1924 (Bevens and others, 1988, v. 3, p. 529–560). During 3 weeks of phreatic explosions the floor of Halemaʻumaʻu continued to collapse, accompanied by more widespread subsidence of Kīlauea's summit, leaving a greatly enlarged and empty pit 400 m deep (Finch, 1926b; Jaggar and Finch, 1924). The estimated volume by which Halemaʻumaʻu was enlarged exceeded the amount of material erupted, in a ratio of 253:1 (Jaggar, 1925). Later investigation of the products showed that no juvenile magma was involved (Finch, 1947), although some of the lithic ejecta were observed to glow at night, and molten lava was observed dripping from sills in the walls of the newly emptied Halemaʻumaʻu Crater.

Contemporary interpretations of the eruption were internally consistent, all invoking draining of magma from Kīlauea's summit into the east rift zone, followed by interaction of groundwater with the heated rocks surrounding the magma conduit. Harold Stearns (1925) ascribed the faulting at Kapoho to draining of magma emplaced in the rift during eruptions of 1922 and 1923. He further hypothesized that the lava column subsided more than 1,100 m (3,600 ft) during the subsidence preceding the explosions, (that is, down to about sea level and well below the local water table), permitting groundwater to enter the hot conduit. Steam accumulated under the plug of talus until explosive pressures were attained. Stearns explained the regularity of the steam blasts by a mechanism similar to that observed in geyser fields. John Dvorak (1992) has added to the interpretation of the 1924 events, concluding that withdrawal of magma alone is not sufficient to trigger explosions. He considers that the explosions were initiated at very shallow levels by interaction of perched ground water brought suddenly in contact with rocks preheated during extended lava lake activity. He also revises the amount of magma removed from Kīlauea's summit to 0.4 km^3, twice the figure estimated by Finch and Jaggar.

The events of 1924 were the most dramatic at Kīlauea since 1790 and 1868. The 1924 eruption, which contained only lithic material, contrasts with the explosive activity in 1790, which was marked by eruption of juvenile ash up to the final explosive surges. Jaggar claimed to have forecast the breakdown in 1924 on the basis of his belief in 130-year cycles of activity, the cycle beginning in 1790 (Bevens and others, 1988, v. 2, p. 721). Whether or not one subscribes to the cyclical theory, the 1924 events marked a change from dominant activity at Kīlauea's summit to increased activity on Kīlauea's east rift zone, a pattern borne out by geologic mapping and dating of Kīlauea flows (Holcomb, 1987). Modern study has also shown that the 1924 eruption represents a geochemical boundary; isotope signatures of lavas erupted before 1924 differ from those erupted after (Pietruszka and Garcia, 1999).

Ground Deformation

Surveys of Kīlauea's summit by triangulation and leveling were conducted in 1912, 1921, and 1926 (Mitchell, 1930; Wilson, 1935). Wilson showed that during both time periods, 1912–21 and 1921–26, the deformation field was very broad, extending more than half the distance to Hilo, at least 20 km from Halemaʻumaʻu Crater (see text table below). The Volcano House on the rim of Kīlauea Caldera rose more than 0.3 m (1 foot) between 1912 and 1921 and dropped 0.5 m (1.7 feet) between 1921 and 1926. The absolute elevation of the rim of Kīlauea Caldera attained in 1921 has not been exceeded as of 2010. Wilson's map also shows that stations within the caldera subsided, presumably in response to draining of magma to feed the 1919 caldera eruption and the 1919–20 Mauna Iki eruption. Wilson questioned the rod corrections applied to the 1912 data and concluded that if the rod corrections were discarded it was possible that the Volcano House benchmark had remained stable through the triangulation of 1921.

The tiltmeter installed in the Whitney Vault began systematic recording in 1913. As explained in appendix A, the Whitney Vault is more than 3 km from Halemaʻumaʻu and will not record draining of lava from Halemaʻumaʻu or from a source at less than 1-km depth beneath Halemaʻumaʻu, whereas it will record inflation or deflation from a source deeper than 3 km. Interpretation of the tilt and triangulation data for the period between 1913 and 1925 are given in the following text table.

Interpretation: The Youngest Summit Shield-Building Period Through May 1924

We interpret the extensive summit lava lake and the high caldera filling rates at the time of the western missionaries' arrival in 1823 (Ellis, 1825) as representing rebound from a massive draining of magma in 1790 that may be represented in Puna by 1790 lava (Trusdell and Moore, 2006) and a possible accompanying intrusion that may have extended beneath the undersea extension of Kīlauea's east rift zone. The eruption of 1790 was not a caldera-forming event (Dibble, 1843; Swanson and Christiansen, 1973; Swanson and Rausch, 2008), but it did drain Kīlauea's entire magmatic system and triggered a very high rate of resupply through lowered pressure above the

Survey Data Relevant to 1924 Subsidence

Original values (elevations in feet)

Date Location	1912	1922	Δ 1912–1922	1927	Δ 1922–1927
Volcano House	3973.090	3974.107	+1.017	3972.541	-1.566
Kea'au	1266.425	1267.229	+ 0.804	1266.477	-0.752

Corrected (Wilson, 1935) values (elevations in feet)

Date Location	1912	1922	Δ 1912–1922	1927	Δ 1922–1927
Volcano House	3976.103	3976.103	0.000	3972.541	-3.562
Kea'au	1266.435	1266.673	+ 0.238	1266.477	-0.194

Whitney tilt values (arc-seconds)[1]

Date begin	Date end	Magnitude	Azimuth
10/22/1918	2/27/1919	85.2	354
4/28/1924	7/3/1924	64.1	198
7/3/1924	12/29/1924	5.3	127

[1]Conversion factors are: 1 arc-second = 4.8468 μrad; 1 μrad = 0.2063 arc-second

magmatic system. We interpret the decline in filling rate after 1840 to be in part a return to a low equilibrium rate (low compared to Kīlauea's post-1950 history) and possibly in part a result of the increased activity at the adjacent Mauna Loa volcano (see chapter 8). We also infer from the occurrence of east rift eruptions and intrusions that, from at least 1790 onward, rift dilation becomes important in controlling the equilibrium magma supply rate (Wright and Klein, 2008).

Following the disappearance of lava in 1894 we interpret Kīlauea to have three different magma sources or reservoirs, as shown in figure 2.4. (1) A shallow magmatic system that fed the lava lakes in Kīlauea Caldera and Halema'uma'u lies at depths of less than a few hundred meters beneath the caldera. This source is evident in the occasional appearance of magma at the bottom of Halema'uma'u preceding the continuous refilling that began in 1907 and the observations cited above regarding the 1919 caldera and 1919–20 southwest rift eruptions. (2) The magma reservoir identified by later instrumental monitoring at between about 2 and 6 km depth beneath Kīlauea's summit (see, for example, Eaton, 1962, and many subsequent studies). (3) A deeper magma system, beneath the shallow summit reservoir at 2–6-km depth, and possibly above the decollement (Delaney and others, 1990). Sources at 3 depths (0.8, 3.5, and deeper than 10 km) are suggested by our modeling of levelling surveys surrounding the 1924 collapse (appendix I). We reinterpret Mogi's source beneath Kīlauea Caldera at 20-km depth (Mogi, 1958) as our source 3, distant from Kīlauea's summit and not directly beneath the caldera.

The period between the end of the large inflation in 1918–19 (following the first overflow of Halema'uma'u in 1918) and the eruption/intrusion of 1922 shows many fluctuations of lava lake level and little correlated tilt change at the Whitney Vault. We interpret this period to be dominated by removal of magma from and drainback into reservoir 1. The changing level of lava in Halema'uma'u associated with intrusion beneath the two rift zones may represent an additional contribution from draining of source 2. For example, a very rapid (3-day duration) episode of draining of the lava lake that followed the end of the Postal Rift eruption (28–30 November 1919) was accompanied by a small east rift earthquake swarm (fig. 2.3*A*) and deflationary tilt (azimuth 198, magnitude 16.2 μr) that suggested intrusion beneath the east rift zone. A second earthquake swarm beneath the southwest rift that took place at the beginning of the Mauna Iki eruption on 15 December 15 1919 showed a 34-m (113-foot) drop in lake level and no tilt response and is interpreted as breaking of rock as magma was draining along the shallow path connecting the summit and Mauna Iki. A slower draining of 90 m (300 feet) between 22 December 1919 and 18 January 1920 and deflationary tilt between 28 and 31 December 1919 (azimuth 194, magnitude 8.2 μrad) suggest a delayed response of the deeper source 2 to the southwest rift zone eruption.

The sharp deflation in 1922 (24 May to 6 June; azimuth 195, magnitude 75 μrad) associated with an east rift eruption and intrusion is correlated with large draining of the lava lake and are consistent with draining magma reservoir 2. The deflation began 4 days before the eruption began, indicating that the eruption was fed from the summit reservoir, consistent wth the absence of an east rift zone eruption since 1840. The 1923 eruption/intrusion, following closely on the similar event of 1922, shows a significant draining of the lava lake accompanied by a very small tilt change (15 μrad deflation at an azimuth of 260 between 25 and 27 August 1923). This azimuth is an unusual one for deflation, and the small tilt magnitude may suggest that eruption of new lava to the surface and further intrusion in 1923 are best interpreted as continued movement of magma that traveled out of the summit reservoir in 1922. Consistent with the idea of dual sources, there is a suggestion that, during intrusions beneath either rift zone, lava first disappears from Halema'uma'u during evacuation of source 1, while continuing to be withdrawn from source 2, much as Harold Stearns suggested for the 1924 collapse (Stearns, 1925).

The large increase in tilt magnitude from 22 October 1918 to 27 February 1919 (413 µrad of inflation at an azimuth of 354) and the great deflation between 28 April and 29 December 1924 (320 µrad of deflation at an azimuth of 194) show tilt azimuths at variance with the other centers of inflation and deflation (appendix A, fig. A2; appendix B, table B3) and are interpreted to have involved source 3. The azimuth of the 1918–19 inflation (fig. 2.3*A*) corresponds to inflation east of the centers defined for the period preceding the 1967–68 eruption (appendix A, fig. A2; appendix B, table B3). The magnitude of the 1918–19 inflation cannot be quantitatively compared to the elevation changes because the 1911–12 leveling took place before the tiltmeter was installed. The azimuth of deflationary tilt up to July 1924 through the period of explosions is similar to the 1922 deflation, consistent with draining from source 2. However, an additional deflation of 25.8 µrad at azimuth 127 between 3 July and 29 December 1924 indicates a more easterly source.

The broad 1924 deflation indicates a source well below the depth of Halema'uma'u Crater. The Japanese geophysicist Kiyoo Mogi modeled the 1924 collapse using the tilt values calculated from Wilson's triangulation to define centers of deflation (Mogi, 1958). Mogi's modeling suggested that two magma reservoirs were activated during the 1924 collapse, one at a depth of about 5 km, within current estimates of Kīlauea's shallow magma reservoir and appoximately consistent with our source 2 above, and one at about 25 km that has not been supported by geodetic and seismic instrumental measurements made in the modern era. Modern data show inflation and deflation confined to Kīlauea Caldera, with little or no change in benchmarks at or beyond the caldera rim. Our fit to the 1921–26 level changes requires two Mogi centers to fit the level changes inside the caldera and three sources to fit the broad caldera subsidence observed along the Kīlauea to Hilo level line (appendix I, fig. I6 and table I1). Mogi center depths of 0.8 and 3.5 km are required to fit the caldera level contours. These correspond to inflation-deflation sources 1 and 2, respectively. If one ascribes the altitude changes seen between the Volcano House and Hilo to a Mogi source, a source geometry which is unconstrained, an additional deflation at 30 km depth is required. John Dvorak's study (Dvorak, 1992) disregarded Wilson's conclusion of a large regional subsidence in 1924 (Wilson, 1935, figure 7), instead assuming the Volcano House benchmark as a fixed datum, citing Wilson's worries about rod corrections applied to the 1912 survey period (Wilson, 1935, p. 41 and following). Yet Wilson's corrections affect only the 1912–21 period and increase the amount that the Volcano House subsided in 1924 to more than 1 m (3.5 feet; see text table above). The large change of Whitney Vault tilt between 1918 and 1919 (figure 2.3*A*) argues for accepting Wilson's uncorrected data, including substantial elevation changes at Volcano House, in order to define a broad regional uplift between 1912 and 1921 and an equal volume of regional subsidence between 1921 and 1926, the latter assumed to have occurred mainly during the events of 1924. If accepted, this increases the volume of subsidence to more than the 0.4 km^3 calculated by Dvorak.

We conclude that the large Whitney Vault tilt changes in 1918–19 and 1924–25 are consistent with Wilson's initial conclusion of a 0.3-m (1-foot) increase in the altitude of the Volcano House benchmark before the 1921 triangulation and an even greater decrease in altitude across the 1924 collapse. The 1912–22 rise and 1922–27 fall of the Kea'au benchmark, about 40 km from Kīlauea's summit along the road to Hilo, argue that its change is due to inflation and deflation of a deep magmatic source, rather than an irreversible tectonic change resulting from a source like south flank spreading. The large changes, extending at least as far as Kea'au, amount to a regional inflation/deflation extending more than 20 km from Halema'uma'u. A possible alternative to the 30-km-deep Mogi source modeled in appendix I, table I1, is additional draining of the deeper magmatic system of source 3, which lies above the decollement and below the east rift zone at depths of 10–12 km but is located at a lateral distance of several tens of kilometers from Kīlauea's summit. Unlike more recent collapses of Kīlauea's summit, the immediate recovery from the 1924 collapse, as measured by tilt, was only a fraction of the total collapse (fig. 2.3*A*), consistent with the need to refill a much larger magma volume than that occupied by sources 1 and 2 alone.

Finally, we test the hypothesis that a large earthquake could explain the 1918–19 tilt change at Kīlauea's summit. Late on the evening 1 November 1918 an earthquake of magnitude 6.4 occurred beneath Mauna Loa's Ka'ōiki Fault Zone (Klein and Wright, 2000, and references cited therein). A similar earthquake of magnitude 6.6 occurred on 16 November 1983 (Buchanan-Banks, 1987; Maley, 1986, Jackson and others, 1992) and produced a large instantaneous and permanent offset in the Uwēkahuna tiltmeter. There was extensive damage around the caldera rim.

The 1918 earthquake produced cracking on the floor of Kīlauea Caldera (Anonymous, 1918) but little damage around the rim (Bevens and others, 1988, v. 2, p. 840) and a small offset of the clinometer in the Whitney Vault (Bevens and others, 1988, v. 2, p. 843). However, the earthquake was an unlikely trigger for the inflation that had begun 10 days before (Anonymous, 1918; Bevens and others, 1988, v. 2, p. 836) and which continued for more than three months (see inset table above and fig. 2.3).

Supplementary Material

Supplementary material for this chapter appears in appendix B, which is only available in the digital versions of this work—in the DVD that accompanies the printed volume and as a separate file accompanying this volume on the Web at http://pubs.usgs.gov/pp/1806/. Appendix B contains the following figures and tables:

Tables B1a and B1b contain earthquakes of magnitudes between 5 and 6 for the same time periods as text tables 2.1 and 2.2, respectively.

Table B2 shows azimuth and distance from Uwēkahuna and Whitney Vaults to points within Kīlauea Caldera shown in text figure 2.4.

Table B3 contains Whitney Vault tilt data to support text figure 2.3.

Figure B1 plots occurrence of earthquakes designated as "south Hawai'i" on a timeline that also shows (1) Kīlauea eruptions and intrusions, (2) Mauna Loa eruptions, and (3) Kīlauea earthquake swarms.

Kīlauea Caldera at the time of missionary William Ellis' vist in 1823, showing activity extending over the entire caldera floor (Ellis, 1825 [1827 edition], plate facing p. 226). Image courtesy of Bishop Museum.

May 1924 explosive eruption in Halemaʻumaʻu viewed by tourists. Image courtesy of Bishop Museum.

Chapter 3

Eruptive and Intrusive Activity, 1925–1953

Kīlauea recovers from the 1924 deflation under a regime of increased magma supply rate and initiation of significant seaward spreading of the volcano's south flank, preceding the first documented long Halema'uma'u eruption in 1952.

Kīlauea experienced seven eruptions and intrusions and large earthquakes from the time of the 1924 collapse through the Halema'uma'u eruption of May-November 1952. These volcanic events and large earthquakes are summarized in table 3.1. Significant events before the beginning of reinflation in early 1950 include seven small eruptions between July 1924 and November 1934 confined to Halema'uma'u (fig. 3.1), the formation of the "Puhimau thermal area," circa 1937, and two upper east rift intrusions in May and August 1938 (fig. 3.2). A possible summit intrusion in December 1944 (Finch, 1944; Klein and Wright, 2000) and several smaller events identified by shallow earthquake swarms beneath Kīlauea's summit are also shown in figure 3.1. During this time the volcano remained deflated, as the Whitney Vault tilt records an actual net drop of an additional 97 microradians (µrad) or 20 seconds of arc between 1925 and 1950.

Between 1925 and 1950 the Bosch-Omori tiltmeter in the Whitney Vault recorded cyclic variations, attributed by early authors to "seasonal" variations, but shown by us to be affected by periods of heavy rainfall[12]. Beginning in March 1950 up to the abandonment of the Whitney Vault in 1962, the Whitney tilt records show rapid deflations, inflations, and longer term trends that correlate well with volcanic events and are typical of modern water-tube tiltmeter observations.

During this time period the accurate location of intrusions and large earthquakes was still limited by the absence of a full seismic network and the short-term affects of rainfall on the Whitney tiltmeter. Major intrusions beneath the rift zones can be located from contemporaneous ground cracking. South flank activity is much less closely defined (Klein and Wright, 2000), and some felt events are classified as unsubdivided "south flank." There is also potential ambiguity in the earthquake swarms accompanying rift intrusion. Absent a large felt event, some of the swarm earthquakes may be a south flank response, but at this time the seismic network cannot discriminate between rift and flank.

[12]The Whitney tiltmeter is discussed in appendix A.

Halema'uma'u Eruptions, 1924–1934

In the decade following the 1924 collapse, Kīlauea erupted seven times at the bottom of the newly deepened Halema'uma'u Crater (table 3.1, figure 3.1). Eruptions in 1930, 1931–32 and 1934 were more voluminous than the others. Whitney Vault tilt signals near the times of eruption are not well-correlated with eruption times. It is likely that the eruptions of 1924–34 were fed from shallow pockets of magma remaining from the lava lake activity before the 1924 collapse (source 1 defined in fig. 2.4) with minor inflation/deflation of source 2. Changes in the shallow magma chamber are too far away to register at the Whitney tiltmeter. Unfortunately, this cannot be verified by the chemistry and petrography of the erupted lavas because only one sample of this series of Halema'uma'u eruptions was collected—the 1931 eruption was sampled in conjunction with recovery of victims of a double suicide from the bottom of Halema'uma'u (Fiske and others, 1987, Volcano Letter 388).

East Rift Intrusions in 1936–1938

Before the last eruption in Halema'uma'u in 1934, the seismicity along the magma supply path and beneath the rift and south flank appeared to increase beneath several regions of the volcano (fig. 3.1, fig. C1). The seismicity increase may be believable, because the seismic network and presence of a staff seismologist were consistent through the 1930s (Klein and Wright, 2000). Earthquakes at shallow (0–5 km) and intermediate (5–15 km) depth beneath Kīlauea's summit and beneath the Koa'e Fault Zone increased in the latter half of 1936 (fig. C1). In the Volcano Letter for May 1938 the HVO staff noted that steam appeared near Kōko'olau Crater accompanied by wilting of vegetation "within the last year or two" (Fiske and others, 1987, Volcano Letter 459. p. 2, 4). The steaming area was referred to as the "Kōko'olau hot area" (later called Puhimau thermal area) and identifies an area of intrusion beneath the east rift zone. Widening of

Figure 3.1. Graphs showing Kīlauea activity, 1925–1935. Dates on the time (x) axis are centered at the beginning of the year, as in figure 2.2. For example, the tick mark at 1926 is at 1 January 1926. Halema'uma'u eruptions following 1924 collapse (first eruption in July 1924 not shown). Seismicity (bottom six panels) and Whitney tilt magnitude (second panel from top) related to times of eruption given in the top panel are emphasized by vertical dotted lines. Tilt magnitudes are given in arc-seconds (left axis) and microradians (right axis). Seismicity is plotted, from bottom to top, for the regions defined in appendix A, table A3 corresponding to the magma supply path, rift zones and Koa'e, and south flank. Earthquakes per day (eq/day) and magnitudes (*M*) greater than or equal to 4.0 are given for each region. Region identifications for south flank earthquakes in this period are not well determined. Eruptions are associated with increase in shallow caldera earthquakes but are not correlated with changes in Whiney tilt. Additional times of of enhanced shallow summit seismicity are not accompanied by eruption. See text for further discussion.

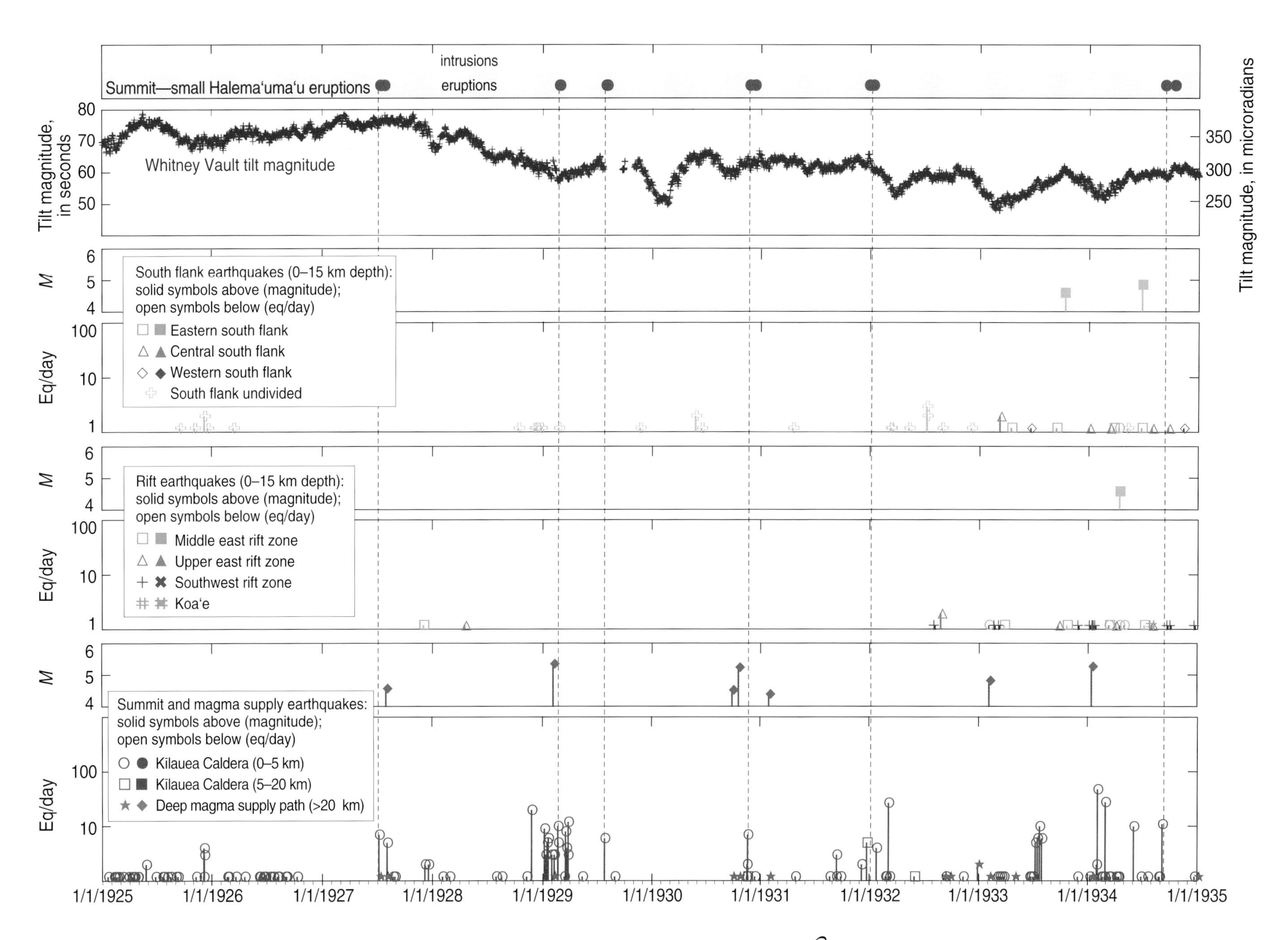
Summit—small Halema'uma'u eruptions
intrusions
eruptions
Whitney Vault tilt magnitude
Tilt magnitude, in seconds
Tilt magnitude, in microradians
South flank earthquakes (0–15 km depth):
solid symbols above (magnitude);
open symbols below (eq/day)
Eastern south flank
Central south flank
Western south flank
South flank undivided
Rift earthquakes (0–15 km depth):
solid symbols above (magnitude);
open symbols below (eq/day)
Middle east rift zone
Upper east rift zone
Southwest rift zone
Koa'e
Summit and magma supply earthquakes:
solid symbols above (magnitude);
open symbols below (eq/day)
Kilauea Caldera (0–5 km)
Kilauea Caldera (5–20 km)
Deep magma supply path (>20 km)
M
Eq/day
1/1/1925
1/1/1926
1/1/1927
1/1/1928
1/1/1929
1/1/1930
1/1/1931
1/1/1932
1/1/1933
1/1/1934
1/1/1935

Table 3.1. Kīlauea eruptions, intrusions, earthquake swarms, and earthquakes $M \geq 4$, 1925–1953.

[In rows with multiple entries text applies down to the next entry; data for eruptions and traditional intrusions are emphasized by grey shading; dates in m/d/yyyy format; closely related events nearby in time are grouped together]

Start date	End date	Location[1]	Activity[2]	Comment	References[3]
07/07/1927	07/20/1927	Hm	E	Eruption in Halema'uma'u Crater ; no precursory seismicity	ESPHVO , v. 3, p. 989 –1007
07/31/1927		KC?	EQ	*M* 4.58 deep beneath Kīlauea	KW; ESPHVO, v. 3, p. 1185
01/10/1928	01/11/1928	KC	EQS	7 precursory events 22:43, 01/10–00:26, 01/11/1928	ESPHVO, v. 3, p. 1059–1066
01/11/1928	01/11/1928	Hm	E	Lava from the 1927 eruption reactivated	
02/05/1929		KC?	EQ	*M* 5.4 deep beneath Kīlauea	KW; ESPHVO, v. 3, p. 1185
02/19/1929	02/20/1929	KC	EQS	15 precursory events 22:42, 02/19–00:46, 02/20/1929	ESPHVO, v. 3, p. 1184–1189
02/20/1929	02/21/1929	Hm	E	Eruption in Halema'uma'u Crater;	
07/25/1929	07/28/1929	KC	EQS	6 precursory events 4:20–5:31, 07/25/1929	ESPHVO, v. 3, p. 1209–1217
07/25/1929		Hm	E	Tilt data missing from 07/22–09/19/1929	
09/28/1930		KC?	EQ	*M* 4.53 deep beneath Kīlauea	KW; VL 301, p. 4
10/20/1930		KC?	EQ	*M* 5.25 deep beneath Kīlauea	KW; VL 305, p. 3
11/19/1930	11/19/1930	KC	EQS	83 precursory events at depths of 0–5 km	VL 309, p. 2–4
11/19/1930	12/07/1930	Hm	E	Eruption in Halema'uma'u Crater;	
01/29/1931		KC?	EQ	*M* 4.4 deep beneath Kīlauea?	KW; VL 364, p.3
12/23/1931	12/23/1931	KC	EQS	5 precursory events at depths of 6–8 km	VL 366–367
12/23/1931	01/05/1932	Hm	E	Eruption in Halema'uma'u Crater;	
03/04/1932	03/04/1932	Hm	EQS	37 events 0–5 km depth—no Whitney tilt signal	VL 376
02/04/1933		sf offshore?	EQ	*M* 4.83 deep beneath Kīlauea's offshore south flank	KW
01/13/1934		sf offshore?	EQ	*M* 5.28 deep beneath Kīlauea's offshore south flank	KW; VL 407, p. 2
02/02/1934	02/02/1934	KC	EQS	47 events at 0–5 km depth	VL 408
09/06/1934	09/06/1934	KC	EQS	9 precursory events at 0-5 km depth	VL 415–416
09/06/1934	10/08/1934	Hm	E	Eruption in Halema'uma'u Crater;	
01/02/1935		KC	EQ	*M*5.15; 30 km beneath Kīlauea Caldera	KW; VL 419, p. 2
06/25/1935		KC	EQ	*M*4.16; 10–20 km beneath Kīlauea Caldera	KW; VL 424, p. 2
04/15/1936		KC	EQ	*M*4.66; 30 km beneath Kīlauea Caldera	KW; VL 434, p. 2
05/28/1938	06/01/1938	ERZ	EQS/I	>88 precursory events (many felt) between 6:05 and 13:00; ground cracking near Devil's Throat	VL 459, p. 2–5; KW
05/28/1938		KC	EQ	*M*4.18; 30 km beneath Kīlauea Caldera	KW; VL 528, p. 2–5
06/01/1938		sf	EQ	*M*4.31	KW; VL 529, p. 3
06/02/1938		KC	EQ	*M*4.05 24 km beneath Kīlauea Caldera	KW; VL 529, p. 3
08/08/1938	08/13/1938	ERZ	EQS/I	>353 precursory events (many felt) 16:38–24:00, 08/08/1938; ground cracking near Devil's Throat	VL 462, p. 2–5
08/08/1939		KC?	EQ	*M*4.55; 30 km beneath Kīlauea Caldera	KW; VL 462, p. 3
10/25/1938		KC	EQ	*M*4.10; 25.6 km depth beneath Kīlauea Caldera	KW; VL 462, p. 5
04/12/1939		KC	EQ	*M*4.12; 28.8 km depth beneath Kīlauea Caldera	KW; VL 464, p. 5
05/24/1939		KC	EQ (2)	*M*5.39 and 4.19; 30 km beneath Kīlauea Caldera	KW; VL 464, p. 6
06/12/1939		KC	EQ	*M*4.65; 20.8 km beneath Kīlauea Caldera	KW; VL 464, p. 6
06/19/1939		KC	EQ	*M*4.29; 24 km beneath Kīlauea Caldera	KW; VL 464, p. 7
07/01/1939		KC	EQ	*M*4.39; 17.6 km beneath Kīlauea Caldera	KW

Table 3.1. Kīlauea eruptions, intrusions, earthquake swarms, and earthquakes $M \geq 4$, 1925–1953.—Continued

[In rows with multiple entries text applies down to the next entry; data for eruptions and traditional intrusions are emphasized by grey shading; dates in m/d/yyyy format; closely related events nearby in time are grouped together]

Start date	End date	Location[1]	Activity[2]	Comment	References[3]
04/20/1941		KC	EQ	*M*4.53; 25 km beneath Kīlauea Caldera	KW; VL 472, p. 3
02/18/1942		KC	EQ	*M*4.05; 12.8 km beneath Kīlauea Caldera	KW; VL 475, p. 2
11/12/1944		KC?	EQ	*M*4.58; 30 km beneath Kīlauea Caldera	KW; VL 486, p. 3
11/12/1944	12/06/1944	KC	EQS	Scattered earthquakes of decreasing depth after 11/15	Finch, 1944,
12/06/1944	12/07/1944		EQS/I	16 shallow events (some felt) beneath Kīlauea Caldera associated with subsidence near Halema'uma'u	Finch, 1944, KW
01/24/1945		KC	EQ	*M*4.32; 21 km beneath Kīlauea Caldera	KW; VL 487, p. 5
08/18/1947		KC	EQ	*M*4.27; 21 km beneath Kīlauea Caldera	KW; VL 497, p. 3
09/30/1947		KC	EQ	*M*4.74; 25 km beneath Kīlauea Caldera	KW; VL 497, p. 3
12/14/1947		KC	EQ	*M*4.69; 32 km beneath Kīlauea Caldera	KW; VL 498, p. 3
03/19/1948		KC	EQ	*M*4.35; 25 km beneath Kīlauea Caldera	KW; VL 499, p. 3
07/29/1948	07/31/1948	KC	I?	6 felt *M*2.8–4.1 on 7/30 at 5–15 km depth	KW; VL 501, p. 3
08/02/1950	08/02/1950	KC	I	3 felt out of 241 recorded; Whitney tilt 8/1–5: 5.8 µrad at azimuth 33.1	KW; 509, p. 6
12/08/1950	12/14/1950	Koa'e[4]	EQS	*M*3.90 recorded in Honolulu; 16 *M*< 4	Finch, 1950;
12/09/1950			EQS	*M*4.40 and 5.12 recorded in Honolulu; 48 *M*< 4	Finch and Macdonald, 1953,
12/10/1950			EQS	*M*4.70 (3), 4.98, 5.26 recorded in Honolulu; 66 *M*< 4	KW: VL 510, p. 3–4
12/11/1950			EQS	*M*4.12 recorded in Honolulu; 36 *M*< 4	
12/12/1950			EQS	8, 6 and 6 *M*< 4 over 3 days	
04/22/1951		KC	EQ	*M*6.23 at 14:52 preceded by a *M*4.21 foreshock at 04:21 and followed by many aftershocks. Location east of Kīlauea Caldera is questionable; Interpreted as 35 km deep along the magma supply path	Macdonald, 1951; Macdonald and Wentworth, 1954; VL 512, p. 1–5
03/13/1952	04/21/1952	SF	EQS	Intense earthquake swarm beneath Kīlauea's offshore south flank. Magnitudes range from 6.2 (2 events), 5–6 (19 events), 4–5 (36) and hundreds at *M*< 4. "Suspected deep intrusion"?	Macdonald, 1952, 1955; KW; VL 515, p.1–7
04/03/1952	04/12/1952	ERZ	EQS?	15 events beneath the onshore east rift zone	KW; VL 516, p. 7
04/12/1952		KC	EQ	*M*4.52 10–20 km beneath Kīlauea's summit	KW; VL 516, p. 7
05/10/1952		KC	EQ	*M*4.04 5–10 km beneath Kīlauea's summit	KW; VL 516, p. 7
06/19/1952		KC	EQ	*M*4.86 25 km deep beneath Kīlauea's summit	KW; VL 516, p. 8
		SWR	EQ	*M*3.97, 4.03 beneath southwest rift zone	
06/27/1952	06/28/1952	KC	EQS	Precursory swarm 0–5 km	KW; VL 516, p. 8;
	11/09/1952	Hm	E	Eruption in Halema'uma'u Crater	Macdonald, 1952, 1955; VL 518, p. 1–10

[1]Location abbreviations correspond to regions shown on chapter 1, figure 1.1A: KC, Kīlauea Caldera; Hm, Halema'uma'u Crater; ERZ, East rift zone; SWR, Southwest rift zone; SF, South flank.

[2]Event abbreviations: E, Eruption; I, intrusion; EQ, Earthquake *M*5; EQS, Earthquake swarm; C, Sharp summit tilt drop indicating transfer of magma to rift zone.

[3]ESPHVO, Bevens and others, 1988, Early serial publications of the Hawaiian Volcano Observatory; KW, Klein and Wright, 2000; VL number, Compilation of The Volcano Letter by Fiske and others, 1987.

[4]Epicenters of these earthquakes are across the Koa'e fault zone from Kōko'olau (east rift zone) to Kamakai'a Hills (southwest rift zone). The larger events were recorded on O'ahu. We interpret at least some to lie deeper than 20 km within the magma supply path. See second plot from bottom of figure 3.4.

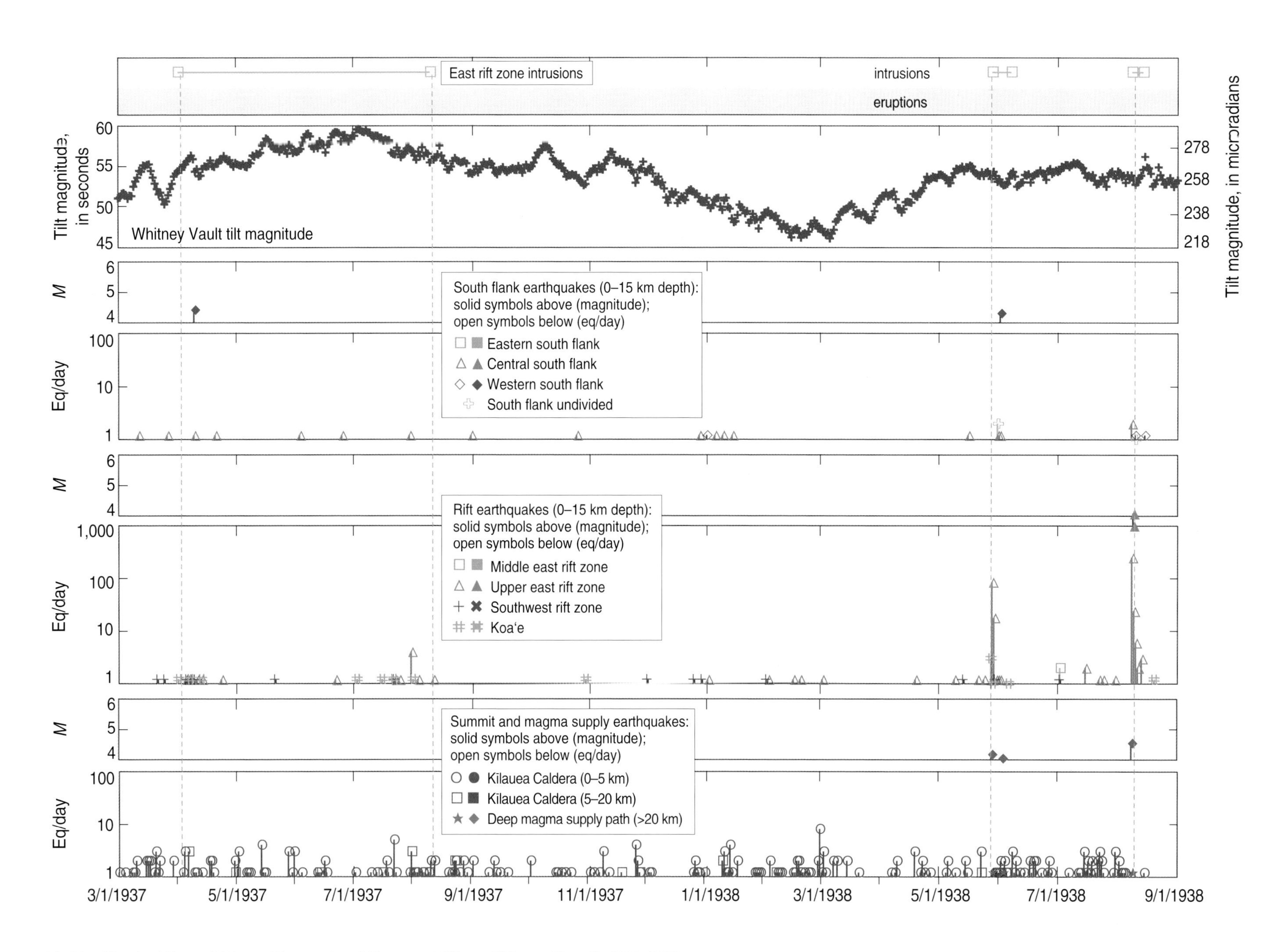
East rift zone intrusions
intrusions
eruptions
Tilt magnitude, in seconds
Whitney Vault tilt magnitude
Tilt magnitude, in microradians
60
55
50
45
278
258
238
218
M
6
5
4
Eq/day
100
10
1
South flank earthquakes (0–15 km depth):
solid symbols above (magnitude);
open symbols below (eq/day)
Eastern south flank
Central south flank
Western south flank
South flank undivided
1,000
Rift earthquakes (0–15 km depth):
solid symbols above (magnitude);
open symbols below (eq/day)
Middle east rift zone
Upper east rift zone
Southwest rift zone
Koa‘e
Summit and magma supply earthquakes:
solid symbols above (magnitude);
open symbols below (eq/day)
Kīlauea Caldera (0–5 km)
Kīlauea Caldera (5–20 km)
Deep magma supply path (>20 km)
3/1/1937
5/1/1937
7/1/1937
9/1/1937
11/1/1937
1/1/1938
3/1/1938
5/1/1938
7/1/1938
9/1/1938

the area of steaming has occurred during the current post-1983 Kīlauea eruption (http://hvo.wr.usgs.gov/volcanowatch/2003/03_02_06.html), indicating that it remains a preferred area of intrusion and potential pit crater formation. The intrusion may have coincided with increased seismic activity beneath the upper east rift zone and Koa‘e Fault Zone (fig. 3.2). The inferred intrusion was not accompanied by ground cracking, indicating that emplacement was below 2-km depth.

In 1938 there were two intrusions into the east rift zone in the vicinity of Devil's Throat, accompanied by major earthquake swarms (figs. 3.2 and 3.3, table 3.1) and by cracking and uplift of the Chain of Craters Road between Devil's Throat and Aloi Craters (Fiske and others, 1987 Volcano Letters 459–460, 462). Apparent Whitney Vault tilt deflations were offset from the time of intrusion and smaller than the measurement uncertainty. The intensity of the ground deformation suggests that the intrusions were too shallow to be detected at Whitney and may have been fed in part from the deeper and earlier formed storage zone beneath the Puhimau thermal area. Cracks that opened in August were staked and the amount of opening measured. The movement continued for at least 6 months following the August intrusion and suggests that the intrusion was still being fed.

Figure 3.2. Graphs showing Kīlauea activity, 1 March 1937–1 September 1938. East rift intrusions including hypothesized formation of Puhimau thermal area. Panels as in figure 3.1. Dates on figure in m/d/yyyy format. Intrusion times are emphasized by dotted vertical lines. Formation of the Puhimau thermal area, probably sometime during 1937, is presumed correlated with occurrence of earthquakes in the upper east rift zone and Koa‘e regions. Whitney tilt is not correlated with either the Puhimau intrusion or the two 1938 intrusions. Earthquake locations are shown in figure 3.3. See text for further discussion.

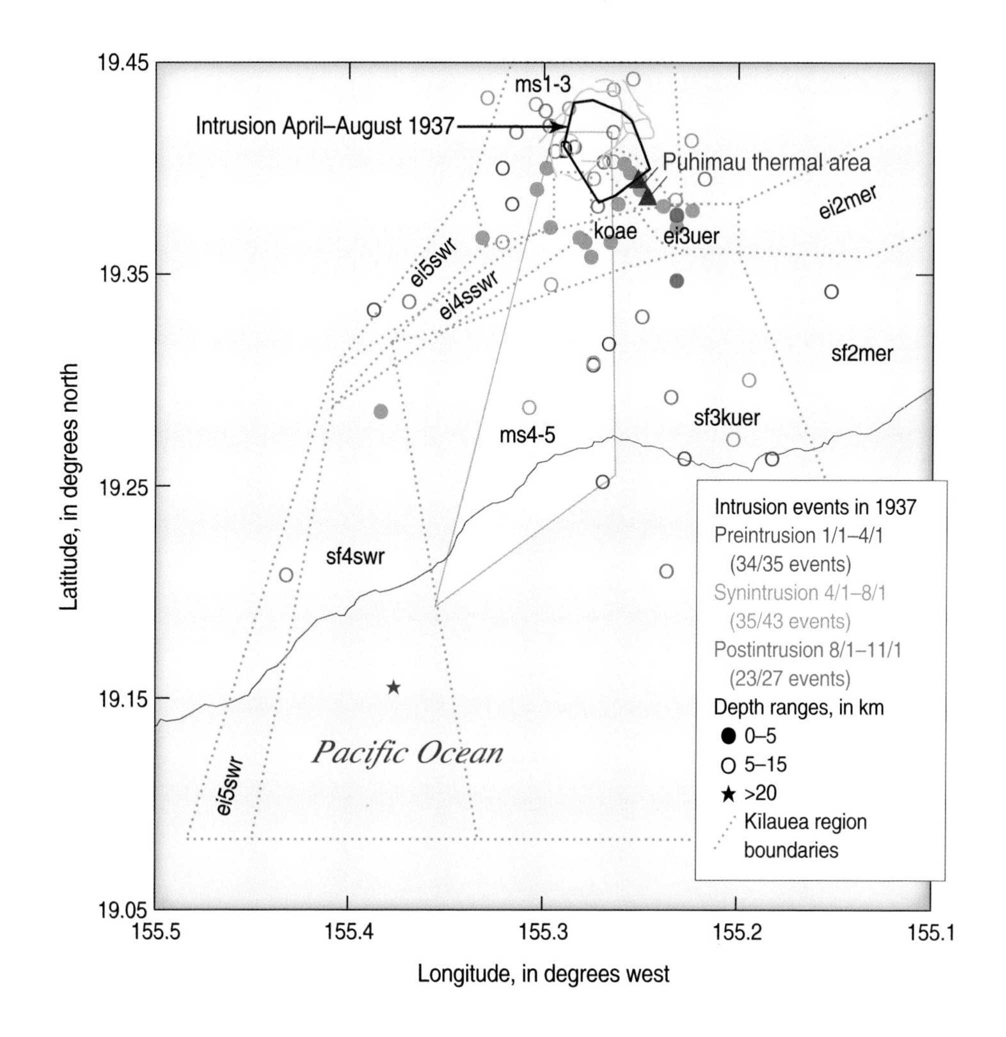

Figure 3.3. Map showing locations of earthquakes between 1 April and 1 August 1937 possibly associated with the formation of the Puhimau thermal area (fig. 3.2). Red triangles mark the location of Puhimau and Kōko‘olau Craters, near which vegetation died within 1–2 years before the May 1938 intrusion. Shallow summit earthquakes for periods before, during, and after the hypothesized intrusion occur within the area outlined in black. Dates on figure in m/d/yyyy format. Earthquake regions are defined in appendix A and shown in appendix A, figure A4. The black polygon outlines the region in which earthquakes presumed to be at depths less than 5 kilometers are present. Within this region there were swarms of events too small to precisely locate. The number of those events presumed to be within the black polygon are given on the figure as a fraction of the total number of events recorded. See text for further discussion.

Summit Intrusions of December 1944 and August 1950

A shallow earthquake swarm of 16 events on 6–7 December 1944 (table 3.1) was preceded by 4.7 µrad of inflation between 5 and 7 December and followed by deflation of 19 µrad ending on 17 December. Rainfall was low during both inflation and deflation, 0.35 cm/day (0.14 inches/day) and 0.25 cm/day (0.10 inches/day), respectively, so the Whitney tilt data can be trusted. Tiltmeters on the crater floor around Halema'uma'u showed marked inflation before the earthquake swarm and deflation during the earthquake swarm, such that they went off scale (Finch, 1944). The data suggest a shallow intrusion beneath Halema'uma'u Crater.

An earthquake swarm on August 2, 1950 consisted of 243 recorded events identified as being shallow beneath Kīlauea's summit (table 3.1, figure 3.4). Three earthquakes in the swarm were felt at the summit. This is probably an intrusion, too shallow to affect the Whitney tiltmeter.

Events Preceding the Return of Lava to Halema'uma'u in June 1952

Summit Inflation

March 1950 marked the onset of continuous inflation measured at the Whitney Vault (fig. 3.4), easily separated from the preceding seasonal variation by lower signal noise and continuity over several months instead of days or weeks. The inflation was interrupted by an abrupt deflation in December 1950, followed almost immediately by reinflation. The reinflation continued without interruption through a *M*6.2 deep Kīlauea "magma supply" earthquake on 22 April 1951 and a major swarm of earthquakes beneath the offshore part of Kīlauea's south flank during March and April 1952 (table 3.1, fig. 3.4). Eruption broke out in Halema'uma'u on 27 June 1952 (Macdonald, 1952).

The Koa'e Crisis of 1950

A major deflation of Kīlauea's summit in December 1950 was accompanied by an earthquake swarm beneath the western Koa'e Fault Zone (table 3.1, fig. 3.5; Finch, 1950; Finch and Macdonald, 1953). This was the first event since the beginning of observation of the volcano in the early 19th century, and certainly since HVO's founding in 1912, in which the Koa'e fault system was noticeably affected. The earthquake swarms of the 19th century were not precisely located, but the available data, including observations of ground cracking, suggest that they were associated with intrusion beneath the rift zones rather than activity between the rifts. The lack of uplift, earthquakes, and ground cracking on the Chain of Craters road suggests that the 1950 intrusion was considerably deeper and involved the east rift less than the intrusions of 1938. Apparently in 1950 no one walked across the Koa'e Fault Zone along the earthquake zone to check for additional cracking and fault movement within the Koa'e.

We don't fully understand how the patterns of the 1950 Koa'e earthquake swarm fit into other events in Kīlauea's history. Many of the earthquakes lie along Kīlauea's magma supply path (Wright and Klein, 2006), and those with magnitudes greater than 4 were recorded on O'ahu, suggesting that these larger events may be deeper than 20 km. The only comparable event in Kīlauea's modern history is the Christmas Eve 1965 eruption and intrusion (Fiske and Koyanagi, 1968). In 1965, the seismicity was shallow and the maximum earthquake magnitude was 3.6 (Bosher and Duennebier, 1985). Observed ground movements in 1965 were far more intense than in 1950 and included nearly 2 m of offset on the Kalanaokuaiki Fault separating the Koa'e from the south flank (Fiske and Koyanagi, 1968, figure 8).

The Earthquake of 22 April 1951

Not long after the 1950 Koa'e crisis a *M*6.2 earthquake occurred at a depth of more than 20 km beneath Kīlauea. Eissler and Kanamori (1986) estimated the depth as 35–50 km. The epicenter of this event was placed near and just northeast of Kīlauea Caldera, and aftershocks were located beneath the entire area of Kīlauea (Macdonald, 1951; Macdonald and Wentworth, 1954). It is likely that these locations have large errors, because modern earthquakes of similar magnitude and depth fall along the magma supply path and aftershock epicenters occur over a much smaller area.

Figure 3.4. Graphs showing Kīlauea activity, 1 March 1950–1 August 1952. Buildup to 1952 Halema'uma'u eruption. Panels as in figure 3.1. Dates in m/d/yyyy format. Inflation of Kīlauea's summit begins circa 1 March 1950. There was a shallow summit earthquake swarm on 2 August 1950 (Klein and Wright, 2000), suggesting a summit intrusion. A large and intense seismic swarm beneath the Koa'e Fault Zone in December 1950, including many earthquakes large enough to be recorded in Honolulu, was accompanied by a major deflation, after which inflation gradually resumed until beginning of Halema'uma'u eruption on 27 June 1952. A deep magma supply earthquake occurred on 22 April 1951. A very large and prolonged earthquake swarm off the south shore of Kīlauea located beneath the midslope bench (chap. 1, fig. 1.1*C*) took place in March and April 1952. Locations of earthquakes are shown in figure 3.5.

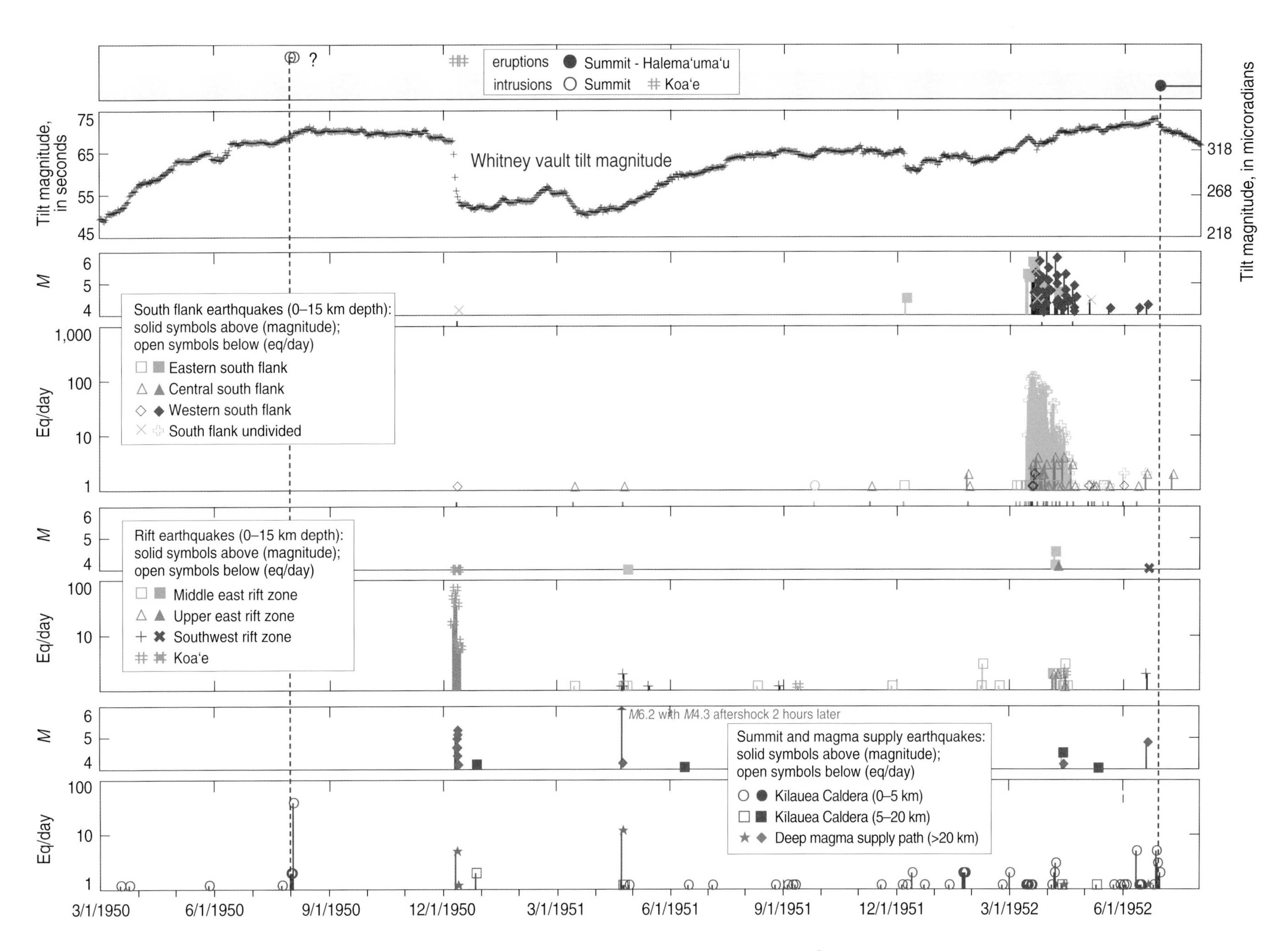

eruptions
Summit - Halema‘uma‘u
intrusions
Summit
Koa‘e
?
Tilt magnitude, in seconds
75
65
55
45
Whitney vault tilt magnitude
318
268
218
Tilt magnitude, in microradians
M
6
5
4
South flank earthquakes (0–15 km depth): solid symbols above (magnitude); open symbols below (eq/day)
Eastern south flank
Central south flank
Western south flank
South flank undivided
Eq/day
1,000
100
10
1
Rift earthquakes (0–15 km depth): solid symbols above (magnitude); open symbols below (eq/day)
Middle east rift zone
Upper east rift zone
Southwest rift zone
Koa‘e
M6.2 with M4.3 aftershock 2 hours later
Summit and magma supply earthquakes: solid symbols above (magnitude); open symbols below (eq/day)
Kilauea Caldera (0–5 km)
Kilauea Caldera (5–20 km)
Deep magma supply path (>20 km)
3/1/1950
6/1/1950
9/1/1950
12/1/1950
3/1/1951
6/1/1951
9/1/1951
12/1/1951
3/1/1952
6/1/1952

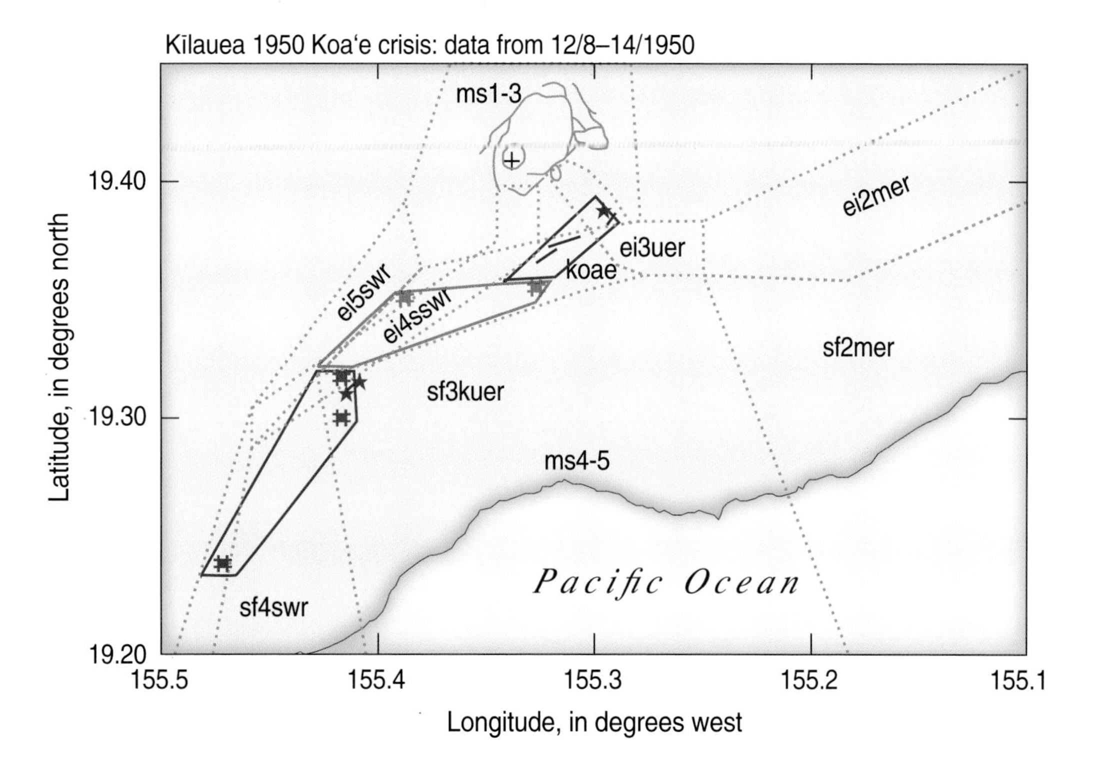

★ Kōko‘olau Crater—12/8–12/14

★ Kamakai‘a Hills—center of main swarm activity

Cracks opened during subsidence—dates uncertain

12/8: more than133 events

12/9–12/11: more than 461 events—only two events are located

Events felt in Honolulu M4.0–5.1—presumed depth >20 km

12/11-12/14: 62 events—locations unknown, not plotted

Kīlauea region boundaries

Figure 3.5. Map showing locations of earthquakes for the Koa‘e crisis of December 1950. Locations of earthquakes and ground cracking during the seismic swarm in December 1950 that was accompanied by dramatic subsidence of Kīlauea's summit (Finch, 1950; Finch and Macdonald, 1953; Klein and Wright, 2000). The swarm propagated from Kōko‘olau Crater on Kilauea's east rift zone across the Koa‘e to the vicinity of Kamakai‘a Hills, then down the southwest rift zone and adjacent flank. Nine earthquakes of M4.16–M5.12 were recorded in Honolulu and were widely felt on the Island of Hawai‘i. These earthquakes are assumed to be deeper than 20 km, hence located within the deep magma supply path. Earthquakes of M<4 are counted as belonging to the Koa‘e Fault Zone at depths sufficiently shallow to explain the observance of new cracking. Single earthquake location errors are unknown and probably large, but the general location of the swarm is well documented. Dates on figure in m/d/yyyy format.

The South Flank Seismic Crisis of March and April 1952

A few months before the 1952 Halema‘uma‘u eruption a great earthquake swarm began beneath the offshore part of Kīlauea's south flank (table 3.1). Epicenters were located along a broad east-west-trending band extending from south of Kīlauea's middle east rift zone to the southwest rift zone (fig. 3.6). The principal swarm lasted for about 2 weeks, and many of the larger events were recorded on O‘ahu. Earthquakes continued at relatively high numbers through April and gradually declined up to the time of the eruption in Halema‘uma‘u. The last confirmed earthquake in this swarm occurred on 16 November 1952, a week after the end of the Halema‘uma‘u eruption. The net tilt change across the earthquake swarm reflected continued inflation leading up to eruption (fig. 3.4). This may qualify as an extreme example of a "suspected deep intrusion."

The Return of Lava to Halema‘uma‘u in June 1952

On 27 June 1952 lava returned to Halema‘uma‘u for the first time since 1934 (table 3.1; fig. 3.3). The eruption was immediately accompanied by a small earthquake swarm beneath Kīlauea Caldera and preceded by about 2 weeks of heightened shallow seismicity beneath the caldera. The summit eruption in Halema‘uma‘u Crater lasted for nearly 6 months, ending on 9 November 1952. During the time of eruption there was a net deflation of 6.21 seconds (30 μrad) measured at the Whitney Vault, indicating that during the eruption additional magma was transferred to the rift zone (table 3.2).

1925–1952: Interpretation

This is a critical period in Kīlauea's history as it involves recovery from the 1924 collapse and the first of several increases in magma supply rate extending to the present. Magma supply calculations follow the procedures outlined in appendix A. The full history of magma supply rate for the entire period from 1950 to present is discussed in chapter 8.

Recovery from the 1924 Collapse

Analysis of the recovery from the 1924 collapse requires an evaluation of the volume lost during the collapse and how that was made up in subsequent years. It is important also to note that the Whitney tilt level stayed low after 1924, in contrast to rapid reinflation following many deflations observed during the modern era. The 1924 collapse created a reduction in summit volume of 0.4 km^3 to be filled (Dvorak, 1992). That volume divided by the 26 years between 17 June 1924 and 5 March 1950 yields a magma supply rate of less than 0.02 km^3/yr. This rate is less than the filling rate before 1924 and it seems unreasonably low.

If we calculate a collapse volume taking into account the deeper, broader source discussed in the last chapter, we obtain a volume of 1.26 km^3, more than three times that assumed by Dvorak[13]. Making up this volume in 26 years yields a magma supply rate of 1.26/25.714 = 0.05 km^3/yr, a magma supply rate that is higher than rates before 1924 and less than a

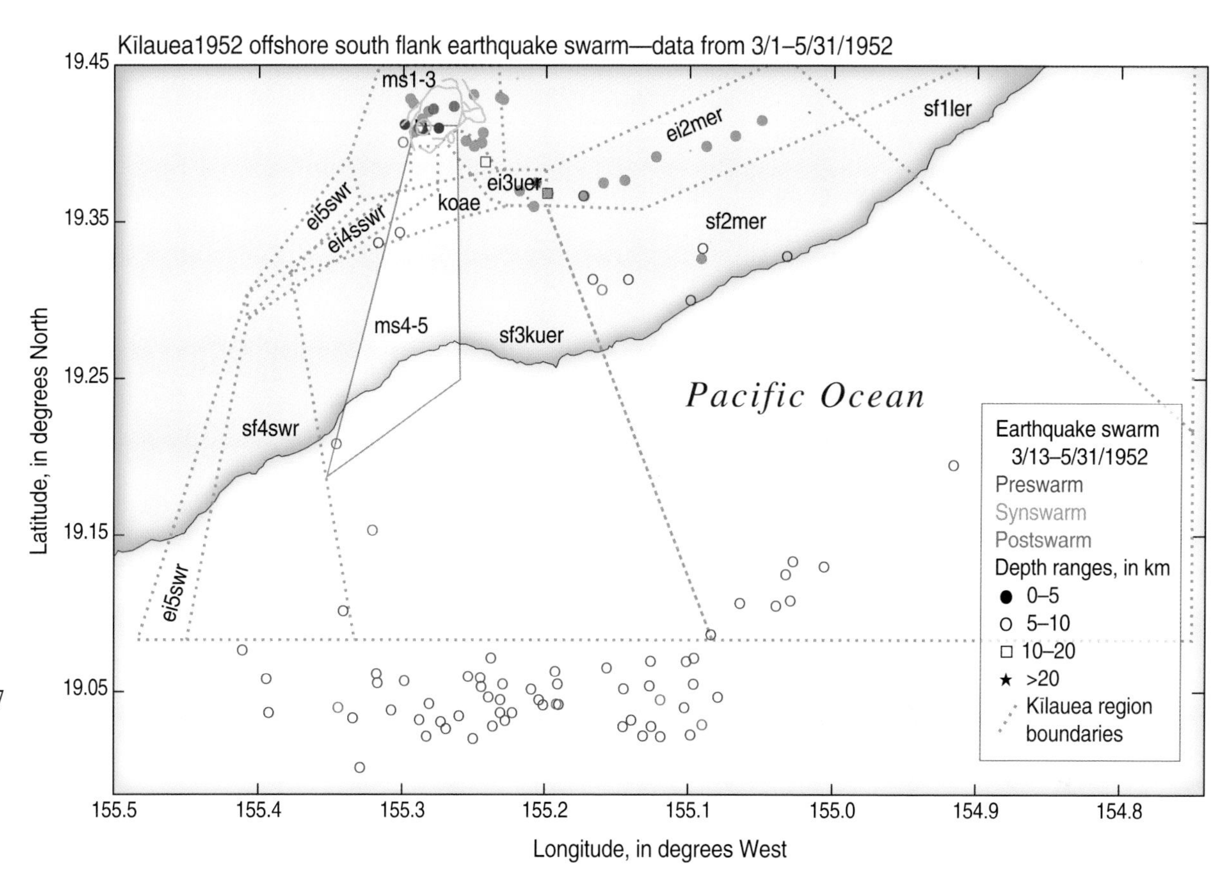

Figure 3.6. Map showing locations of earthquakes during March and April 1952 across a broad region of the offshore south flank (Fiske and others, 1987, Volcano Letter 515, fig. 1, p. 2). The smaller quakes are counted here as south flank undivided because precise locations are not known. The earthquakes of $M \geq 4$ are plotted at their given location. The linear trend is possibly an artifact of the limited seismic network, in which distance from Kīlauea's summit constituted the only reliable measurement. Earthquake locations are color-coded for pre-, syn- and postintrusion/eruption. Dates on figure in m/d/yyyy format. Only epicenters of the larger events in the offshore earthquake swarm are given in table 2 of Macdonald (1955). Additional earthquakes are given a time and a description ("off south shore of Hawai'i). For these events locations are randomly generated to match the published distribution of epicenters (Macdonald, 1955, fig. 5). The occurrence of many east rift earthquakes during the south flank swarm suggests that an intrusion may have occurred at the same time.

[13]We have calculated the volume of subsidence as a two-part cone, with very gentle slope over a broad area (almost disk-like in shape) and much steeper slope in the central area within 1 km of Halema'uma'u.

Table 3.2. History of tilt and volume changes prior to and associated with the 1952 eruption.

Cycle	Start date	End date	Time (years)	Tilt magnitude (μrad)	Tilt azimuth (degrees)	Tilt volume[1] (km^3)	Eruption volume[2] (km^3)	Magma supply rate (km^3/year)	Comment msr = magma supply rate
	2/23/1918	2/9/1924	5.960			0.1425		0.0239	1918-1924 filling rate from table 2.5
1924-1934	6/1/1924	10/9/1934	10.36	flat			0.0211	0.0020	Filling rate for 1924–1934 eruptions
1924-1950	6/17/1924	3/5/1950	25.17	flat		1.26[3]		0.0490[3]	Filling of the 1924 collapse volume
1950-1952	3/5/1950	6/27/1952		135.5	27.85[4]	0.0870			Total inflation
	9/15/1950	12/8/1950		13.7	218.1	0.0088			Magma transfer to rift zone
	12/8/1950	12/16/1950		78.9	209.98	0.0507			Magma transfer to rift zone
Total			**2.313**			**0.1465**	**0.0021**	**0.0642**	**Pre-eruption magma supply rate**
1952	6/27/1952	11/9/1952	0.364				0.0370	0.1016	Filling rate 1952 Halema‘uma‘u
							0.0234		Volume added at preeruption msr
							-0.0136		Added volume minus erupted volume
	6/27/1952	11/9/1952		30.1	195.27	0.0193			Syneruption deflation
			0.364				0.0058	0.1176	Syneruption magma transfer to rift
							0.0428		Syneruption magma supply rate

[1]Tilt volume in cubic kilometers = Whitney tilt magnitude × 0.00064 after conversion from arc-seconds to microradians (see text for explanation).
[2]Equivalent magma volume obtained by multiplying published lava volumes by 0.8 to account for 20 per cent vesicles.
[3]Volume assumed; minimum magma supply rate calculated as described in text.
[4]Corrected for offset from the 1951 *M*6 earthquake.

rate calculated between 1950 and 1952 (table 3.2 and discussion below). This allows for a rapid recovery during a short time following the 1924 collapse, for example, the first year, followed by a gradual decrease in supply rate over the 1925–50 time period. The net deflationary tilt change of more than 20 arc-seconds (97 μad) from 1925 to 1950 cannot be explained by meteorological variations, and most likely represents continued regional subsidence[14].

We interpret that the discrepancy in volume during 1925–50 was made up through refilling of the deep magma system beneath the rift zone (source 3 in fig. 2.4), recovery being complete when the summit began to inflate in March 1950. The recovery of source 3 is consistent with the continuation of regional subsidence and the delayed reinflation of source 2.

[14]Long-term drift of the tiltmeter is not possible to confirm.

History of Magma Supply, 1925–1953

Estimates of magma supply for the entire period from 1925 to 1953 are shown in table 3.2. Magma for eruption and intrusion during this period came from the three sources shown in figure 2.4. We consider that the small eruptions in Halema‘uma‘u used up residual magma left at shallow depth beneath the caldera (source 1). The continued recovery after 1934 produced at least four east rift intrusions during 1937–38, all of which occurred with little tilt change (table 3.2; fig. 3.2). Only when all the volume lost in the 1924 collapse is restored to reservoir 3 (see above) does reservoir 2 become pressurized and show measurable inflation in the manner observed during post-1952 cycles of inflation and deflation. This hypothesized style of refilling following the massive and unusual 1924 collapse is similar to that invoked to explain gravity changes during the recovery from the 1975 earthquake (Dzurisin and others, 1980).

The filling rate from 1918 to 1924 of 0.024 km^3/yr (table 2.5) contrasts with the filling rate during the eruptions in Halemaʻumaʻu following the 1924 collapse, which are lower by a factor of 2, and during the 1952 eruption , which are higher by a factor of more than 4. The magma supply rate from March 1950 to the beginning of eruption in 1952, calculated from the volume derived from the Whitney Vault inflationary tilt (see appendix A), is 0.06 km^3/yr, a value falling between the filling rate for 1918–24 and the filling rate during the 1952 eruption, and somewhat greater than the estimated average rate for the entire 1925–53 period (table 3.2).

The low rate of filling between 1924 and 1934 can be explained in a number of ways. Perhaps the most likely is that magma was diverted to the east rift zone from below source 2 as part of the restoration of source 3. This hypothesis is consistent with the calculation of a higher overall magma supply rate for the entire period between 1925 and 1953, as discussed in the last section. Less likely would be a diversion of magma at this time to Mauna Loa in anticipation of its large southwest rift eruption in 1950.

Increased Kīlauea magma supply probably triggered the Koaʻe crisis of December 1950 and may have triggered the later occurrence of the deep *M*6.2 earthquake of 22 April 1951 that occurred near the postulated magma supply conduit from the mantle. Deep earthquakes near Kīlauea's conduit are caused by stress from the flexure of the lithosphere (Klein, 2007; Klein and others, 1987), but their timing could be influenced by increases in magma supply. Later the higher magma supply may have triggered the offshore south flank earthquake swarm of March and April 1952. We interpret the offshore swarm to have unlocked the south flank, potentially marking the beginning of much less constrained seaward spreading. We also can consider the south flank swarm as a massive suspected deep intrusion associated with an increase in rift dilation that in turn could provide a reason for deflation of about 50 µrad during the period of the 1952 eruption (table 3.2). Using a postearthquake dilation rate of ~0.005 km^3/yr , a calculated preeruption magma supply rate of ~0.07 km^3/yr can be compared to a syneruption filling rate of about 0.10 km^3/yr. The additional transfer of 0.006 km^3 of magma to the rift zone calculated from the deflation during the 1952 eruption yields a syneruption magma supply rate of 0.118 km^3/yr, which is larger than the filling rate. This pattern was repeated before and during the 1967–68 Halemaʻumaʻu eruption, at an assumed constant spreading rate, with a preeruption magma supply rate of 0.05km^3/yr, an eruption rate of 0.11 km^3/yr, and a syneruption magma supply rate of 0.12 km^3/yr (see chap. 5).

We interpret the differences among preeruption and posteruption magma supply rates, summit eruption filling rate, and syneruption magma supply rate to the effects of a confining pressure and its release on the rate of magma supply. Before eruption (1925–52) the magma is confined and the supply is throttled. As soon as eruption begins, magma is released at a rate compatible with the ability of the plumbing above the storage reservoir to deliver magma to the surface. When the magma supply rate exceeds the capacity of the delivery system, magma either continues to inflate the storage reservoir (inflationary tilt) or is released to the rift zone as part of the spreading process (deflationary tilt). If either value is greater than 0 the syneruption supply rate will exceed the Halemaʻumaʻu filling rate. When the eruption ends, the magma is again confined and the magma supply rate is temporarily reduced during subsequent inflation. The relationship between an open and closed magma supply system agrees with ideas previously expressed (compare Dvorak and Dzurisin, 1993).

Conclusions: (1) the Kīlauea magma supply rate from 1840 to 1950 (<0.03 km^3/yr) was far less than what we observe from 1950 and after; (2) in the period between the 1924 collapse and the March 1950 inflation, the volume lost in the collapse was regained at a rate still uncertain. The net magma supply rate before 1950 must have decreased as the pressure deficit caused by the 1924 collapse was equalized to a value greater than the 1918–24 filling rate; (3) summit inflation, driven by increased magma supply, began when the volume lost in the 1924 collapse was recovered, possibly augmented by a shift in the deep mantle plumbing to favor eruption at Kīlauea over eruption at Mauna Loa (see chapter 8).

Supplementary Material

Supplementary material for this chapter appears in appendix C, which is only available in the digital versions of this work—in the DVD that accompanies the printed volume and as a separate file accompanying this volume on the Web at http://pubs.usgs.gov/pp/1806/. Appendix C includes the following supplementary material:

Table C1 contains earthquakes of magnitudes between 5 and 6 for the same time periods as text table 3.1.

Table C2 contains Whitney tilt data to support text figure 3.2.

Figures C1 and C2 show time series plots of earthquakes associated with eruptions and intrusions at Kīlauea from 1925 to 1953.

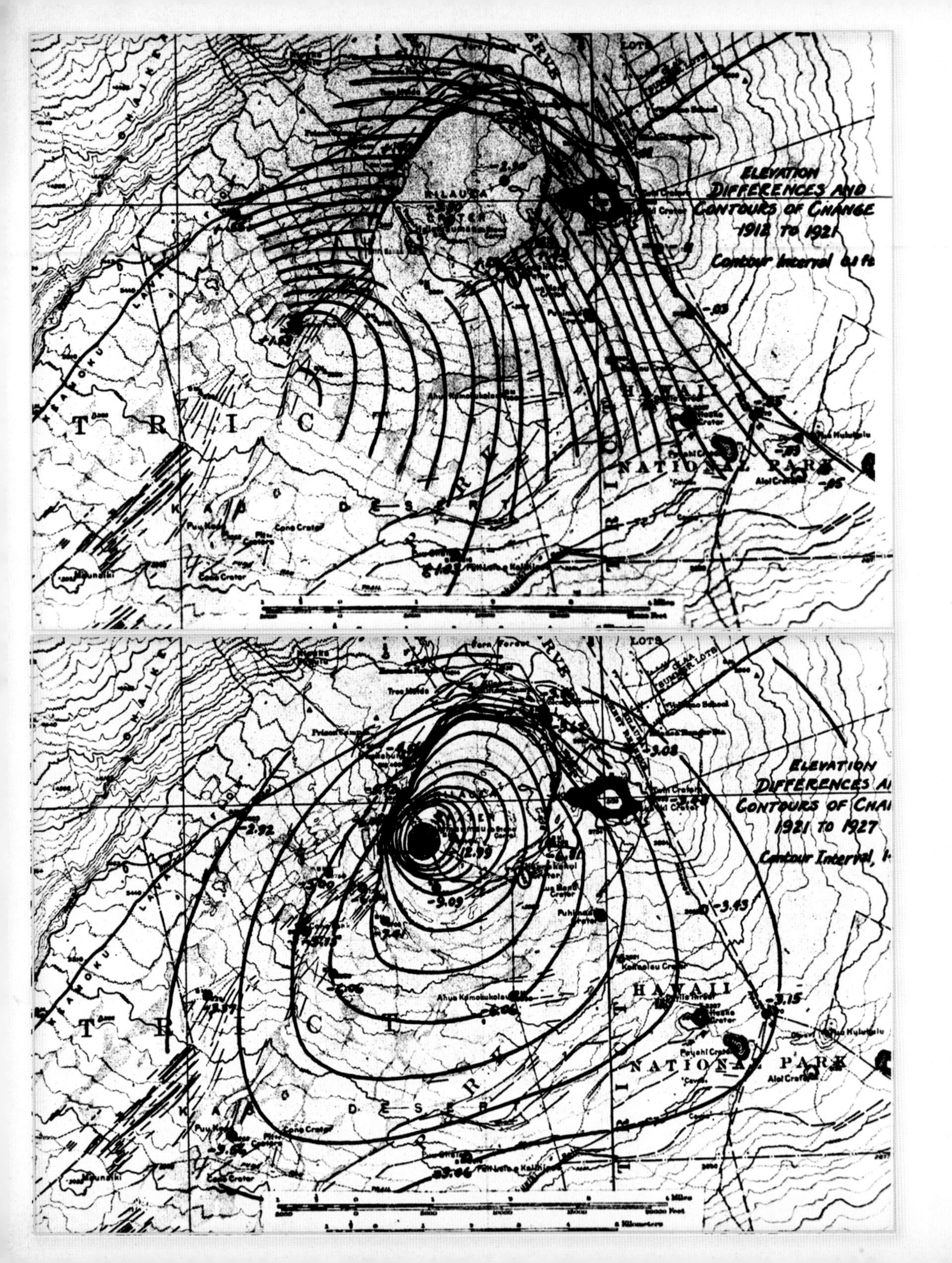

Elevation changes at Kilauea: (top) upward movement, 1912–1921; (bottom) subsidence, 1921–1927, associated with the great collapse of Halema'uma'u in May 1924. From Wilson (1935).

Chapter 4

Eruptive and Intrusive Activity, 1953–1967

In the interval between the Halema'uma'u eruptions of 1952 and 1967–68, a unique eruption in Kīlauea Iki occurred in 1959. Rift eruptions between 1955 and 1965 are shown to be mixtures of fractionated magma residing in the rift zone and three magmas moving through the east rift zone earlier than, but in the same order as, eruptions of equivalent chemistry at the summit.

Lava lake formed in Kilauea Iki Crater during eruption in 1959. USGS photo by J.P. Eaton.

The period 1953–67 was a very active time in Kīlauea's history, with pioneering research and monitoring and great leaps in understanding achieved at HVO. Kīlauea eruptions, intrusions, and large earthquakes following the 1952 eruption and through the end of 1960 are summarized in table 4.1 and figures 4.1. The 1961–67 period is covered later in this chapter. Additional seismic analysis for the period 1963–67 has been published in Klein and others (1987), and figures from that paper are cross-referenced in table 4.1.

The period following the 1952 eruption is unique in Kīlauea's known history because it involves many different eruption styles and eruption sequences, including (1) a short eruption in Kīlauea Caldera, (2) three longer eruptions in Halema'uma'u, (3) one episodic eruption in Kīlauea Iki Crater, previously inactive for nearly a century, (4) two eruptions on Kīlauea's lower east rift zone, not active since the great eruption of 1840, and (5) a series of short east rift eruptions uninterrupted by summit activity.

It was also a time of great advances in HVO's ability to effectively monitor volcanic and seismic activity. Expansion and upgrading of the HVO seismic network (Klein and Koyanagi, 1980) began following the 1952 eruption with the arrival of Jerry Eaton. The expanded seismic network allowed earthquake locations to be determined with far more accuracy than in preceding periods. The arrival of Jerry Eaton at HVO also led to the installation of a water-tube tiltmeter permanently mounted in Uwēkahuna Vault and a network of portable water-tube tiltmeters around Kīlauea's summit (Eaton, 1959). Thus, for the first time HVO was able to verify, through monitoring of ground tilt and leveling, the location of a magma chamber at depths of 2–6 km beneath Kīlauea's summit (source 2 of fig. 2.4), from which all eruptions and intrusions were fed.

Tilt and Leveling Data[15]

During the time between the end of the 1965 east rift eruption and the beginning of the 1967–68 Halema'uma'u eruption the leveling network in and near Kīlauea Caldera was measured numerous times, yielding a valuable record of changing epicenters of deformation during an extended period of inflation (fig. 4.2; Fiske and Kinoshita, 1969, figure 5). Deformation centers of nearly all subsequent inflations and deflations determined by a variety of methods fall within the grouping of centers for the 1965–68 period reported by Fiske and Kinoshita (fig. 4.2; Fiske and Kinoshita, 1969, figure 5). The range of deformation centers indicated by the long-base water-tube network (fig. 4.3) lie within a circle of approximately 1 km diameter in the south caldera.

Eruptions and Intrusions

The Eruption of 31 May–3 June 1954

The 31 May–3 June 1954 eruption in Halema'uma'u (Macdonald and Eaton, 1957) was a short eruption during a long period of inflation that began following the end of the 1952 eruption and ended with the lower east rift eruption at the end of February 1955 (fig. 4.1). Inflation continued during the May 1954 summit eruption, with only a small deflection on the Whitney tilt record (fig. 4.4*A*). The eruption on 31 May 1954 was preceded on 30 March 1954 by an earthquake of *M*6.4 beneath Kīlauea's south flank (fig. 4.5) that did not seem to be directly connected to the summit eruption. However, immediately following the earthquake seismicity picked up beneath the lower east rift (fig. 4.4*A*), initially occurring at a rate of about 25 events per month (Macdonald and Eaton, 1964, p. 116, 120).

The Eruption of 28 February–26 May 1955

The 1955 eruption (Macdonald and Eaton, 1964) was the first historically documented eruption on Kīlauea's lower east rift zone since 1840 and is unique in the following respects, when compared to later eruptions: (1) the erupted lavas have a highly fractionated chemistry, (2) deflation of Kīlauea's summit that normally accompanies rift eruptions was delayed more than 1 week after the beginning of the eruption (fig. 4.4*B*), and (3) south flank seismicity adjacent to the eruption site was absent (table 4.1). Beginning two months before the 1955 eruption the number and magnitude of shallow earthquakes beneath the lower east rift zone increased from 1 per day to more than 1,000 per day in the week preceding eruption (fig. 4.4). A newly installed north-south tiltmeter at the Pāhoa schoolhouse registered 97 µrad (~20 arc-seconds) of inflation of the east rift zone beginning 2 days before lava reached the surface (fig. 4.4*A*).

The eruption took place in four stages punctuated by temporary cessation of eruptive activity and (or) shift of vent location (figs. 4.4, 4.5). Toward the close of stage 2, on 5 March 1955, an intrusion occurred near Kalalua Crater on the middle east rift zone (Macdonald and Eaton, 1964, p. 122). The rift deflated at the time of the Kalalua intrusion and resumed inflation through the opening of new vents associated with stage 3 (5 March–7 April 1955) as measured by the Pāhoa tiltmeter (fig. 4.4*A*). The rift zone continued to inflate intermittently before stabilizing within stage 4, a total of 291 µrad (60 arc-seconds) of change (fig. 4.4*A*). Deflation at

[15] A comprehensive review of tilt measurements, including discussion of uncertainties regarding locations of inflation and deflation centers and calculation of magma volume from tilt measurements, is given in appendix A.

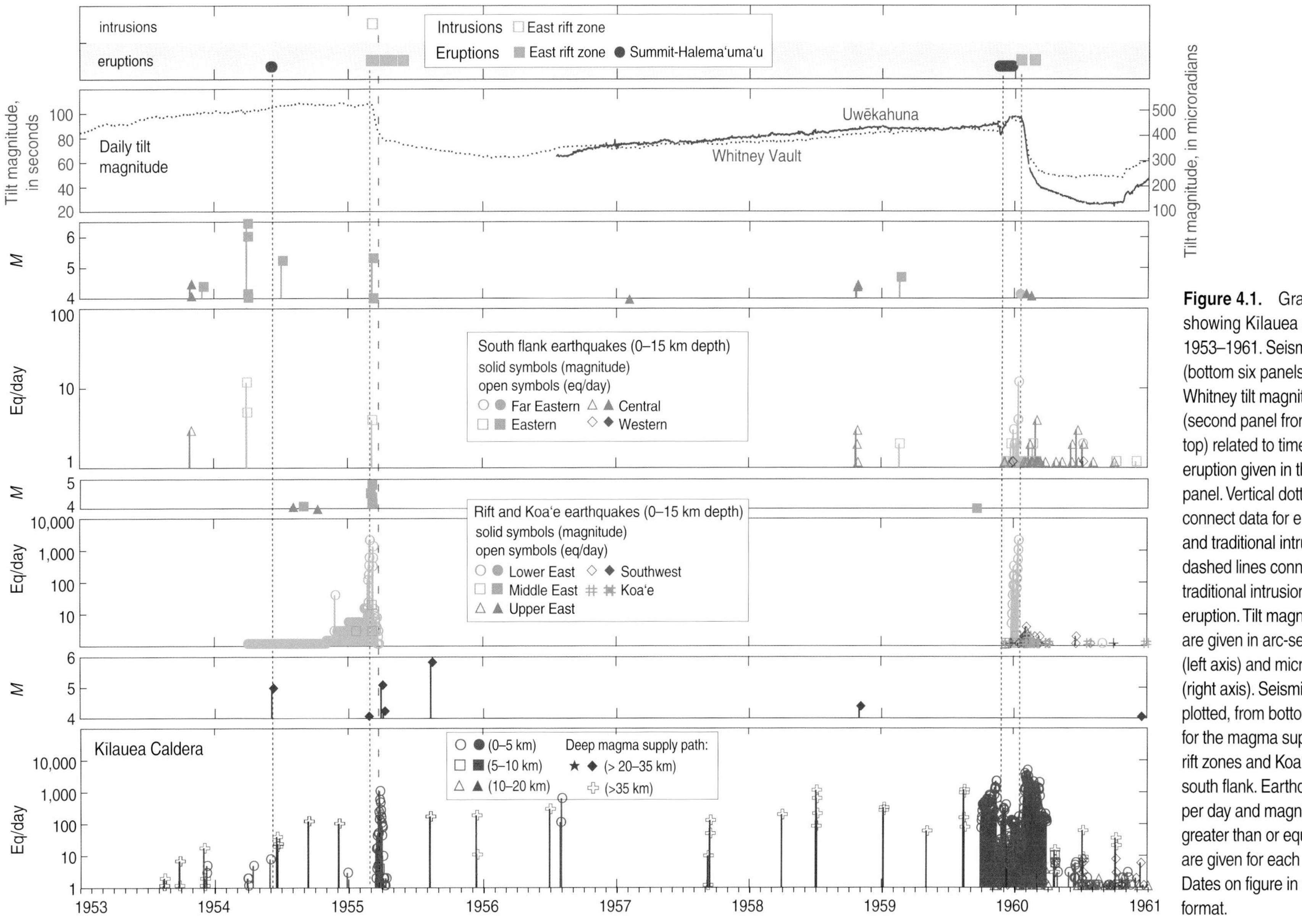

Figure 4.1. Graphs showing Kīlauea activity, 1953–1961. Seismicity (bottom six panels) and Whitney tilt magnitude (second panel from top) related to times of eruption given in the top panel. Vertical dotted lines connect data for eruptions and traditional intrusions; dashed lines connect traditional intrusions without eruption. Tilt magnitudes are given in arc-seconds (left axis) and microradians (right axis). Seismicity is plotted, from bottom to top, for the magma supply path, rift zones and Koaʻe, and south flank. Earthquakes per day and magnitudes greater than or equal to 4.0 are given for each region. Dates on figure in m/d/yyyy format.

Table 4.1. Kīlauea eruptions, intrusions, earthquake swarms and earthquakes $M \geq 4$, 1953–1967.

[In rows with multiple entries text applies down to the next entry; dates in m/d/yyyy format; data for eruptions and traditional intrusions emphasized with gray shading; do, ditto (same as above)]

Time and Date		Region[1]	Event Type[2]	Number of events[3]	Tilt[4]		Lag[5]	Comment[6]	Figures[7]	References[8]
Start	End				Mag	Az				
07:50 9/23/1953	9/26/1953	ms5gln	EQS	15				First of several 40-60 km deep swarms north of Kīlauea Caldera; precursory to 1959 eruption and continuing to late 1960 as shown on figure 4.7	**4.1, 4.7,** table D1	1; VL 521, p. 6
06:40 3/30/1954	08:42 3/31/1954	sf2 sf2	EQ EQ		flat tilt			*M*6.03 foreshock; 3 aftershocks *M*6.45 mainshock, 17 as to 3/31	**4.5**	1, 2; VL 523, p. 7
4//1/1954	2/1/1955	eiller	EQS	~275				Precursors to 1955 eruption ~25/month, no date/time	**4.4**. D5	1, 3; VL 524, p. 10
03:42 5/31/1954	03:50 5/31/1954	ms1	EQS	7				Followed by 1 ms2 at 03:54	**4.1–4.4**, D5, 10	1, 4, 2; VL 524, p. 1–9
04:10 5/31/1954	03:00 6/3/1954	KC	E					1954 eruption in Kīlauea Caldera		
14:26 8/07/1954		ms4	EQ					*M*5.0, 25 km beneath Kīlauea Caldera	**4.1**	1; VL 525, p. 7
15:06 2/24/1955	22:30 2/24/1955	eiller	EQS	14				3 located; others no time	**4.1**	1; VL 527, p. 4–5
01:34 2/25/1955	22:10 2/25/1955	eiller	EQS	127				6 located; others no time	**4.1**	do
07:30 2/26/1955	18:56 2/26/1955	eiller	EQS	344				6 located; others no time	**4.1**	do
00:37 2/27/1955	23:12 2/27/1955	eiller	EQS	594				8 located; others no time	**4.1**	1, 3; VL 527, p. 4–5
07:45 2/28/1955	07:45 2/28/1955	eiller	EQS	691				2 located; others no time	**4.1, 4.4A–4.5**	3; VL 529–530, p. 1–10
08:00 2/28/1955	14:00 3/4/1955	LERZ	E					1955 eruption stage 1	**4.1, 4.4A–4.5**	
14:15 3/2/1955	06:00 3/7/1955		E					1955 eruption stage 2	do	do
12:39 3/5/1955	11:45 3/6/1955	ei2mer	EQS/I	23				*M*3.52-4.79 near Kalalua Crater	**4.1, 4.4***A*	do
22:21 3/7/1955	23:32 3/7/1955	sf2	EQS	4				*M*5.32, 3.95, 4.02, 4.00;	**4.1, 4.4***A*	1, 3; VL 527, p. 5–6
3/7/1955	3/30/1955				151.7	211		Summit def	**4.1, 4.4**	do
00:00 3/8/1955	23:59 3/8/1955	eiller	EQS	326				No time; 1 located at 20:03	**4.4**	do
00:00 3/9/1955	23:59 3/9/1955	eiller	EQS	601					do	
3/9/1955		sf2		2				09:16, 21:08, *M*3.5, 3.8		
00:00 3/10/1955	23:59 3/10/1955	eiller	EQS	1385				No time	do	do
00:00 3/11/1955	23:59 3/11/1955	eiller	EQS	76				No time	**4.4**	3; VL 527, p. 5–6
00:00 3/12/1955	23:59 3/12/1955	eiller	EQS	11				No time		1; VL 527, p. 5–6
17:05 3/12/1955	18:30 4/7/1955	LERZ	E					1955 eruption stage 3	**4.4–4.5**	1, 3; VL 527, p. 5–6
19:01 3/15/1955	19:02 3/15/1955	ms1	EQS	2				Initiation of summit subsidence	**4.1, 4.4**	1; VL 527, p. 5–6
00:00 3/16/1955	23:59 3/16/1955	ms1	EQS	21				No time	do	do
01:00 3/17/1955	13:35 3/17/1955	ms1	EQS	9				6 located; 3 unlocated eiler	do	do
09:54 3/18/1955	22:43 3/18/1955	ms1	EQS	48				3 located	do	do
00:00 3/20/1955	23:59 3/20/1955	ms1	EQS	6				1 located at 18:10	do	do
00:00 3/21/1955	23:59 3/21/1955	ms1	EQS	51				No times	do	do
00:00 3/22/1955	23:59 3/22/1955	ms1	EQS	52				No times; 2 eiler at 00:42, 20:46	do	do
00:00 3/23/1955	23:59 3/23/1955	ms1	EQS	324				2 at 00:10, 07:54	do	do

Table 4.1. Kilauea eruptions, intrusions, earthquake swarms and earthquakes $M \geq 4$, 1953–1967.—Continued

[In rows with multiple entries text applies down to the next entry; dates in m/d/yyyy format; data for eruptions and traditional intrusions emphasized with gray shading; do, ditto (same as above)]

Time and Date		Region[1]	Event Type[2]	Number of events[3]	Tilt[4]		Lag[5]	Comment[6]	Figures[7]	References[8]
Start	End				Mag	Az				
3/23/1955	3/23/1955	eiller		3				3 at 02:06, 03:42, 12:08		
06:56 3/24/1955	10:27 3/24/1955	ms2		4						
23:26 3/24/1955	23:26 3/24/1955	ms3		1						
05:23 3/24/1955	11:02 3/24/1955	eiller		2						
04:27 3/25/1955	08:55 3/25/1955	ms1	EQS	590				8 located; 1 ms5 at 20:17	**4.1, 4.4**	1, 3; VL 527, p. 5–6
3/26/1955	3/26/1955	ms1	EQS	1103				No times	**4.1, 4.4**	1; VL 527, p. 5–6
3/27/1955	3/27/1955	ms1	EQS	482				1 at 09:23; 1 ms29 at 16:02	do	do
3/28/1955	3/28/1955	ms1	EQS	278				1 at 20:39	do	do
3/29/1955	3/29/1955	ms1	EQS	165				No time	do	do
3/30/1955	3/30/1955	ms1	EQS	111				End of summit subsidence	do	do
00:00 3/31/1955	23:59 3/31/1955	ms1	EQS	44				No time	do	do
04:24 4/1/1955		ms4	EQ					*M*5.1 at 30 km; *M*3.6 at 04:35	do	do
00:00 4/1/1955	23:59 4/1/1955	ms1	EQS	88				No time		
00:00 4/2/1955	23:59 4/2/1955	ms1	EQS	77				No time; 2 at 05:09, 06:37	do	1; VL 528, p. 5
00:00 4/3/1955	23:59 4/3/1955	ms1	EQS	10				End of intense ms1 EQS	do	do
13:28 4/4/1955	12:49 4/12/1955	ms1	EQS					8 located; intermittent activity to 6/1/1955		
01:27 4/07/1955		ms4	EQ					*M*4.25 at 30 km beneath Kīlauea Caldera	**4.1**	do
15:00 4/24/1955	11:15 5/26/1955	LERZ	E					1955 eruption stage 4; subsidence ended	**4.1, 4.4**	1, 3; VL 528, p. 5
02:28 8/14/1955		ms4	EQ					*M*5.84 at 25 km beneath Kīlauea Caldera Isoseismal map in Wyss and Koyanagi, 1992		1; VL 529–530, p. 12
10/1/1959	10/31/1959	ms1	EQS	12,543				Precusory swarm; 75-800 per day	**4.1, 4.4, 4.6**	10
11/1/1959	11/14/1959	ms1	EQS	12,323				Swarm continued; 90-2200 per day		
06:45 11/7/1959	07:53 11/7/1959	ms3	EQS	5				11/4, 6 2, 9, 13; 12.7-16.6 km depth	**4.6**	11
20:08 11/14/1959	08:00 12/20/1959	KI	E					Kīlauea Iki Crater eruption	**4.1-4.3, 4.6,** D6	7
20:08 11/14/1959	19:25 11/21/1959							Episode 1		
00:30 11/26/1959	16:35 11/26/1959	KI	E					Episode 2	do	do
16:30 11/28/1959	21:47 11/29/1959	KI	E					Episode 3	do	do
01:00 12/4/1959	09:27 12/5/1959	KI	E					Episode 4	do	do
14:40 12/6/1959	00:23 12/7/1959	KI	E					Episode 5	do	do
15:30 12/7/1959	02:45 12/8/1959	KI	E					Episode 6	do	do
13:00 12/8/1959	20:12 12/8/1959	KI	E					Episode 7	do	do
15:15 12/10/1959	11:40 12/11/1959	KI	E					Episode 8	do	do
10:07 12/12/1959	12/15/1959	ms5gln	EQS	56				Located; 43.57±2.24 km depth	do	do
05:08 12/13/1959	13:40 12/13/1959	KI	E					Episode 9	do	do
07:45 12/14/1959	15:36 12/14/1959	KI	E					Episode 10	**4.1, 4.6**	7
06:11 12/15/1959	10:25 12/15/1959	KI	E					Episode 11	do	do
19:30 12/15/1955	21:30 12/15/1959	KI	E					Episode 12	do	do
13:35 12/16/1959	17:19 12/16/1959	KI	E					Episode 13	do	do
11:10 12/17/1959	15:32 12/17/1959	KI	E					Episode 15	do	do

Table 4.1. Kīlauea eruptions, intrusions, earthquake swarms and earthquakes $M \geq 4$, 1953–1967.—Continued

[In rows with multiple entries text applies down to the next entry; dates in m/d/yyyy format; data for eruptions and traditional intrusions emphasized with gray shading; do, ditto (same as above)]

Time Date		Region[1]	Event type[2]	Number of events[3]	Tilt[4]		Lag[5]	Comment[6]	Figures[7]	References[8]
Start	End				Mag	Az				
02:40 12/19/1959	06:16 12/19/1959	KI	E					Episode 16	do	do
20:45 12/19/1959	08:00 12/20/1959	KI	E					Episode 17	do	do
12/25/1959	12/31/1959	ei1ler	EQS	635				Increase from 5/ to 320/day	**4.6**	13
1/1/1960	1/7/1960	ei1ler	EQS	317				Decrease from 160 to 1/day	**4.6**	14
1/8/1960	1/13/1960	ei1ler	EQS	5100				Increase from 65 to 2,080/day	**4.6**	14
15:29 12/28/1959	17:32 1/12/1960	sf1/sf1os	EQS	8/9				South flank precursor to 1960 eruption; os, offshore	**4.6, 4.8b**	11
14:41 1/13/1960	20:13 1/13/1960	sf1/sf1os	EQS	2/10				Do; *M*4.13 at 16:30		
19:35 1/13/1960	14:00 2/19/1960	LERZ	E					Kapoho eruption	**4.6, 4.8,** D6	7
08:30 1/14/1960	08:30 1/21/1960				21.1	104		Summit subsidence Uwēkahuna tilt		
08:30 1/21/1960	08:30 1/26/1960				43.5	113		1/14-7/9/1960—Clockwise rotation		
08:30 1/26/1960	08:30 7/9/1960				299.8	127		1 arc-second = 4.848 microradians (μr)		
16:56 4/20/1960	08:32 4/22/1960	ms5gln	EQS	25				Continuation of pre-1959 swarms, now coincident with the beginning of 30 km deep swarms beneath Kīlauea Caldera	**4.1**, D6, 11	11
06:31 4/21/1960	09:03 4/22/1960	ms2	EQS	16				6 more on 5/27 (2), 6/7, 8, 7/6, 7		
10:31 4/21/1960	03:19 4/22/1960	ms4	EQS	10				3 more 5/8 2, 5/19		
21:26 4/21/1960	10:36 04/22/1960	ms3	EQS	4				First overlap ms5gln-ms4		
								Second overlap ms5gln-ms4	**4.1**, D6, 11	11
21:47 7/6/1960	01:13 7/8/1960	ms5gln	EQS	55				Continuation of pre-1959 swarms 40–60 km deep swarms; second overlap of ms5gin and ms4		
15:10 7/7/1960	13:53 7/8/1960	ms4	EQS	13				More events on 7/6 18:51, 19:11, 7/8-9 (4)		
17:17 7/8/1960	08:19 7/9/1960	ms5gln	EQS	13				1 more on 7/9 17:02		
02:32 10/5/1960	20:51 10/6/1960	ms5gln	EQS	64				End of pre-1959 40-60 km deep swarms	**4.1**, D6, 11	11
19:48 10/5/1960	03:52 10/6/1960	ms4	EQS	6				2 more on 10/6; last overlap ms5gln-ms4		
07:20 2/24/1961	15:08 2/24/1961	Hm[9]	E					Halema'uma'u Crater	**4.9**, D7	14
22:00 3/3/1961	20:00 3/25/1961	Hm[9]	E					Halema'uma'u Crater	**4.9**, D7	14
02:48 6/29/1961	02:45 6/30/1961	ms4	EQS	17				8 more on 6/30-7/3 (5), 7/9, 10, 14 (2)	**4.9**, D7, 19	11
20:15 7/10/1961	12:45 7/17/1961	Hm[9]	E					Halema'uma'u Crater	**4.9**, D7	14
04:19 7/23/1961	07:50 7/24/1961	ms4, 5	EQS	25				*M*4.07 on 7/23 05:24; 5 > 35 km deep	**4.9**, D7,20	11
13:17 9/21/1961	14:33 9/21/1961	sf3		4				South flank anticipation; 1 more on 9/22	**4.9**, D7	11
04:30 9/22/1961	05:00 9/25/1961	MERZ	E		173.6	136	-0h 54m	Heiheiahulu vicinity[10]	**4.9–10**, D12	11, 14
08:30 9/21/1961	08:30 9/22/1961				93.6	129				
08:30 9/22/1961	08:30 9/25/1961				72.6	143				
08:30 9/25/1961	08:30 10/1/1961				8.8	157				

Table 4.1. Kilauea eruptions, intrusions, earthquake swarms and earthquakes $M \geq 4$, 1953–1967.—Continued

[In rows with multiple entries text applies down to the next entry; dates in m/d/yyyy format; data for eruptions and traditional intrusions emphasized with gray shading; do, ditto (same as above)]

Time Date		Region[1]	Event Type[2]	Number of events[3]	Tilt[4]		Lag[5]	Comment[6]	Figures[7]	References[8]
Start	End				Mag	Az				
19:29 9/24/1961	21:39 9/24/1961	sf2	EQS	11				*M*4.0, 4.32, 9/24 (2), 25, 26 (2), 27 (2)	**4.9**, D7	11
19:50 9/24/1961	21:14 9/24/1961	ei2mer		2				1 on 9/23	**4.9**, D7	11
16:29 11/23/1961	16:59 12/9/1961	ms4	EQS	48				Broken swarm; *M*4.19 12/2, 35.7 km	**4.9**, D7, 21	11
08:52 12/31/1961	23:21 12/31/1961	ms4	EQS	11				*M*4.10 28.34 km occurs as first EQ in swarm	D22	11
19:07 1/1/1962	16:19 1/3/1962	ms4	EQS	6				Broken swarm; diminished deep seismicity	D22	11
21:04 5/9/1962	06:37 5/11/1962	ms4.5	EQS	14					D23	11
19:11 12/6/1962	23:57 12/8/1962	sf3	EQS	10				South flank anticipation/response	**4.9**, D8	11
01:10 12/6/1962	15:00 12/9/1962	UERZ	E		20.5	130	-3h 20m	Aloi Crater; tilt 12/6–7	**4.9–10**, D8, 13	19
19: 31 12/6/1962	21:26 12/7/1962	koae	EQS	4				Koa‘e response		
07:13 12/7/1962	19:56 12/8/1962	sf2	EQS	7				South flank response		
13:12 12/10/1962	13:18 12/17/1962	sf3.4	EQS	12				South flank response shift to west		
07:56 1/8/1963	15:55 1/8/1963	ms4.5	EQS	10				*M*4.46, *M*4.11; 1/9 (3), 10, 11 (2), 12	**4.9**, D24	11
20:54 5/9/1963	20:01 5/12/1963	sf3	EQS	114				South flank anticipation/response; 5/13 (4)	43.21	8, 22
21:19 5/9/1963	19:32 5/11/1963	koae	I	62	35.0	126	-2h 56m	Koa‘e crisis; er3 on 5/10 (3), 11, 12; tilt 5/8–12		11, 15
23:51 5/9/1963	08:04 5/11/1963	sf4	EQS	5				Broken swarm		
04:31 5/10/1963	16:49 5/12/1963	ei4.5	I	5				Scattered	**4.9**, 11a	
03:56 5/12/1963	19:51 5/12/1963	koae	EQS	7				Continued tilting within the Koa‘e		
06:06 7/2/1963	08:04 74/1963	sf3	EQS	11				South flank anticipation/response	43.21	22
07:15 7/2/1963	06:38 7/31963	koae	EQS/I	15	19.0	122	-1h 55m[11]	Koa‘e crisis; tilt 7/1-2	**4.9**, 11b, D8	11
05:04 7/4/1963	20:32 7/4/1963	koae	EQS/I	8				Koa‘e crisis continued; tilting within the Koa‘e		
08:39 7/4/1963	16:56 7/4/1963	ei3kuer	EQS	3				East rift response		
17:01 7/4/1963	20:06 7/4/1963	sf3	EQS	3				South flank response continued		
18:15 8/21/1963	08:10 8/23/1963	MERZ	E		11.5	112	-0h 4m[12]	‘Alae Crater: lava lake; tilt 8/20–22	**4.9**, D8, 14	17
03:32 10/5/1963	05:44 10/5/1963	sf2/sf3	EQS	9/3				South flank anticipation/response		11
05:25 10/5/1963	10:00 10/6/1963	MERZ	E		75.7	121	-0h 4m[13]	Nāpau Crater; tilt 10/4-7, clockwise rotation	43.23; **4.9–10**	22, 18
20:43 10/5/1963	21:40 10/5/1963	koae	EQS	3				2 more on 10/6, 7	D8, 15	
00:38 10/6/1963	13:04 10/8/1963	sf2	EQS	25				South flank response		
04:06 10/6/1963	23:10 10/8/1963	ei2mer	EQS	7				East rift response; scattered; 10/6 (3), 7, 8 (3)		
21:47 10/6/1963	22:21 10/7/1963	sf3	EQS	15				South flank response continued		
08:35 10/8/1963	03:15 10/9/1963	sf3	EQS	7				South flank response continued		
18:45 10/9/1963	14:49 10/22/1963	sf3/sf2	EQS	18/7				South flank response continued		
22:29 12/2/1964	02:18 12/3/1964	ms4	EQS	12				*M*4.45; as on 12/3, 4 (4), 6 (2)	**4.9**, D25	11
09:16 3/5/1965	17:56 3/5/1965	sf2	EQS	8				South flank anticipation/response	**4.9**, D9	11
09:23 3/5/1965	23:00 3/15/1965	MERZ	E		87.8	128	-0h 28m[14]	Makaopuhi Crater: lava lake; tilt 3/5-9	43.24; D16	22, 19
18:50 3/5/1965	06:22 3/7/1965	sf3	EQS	6				South flank response; broken swarm		
13:06 3/6/1965	19:03 3/8/1965	sf2	EQS	18				South flank response continued; broken swarm		
23:53 8/21/1965	9:46 8/26/1965	sf3	EQS	9	3.4	122		Broken swarm; tilt 8/23–25	43.25; **4.9, 4.12**	22, 11

Table 4.1. Kīlauea eruptions, intrusions, earthquake swarms and earthquakes $M \geq 4$, 1953–1967.—Continued

[In rows with multiple entries text applies down to the next entry; dates in m/d/yyyy format; do, ditto (same as above).]

Time Date		Region[1]	Event type[2]	Number of events[3]	Tilt[4]		Lag[5]	Comment[6]	Figures[7]	References[8]
Start	End				Mag	Az				
00:15 8/25/1965	12:58 8/25/1965	sf2	SDI	32				Also 8/23, 24; first suspected deep intrusion	D9	
22:35 8/25/1965	13:25 8/26/1965	sf2	SDI	5				8/26 (10), 27 (2), 28 (2), 30, 9/1 (2), 2 (2)		
00:31 8/25/1965	06:02 8/26/1965	ei2mer	I	4	3.4	121	+46h	I-A: 4 μrad deflation 8/23 02:11-8/26 08:34		
07:29 11/13/1965	18:21 11/13/1965	sf3	SDI	7				Additional earthquakes on 11/12 (3), 14 (2)		
13:04 12/25/1965	06:52 12/30/1965	sf3[15]	EQS	175				South flank anticipation/response	D9	21
13:53 12/25/1965	06:39 12/30/1965	koae	I	131				Koa'e: tilting and ground cracking		21
21:30 12/24/1965	05:30 12/25/1965	UERZ	E		47.8	98	-0h 1m	Aloi Crater; tilt 12/23-28[16]	**4.2**; D9, 17	20, 11
11:18 12/28/1965	03:04 12/29/1965	sf3	EQS	7				USGS locations; south flank response		
11:35 12/28/1965	12:21 12/30/1965	koae	EQS	4				do; scattered		
15:16 12/28/1965	16:18 12/28/1965	sf4[15]	EQS	3				do; 12/25 (6), 26 (2), 27 (2), 28		
17:57 7/5/1966	01:34 7/6/1966	sf3	SDI?	6				Broken swarm	**4.9, 4.12**; 43.26	11, 22
23:52 7/5/1966	13:25 7/6/1966	sf2	SDI?	6						
23:59 7/5/1966	01:12 7/6/1966	ei3kuer	I?	4	3.9	132	+22h	I-A: sharp def of 3 μrad on 7/4-6		

[1]Earthquake classification abbreviations are given according to the classification in table A3A, and locations are shown on figure A4. Eruption location abbreviations correspond to regions shown on chapter 1, figure 1.1*A*; **KC**, Kīlauea Caldera; **SWR**, Southwest rift zone; **UERZ**, upper East rift zone, **MERZ**, middle East rift zone, **LERZ**, lower East rift zone; **KI**, Kīlauea Iki Crater.

[2]Event type abbreviations: **E**, Eruption; **I**, traditional intrusion; SDI, suspected deep intrusion; EQ, Earthquake ≥*M*5; EQS, Earthquake swarm.

[3]Minimum number of events defining a swarm: 5 for south flank; 3 for all other regions (see appendix table A3).

[4]Magnitude in microradians and azimuth of daily tilt measurements from the Bosch-Omori tiltmeter in Whitney Vault (1953–1957); water-tube tiltmeter in Uwēkahuna vault (1957–1967). 1 second of arc = 4.848 microradians (μrad).

[5]Lag times compare onset of tilt deflection and the beginning of an eruption or earthquake swarm. (+) tilt leads, (-) tilt lags. Readings from Press-Ewing seismometer in Uwēkahuna Vault in plain text, **bold** across eruptions; readings from Ideal-Arrowsmith tiltmeter in Uwēkahuna Vault in *italic* text, ***bold/italic*** across eruptions.

[6]Abbreviations as follows: *M*, earthquake magnitude; fs, foreshock; as, aftershock; ant, anticipation; resp, response; inf, inflation; def, deflation; cw, clockwise; ccw, counterclockwise.

[7]Text figures **bold** text; Appendix figures plain text; 43.xx = figures in Klein and others, 1987

[8]Reference codes (for all tables in the report) as follows: 1; Klein and Wright, 2000, additional references given, 2; Macdonald and Eaton, 1957, 3; Macdonald and Eaton, 1964, 4; 8Macdonald and Eaton, 1954, 5; Macdonald and Eaton, 1956, 7: Richter and others, 1970, 8; Eaton and Fraser, 1957, 9; Eaton and Fraser, 1957, 10; HVO, unpub., 11; HVO seismic catalog, 12; Nakata, 2007, 13; Nakata, 2007, 14; Richter and others, 1964, 15; Kinoshita, 1967, 16; Peck and Kinoshita, 1976, 17; Moore and Koyanagi, 1969, 18; Moore and Krivoy, 1964, 19; Wright and others, 1968, 20; Fiske and Koyanagi, 1968, 21; Bosher, 1981, 22; Klein and others, 1987.

[9]Seismic summaries missing for six quarters between 4/1/1960-10/1/1961. Larger earthquakes were located by hand. There is no located seismicity for the three Halema'uma'u eruptions.

[10]Graben formed during eruption; tilt vectors in this row record the total deflationary tilt; the three succeeding rows record a clockwise rotation during deflation.

[11]Tremor and earthquake counts begin at 7/1 21:50, nearly 11 hours before first located earthquake—tilt lag calculated from the earlier time.

[12]Earthquake counts and tremor begin at 13:46 8/21/1963; tilt lag calculated from that time.

[13]Earthquake counts and tremor begin at 03:16 10/5/1963; tilt lag calculated from that time.

[14]Earthquake counts and tremor begin at 08:02 3/5/1965; tilt lag calculated from that time.

[15]Data from Bosher MS thesis cited in Bosher, 1985.

[16]Earthquake counts and tremor begin at 19:29 12/24/1965; tilt lag calculated from that time.

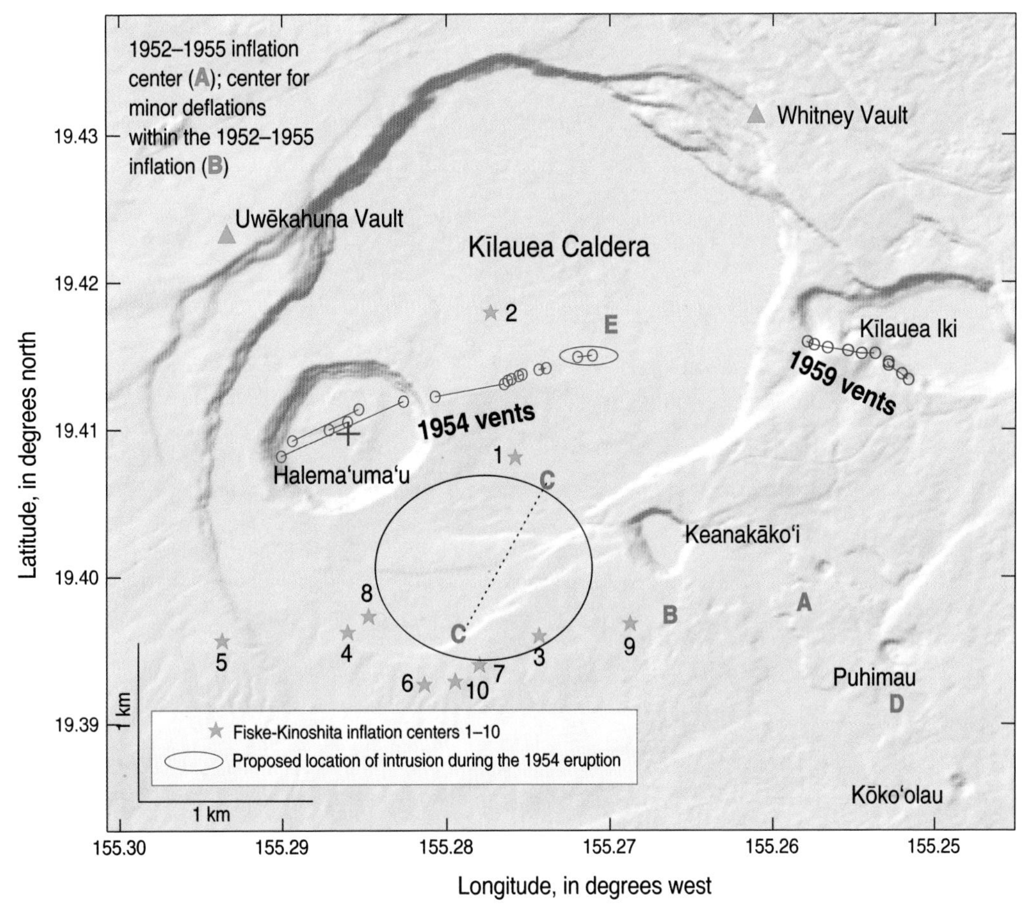

Figure 4.2. Map of Kīlauea activity, 1952–1959. Location of centers of inflation/deflation shown as dark green capital letters. Deformation centers determined from Whitney tilt azimuths are shown for: A, The center of inflation between 27 June 1952 and 27 February 1955 estimated from the tilt azimuth from Whitney to be between Halema'uma'u and the east rift zone. B, Centers of small deflations within deflation period A estimated in the same way as for A and located near the eastern boundary of the Fiske-Kinoshita array. Location of the 1954 vents (small circles) and the oval surrounding the presumed 1954 summit intrusion are shown in purple. Vents for the 1959 eruption are shown as red circles. Fiske-Kinoshita centers of inflation are shown as light green stars, and upper east rift craters are labeled. C, Net deflation between the beginnings of the 1955 and 1959 eruptions. D, The major deflation accompanying and following the 1955 eruption. E, Inflation on the first day of the 1959 eruption. Tilt vectors are given in appendix D, table D3. Deflation center B lies east of the Fiske-Kinoshita array, C lies within the array, and A lies within the upper east rift zone. The black oval outlines the extent of deformation centers determined from the long-base watertube tilt network shown in appendix I, figure I5 . See text for further discussion.

Kīlauea's summit began 2 days after the intrusion. The intrusion triggered a strong south flank response, including a *M*5.3 flank earthquake later the same day (fig. 4.4).

The Eruptions of 1959 and 1960

The 1959 eruption in Kīlauea Iki was unusual for (1) its episodic eruption style, with 17 episodes (table 4.1, fig. 4.6), most lasting less than 24 hours, (2) fountains higher than seen in Hawai'i before or since (Richter and others, 1970), and (3) its hybrid chemistry (Wright, 1973). The eruption filled Kīlauea Iki Crater to a depth of about 110 m, forming the first of several "passive" lava lakes[16] formed during eruption (Helz, 1993). Subsequent drilling of the lava lake showed the lake to be considerably deeper than originally estimated, indicating a syneruption collapse of the crater floor amounting to more than 30 m (Helz, 1993, p. 12–14, figure 4).

Precursory to the 1959 eruption was the occurrence of deep earthquake swarms at 40–60 km depth located north of Kīlauea Caldera (fig. 4.7). Such swarms were first documented in September 1953 (fig. 4.7; Klein and Wright, 2000). These deep swarms were considered to be related to magma movement from a source region in the mantle (Eaton, 1962, p. 21; J.P. Eaton, unpub. data).

The 1959 erupted products were composed of mixtures of two kinds of magma defined by samples

[16] A passive lava lake is one formed during filling of a pit crater. The filling ends when the eruption ends. Active lava lakes are those within a pit crater or at the top of a shield that are connected to the feeding source and thus continuously filled from below.

Figure 4.3. Map of Kīlauea activity, 1959–1960. Locations of centers of inflation/deflation are shown as dark green capital letters and continue the sequence begun in figure 4.2. The small purple oval indicates the area of the 1954 intrusion from figure 4.2. Deformation centers, unless otherwise indicated, are determined from the intersection of tilt vectors from Whitney and Uwēkahuna. (F) Deflation during episode 1 of the 1959 eruption followed by (G) inflation between episode 1 and the beginning of the 1960 eruption. Locations F and G are determined from intersection of tilt vectors from Uwēkahuna and Whitney Vaults (see appendix A). (H) Deflation during 21–24 November 1959 at the end of episode 1 recorded at Whitney Vault only. Using the Whitney azimuth, center H is placed at the west end of Kīlauea Iki and records the collapse of the crater at the end of episode 1. See text for further explanation. The center of deflation during the 1960 eruption is determined from (I) the intersection of tilt vectors from Uwēkahuna and Whitney Vaults, (J) the intersection of azimuths from the array of long-base water-tube tiltmeters, and (K) unpublished leveling data (appendix I, fig. I3; J.P. Eaton, written commun., 1999). Deflation most likely occurred near Fiske-Kinoshita center 1 and within the region defined by the three estimates I, J, and K. The scattering of these deflation center estimates (an area comparable to the size of Halema'uma'u) is an estimate of the minimum error in each location determination. (L) Post-1960 inflation, as defined by intersection of azimuths of short-base tiltmeters located at Uwēkahuna and Whitney Vaults, still lies east of the Fiske-Kinoshita array. Tilt vectors are given in appendix D, table D3. The black oval covers deformation centers determined from the long-base water-tube tilt network. See text for further discussion.

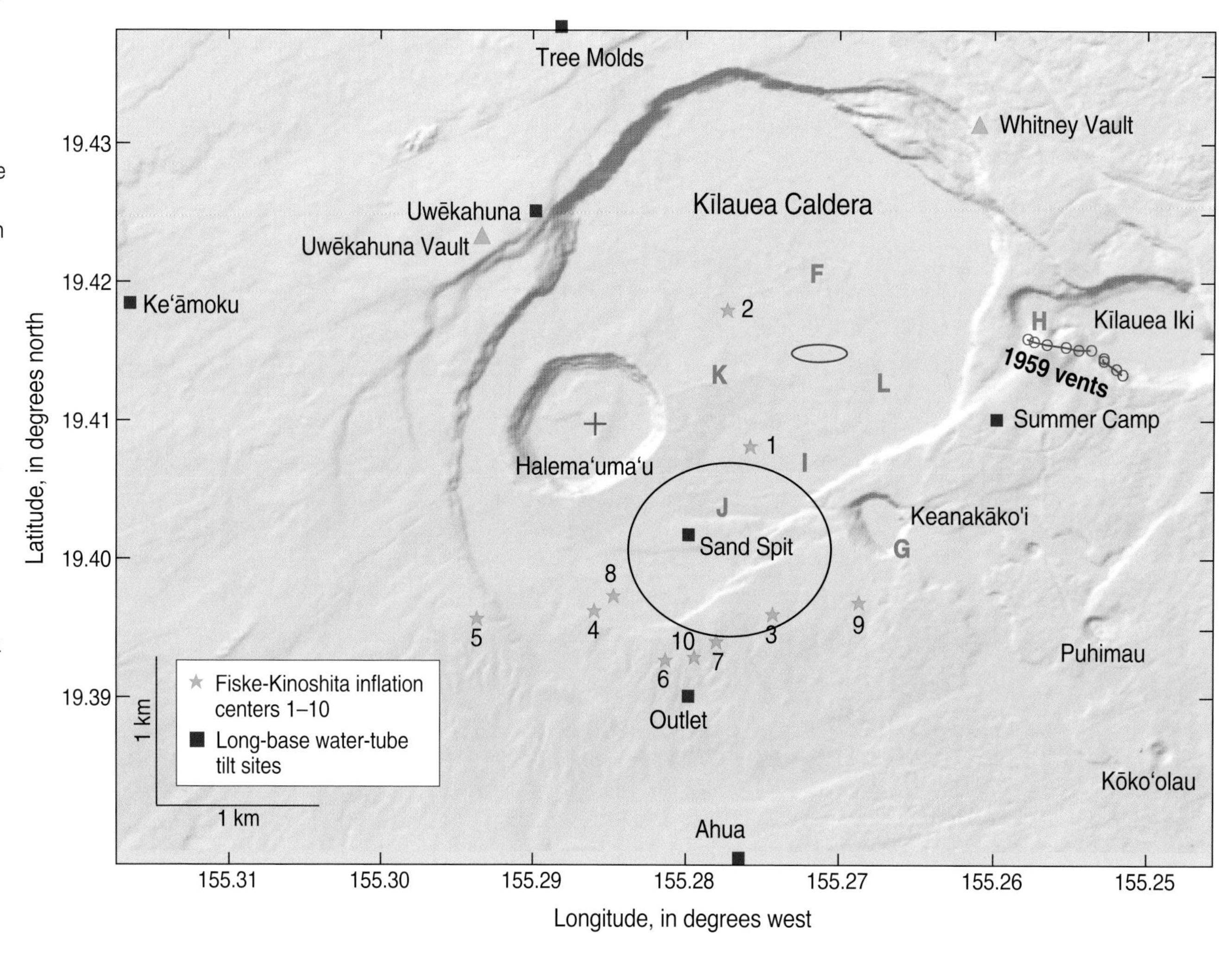

collected at the east and west ends of the initial line of erupting vents in episode 1[17]. The sample collected at the western end closer to the caldera (variant 1) had few olivine phenocrysts and a composition identical to that of magma erupted within Halema'uma'u and Kīlauea Caldera in 1954. The sample at the eastern end (variant 2) had an unusual composition unmatched, for example, in its high ratio of CaO to MgO, by any other historical Kīlauea eruption. The eruption products also contained abundant phenocrysts of deformed and twinned olivine, indicating an origin considerably deeper than other Kīlauea eruptions (Helz, 1987). Products of subsequent episodes had chemistry that could be modeled as mixtures of variants 1 and 2 with varying amounts of additional olivine (Wright, 1973). Variant 2 is interpreted as related to the deep earthquake swarms originating north of the caldera as described above. Variant 2 later found its way into the east rift, where it was identified as a mixing component in hybrid lavas from the latter part of the 1960 eruption (Wright and Helz, 1996).

From 17 January through 19 February 1960 Kīlauea erupted again in the lower east rift zone to the east of the vents of the 1955 eruption (fig. 4.6; Richter and others, 1970). Lavas from this eruption destroyed the village of Kapoho. Precursory seismicity included many earthquakes beneath the south flank southeast of the 1960 vents, as well as shallow earthquakes beneath the eventual eruption site (fig. 4.8). Seismic network coverage of this eruption in this area was poor. The shallow earthquakes below the eruption site ended as the eruption began, because stress was relieved when magma escaped to the surface. As in 1955, the 1960 collapse of Kīlauea's summit occurred after a time delay, 4 days (instead of 8, as in 1955) following the beginning of eruption (fig. 4.6). The succeeding summit earthquake swarm also developed 8 days later, similar to the delay in 1955. The summit subsidence was the largest observed in the post-1924 tilt record and extended into midsummer of 1960.

During the period of maximum deflation, there was a dramatic collapse of Halema'uma'u Crater itself, described in detail in the narrative report of the two eruptions (Richter and others, 1970, p. E68–E73, figures 76–81). The 1960 collapse of Halema'uma'u took place in three stages, on 7 and 9 February and 11 March 1960 (fig. 4.6). During the first stage, the lava lake emplaced in 1952 first drained and then, as collapse proceeded, some of the 1952 lava was reextruded on the crater floor. The 7 February 1960 collapse was accompanied by a plume of explosively pulverized rock, and there was considerable apprehension that the collapse might develop into a major phreatic eruption, such as occurred in 1924. The second collapse, on 9 February, was heralded by a felt earthquake at shallow depth beneath Kīlauea Caldera and was succeeded by more felt earthquakes, some of which were located in the adjacent Koa'e fault system and south flank (fig. 4.8). The third collapse, on 11 March, was again accompanied by a plume of dust but was otherwise uneventful. Fortunately the 1924 experience was not repeated[18].

Deep (45–55 km) earthquake swarms north of Kīlauea Caldera continued beyond the end of the 1960 eruption, the last one being recorded in October 1960. During this time interval earthquake swarms at 20–35 km depth began, marking a transition to frequent earthquake swarms at that depth beneath Kīlauea Caldera beginning in 1961 (fig. 4.9). The 1960 earthquake swarms shallower than 35 km are still north of the caldera, but with locations closer to the caldera than the earthquakes deeper than 35 km.

The Eruptions of 1961 Through 1965

The 1960 eruption was followed by three short eruptions in Halema'uma'u (table 4.1, fig. 4.9), in turn followed by a series of six eruptions on the east rift zone without intervening summit activity (Fiske and Koyanagi, 1968; Moore and Koyanagi, 1969; Moore and Krivoy, 1964; Peck and Kinoshita, 1976; Richter and others, 1964; Wright and others, 1968). Vents for the east rift eruptions are shown in figure 4.10. Earthquake swarms at 20–35 km depth beneath Kīlauea Caldera also continued in 1961 and extended through and beyond 1965 at decreasing frequency and intensity (fig. 4.9).

No earthquakes associated with the 1961 summit activity were documented because they fall within the gap from 1 April 1960 to 1 October 1961, when no seismic summaries were published and no data for earthquakes with magnitudes less than 2 were recorded with information sufficient for later processing (J. Nakata, oral commun., 1992). The September 1961 rift eruption was preceded by an intense swarm of felt earthquakes originating near Nāpau Crater (fig. 4.9; see also Richter and others, 1964, p. D19). The larger events were located from unpublished seismic data that fell within the seismic summary gap mentioned above.

Intrusions into the eastern end of the Koa'e Fault Zone (Klein and others, 1987) accompanied eruptions in December 1962, October 1963, and December 1965 (fig. 4.9). Koa'e intrusions without eruption also occurred in May (figs. 4.9, 4.11*A*; Kinoshita, 1967) and July 1963 (figs. 4.9, 4.11*B*).

Suspected deep intrusions occurred in August and November 1965 and July 1966 (Fig. 4.12). From 25 February 1962 to 2 April 1963 similar small swarms of earthquakes were detected from the eastern south flank and were labeled "KT" (for

[17] Referred to colloquially as a "curtain of fire."

[18] Because the smaller summit collapses in the decades after 1960 did not cause a Halema'uma'u floor collapse, we infer that summit caldera elevation drops of a meter or less can be sustained without completely evacuating magma from under Halema'uma'u.

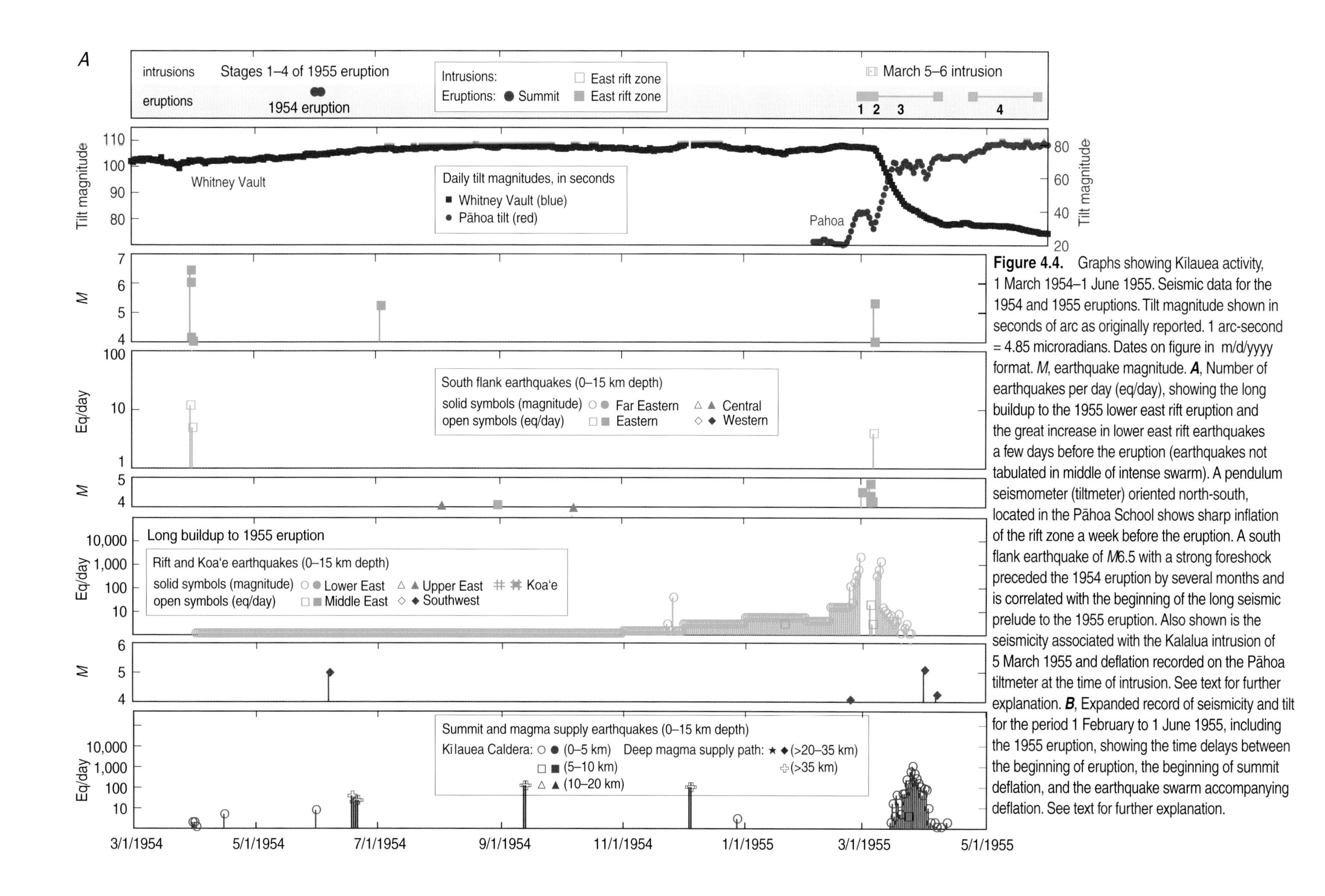

Figure 4.4. Graphs showing Kīlauea activity, 1 March 1954–1 June 1955. Seismic data for the 1954 and 1955 eruptions. Tilt magnitude shown in seconds of arc as originally reported. 1 arc-second = 4.85 microradians. Dates on figure in m/d/yyyy format. *M*, earthquake magnitude. ***A***, Number of earthquakes per day (eq/day), showing the long buildup to the 1955 lower east rift eruption and the great increase in lower east rift earthquakes a few days before the eruption (earthquakes not tabulated in middle of intense swarm). A pendulum seismometer (tiltmeter) oriented north-south, located in the Pāhoa School shows sharp inflation of the rift zone a week before the eruption. A south flank earthquake of *M*6.5 with a strong foreshock preceded the 1954 eruption by several months and is correlated with the beginning of the long seismic prelude to the 1955 eruption. Also shown is the seismicity associated with the Kalalua intrusion of 5 March 1955 and deflation recorded on the Pāhoa tiltmeter at the time of intrusion. See text for further explanation. ***B***, Expanded record of seismicity and tilt for the period 1 February to 1 June 1955, including the 1955 eruption, showing the time delays between the beginning of eruption, the beginning of summit deflation, and the earthquake swarm accompanying deflation. See text for further explanation.

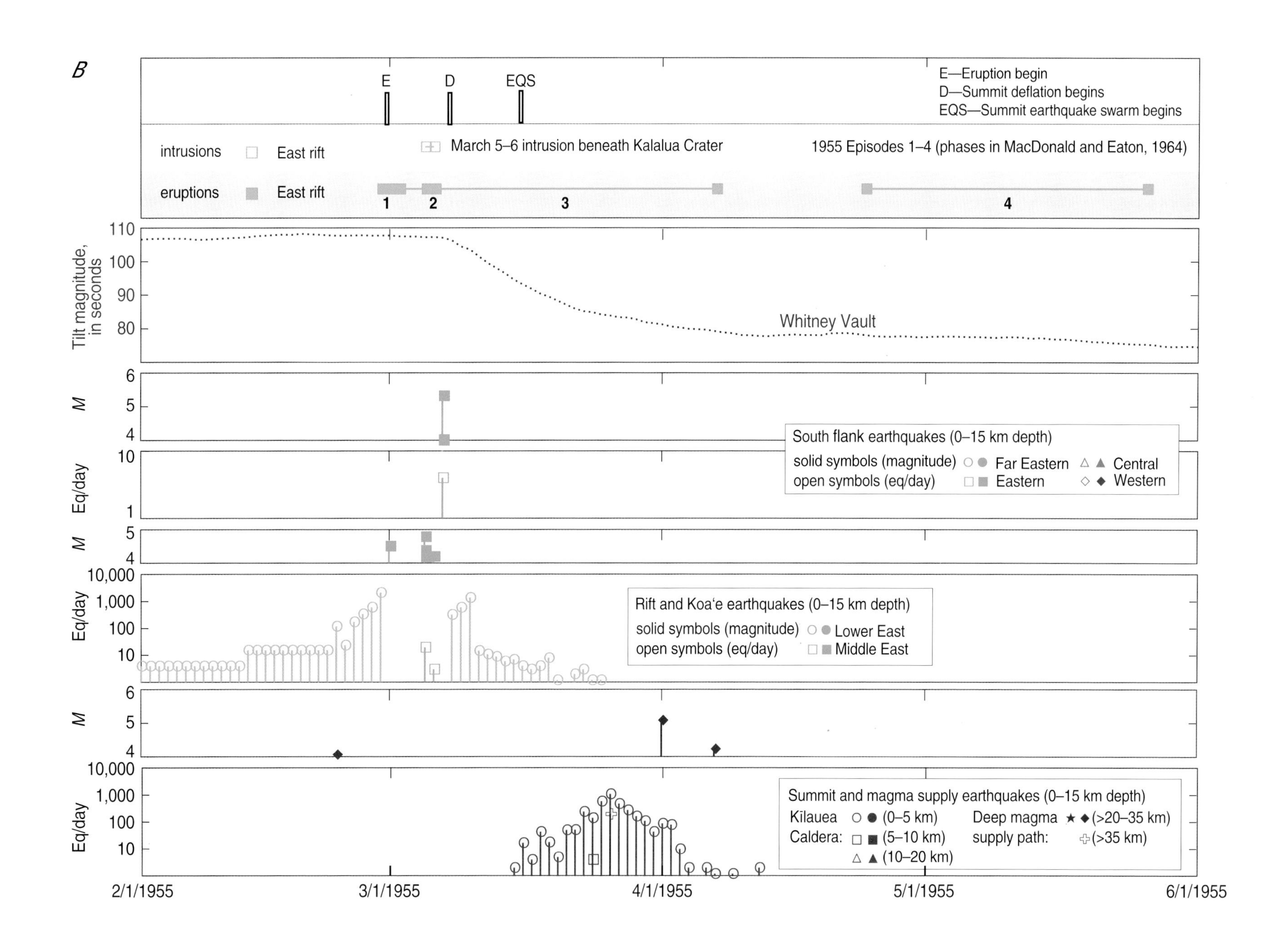
B
E
D
EQS
E—Eruption begin
D—Summit deflation begins
EQS—Summit earthquake swarm begins
intrusions
East rift
March 5–6 intrusion beneath Kalalua Crater
1955 Episodes 1–4 (phases in MacDonald and Eaton, 1964)
eruptions
East rift
1
2
3
4
Tilt magnitude, in seconds
Whitney Vault
M
Eq/day
South flank earthquakes (0–15 km depth)
solid symbols (magnitude)
open symbols (eq/day)
Far Eastern
Eastern
Central
Western
Rift and Koa'e earthquakes (0–15 km depth)
solid symbols (magnitude)
open symbols (eq/day)
Lower East
Middle East
Summit and magma supply earthquakes (0–15 km depth)
Kīlauea
Caldera:
(0–5 km)
(5–10 km)
(10–20 km)
Deep magma supply path:
(>20–35 km)
(>35 km)
2/1/1955
3/1/1955
4/1/1955
5/1/1955
6/1/1955

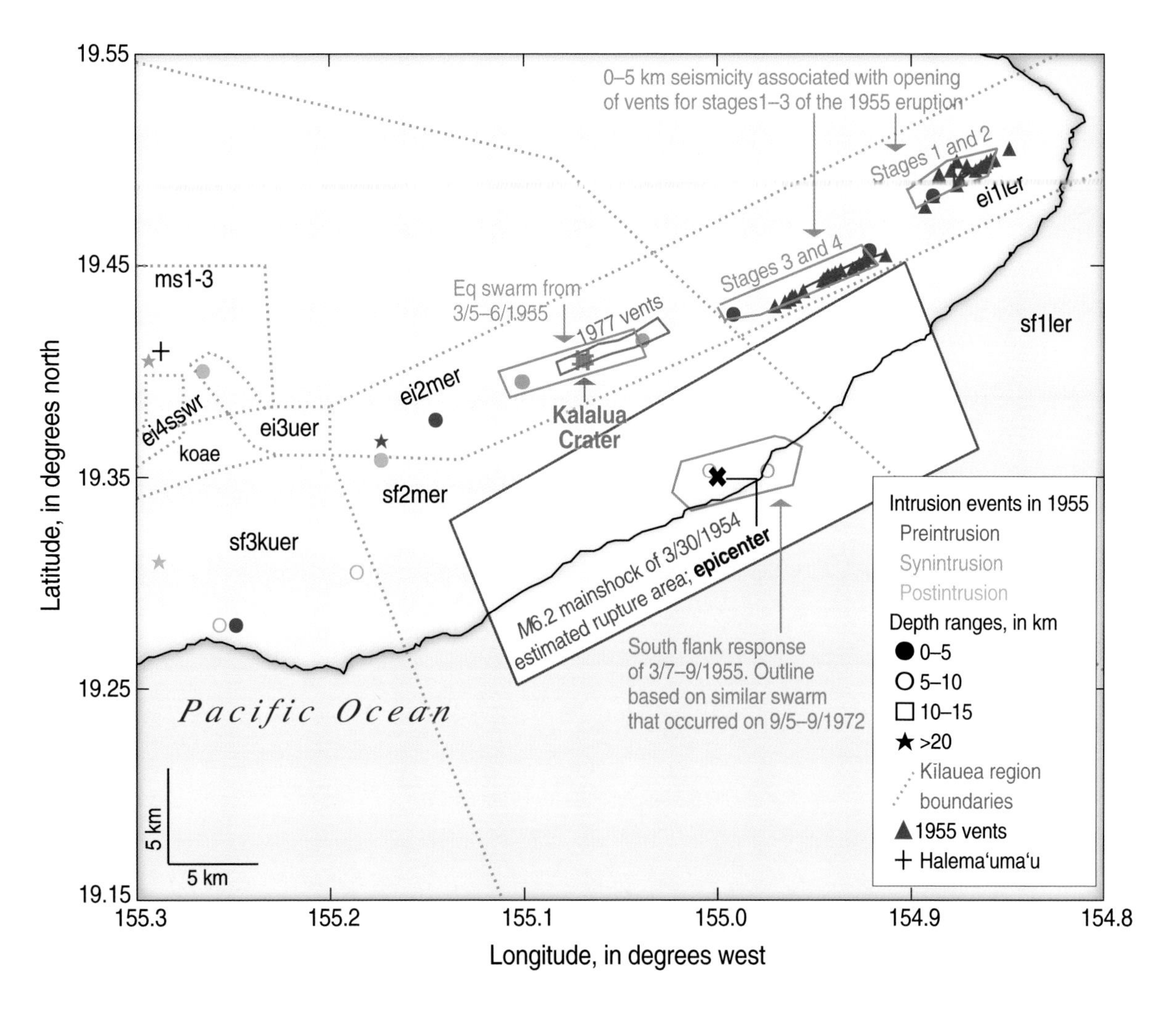

Figure 4.5. Map showing locations of earthquakes (circles and dots) associated with the 1955 Kīlauea lower east rift eruption and Kalalua intrusion. Vents for that eruption are shown as solid red triangles and three rift and one south flank swarms are approximately outlined in orange. Shown for comparison is the area of the 1977 vents. A parent magma for fractionated lavas erupted in 1977 is inferred to correspond to magma emplaced in the 1955 Kalalua intrusion. The south flank region activated in response to the March 1955 Kalalua intrusion is compared with the estimated rupture zone of the 30 March 1954 south flank earthquake. Dates on figure in m/d/yyyy format.

Kalapana Trail) in the HVO seismic summaries (Koyanagi and others, 1963; Krivoy and others, 1964; Okamura and others, 1963, 1964). The earthquakes were all assigned to a single approximate location listed in the HVO seismic summaries and shown in figure 4.12.

Interpretations 1952–1967

Events of this period, bracketed by the 1952 and 1967–68 eruptions in Halema'uma'u Crater, can be better interpreted than was possible for the events in the preceding period because of (1) the expansion of seismic and ground deformation networks and (2) the acquisition of high-quality chemical data for every eruption. The relative timing of earthquake swarms and summit tilt deflections relative to the beginning of eruption and intrusion can now be calculated for this period, as can magma supply rates and eruption efficiencies for the intervals separating eruptions and intrusions. Petrologic study has established that magmas erupted at Kīlauea's summit enter the summit reservoir in their order of eruption and move through the rift plumbing to (1) serve as parents for fractionated lavas and (2) appear as components of mixed magma eruptions (Helz and Wright, 1992; Wright and Fiske, 1971; Wright and Helz, 1996; Wright and others, 1975; Wright and Tilling, 1980). Compositional differences among the summit magmas are small, but they can be clearly discriminated using petrologic mixing calculations (Wright and Fiske, 1971). It is evident that these "magma batches" retain both their chemical identity and low phenocryst content as they move through the Kīlauea plumbing, as shown in figure 4.13. In this section we interpret each eruption in terms of patterns of seismicity (figs. 4.1, 4.9), ground

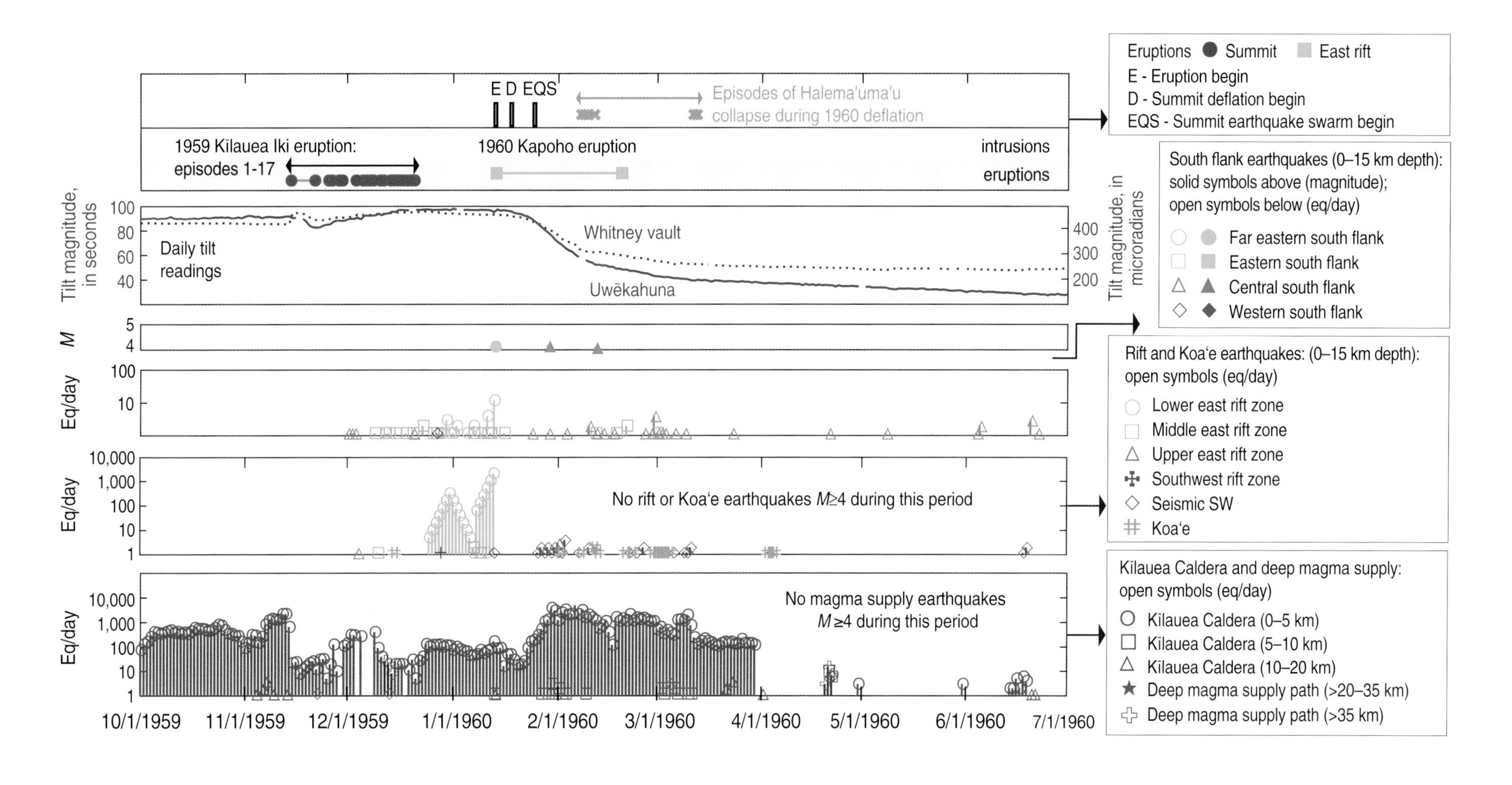

Figure 4.6. Graphs showing seismicity and tilt changes associated with the 1959 Kīlauea Iki and 1960 Kapoho eruptions. Earthquake counts and earthquakes of *M*>4 are shown for all regions and compared with eruption times and tilt changes measured at both the Uwēkahuna and Whitney Vaults. The 1959 eruption in Kīlauea Iki Crater had 17 short episodes. The mismatch between the Whitney and Uwēkahuna tilt reflects a late surge of magma to shallow depth beneath Kīlauea Iki and a later syneruption collapse of the crater floor, both evident only on the Whitney instrument. See text for further explanation. The 1960 eruption shows similar but shorter delays separating the beginning of eruption, summit collapse, and summit earthquake swarm compared to the 1955 sequence shown in figure 4.4*B*. Earthquake swarms at 40-60 km depth north of Kīlauea Caldera continued through both eruptions. Unlike the 1955 eruption and more similar to later rift eruptions, south flank earthquakes occur as a precursor to and, to a lesser extent, as a delayed response to the 1960 eruption. Dates on figure in m/d/yyyy format.

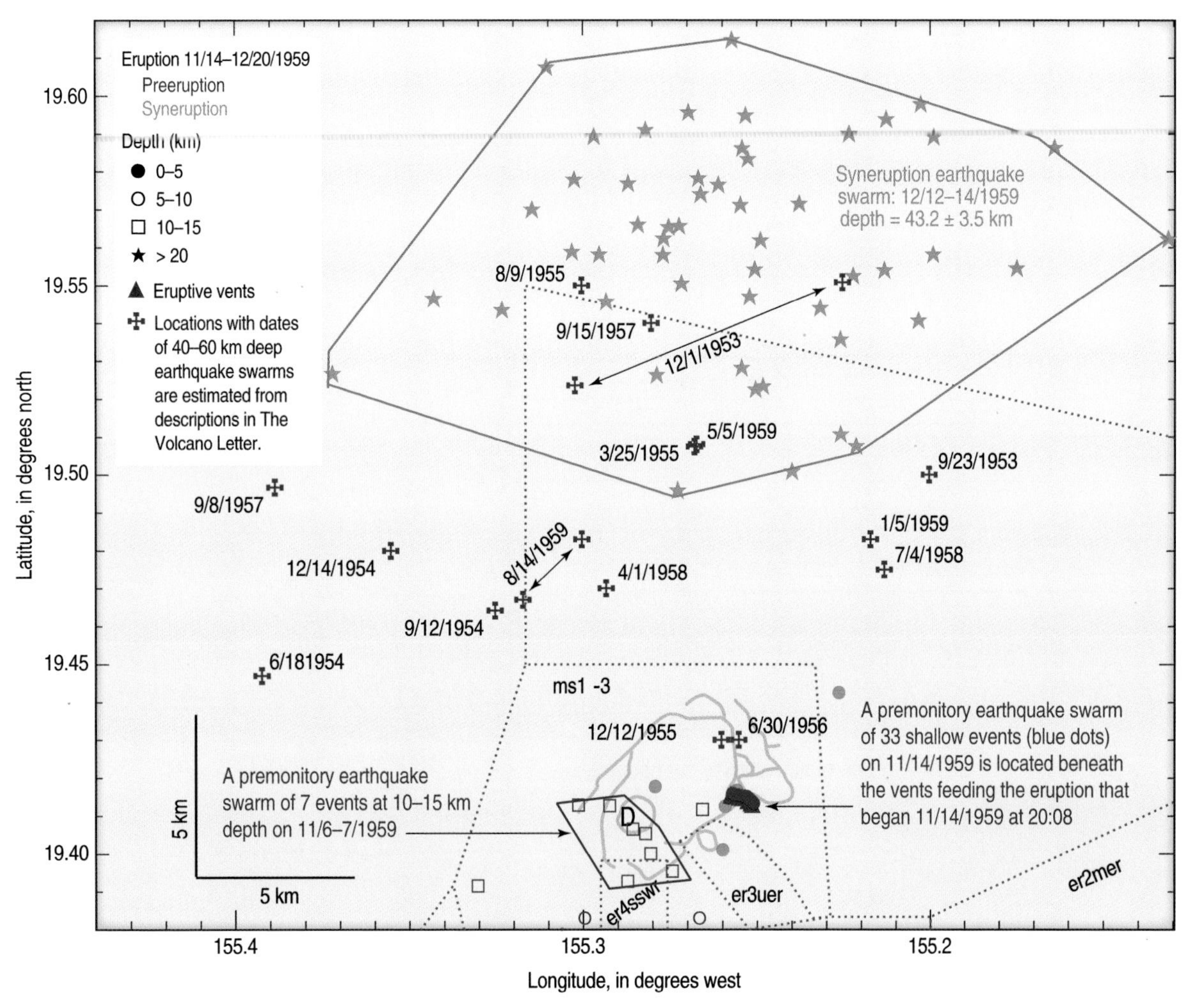

Figure 4.7. Map showing locations of earthquakes associated with the 1959 Kīlauea Iki eruption. Most swarms before December 1959 were reported as consisting of several hundred events at depths of 40–60 km, only approximately located by hand as less than 10 km north (toward Mauna Loa) from the Uwēkahuna seismometer. The orange polygon outlines the computer-located deep earthquake swarm of 12–14 December 1959 and is a likely representation of the spread of epicenters for the larger swarms before 1 October 1959 that are labelled with the date of the swarm next to a blue cross. Dates on figure in m/d/yyyy format.

deformation, and eruption chemistry for the entire sequence of eruptions and intrusions. A quantitative evaluation of where magma is stored beneath Kīlauea's summit and how it moves into the east rift plumbing is given in chapter 8.

1954 Eruption and Intrusion

The 1954 eruption differed from eruptions before and after. Between 1952 and 1955, the upper east rift zone inflated, suggested by the azimuth of the Whitney Vault tilt vector pointing east of the pre-1967 array of inflation centers (fig. 4.2; location "A" assumed). Inflation of the upper east rift zone contrasted with the summit inflation preceding the 1952 eruption, and with post-1960 inflations and deflations (table 4.1). Minor deflations within the overall 1952–55 inflation also lie east of the principal center of inflation and deflation (fig. 4.2; location "B" assumed). The line of 1954 vents extends within the caldera to a point north of Keanakākoʻi (location "E" in fig. 4.2).

The relation of the 1954 magma to the Kīlauea plumbing is problematic. The eruption was of small volume, and the magma does not appear as a mixing component of any subsequent rift eruptions (see, for example, Wright and Fiske, 1971). It is possible that the eruption was fed from the postulated shallow intrusion at the eastern end of the line of vents, which may have been emplaced at the top of a path from the mantle closer to the east rift zone that bypassed the reservoir feeding the larger Halemaʻumaʻu eruptions (fig. 4.13*C*). The line of vents on the caldera floor opened minutes after the fountaining in Halemaʻumaʻu (Macdonald and Eaton, 1954), but the initiation of eruption in Halemaʻumaʻu may have been because the floor of Halemaʻumaʻu was lower in altitude.

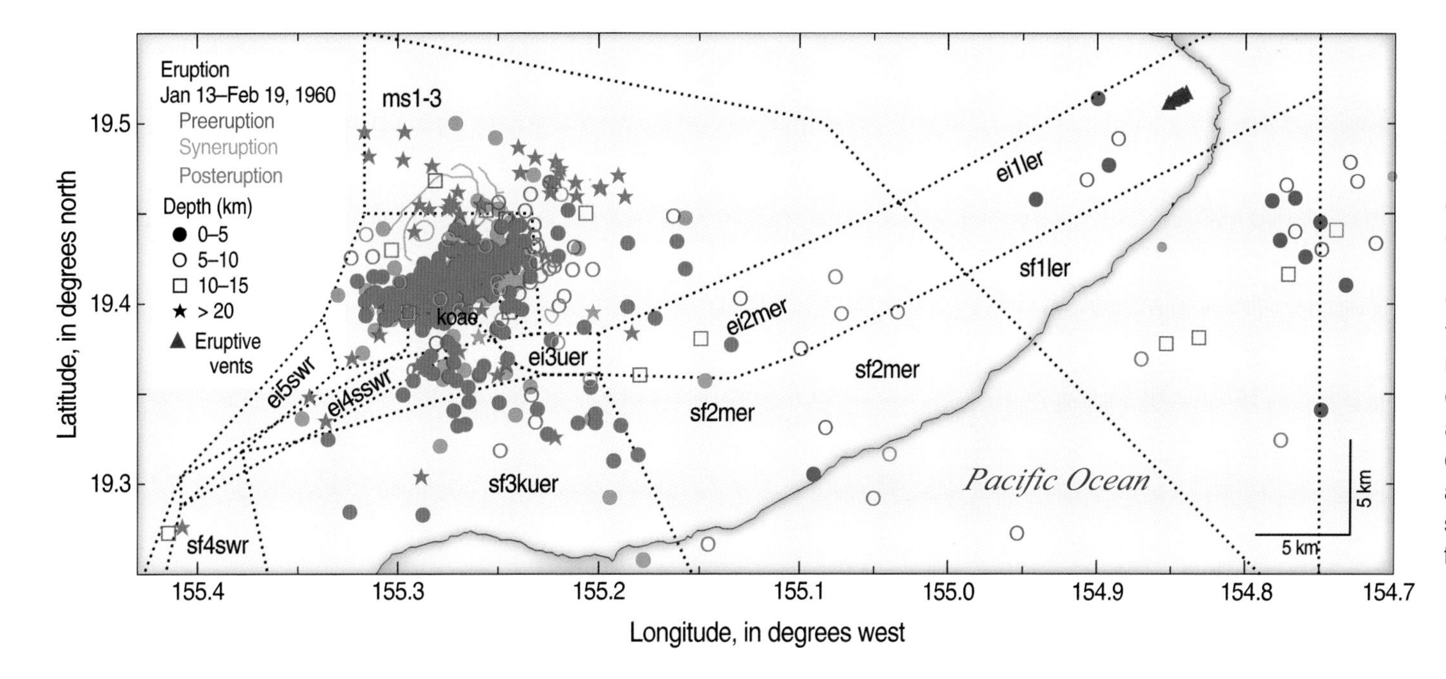

Figure 4.8. Map showing locations of earthquakes and earthquake swarms associated with the 1960 Kapoho eruption. A few south flank earthquakes before the eruption (blue dots and circles) are located southeast of the eruption site, but poor earthquake location precision this far from Kīlauea Caldera means that their location closer to the vents can't be ruled out. The magma supply seismicity at 5–10 km depth occurring during the eruption, as well as the south flank earthquakes during and following the eruption, are associated with the collapse of Kīlauea's summit, beginning about 10 days following the beginning of eruption.

We interpret the 1954 summit eruption as a minor, and almost accidental, event between the 1952 summit eruption and the 1955 east rift eruption. We propose that the 1954 eruption occurred when the edifice above the summit magma reservoir cracked during a time of inflation to allow an eruption, but was quickly healed because the small volume of stored magma was insufficient to sustain a longer eruption. In the longer historical context, we reject the idea of a pairing of the 1954 summit and 1955 flank eruptions (compare Macdonald, 1959, p. 4; Macdonald and Abbott, 1970). The failure of consistent pairing or alternation of summit and flank eruptions is demonstrated by a statistical runs test (Klein, 1982); some summit/flank eruption pairs like 1954/1955 and 1959/1960 may be linked by the volcano's response to gradual increase of its current stress state, but pairing is not a feature of the historical record.

1955 Eruption

The 1955 eruption differed from all previous Kīlauea eruptions in that the erupted magma was highly fractionated and therefore not fed directly from the summit magma reservoir. Instead, it was fed from the body of cooling magma emplaced below the lower east rift zone during the intrusion of April 1924. The erupted magma represents a liquid composition obtained by ~50-percent crystallization of a magma similar in chemistry to what was erupted at Kīlauea's summit in 1924 (table 4.2). Left alone, this body of magma would have cooled and crystallized past the point where it could erupt (compare Marsh, 1981). Instead, the *M*6.5 south flank earthquake of 30 March 1954 was instrumental in making possible the subsequent 1955 lower east rift eruption[19]. The accelerating preeruption lower east rift seismicity may have been caused by slow, downrift magma flow to fill rift voids left by stress release from the 1954 earthquake. The leveling off of summit tilt and inferred stop of summit magma accumulation, after a period of inflation and before the 1955 eruption (fig. 4.4) suggest that magma slowly entered the rift zone to increase pressure on the 1924 magma already in residence. At the same

[19]Flank earthquakes sometimes encourage subsequent eruptions (see, for example, Walter and Amelung, 2006), but at other times they promote unclamping of the rift zone, as happened during the 1975 Kalapana earthquake, thus encouraging intrusions rather than eruptions (Klein, 1982).

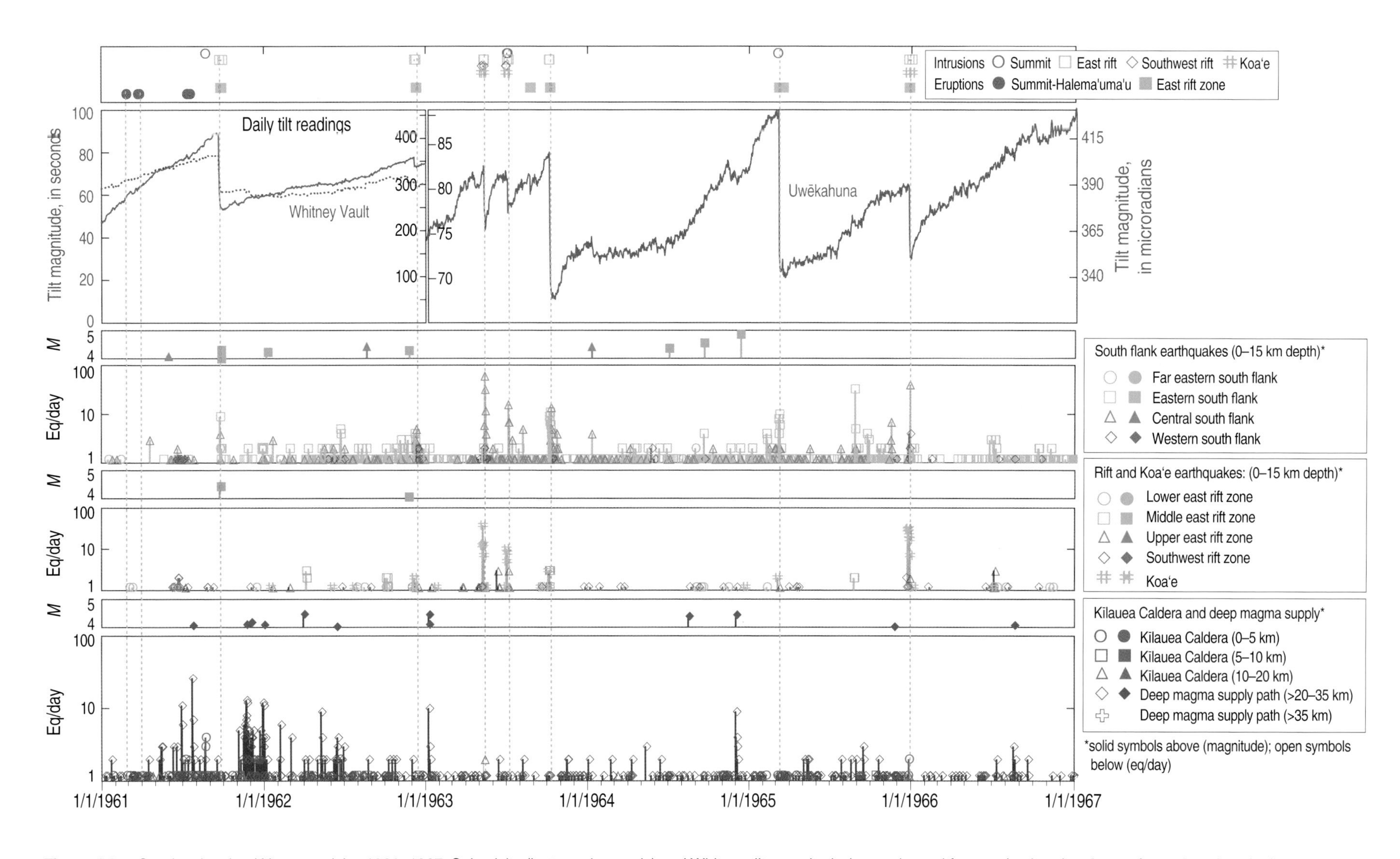

Figure 4.9. Graphs showing Kīlauea activity, 1961–1967. Seismicity (bottom six panels) and Whitney tilt magnitude (second panel from top) related to times of eruption given in the top panel. Vertical dotted lines connect data for eruptions and traditional intrusions; dashed lines connect traditional intrusions without eruption. Tilt magnitudes are given in arc-seconds (left axis) and microradians (right axis). Seismicity is plotted, from bottom to top, for the magma supply path, rift zones and Koa'e, and south flank. Earthquakes per day (eq/day) and magnitudes (*M*) greater than or equal to 4.0 are given for each region. The tilt plot is divided at 1 January 1963 following abandonment of the Whitney tilt site. The scale is also changed at that date to emphasize the smaller deflations associated with eruptions and intrusions after 1961. Dates on figure in m/d/yyyy format.

Table 4.2. Parentage of 1955 and 1977 differentiated lavas.[1]

[Fo = forsterite content of olivine; An = anorthite content of plagioclase; MgO in weight percent]

	Parent	Differentiate	Minerals removed from parent				Residuals	
			Olivine[2]	Augite[3]	Plagioclase	Ilmenite	Max.	Min.
1955 components								
Name	1924	1955E: MgO = 5.39						
Wt %	100	49.84	5.24	23.10	20.52	1.29	0.053	0.003
Composition			Fo 77.1		An 66.1			
1977 components using least magnesian differentiate								
Name	1961	1977: MgO = 5.43						
Wt %	100	62.97	3.85	17.31	15.29	0.58	0.006	0.001
Composition			Fo 77.1		An 68.4			
1977 components using most magnesian differentiate								
Name	1961	1977: MgO = 5.89						
Wt %	100	73.39	3.20	12.80	10.31	0.29	0.010	0.001
Composition			Fo 77.1		An 69.8			

[1]Petrologic mixing calculations modified from a program described by Wright and Doherty, 1970.

[2]Olivine chemistry fixed.

[3]Augite chemistry averaged from two analyzed augites from similar rocks.

time the earthquake slip on the flank decollement acted to both reduce the confining stress on the rift zone and to reduce the pressure above the stored magma body below, thus promoting eruption. The eruption occurred while the summit was still inflating. The earthquake location beneath the eastern south flank also relieved confining stress on the rift near the March 1955 Kalalua intrusion (fig. 4.5).

Seismicity and ground deformation before, during, and following the 1955 eruption differ markedly from subsequent east rift eruptions and intrusions. At that time, it was not possible for HVO to locate earthquakes with magnitudes less than 3, or discriminate rift earthquakes associated with dike formation from triggered flank events. Therefore we cannot plot earthquakes on the map (fig 4.5) but can plot daily counts (figs 4.4). The occurrence in time of earthquakes in the days before the eruption started are typical of earthquake swarms accompanying dike propagation. The March 1955 Kalalua intrusion was accompanied by an intense swarm

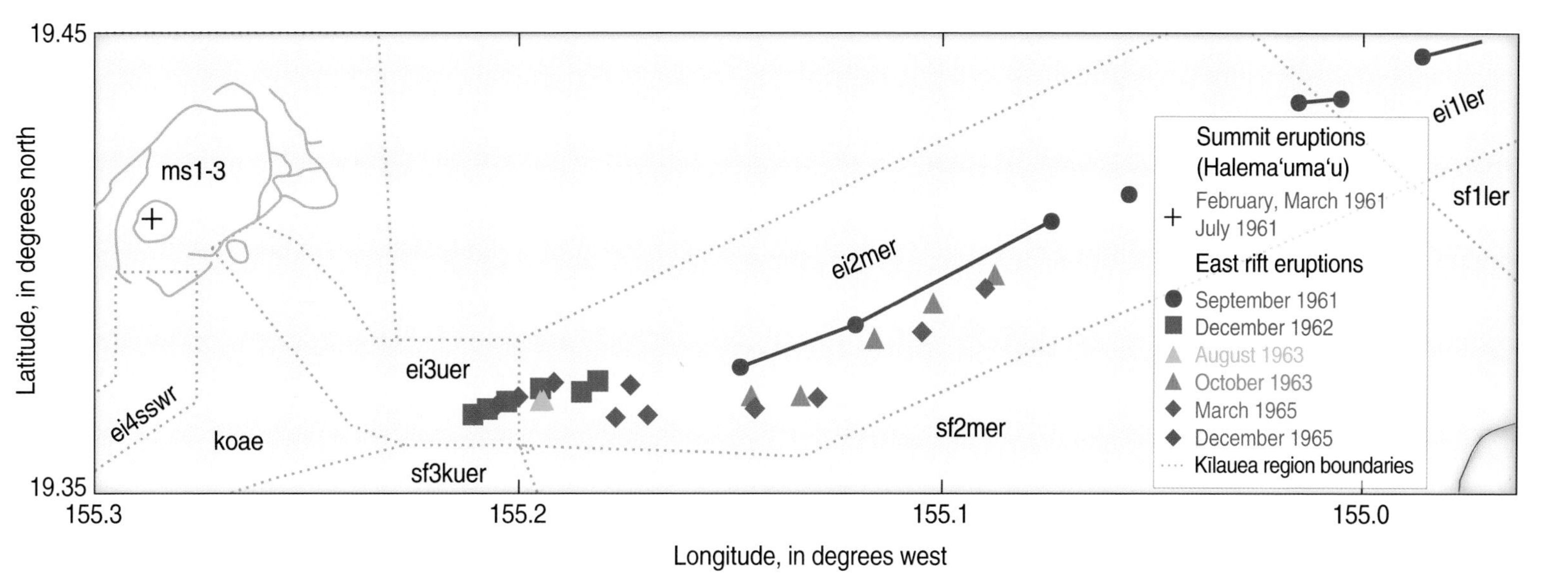

Figure 4.10. Map showing Kilauea activity, 1961–1965. Vent locations for 1961–1965 eruptions on Kilauea's east rift zone. Many eruptions took place at several locations along the rift zone.

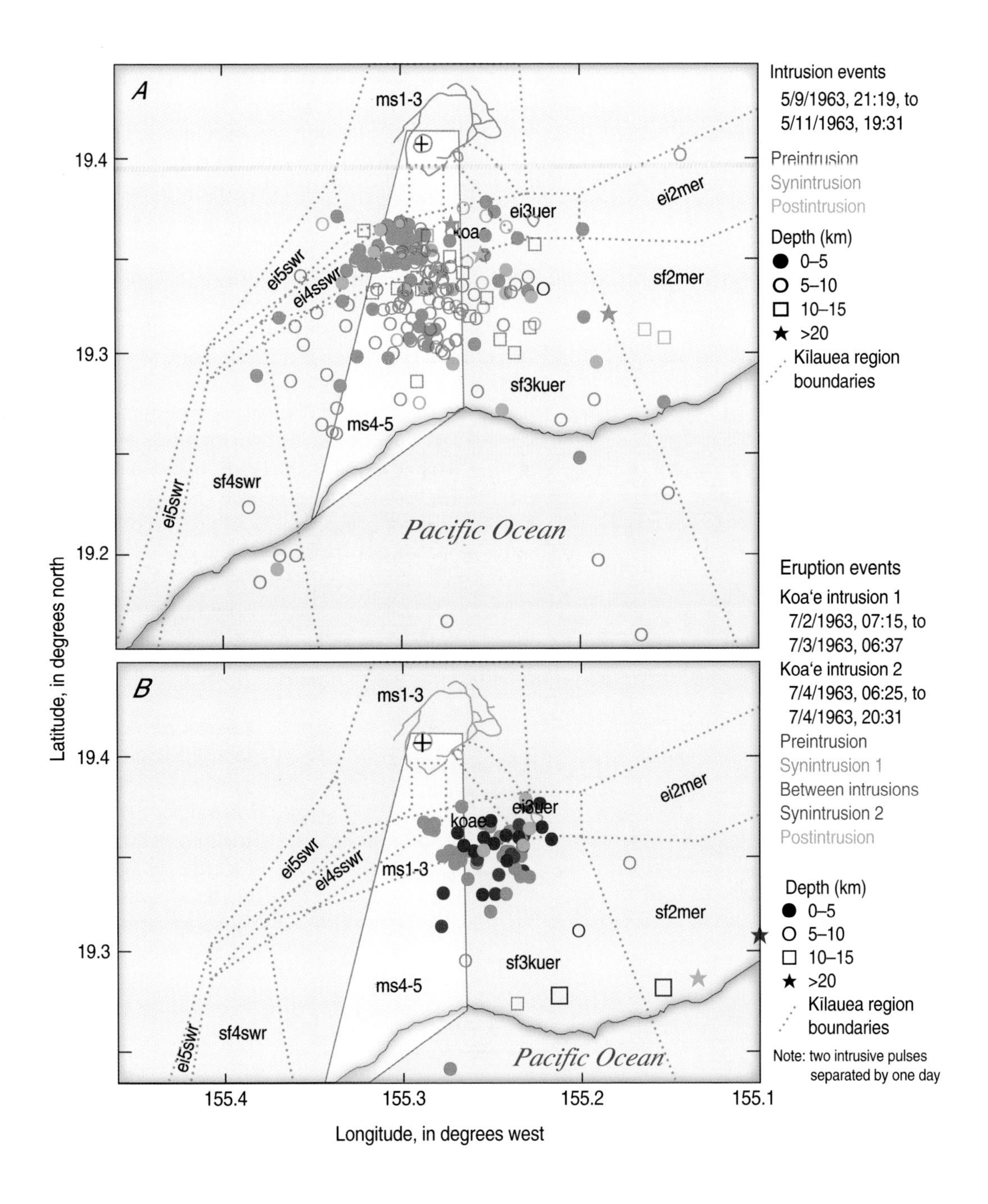

Figure 4.11. Maps showing seismicity associated with 1963 Koa'e intrusions. Symbols for pre-, during-, and postintrusion are explained in text preceding figure 4.7*D*. Dates on figure in m/d/yyyy format. Times given represent the beginnings and endings of intrusions. ***A***, Locations of earthquakes associated with the May 1963 Koa'e intrusion. ***B***, Locations of earthquakes associated with the July 1963 Koa'e-east rift intrusion. This intrusion occurred to the east of the May event, filling in the relatively aseismic area left from the May intrusion. Both intrusions have a strong south flank response.

of rift earthquakes of *M*3 to *M*>4 and a delayed south flank response with one earthquake of *M*5.5 about 2 days after the intrusion began (fig. 4.4). The early Kalalua intrusion phase was neither preceded nor accompanied by draining of Kīlauea's summit magma reservoir, as has been the case for every rift eruption beginning in 1961. However, the east rift zone did later deflate about 10 arc-seconds (50 μr), as shown by the Pāhoa tiltmeter record (fig. 4.4*A*). The large summit deflation that followed the beginning of eruption showed a deformation center somewhat east of the Fiske-Kinoshita array (fig. 4.2, point labeled "D").

Petrologic calculations made for the fractionated magmas of the 1977 eruption indicate that the chemistry of the parent magma, assumed to be emplaced beneath the 1977 eruption site in 1955, matches the chemistry of lavas erupted at Kīlauea's summit during 1961 (table 4.2). We infer that magma of 1961 composition moved upward beneath Kalalua in 1955 to cause the deflation coincident with intrusion observed on the Pāhoa tiltmeter in 1955 (fig. 4.4*A*). Subsidence of Kīlauea's summit began 2 days after the Kalalua intrusion, and 8 days after the beginning of the 1955 eruption (fig.4.4*B*). This implies the 1955 eruption was initally fed from magma stored within the rift. After another 8 days a major earthquake swarm began beneath Kīlauea's summit, with epicenters extending to 10–km depth. We associate these summit earthquakes with caldera subsidence and magma flow from the summit magma reservoir. Notably, the deflation was not accompanied by seismicity beneath the east rift zone, indicating that magma moved aseismically downrift below the seismic zone of shallow intrusion without breaking rock along the path.

Petrologic study indicates that lava erupted during the latter part of the 1955 eruption had mixed with magma having a chemistry identical to that of Kīlauea's 1952 summit eruption (Helz and Wright, 1992). The timing and chemistry of magmas emplaced in 1955 are summarized in Wright and Helz (1996, table 3a). The relative ages and inferred position of magmas within Kīlauea's plumbing during the 1955 eruption are shown in figures 4.13*D* and 4.13*E*. We infer that the 1952 magma occupied only that part of the rift zone east (downrift) of Kalalua. Mixing of 1952 with the early 1955 magma began before pressure was applied from summit deflation and this mixture was the only magma occupying the rift zone beneath the site of the 1955 eruption.

We interpret the delay in summit collapse in 1955 to the distance between the eruption site and Kīlauea's summit. The loss of magma from the 1955 section of the lower rift was too far away to be immediately detected at the summit, but the pressure drop drove eventual replacement from the summit storage reservoir, which was then recorded by the tilt drop.

We can compare the volume erupted in 1955 (0.07 km^3) with the volume of magma moving into the rift plumbing from the summit magma reservoir. Before the 1955 eruption, Kīlauea's summit reservoir received an estimated volume of magma of about 0.036 km^3, based on inflation of 73 µrad (16.3 arc-seconds) measured at the Whitney Vault between the beginning of the 1952 eruption and the beginning of the 1955 eruption. The volume of subsidence calculated from the Whitney tilt record is 0.0952 km^3. The difference of these two values represents net loss from the summit of 0.059 km^3, and within measurement error this is equal to or less than the magma volume of 0.07 km^3 for the 1955 erupted lava after correcting the lava volume for 20 percent vesicles. The volume comparison, which implies no additional intrusion over and above replacing what was lost during eruption, is consistent with the scarcity of south flank seismicity during magma recharge.

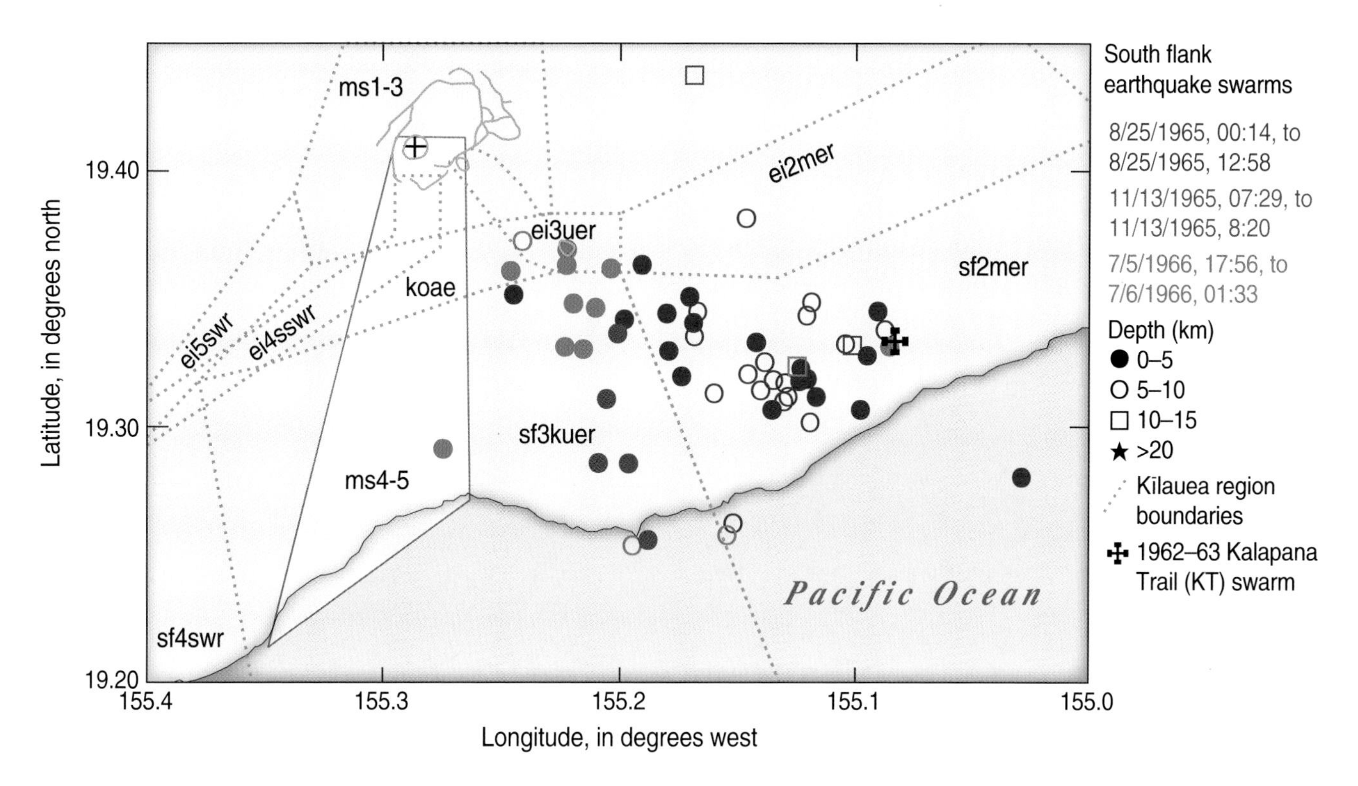

Figure 4.12. Map showing seismic activity in 1965–1966 associated with two suspected deep intrusions and one minor rift intrusion with a similar pattern in its south flank response. The patterns shown fit later documented examples of suspected deep intrusions in which earthquakes are distributed at a high angle to the east rift zone. Also shown are the imperfectly located earthquakes designated "Kalapana Trail" or "KT" in the HVO seismic summaries for 1962, which may be the earliest cataloged occurrence of south flank earthquake swarms without accompanying rift activity, and could be the earliest examples of suspected deep intrusions. Dates on figure in m/d/yyyy format.

1959 Eruption

Tilt data between the beginning of the 1955 eruption and the beginning of the 1959 eruption failed to recover the net deflation located beneath the upper east rift zone (fig. 4.2, point labeled "C"). This was the last time that an inflation or deflation vector lay outside of the Fiske-Kinoshita array. The 1959 eruption has been interpreted as a mixed-magma eruption, and different models have been offered to explain the origin and mixing of the two magmas erupted in 1959 (Helz, 1987; Wright, 1973). As indicated in the section interpreting the 1954 eruption, a postulated intrusion below the eastern end of the line of 1954 vents lies close to the westward projection of the initial line of vents for the 1959 eruption (purple oval in figs. 4.2 and 4.3). As the altitudes of the 1959 vent openings at Kīlauea Iki were lower than the altitude of the caldera floor we infer that the opening of the Kīlauea Iki fracture could draw the magma erupted at the western end of the 1959 vents (variant 1) from this caldera source, even though magma erupted at the eastern end of the 1959 vents (variant 2) came from a deeper, distinctive source (see below).

An additional observation suggests the 1954 and 1959 magmatic plumbing systems were connected. On the day before the 1959 eruption began, the Whitney tilt magnitude showed a sharp upward excursion at an azimuth of about 31° east of north (tilt down to the northeast). Intersection with a smaller inflation vector at Uwēkahuna indicates inflation of a source close to the inferred location of the magma intruded in 1954 (fig. 4.2, point labeled "E"). This may have been the final magma-pressure increase that caused the 1959 fractures to open.

Meanwhile the magma component of the 1959 eruption associated with the deep (40–60 km) earthquake swarms north of Kīlauea Caldera (variant 2) must have been accumulating from the time of initial detection of deep seismicity in 1953. This component is inferred to have flooded the shallow caldera reservoir containing the magma of 1954 chemistry to produce the hybrid magma compositions seen during the 1959 eruption (fig. 13*F*). Upward movement of this component is presumed to have occurred rapidly to preserve the deformed and twinned olivine.

The collapse of Kīlauea Iki Crater during the 1959 eruption can also be detected in the tilt record at the Whitney Vault. During 1959 episode 1 both the Whitney and Uwēkahuna isntruments recorded deflation at a point less than 1 km north of the proposed 1954 intrusion (fig. 4.3, label "F"). We conclude that this represents removal of magma from the 1954 intrusion. In the pause between eruptive episodes 1 and 2, the Whitney tilt difference for 21–24 November 1959 shows a 1.1 arc-second (5.3 µrad) deflation at an azimuth of 161° pointing toward the west end of the Kīlauea Iki vents (fig. 4.3, point labeled "H"), whereas the Uwēkahuna vector over the same interval shows inflation. This is the only evidence to corroborate the collapse of the floor of Kīlauea Iki documented above, and the timing was logical, occurring at the end of the long and large-volume initial episode of the Kīlauea Iki eruption (Richter and Moore, 1966). This shallow collapse was too far away from the Uwēkahuna Vault to be recorded by the Uwēkahuna tiltmeter and barely above the noise level at Whitney.

Tilt during and following the 1959 eruption signifies the end of inflation and deflation located close to the upper east rift zone. The intersection of Uwēkahuna and Whitney tilt vectors for the period between the onset of the Kīlauea Iki eruption on 14 November 1959 and the onset of the 1960 eruption falls within the caldera (fig. 4.3, point labeled "G") and is consistent with a source within the summit reservoir rather than a shallow center in the eastern caldera or beneath Kīlauea Iki.

1960 Eruption

The January-February 1960 Kapoho eruption is linked chemically and petrographically to the 1955 lower east rift eruption and to the 1959 eruption in Kīlauea Iki. During the summit deflation following the 1955 eruption, most but not all of the 1952 magma mixed with the unerupted 1955 magma to refill the 1955 storage volume (fig. 4.13*E*), producing a hybrid magma that was erupted at the beginning of the 1960 eruption (fig. 4.13*G*).

The earliest 1960 lavas are chemically identical to those erupted from the east rift in late 1955 but have about 5 percent more phenocrysts formed during the 5 years separating the two eruptions (Helz and Wright, 1992; Murata and Richter, 1966; Wright and Fiske, 1971). The earliest recharge in 1960 involves only magma having the 1952 chemistry, which is interpreted as the last remnant of the 1952 magma that was not intruded in 1955, and which occupied a region east of the easternmost 1955 vents. Subsequent recharge involved the other two summit magma batches in a sequence matching their appearance in later eruptions at Kīlauea's summit, that is, 1952⇒1961⇒1967–68 (fig. 4.13*G*; Wright and Helz, 1996), but carrying more olivine than was seen in the corresponding Halema'uma'u eruptions. A final component of the 1960 recharge matched the eastern (mantle) component of the preceding 1959 eruption, distinguished both by chemistry and by the twinned and deformed olivine (fig. 4.13*G*). Addition of this 1959 component contributed further to the olivine-rich character of the 1960 eruption.

Deformation data indicate that the patterns of inflation and deflation in 1959–61 represent a transition from centers located to the east of the Fiske-Kinoshita array to centers within the Fiske-Kinoshita array. The intersection of tilt vectors and the center of leveling contours for the 1960 collapse (fig. 4.3, points labeled "I," "J," and "K") are located in the central caldera surrounding inflation center 1 of the Fiske-Kinoshita array of inflation sites. The differences in the centers determined for the 1960 collapse arise from different time intervals for the measurements and, in the case of the short-base tilt, from continued influence of the east rift zone on the Whitney tilt azimuths, as well as higher error in tilt vector inversions. The tilt intersection of the pre-1961 inflation (fig. 4.3, label "L") likewise falls near center 1 of the Fiske-Kinoshita array.

The south flank seismicity adjacent to the lower east rift zone preceding and accompanying the 1960 eruption (fig. 4.8), and lack of much seismicity in the middle rift or flank, is consistent with expansion of the 1960 magma reservoir by addition of magma already in the lower east rift zone underneath the reservoir. Finally, the volume of collapse during and following the 1960 eruption, as estimated from the Uwēkahuna tilt (0.1405 km^3), is nearly twice the erupted volume after correction for 20 percent vesicles (0.0906 km^3). This volume difference suggests that significant additional rift intrusion took place, even after refilling of the summit reservoir evacuated during the eruption. In this case the absence of posteruption south flank seismicity is puzzling and may indicate that the intruded volume was small and (or) that the eruption volume was underestimated by the amount that flowed into the ocean (Richter and others, 1970).

History of Magma Mixing in 1955 and 1960

The identification of the sequence of chemical changes associated with magma mixing in the period from the 1952 summit eruption through the 1960 east rift eruption was possible only through the acquisition of very high quality major-oxide chemical data from well-sampled eruptions. The interpretations cartooned in figure 4.13 are rendered credible by the observation that the time sequence of compositions identified as mixing components matches the time sequence of the three chemically distinguishable summit eruptions of 1952, 1961, and 1967–68. The mixing calculations force a further inference that these magmas entered the east rift zone along parallel paths beginning even before their appearance in the three summit eruptions. The mixing calculations also indicate that (1) the 1954 summit magma, also chemically distinct, never moved through the main Kīlauea plumbing feeding the summit reservoir from below and (2) a magma of unique chemistry moved from great depths at a location far from the usual mantle source region to be erupted only in 1959 and 1960. In chapter 8 we augment the interpretations illustrated in figure 4.13 with consideration of volumes and residence times for the three summit magmas in order to specify the nature of magma storage at Kīlauea's summit during this period.

Unique Deep Magma Pathways

The events of the 1954–60 period stand apart from the rest of Kīlauea's recorded history. The vertical changes observed during the 1960 collapse record the integrated movement of summit, east rift, and south flank. The deep (40–60 km) seismicity during 1953–60 and the unique chemistry of the Kīlauea Iki eruption strongly suggest that the magma being supplied was from an alternate deep mantle source above the normal Hawaiian mantle source and that movement of magma from this source to the surface did not follow the typical magma conduit feeding Kīlauea's shallow reservoir. Instead, the deep earthquakes and magma movement toward Kīlauea Iki were independent of Kīlauea's typical plumbing, but instead they were assisting inflation of Kīlauea's summit from a position eccentric to the main plumbing geometry. We thus think of the deep Kīlauea conduit as being multistranded (Wright and Klein, 2006, fig. 10 and discussion, p. 63–64). On arriving at shallow depth beneath the eastern caldera, the deeper magma (variant 2 of the 1959 eruption) mixed with and pushed out a shallow pocket of magma (variant 1 of the 1959 eruption) that also may have followed a different path to be stored some distance away from the primary Kīlauea reservoir. The absence of shallow rift seismicity, the migration of flank seismicity to the far eastern south flank, and the appearance of the deep component of the 1959 Kīlauea Iki eruption during the latter part of the 1960 eruption all suggest that magma pressure was being applied at a deeper (>5 km?) level beneath the east rift zone than in other typical eruptions (2–4 km). Finally, the movement of magma out of the shallow summit reservoir in 1960 was triggered by downrift movement of magma batches within the deeper parts of the rift zone.

Comparison with Other Time Periods

The rift activity between 1961 and 1966 is similar to the earlier 20th century east rift zone eruptions (1922 and 1923) and intrusions (1938). During the time (1956–62) that both Whitney and Uwēkahuna Vaults were in use, tilt measurements were consistent, the respective vectors defining centers of inflation and

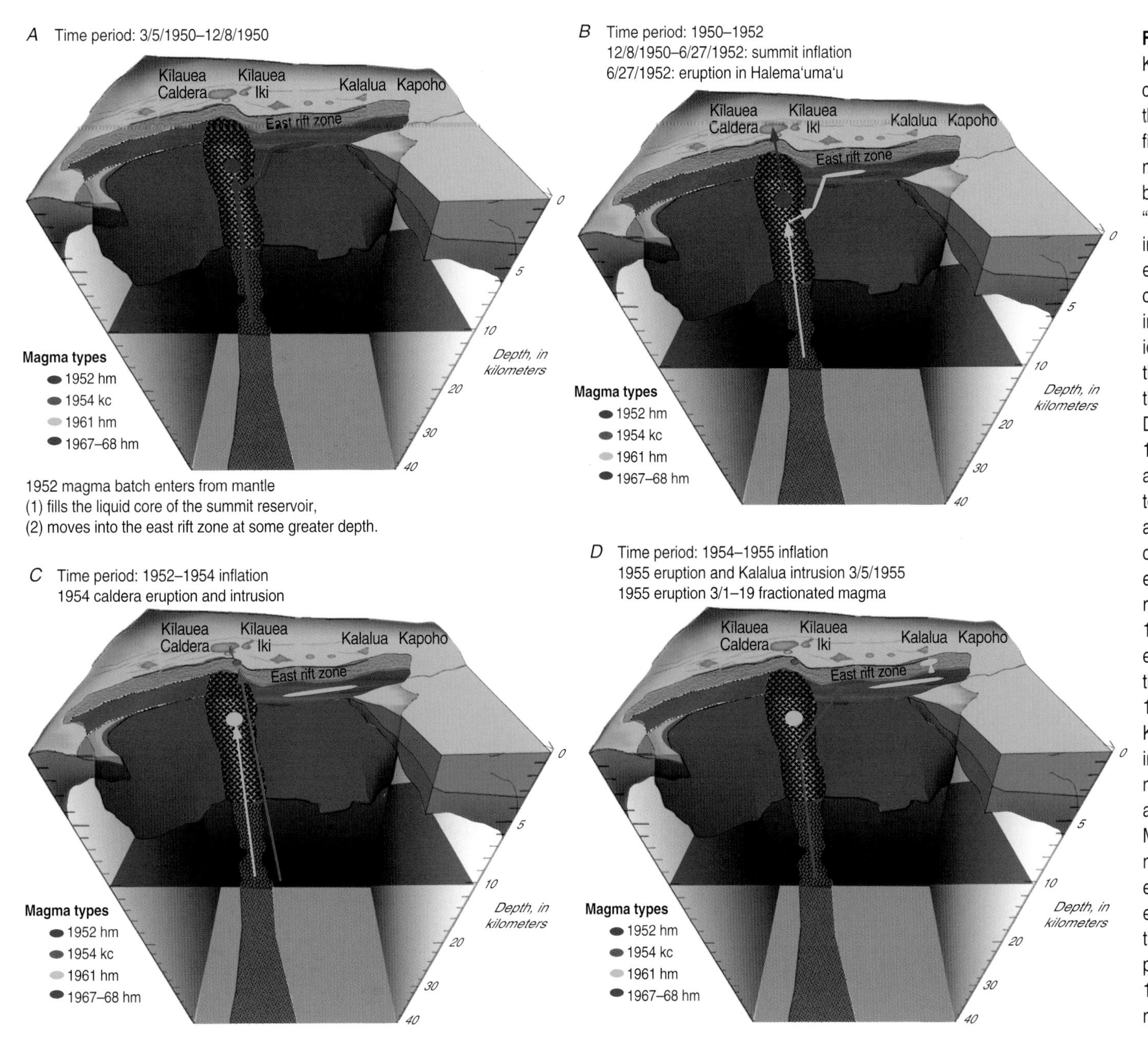

Figure 4.13. Cutaway diagram of Kīlauea showing transport paths of chemically distinct magma batches through the volcano's plumbing. Dates on figure in m/d/yyyy format. Four different magma chemistries are distinguished by color. The "1952 hm", "1961 hm" and "1967-68 hm" magma chemistries erupted in Halema'uma'u and also entered the east rift zone. The "1954 kc" magma chemistry erupted in Kīlauea Caldera in 1954 and its intruded equivalent was identified as a mixing component of the 1959 eruption, but never entered the east rift zone. ***A***, 5 March 1950 to 8 December 1950. Possible distribution of 1952 magma beneath both the summit and east rift zone. ***B***, 8 December 1950 to 27 June 1952. Introduction of 1954 and 1961 magma following the Koa'e collapse in December 1950. The 1952 eruption is fed from the top of the summit reservoir. ***C***, Inflation between 1952 and 1954 eruptions. The 1952 Halema'uma'u eruption used up the 1952 magma in the summit reservoir. The small-volume 1954 magma is erupted and intruded at Kīlauea's summit without any transport into the east rift zone. ***D***, The 1961 magma moves up and outward into the rift zone and is intruded beneath Kalalua on 5 March 1955. Introduction of 1967–68 magma into the east rift zone. ***E***, 1955 eruption. During the latter part of the 1955 eruption, 1952 magma stored beneath the lower east rift zone is mixed with previously stored 1961 magma. ***F***, 1955–1960. The 1959 eruption is a mixture of magma from a mantle source (black oval

at lower left) with magma of 1954 chemistry intruded west of Kīlauea Iki. ***G***, The 1960 eruption on the east rift zone initially erupts the mixed magma produced in 1955, then is mixed with additional rift magma in the following sequence: 1952–1961–1967–1968. A component of the mantle-derived magma is also a component of the latter part of the 1960 eruption (black arrow, unknown path from mantle). ***H***, The 1967-68 eruption and afterward. 1967-68 magma erupts in Halemaʻumaʻu. Subsequent east rift hybrid eruptions are mixed with new magma batches arriving from the mantle at more frequent intervals, illustrated schematically by the solid black arrows.

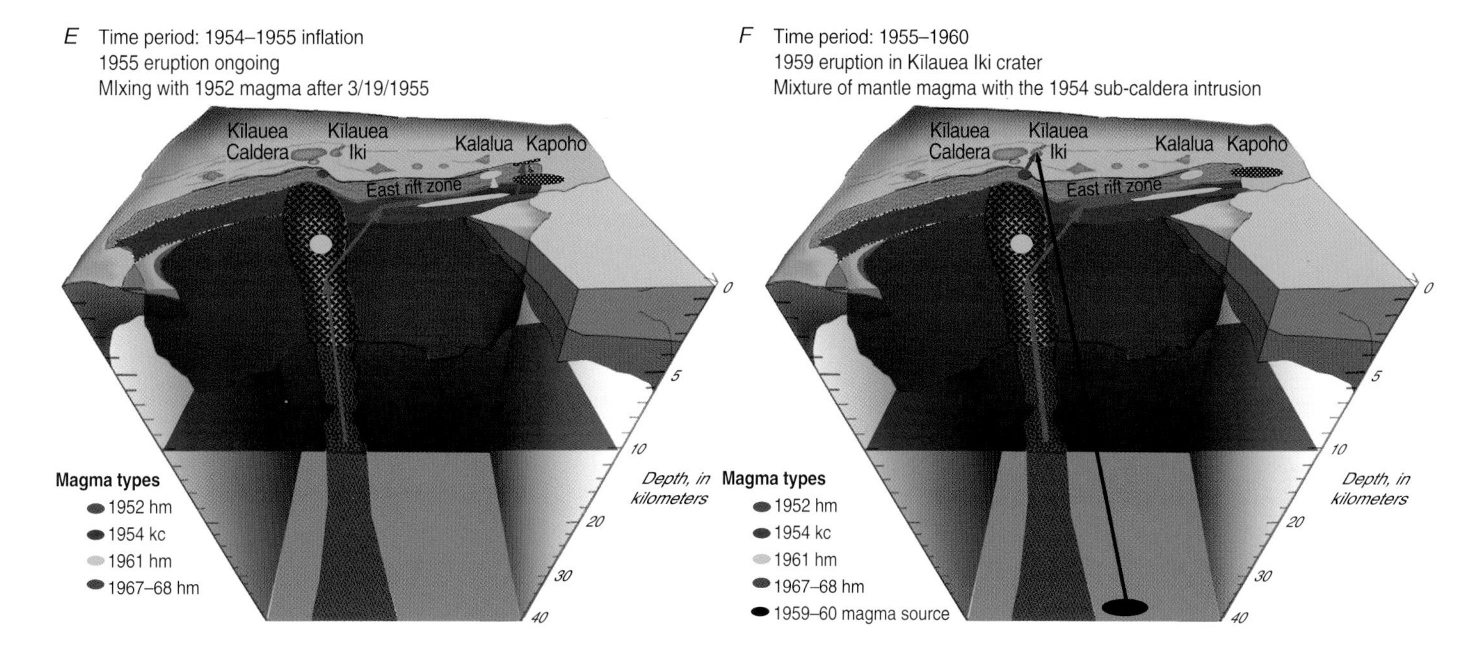

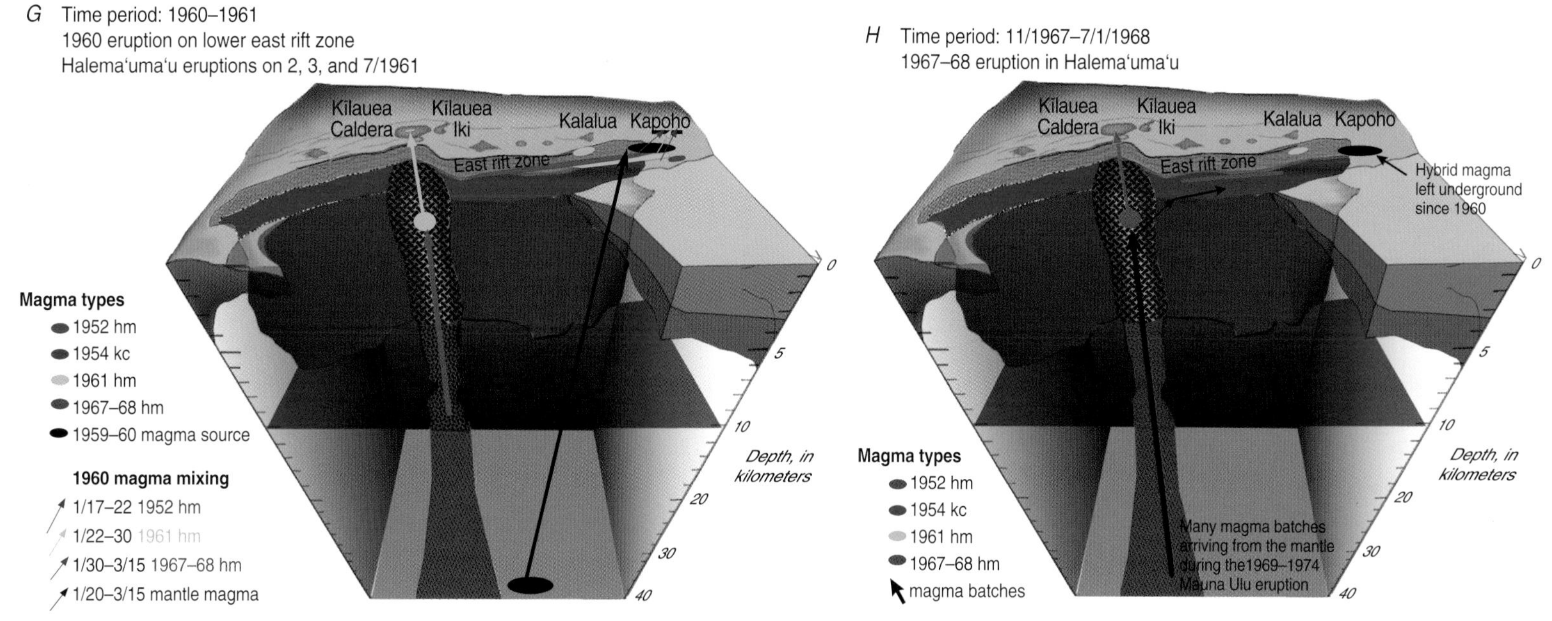

deflation around the Sand Spit tilt station (fig. 4.3). The pattern of inflation before rift eruptions, and deflation associated with rift eruptions and intrusions, of this 1956–1962 period is similar to that of later post-1961 periods in which deformation data were more numerous and locations more accurate.

Deep Magma Supply Seismicity

Earthquake swarms in the 20–35-km depth zone dominantly represent response to variable stresses associated with magma supply, whereas larger single earthquakes at this depth may be triggered by flexural stresses in the lithosphere under the volcanic load and by lateral shear stresses from the south flank. The incidence of earthquake swarms at 20–35-km depth sharply decreased following the December 1962 eruption (fig. 4.9). A similar decrease in earthquake rate only for the year 1963 occurs in the eastern south flank sector at 5–10-km depths adjacent to the middle east rift zone. The rate of seismic release beneath the various rift segments and Koa‘e, however, does not diminish until 1966.

Reading of wet tilt. USGS photograph by R.S. Fiske, January, 1968.

Supplementary Material

Supplementary material for this chapter appears in appendix D, which is only available in the digital versions of this work—in the DVD that accompanies the printed volume and as a separate file accompanying this volume on the Web at http://pubs.usgs.gov/pp/1806/. Appendix D comprises the following:

Table D1 shows additional earthquake swarms and earthquakes of *M*4–6 from 1953 to 1967.

Table D2 summarizes tilt volume, eruption efficiency, and magma supply rate for the period 1950–67

Table D3 summarizes tilt data for deformation centers plotted in text figures 4. 2 and 4.3.

Figures D1 and D2 show earthquake swarms using the same data as in text figures 4.1 and 4.9.

Figure D3*A–N* shows the same data as text figures 4.1 and 4.9 plotted at 1-year intervals.

Figure D4*A–N* shows earthquake swarm data as in figures D1 and D2 plotted at 1-year intervals, but beginning 3 years earlier at 1 February 1950.

Figures D5–9 show time series plots of earthquakes associated with eruptions and traditional intrusions between 1954 and 1965.

Figures D10–17 show map plots of earthquakes associated with eruptions and traditional intrusions between 1954 and 1965.

Figures D18–25 show map plots of deep (20–35 km) earthquake swarms in and near the magma supply path, organized chronologically.

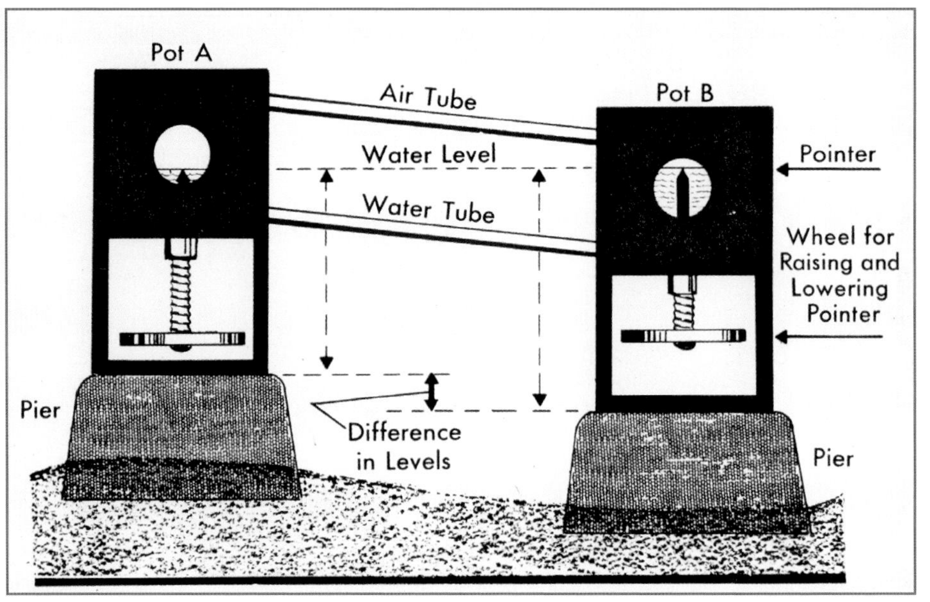

Field measurement of watertube (wet) tilt at Kīlauea as pioneered by Jerry Eaton showing instrument setup (Herbert, D., and Bardossi, F. , 1968, Kīlauea: Case History of a volcano, New York, Harper & Row, p. 106).

Chapter 5

Eruptive and Intrusive Activity, 1967–1975

In the period after 1967–68, more frequent changes in the chemistry of magma entering the system are manifested in the history of the long east-rift eruption at Mauna Ulu. Magma supply continues to increase during and beyond the end of the Mauna Ulu eruption, leading to a large intrusion in the southwest part of Kīlauea, which destabilized the south flank and is interpreted as a proximate cause of the 1975 M7.2 south flank earthquake.

Mauna Ulu lava shield built on Kilauea's east rift zone during the 1969–1974 eruption. USGS photo by J.D. Griggs, June 14, 1989.

The period beginning with the long 1967–68 eruption in Halema'uma'u and ending with the *M*7.2 south flank earthquake in 1975 is an important chapter in HVO's history. Expansion of the seismic network was largely completed, and expansion of ground-deformation networks was accelerating. Improvements in monitoring ground deformation had begun in the 1960's with the nailing of setups and turning points on level lines and the acquisition of electronic distance measuring (edm) instrumentation. Later in the period a "dry" tilt network replaced the water-tube tilt network. In late 1966 a continuously recording Ideal-Arrowsmith (IA) tiltmeter was installed in Uwēkahuna Vault. Although it measured only the east-west component of tilt, it allowed continuous strip-chart recording and precise identification of the onset of summit deflation, filling the same role that the Press-Ewing seismometer installation had filled during the early 1960s.

Along with improved instrumentation and methods came increased challenges to the HVO staff as eruption frequency underwent a dramatic increase. In the years between the two summit eruptions in 1961 and 1967, there were a total of six rift eruptions, although 2 of the 6 years saw no eruptive activity. By contrast, following the summit eruption ending in July 1968, there were four rift eruptions in less than 1 year. The last of these produced the small Mauna Ulu shield that continued in semicontinuous eruption for a total of almost 5 years. As a result, incoming staff members no longer had the luxury of focusing their interpretations on the first eruption observed during their stay, but now had to consider multiple periods of activity.

Kīlauea eruptions, intrusions, and large earthquakes beginning with the 1967–68 Halema'uma'u eruption up to the occurrence of the *M*7.2 south flank earthquake in 1975 are summarized in tables 5.1, 5.3–5.7. The eruptions from 1967 through 1975 are covered in seven papers (Duffield and others, 1982; Jackson and others, 1975; Kinoshita and others, 1969; Lockwood and others, 1999; Swanson and others, 1976b, 1979; Tilling and others, 1987). Additional interpretation of events in this period is covered in two summary papers (Klein and others, 1987; Wright and Klein, 2008). Seismic analysis for the entire period is included in the Klein reference, and figures from that paper are cross-referenced in tables 5.1 and 5.3 to 5.7. In the tables and figures we adopt the definitions of eruption and intrusion given in chapter 1 of this paper, modified from terminology used previously to define intrusions (Wright and Klein, 2008, table 2).

The Mauna Ulu eruption in 1969 is of particular interest for two reasons. First, it is the centerpiece of an epoch of much greater seismic activity than in any previous period in Kīlauea's post-1952 history and takes place within a period of accelerating magma supply that we consider an important factor in the triggering of the 1975 earthquake. Second, it is Kīlauea's first (and failed) 20th century attempt at continuous east-rift eruption, an attempt that finally succeeded with the east-rift eruption that began in 1983 and continues at this writing (2014). The Mauna Ulu eruption has been described in two papers (Swanson and others, 1979; Tilling and others, 1987) and is the primary focus of another paper (Wright and Klein, 2008). In the following we summarize the sequence of activity and amplify and add to the conclusions of the latter paper.

Mauna Ulu Prelude: 1 January 1967–1 April 1969

A long summit eruption in Halema'uma'u occurred from November 1967 to July 1968 (Kinoshita and others, 1969) and was followed by three small east-rift eruptions in 1968 and 1969 (Jackson and others, 1975; Swanson and others, 1976b). Leveling surveys conducted around Kīlauea's summit in the period preceding the summit eruption showed that the center of inflation migrated around Halema'uma'u from northeast to southeast to southwest (chap. 4, fig. 4.2; Fiske and Kinoshita, 1969, figure 5). Parts of this cycle of inflation centers can be identified in the buildup to other eruptions, both before and after 1967–68[20].

The immediate onset for the November 1967 summit eruption was three *M*2.7–2.8 earthquakes shallower than 2 km beneath Kīlauea's summit (fig. 5.1, lower panel). Other activity that might be considered as precursory were magma-supply earthquakes of *M*4+ at 20–35-km depth on 31 December 1966 and 1 July 1967, followed by a *M*4.1 magma-supply earthquake at 10–20-km depth on 8 September 1967—though many such earthquakes occur routinely and not immediately preceding an eruption. Slight increases in the rate of inflation followed each of these earthquakes. The beginning of eruption on 5 November 1967 was marked by a sharp deflation of about 13 μr and, after recovery to a value near the preeruption level, deflation resumed at a lower rate. The initial deflation ended with the occurrence of a swarm of south flank earthquakes on 5–10 January 1968 identified as a multiple suspected deep intrusion labeled SDI 1, SDI 2, and SDI 3 in figure 5.2. All three source areas are seen in subsequent suspected deep intrusions, the most common locus being SDI 2. The end of eruption in Halema'uma'u in July 1968 was preceded by a sharp inflation 5–6 weeks before and by gradually

[20] With reference to chapter 4, figure 4.2, tilt migration for three eruptions out of many can serve as examples: October 1963—south of center 2 to north of center 1 over 2 days; February 1969—south of center 2 to center 1 over 2 days; December 1974—the full cycle from north (center 2) through south (center 3) to west (center 5) over 4 days.

Table 5.1. Kīlauea eruptions, intrusions, and earthquakes, pre-Mauna Ulu (see figs. 5.1, E3).

[In rows with multiple entries text applies down to the next entry; dates in m/d/yyyy format; do = ditto (same as above); data for eruptions and traditional intrusions are emphasized by grey shading]

Time and Date		Region[1]	Event Type[2]	No.[3]	Tilt[4]		Lag[5]	Comment	Figures[6]	References[7]
Start	End				Mag	Az				
02:32 11/05/1967	19:00 7/31/1968	kcal-hm	E		19.3	166.6	no data	Net tilt during eruption	E4	1, 2, 3
06:30 11/5/1967	08:30 11/6/1967				13.4	125		Tilt 11/5-6—initial deflation		
21:44 1/05/1968	20:54 1/08/1968	sf2mer	SDI[8]	29					**5.2**	
06:30 1/07/1968	10:29 1/10/1968	sf3kuer	SDI	49				Slight inflation following SDI		
06:30 1/07/1968	21:50 1/07/1968		EQS	16				Broken earthquakes , of 15 events 1/05/, 02:58–06:11	**5.2**	
15:46 1/07/1968	01:57 1/08/1968		EQS	11						
19:01 1/08/1968	10:29 1/10/1968		EQS	26				5 additional events 1/08, 06:59–11:47		
03:20 8/22/1968	09:39 8/22/1968	koae	EQS	11				Exceptionally few eq in all reions		
03:50 8/22/1968	11:29 8/22/1968	ei3uer		5	56.7	114		No swarm	*43.28*	
06:00 8/22/1968	10:00 8/26/1968	UERZ/koae	E/I				no data	Hiiaka Crater and east; Tilt 8/21–25	E5; *43.28*	4, 8
10:37 10/07/1968	23:27 10/09/1968	sf3kuer	EQS	36				South flank-ant/acc[9]	*43.29*	
10:42 10/07/1968	20:43 10/09/1968	ei2mer	EQS	145						
11:11 10/07/1968	12:04 10/08/1968	sf2mer	EQS	29				South flank acc/resp[9]		
12:32 10/07/1968	00:28 10/08/1968	ms1	EQS	41						
14:35 10/07/1968	04:00 10/12/1968	MERZ	E/I		60.2	116	no data	Nāpau Crater and east; Tilt 10/7–9	**5.3**	4
00:00 10/21/1968	17:00 10/22/1968	MERZ	E				no data	Renewed eruption		
16:33 12/16/1968		sf3kuer	EQ		flat tilt			*M*4.2 with 2 possible foreshocks and >26 aftershocks	E6	
16:25 2/09/1969		sf2mer	EQ					*M*4.3 with 3 possible foreshocks and >30 aftershocks		12
09:56 2/21/1969	14:59 2/22/1969	ms1	EQS	20					*43.31*	
05:23 2/22/1969	09:46 2/22/1969	sf2mer	EQS	17				South flank anticipation/ accompaniment		
06:23 2/22/1969	14:54 2/22/1969	sf3kuer	EQS	6				do		
07:15 2/22/1969	09:29 2/22/1969	ei2mer	EQS	6						
09:50 2/22/1969	03:00 2/28/1969	UERZ	E/I		48.7	124.2	+1h 50m	Aloi Crater and east; tilt 2/21–28	E7	5
12:16 2/22/1969	14:54 2/22/1969	ei3uer	I	5						
02:21 3/21/1969	05:55 3/21/1969	sf3kuer	EQS	15				South flank anticipation/ accompaniment	*43.32*	
02:34 3/21/1969	03:26 3/21/1969	ei3uer	I[10]	11	3.0	101.3	+0h 21m	Tilt 3/20-22	E8	
02:42 3/21/1969	06:04 3/21/1969	sf2mer	EQS	5				South flank-accompaniment/ response		

[1] Earthquake classification abbreviations are given according to the classification in appendix A, table A3, and locations of regions are shown in appendix A, figure A4. Eruption locations are designated in bold type as follows: KC, Kīlauea Caldera; LERZ, lower east rift zone; MERZ, middle east rift zone; UERZ, upper east rift zone; SWR, outhwest rift zone.

[2] E, Eruption; intrusion ("traditional" I; "inflationary" II; "suspected deep intrusion" SDI-see chapter 1; earthquake, EQ; earthquake swarms EQS).

[3] Minimum number of events defining a swarm: 20 for south flank; 10 for all other regions.

[4] Magnitude in microradians and azimuth of daily tilt measurements from the water-tube tiltmeter in Uwēkahuna Vault.

[5] Lag times separating the onset of the earliest earthquake swarm (excluding south flank) for a given event and the beginning of deflation or inflation measured by the continuously recording Ideal-Arrowsmith tiltmeter in Uwēkahuna Vault. (+) tilt leads, (-) tilt lags.

[6] Text figures **bold text**; appendix figures plain text; 43.xx = figures in Klein and others, 1987.

[7] References coded as follows: 1. Fiske and Kinoshita, 1969, 2. Kinoshita and others, 1969, 3. Wright and Klein, 2008, 4. Jackson and others, 1975, 5. Swanson and others, 1976b, 6. Swanson and others, 1979,, 7. Duffield and others, 1974, 8. Klein and others, 1987, 9. Tilling and others, 1987, 10. Nielsen and others, 1977, 11. Lockwood and others, 1999, 12. Klein and others, 2006. Note: This is a master list for all tables in chapter 5. Only some references will be cited in this table.

[8] Suspected deep intrusions are defined by south flank earthquake swarms (10 eq/day minimum) with little rift seismicity or tilt change.

[9] Abbreviations as follows relative to time of intrusion or eruption: ant, anticipation (before); acc, accompaniment (during); resp, response (after).

[10] Traditional intrusions unaccompanied by eruption are defined by at least 5 rift events per day.

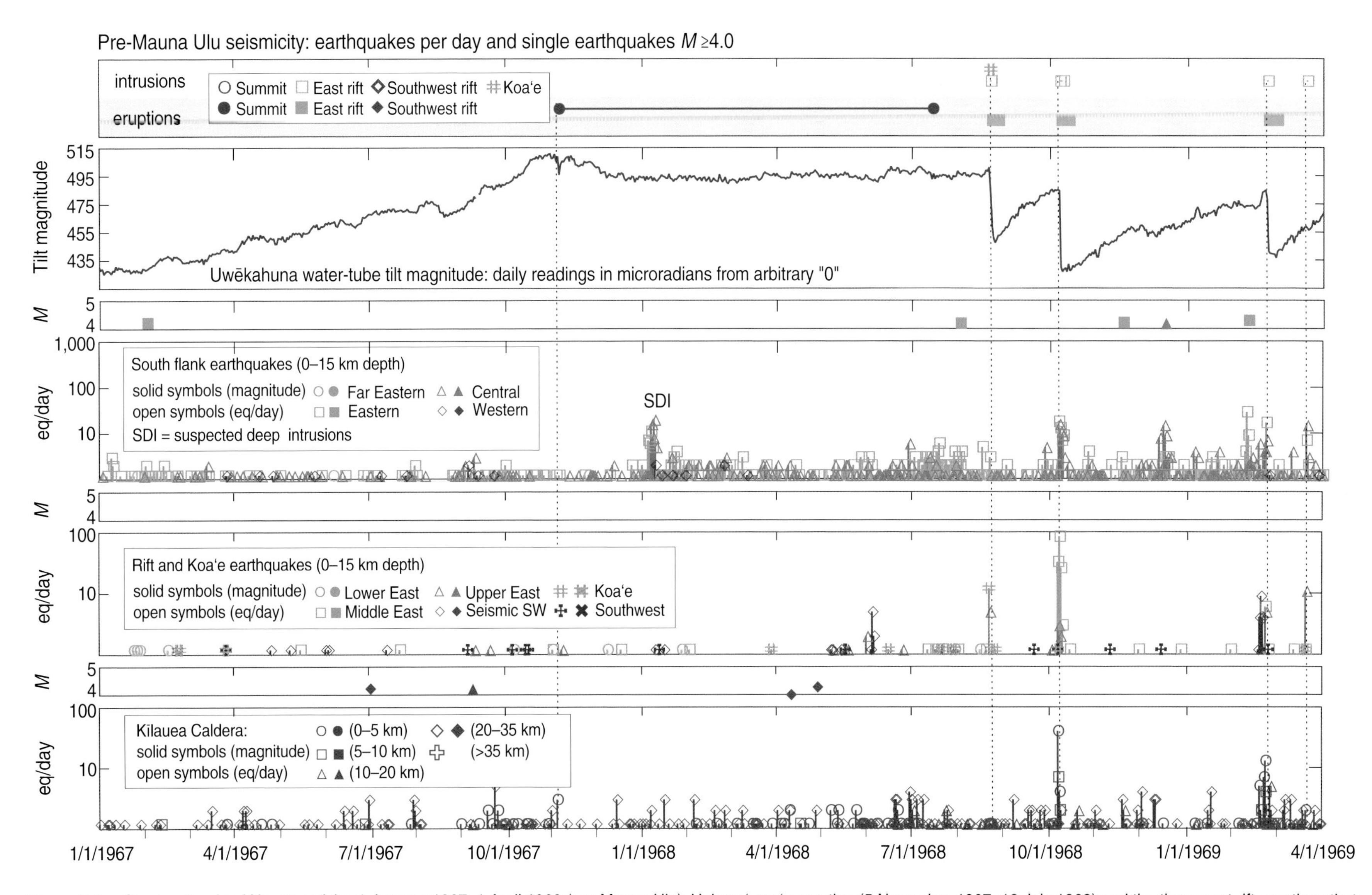

Figure 5.1. Graphs showing Kīlauea activity, 1 January 1967–1 April 1969 (pre-Mauna Ulu): Halema'uma'u eruption (5 November 1967–13 July 1968) and the three east rift eruptions that precede the Mauna Ulu eruption that began on 24 May 1969. Top panel: Times of eruption and traditional intrusion. Second panel from top: Uwēkahuna tilt magnitude related to times of eruption and intrusion emphasized by vertical dotted lines. Tilt magnitudes are given in microradians. Bottom six panels: Seismicity is plotted, from bottom to top, for the magma supply path, rift zones and Koa'e, and south flank. Earthquakes per day (eq/day) and magnitudes (M) greater than or equal to 4.0 are given for each region. Dates on figure in m/d/yyyy format.

increasing seismicity beneath Kīlauea's south flank and the deep magma-supply path (fig. 5.1).

The net volume of summit inflation preceding the 1967–68 eruption can be compared with the volume erupted and with the volume of magma transfer to the rift zone as noneruptive deflations (table 5.2). The preeruption volume of summit inflation estimated from applying a Mogi model to the Fiske-Kinoshita leveling data is ~0.056 km^3. compared with a nearly identical volume of 0.055 km^3 calculated from the daily Uwēkahuna tilt by methods given in appendix A. This represents the volume of magma supplied to the summit reservoir before eruption began. The erupted volume, taken from Kinoshita and others (1969, table 2) after correction for 20 percent vesicles, is 0.078 km^3. Episodes of significant deflation occurring during the 1965–67 cycle, beginning with the deflation associated with the December 1965 eruption/intrusion, represent additional magma transferred to the east rift zone. The initial rapid rate of filling and initial deflation during the 1967 eruption are attributed to lowered pressure on the magmatic system produced by the shift from a capped and throttled magma supply to an open eruption at the surface. We attribute the net deflation during the eruption to additional magma transfer to the rift zone

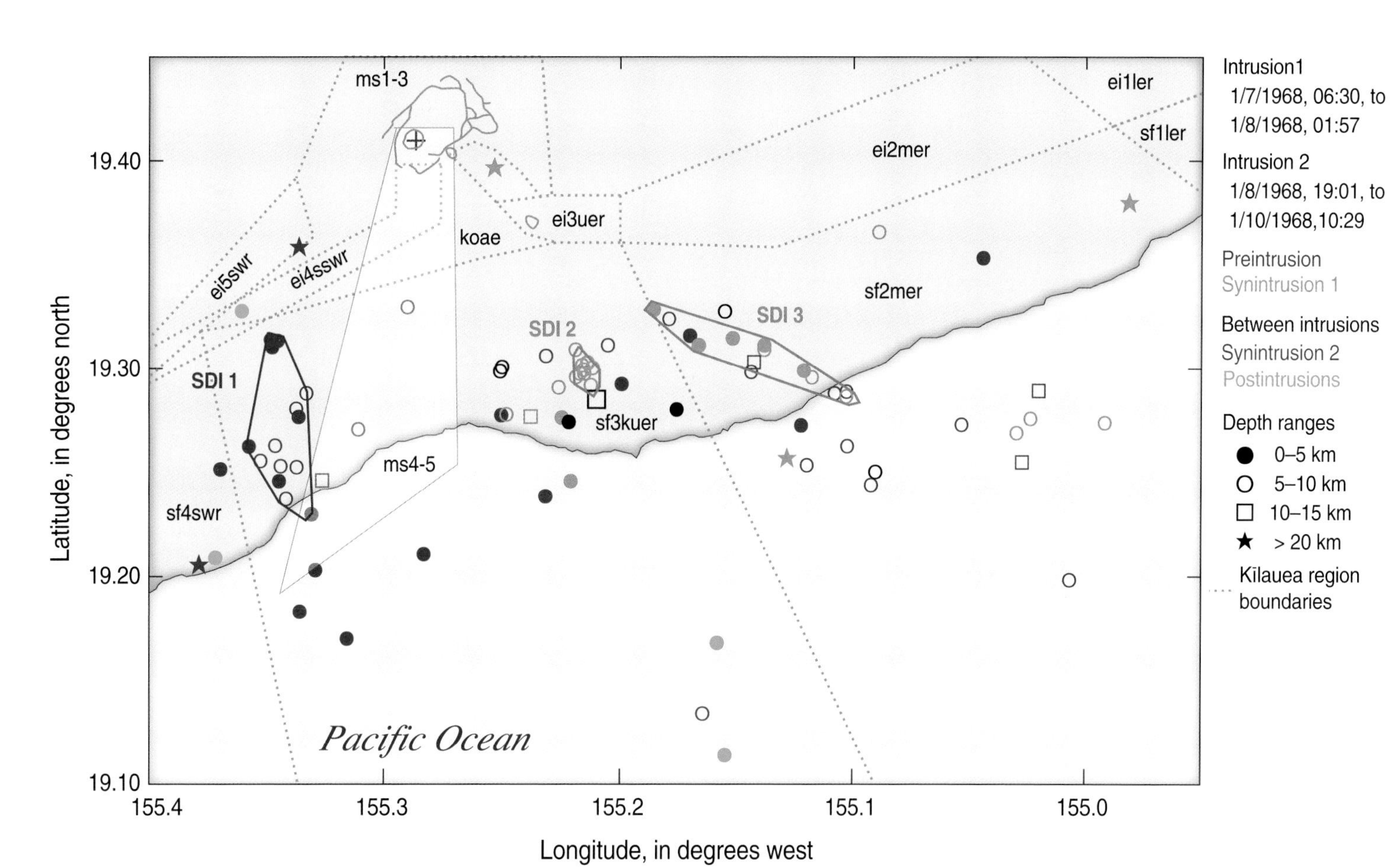

Figure 5.2. Map showing Kīlauea activity, 5–10 January 1968. Earthquake locations are shown for a subset of events covering a period within the 1967–68 Halema'uma'u eruption dominated by suspected deep intrusions. Suspected deep intrusions are distinguished by small swarms of south flank earthquakes oriented at a high angle to the rift zone, in contrast to the south flank response to traditional and inflationary intrusions, in which earthquakes are distributed nearly parallel to the rift zones. Three locations are outlined and labeled SDI 1, SDI 2, and SDI 3. South flank earthquake swarms at these locations appear throughout Kīlauea's history, with earthquakes near SDI 2 being the most common. Dates on figure in m/d/yyyy format.

Table 5.2. Tilt and volume changes associated with 1967–1968 eruption.

Cycle	Event	Date		Δ Time Years	Tilt		Tilt Vol (km^3)[2]	Eruption Volume (km^3)[3]	Magma supply rate msr (km^3/yr)[4]	Comment
		Start	End		Mag[1]	Az				
Preeruption	Net	12/23/1965	11/5/1967	1.8672	123.0	328	0.0555			Total inflation
	Deflation	12/23/1965	12/29/1965		47.4	99	0.0213			Magma transfer to rift zone
	Deflation	10/1/1966	10/7/1966		9.8	140	0.0044			Magma transfer to rift zone
	Deflation	8/9/1967	8/18/1967		8.4	121	0.0038			Magma transfer to rift zone
	Sum Eruption				188.6		0.0850		0.0455	Preeruption magma supply rate
Eruption		11/5/1967	7/15/1968	0.6927				0.0748	0.1083	Filling rate
Syneruption		11/5/1967	7/15/1968					0.0315		Volume added at preeruption msr
								-0.0464		Added volume minus erupted volume
	Deflation	11/5/1967	7/14/1968		4.09	166.57	0.0089			Syneruption deflation
								0.0375		Reservoir gain during eruption
				0.6927				0.1123	0.1208	Syneruption magma supply rate

[1]Tilt magnitude in microradians.

[2]Tilt volume in cubic kilometers = Uwēkahuna tilt magnitude × 0.00045 (see text for explanation).

[3]Equivalent magma volume obtained by multiplying published lava volumes by 0.8 to account for 20 percent vesicles.

[4]Minimum magma supply rate (msr) calculated as described in text.

(table 5.2). The filling rate of 0.109 km^3/yr during the eruption is much greater than the preeruption magma supply rate of 0.046 km^3/yr, and the magma supply rate during the eruption is even higher (0.121 km^3/yr) when the additional transfer of magma to the east rift zone is counted (table 5.2; see also Dvorak and Dzurisin, 1993, discussion on p. 22,263 and following).

The differences in calculated magma supply rate before and after November 1967 may be reflected in the seismicity. The overall rate of deep magma-supply seismicity increased before the eruption from 45 events during all of 1966 to 66 events in the first 10 months of 1967, and it continued to increase to 91 events during the 8-month 1967–68 eruption. South flank seismicity, even excluding south flank events associated with the suspected deep intrusion of January 1968, also increased dramatically during the eruption, 264 events compared to 98 events in the preceding 8-month period. The increase in number of events during this period is much greater than would be expected from improved recording techniques within an expanding seismic network, though the installation of the Develocorder at HVO in March 1967 coincident with the seismicity increase means the numbers quoted above should not be interpreted absolutely.

The 1968–69 eruptions show patterns of seismicity similar to those related to rift eruptions between 1961 and 1965. Ground deformation surveys show that each of these rift eruptions was accompanied by shallow rift intrusion (Jackson and others, 1975, figures 13 and 32; Swanson and others, 1976b, figure 20). The August 1968 eruption was accompanied by intrusion into the Koa‘e Fault Zone (table 5.1). The overall rift seismicity is anomalously low near the eruption site, considering that the east rift segment beneath the eruption site had had no documented intrusions after 1952. We attribute the lack of seismicity to addition of magma to an existing, still molten, dike, or more likely to significant dilation of the rift during the earlier suspected deep intrusion of January 1968 (Wright and Klein, 2008, p. 105). The October 1968 eruption shows a typical pattern of shallow intrusion with strong south flank response (fig. 5.3). A rapid summit inflation occurred just before the February 1969 eruption. The February 1969 eruption/intrusion was accompanied by reduced rift and flank seismicity compared to the October 1968 eruption/intrusion. Another small intrusion occurred on 21 March 1969.

Mauna Ulu Eruption, 1969–1974: Observations

We divide the Mauna Ulu eruption into six parts as follows:

IA: 1 April 1969–31 December 1969 (table 5.3; fig. 5.4). This section discusses the immediate precursors to the beginning of the Mauna Ulu eruption on 24 May 1969 and the 12 episodes of high fountaining ending on 31 December 1969

IB: 1 January 1970–15 June 1971 (table 5.4; fig. 5.5). A period of sustained eruption at two interconnected vents. The primary vent built the Mauna Ulu shield, and a second active vent was beneath the former location of ʻAlae Crater.

Pause: 16 June 1971–3 February 1972 (table 5.5; fig. 5.7). A period of eruption elsewhere on the volcano coincident with temporary cessation of eruption from Mauna Ulu.

II: 4 February 1972–15 June 1974 (table 5.6; fig. 5.9). This period is further divided by the occurrence of a distant earthquake, the *M*6.6 Honumu earthquake off the east coast of Mauna Kea. The period before the earthquake (IIA) was a quiet period of continuous eruption at Mauna Ulu. Following the earthquake (IIB), the eruption was marked by many more intrusions and two eruptions elsewhere on the east rift zone, during which the Mauna Ulu activity temporarily ceased.

Post: 16 June 1974–29 November 1975 (table 5.7; fig. 5.11). This period marks the end of the Mauna Ulu eruption and ends with the *M*7.2 south flank earthquake on 29 November 1975.

Period IA: 24 May 1969–1 January 1970

The month before the Mauna Ulu eruption was marked by heightened seismicity that began at the

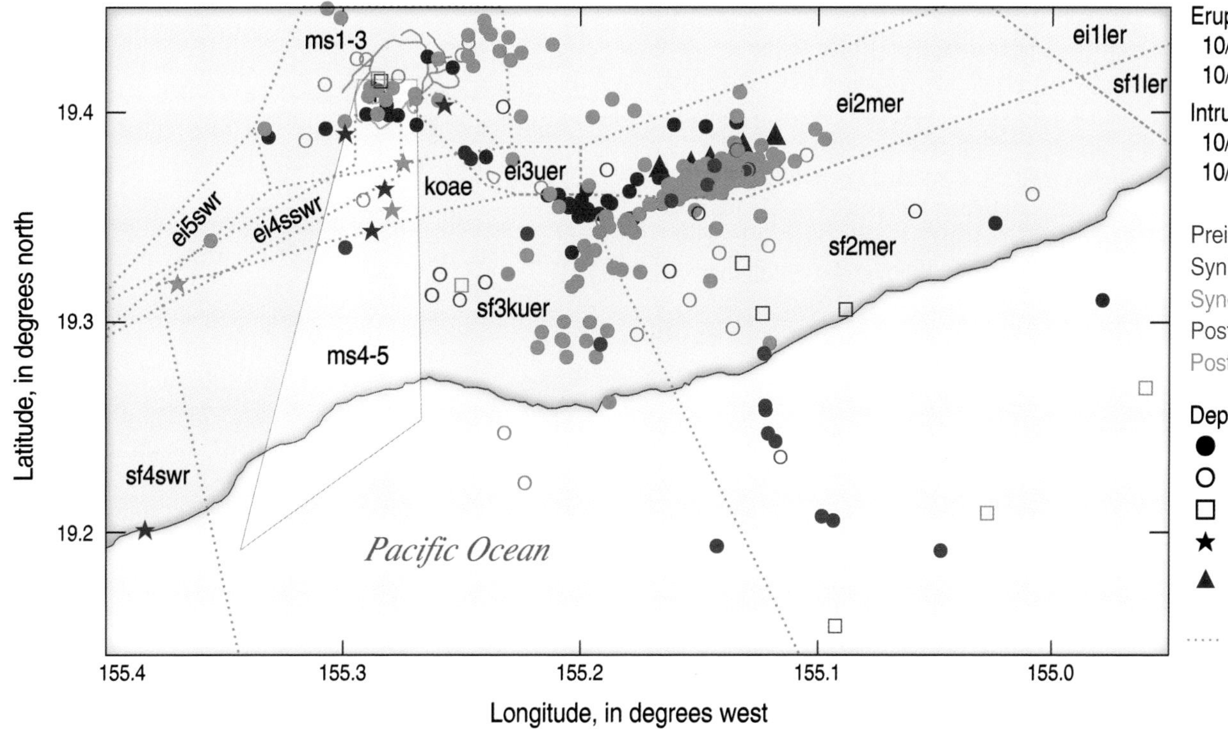

Figure 5.3. Map showing Kilauea activity, 1–19 October 1968. Earthquake locations are shown for a traditional eruption (7–12 October) preceded and accompanied by intrusion (7–10 October). Data are shown for periods before, during, and after the events, and eruptive vents are plotted as red triangles. Precursory shallow rift seismicity is accompanied by a south flank response, both located west (uprift) of syneruption seismicity and near the eruptive vents. The south flank response following the end of intrusion, but still within the period of eruption, includes a possible suspected deep intrusion extending offshore beneath the eastern south flank (region sf2mer). Eruption is preceded and followed by deep magma-supply earthquakes shown by stars. Dates on figure in mm/dd/yyyy m/d/yyyy format.

end of April and culminated with the beginning of the eruption on 24 May 1969. Intrusions beneath Kīlauea's summit occurred on 30 April and in early May. On 21 May intense seismicity beneath Kīlauea's south flank began 3 days before eruption at the surface on 24 May.

Mauna Ulu IA was marked by 12 different high-fountaining episodes over 8 months. All fountaining episodes were accompanied by small summit deflations and, with the exception of episodes 1, 2, 9, and 12, these eruptive episodes were not accompanied by shallow seismic activity beneath Kīlauea's summit or the rift zone adjacent to the eruption site (fig. 5.4). Two periods of intense south flank earthquake swarms (suspected deep intrusions) occurred during this period. Seismicity marking the suspected deep intrusion of 3–9 June occurred beneath the area of figure 5.2 designated as SDI 3. The swarm of 29 September–4 October occurred beneath SDI 1, and the swarm of 4–16 October occurred beneath both sites. The last high-fountaining episode was preceded by both enhanced south flank seismicity characteristic of a suspected deep intrusion and also by an intrusion into the northernmost seismic southwest and east rift zones that anticipated the more intense inflationary intrusions of Mauna Ulu IB (see below).

Period IB: 1 January 1970–14 June 1971

A high level of seismicity continued following the end of episodic high fountaining during period IA. Although lava was continuously visible in Mauna Ulu and at the nearby former location of 'Alae Crater, and overflows fed lava flows that traveled downslope from both locations, there were also several intrusive episodes throughout this period. A possible pair of suspected deep intrusions on 8–14 January was followed by a summit intrusion on 22 January. The first of many inflationary intrusions took place during 4–9 February (fig. 5.6), during which time earthquakes clearly migrated downrift (Klein and others, 1987, figure 43.40). Similar intrusions occurred later in February and in March and April.

A large intrusion beneath Kīlauea's summit, the upper east rift zone, and adjacent Koa'e Fault Zone occurred from 15 to 18 May 1970, with a strong south flank response before, during, and after the intrusion (table 5.4; fig. 5.5). According to Swanson and others (1979, p. 39), this event marked the reopening of the connection between the summit and the east rift zone. Deep seismic activity occurred before the intrusion southwest of Kīlauea's summit and after the intrusion in the mantle beneath the eastern south flank[21].

A large inflationary intrusion occurred between 25 December 1970 and 5 January 1971, with activity concentrated beneath the two main segments of the southwest rift zone. South flank activity, however, was located beneath the central and eastern south flank, focused on the middle segment (SDI 2) associated with the suspected deep intrusions. The end of continuous eruptive activity at Mauna Ulu is arbitrarily placed at 15 June 1971, coincident with the last overflow, although lava remained visible deep in Mauna Ulu Crater until October 1971 (Swanson and others, 1979). As in Mauna Ulu IA, the end of period IB is marked by heightened seismicity, including swarms of deep magma-supply earthquakes between 29 April and 9 May (fig. 5.5). Uwēkahuna tilt shows net inflation up to the end of November 1970, then deflation to February 1971, and renewed inflation up to the end of period IB in June 1971 (fig. 5.5).

[21] Deep earthquakes beneath the eastern south flank, and several kilometers east of the magma supply path previously outlined (Wright and Klein, 2006) are considered to be stress release in the mantle associated with flexure of the Pacific Plate.

Mauna Ulu Pause: 15 June 1971–4 February 1972

During the Mauna Ulu pause two eruptions occurred at and near Kīlauea's summit in August and September 1971 (Duffield and others, 1982), accompanied by strong tilt changes and heightened seismic activity across all sectors of the volcano (table 5.5; fig. 5.7). A noteworthy suspected deep intrusion occurred on 4–5 July beneath area SDI 2 (fig. 5.8). Inflationary intrusions continued—three in July 1971, one preceding the eruption of 14 August, and three preceding the eruption of 24 September. Inflation of more than 100 µrad associated with the September eruption (table 5.5) brought the Uwēkahuna tilt site to its highest level of inflation since 1956. The 24 September tilt azimuth was consistent with intrusion beneath the uppermost seismic southwest rift zone. The earthquake sequence within the south flank resembles the pattern seen during suspected deep intrusions beneath area SDI 2.

Seismic activity beneath Kīlauea's southwest south flank increased relative to preceding periods and a small suspected deep intrusion in November was followed by a major suspected deep intrusion beneath the central and western south flank at the end of December 1971, unaccompanied by eruption or deflationary tilt (fig. 5.7). This event was accompanied by minor activity on the upper section of the east rift zone, the middle section of the seismic southwest rift zone, and an additional concentration of south flank earthquakes south of the uppermost east rift zone located similarly and slightly to the east of the earthquakes associated with the suspected deep intrusion of June 1969.

Return of activity to Mauna Ulu on 4 February 1972 was heralded by a series of east rift inflationary

Table 5.3. Kīlauea eruptions, intrusions, and earthquakes, Mauna Ulu period IA (see figures 5.4, E9).

[In rows with multiple entries text applies down to the next entry; dates in m/d/yyyy format; do, ditto (same as above); data for eruptions and traditional intrusions are emphasized by grey shading]

Time and Date		Region[1]	Event Type[2]	No.[3]	Tilt[4]		Lag[5]	Comment[6]	Fig.[7]	Ref.[8]
Start	End				Mag	Az				
12:35 4/30/1969	13:41 4/30/1969	sf2mer	II	11	5.8	325		South flank anticipation	E10	
18:56 4/30/1969	22:07 4/30/1969	ms1	I	11			+45h 4m	Tilt 4/29-5/4; I–A: inflation 4/29-5/2		
09:16 5/4/1969	00:34 5/6/1969	ms1	II	22	4.4	130	deflation	Tilt 5/4-9; I–A: 5/7-9; 5/4-6/1969	E11	
22:18 5/5/1969	23:57 5/5/1969	ms2	II	10						
15:33 5/9/1969		sf2mer	EQ					*M*4.3; productive aftershock sequence		12
08:40 5/21/1969	12:20 5/22/1969	sf3kuer	EQS	14				South flank anticipation	43.33	8
17:32 5/21/1969	03:49 5/22/1969	ms1	I	8	5.3	311		Tilt: 5/19-22; earthquakes move south 5/21-22	E12	
19:03 5/21/1969	02:37 5/22/1969	ei3uer	I?	3						
04:31 5/22/1969	23:19 5/22/1969	ms2	I?	5						
05:22 5/22/1969	23:36 5/22/1969	ei4sswr	I?	9			-16h 38m			
07:31 5/21/1969	00:24 5/25/1969	sf2mer	EQS	119				South flank anticipation		
14:25 5/23/1969	19:21 5/24/1969	sf3kuer	EQS	52				do		
03:42 5/24/1969	12:36 5/24/1969	ei3uer	I	9			-2h 18m	Precursory intrusion		
04:16 5/24/1969	23:29 5/24/1969	koae	EQS	5				No swarm	43.34	8
04:45 5/24/1969	15:00 5/25/1969	MERZ	E/I		26.0	110		Episode 1; Tilt 5/22-27	E13	6
07:35 5/25/1969	06:00 5/26/1969	sf2mer	EQS EQS	9				South flank accompaniment/response		
20:38 5/25/1969	05:20 5/26/1969	sf3kuer		5				South flank response		
14:27 5/26/1969	19:47 5/26/1969	sf3kuer	EQS	5				South flank anticipation; also 5/25-26 (8)		
16:40 5/26/1969	17:22 5/26/1969	ei3uer	EQS	9				Precursory seismicity		
19:00 5/27/1969	09:00 5/29/1969	MERZ	E/I		6.6	114	+0h 40m	Episode 2; Tilt 5/28-30	E13	6
10:28 6/3/1969	12:09 6/9/1969	sf3kuer	SDI	130					E14	
15:47 6/3/1969	10:48 6/4/1969	sf2mer	SDI	7						
13:30 6/12/1969	11:00 6/13/1969	MERZ	E		15.2	112	-0h 30m	Episode 3; tilt 6/11-13; no rift earthquakes		6
21:45 6/25/1969	07:00 6/26/1969	MERZ	E		20.0	115	-2h 15m	Episode 4; tilt 7/25-26; no rift earthquakes		6

Table 5.3. Kīlauea eruptions, intrusions, and earthquakes, Mauna Ulu period IA (see figures 5.4, E9).—Continued

[In rows with multiple entries text applies down to the next entry; dates in m/d/yyyy format; do, ditto (same as above); data for eruptions and traditional intrusions are emphasized by grey shading]

Time and Date		Region[1]	Event Type[2]	No.[3]	Tilt[4]		Lag[5]	Comment[6]	Fig.[7]	Ref.[8]
Start	End				Mag	Az				
22:31 7/2/1969	09:07 7/4/1969	sf3kuer	EQS	26				South flank anticipation	E15	
06:20 7/3/1969	13:01 7/3/1969	sf2mer	EQS	8				South flank anticipation/accompaniment		
06:36 7/3/1969	22:15 7/3/1969	ei3uer	I	60	2.7	162	inflation	Tilt 7/2-4; I–A tilt 6/26 -7/6; no deformation indicates redistribution of intruded magma	E15; *43.35*	8
06:59 7/3/1969	18:04 7/3/11969	koae	I	43						
10:39 7/3/1969	13:41 7/3/1969	ei2mer	I	5						
03:15 7/15/1969	05:56 7/15/1969	sf3kuer	EQS	3				South flank anticipation		
03:45 7/15/1969	12:20 7/15/1969	MERZ	E		13.0	111	+1h 45m	Episode 5; tilt 7/14-15; no rift earthquakes	**5.4**	6
16:30 7/15/1969	09:48 7/16/1969	sf3kuer	EQS	7				South flank response		
17:15 8/3/1969	00:10 8/4/1969	MERZ	E		12.4	108	-2h 45m	Episode 6; Tilt 8/2-4; few rift earthquakes	E16	6
21:00 8/5/1969	05:45 8/6/1969	MERZ	E		12.7	120	-1h 0m	Episode 7; Tilt 8/5-6; few rift earthquakes	E16	6
00:15 8/22/1969	04:40 8/22/1969	MERZ	E		10.7	109	+0h 15m	Episode 8; Tilt 8/21-23; no rift earthquakes		6
19:30 9/06/1969	4:30 9/07/1969	MERZ	E		30.0	114.4	+0h 30m	Episode 9: Tilt 9/6–7; no rift eqrthquakes		6
23:51 9/29/1969	18:37 10/4/1969	sf3kuer	SDI	61			inflation	Intrusion from 5-10 km deep source	E17	
10:26 10/7/1969	12:52 10/9/1969	sf3kuer	SDI	154			inflation	South flank anticipation	*43.36*	8
15:41 10/9/1969	09:03 10/10/1969	sf2mer	EQS	8				South flank anticipation; SDI continuation?		
18:55 10/9/1969	12:44 10/11/1969	sf3kuer	EQS	34				South flank anticipation/accompaniment		
09:00 10/10/1969	11:00 10/13/1969	MERZ	E		11.1	121	+7h 0m	Episode 10; Tilt 10/8-13; no rift earthquake		6
12:55 10/14/1969	16:57 10/14/1969	sf2mer	EQS	8				South flank response		
15:41 10/14/1969	19:57 10/14/1969	sf3kuer	EQS	5				do		
15:30 10/19/1969	00:24 10/20/1969	sf3kuer	EQS	7				South flank anticipation		
01:00 10/20/1969	08:20 10/20/1969	MERZ	E		21.5	108.0	+1h 0m	Episode 11; tilt 10/19-20; no rift earthquake		6

Table 5.3. Kilauea eruptions, intrusions, and earthquakes, Mauna Ulu period IA (see figures 5.4, E9).—Continued

[In rows with multiple entries text applies down to the next entry; dates in m/d/yyyy format; do, ditto (same as above); data for eruptions and traditional intrusions are emphasized by grey shading]

Time and Date		Region[1]	Event Type[2]	No.[3]	Tilt[4]		Lag[5]	Comment[6]	Fig.[7]	Ref.[8]
Start	End				Mag	Az				
10:22 11/3/1969	11:22 11/3/1969	sf2mer	EQS	3				South flank anticipation	E18; *43.37*	8
11:11 11/3/1969	11:44 11/3/1969	ei3uer	I	16	5.8	117	+3h 11m	Tilt 11/1-4		
11:14 11/3/1969	13:13 11/3/1969	ei2mer		4				No swarm		
11:26 11/3/1969	12:54 11/3/1969	sf3kuer	EQS	4				South flank response		
01:58 12/11/1969	13:28 12/12/1969	ei4sswr	II	14	1.7	121	-32h	Tilt 12/12-13	E19	
01:04 12/23/1969	11:51 12/23/1969	ei4sswr	II	15	7.9	303	inflation	Tilt 12/20-27; I–A tilt 12/18-12/23	E20; *43.38*	8
12:05 12/26/1969	00:03 12/29/1969	ei4sswr	II	11				Broken swarm		
16:59 12/27/1969	08:21 12/29/1969	sf3kuer	SDI	41				South flank accompaniment as SDI	E21	
05:00 12/30/1969	08:25 12/30/1969	MERZ	E		4.7	113		Episode 12a; Tilt 12/30-31; no rift eq		6
10:00 12/30/1969	18:30 12/30/1969	MERZ	E					Episode 12b; I–A: deflation		6
10:41 12/30/1969	04:05 12/31/1969	sf3kuer	EQS	8[9]				South flank accompaniment/response		

[1]Earthquake classification abbreviations are given according to the classification in appendix table A3, and locations are shown on appendix figure A4. Eruption locations are designated in bold type as follows: KC, Kīlauea Caldera; LERZ, lower east rift zone; MERZ, middle east rift zone; UERZ, upper east rift zone; SWR, southwest rift zone.

[2] **E**, Eruption; intrusion ("traditional" **I;** "inflationary" II**;** "suspected deep" SDI (see Wright and Klein, 2008, table 2); south flank earthquake swarm EQS.

[3]Minimum number of events defining a swarm: 20 for south flank; 10 for all other regions.

[4]Magnitude in microradians and azimuth of daily tilt measurements from the water-tube tiltmeter in Uwēkahuna Vault.

[5]Lag times separating the onset of the earliest earthquake swarm (excluding south flank) for a given event and the beginning of deflation or inflation measured by the continuously recording Ideal-Arrowsmith tiltmeter in Uwēkahuna Vault. (+) tilt leads, (-) tilt lags.

[6]Abbreviations as follows: ftn, fountaining; eq, earthquake; eqs, earthquake swarm; fs, foreshock; as, aftershock; ms, mainshock; sf, south flank; inf, inflation; def, deflation; ant, anticipation (preceding event); acc, accompaniment (during event); resp, response (following event); I–A, Ideal-Arrowsmith continuously recording tiltmeter in Uwēkahuna Vault.

[7]Text figures **bold text**; appendix figures plain text; 43.xx = figures in Klein and others, 1987.

[8]References coded as follows: 1. Fiske and Kinoshita, 1969, 2. Kinoshita and others, 1969, 3. Wright and Klein, 2008, 4. Jackson and others, 1975, 5. Swanson and others, 1976b, 6. Swanson and others, 1979, 7. Duffield and others, 1974, 8. Klein and others, 1987, 9. Tilling and others, 1987, 10. Nielsen and others, 1977, 11. Lockwood and others, 1999; 12. Klein and others, 2006. Note: This is a master list for all tables in chapter 5. Only some references will be cited in this table.

[9]Productive aftershock sequences defined in reference (12) above.

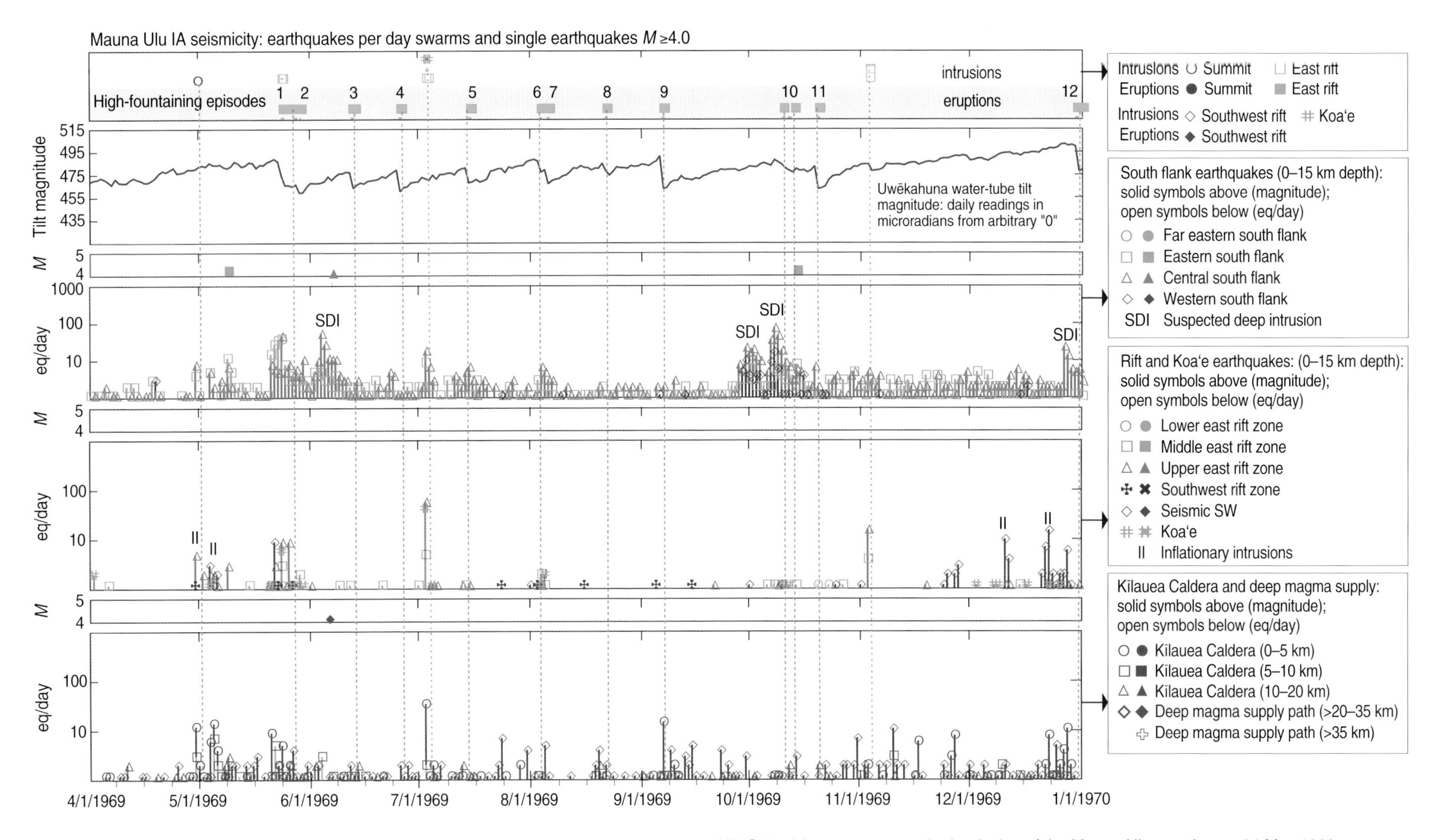

Figure 5.4. Graphs showing Kīlauea activity, 1 April 1969–1 January 1970 (Mauna Ulu period IA): Seismicity precursory to the beginning of the Mauna Ulu eruption on 24 May 1969, continuing through the end of episodic high fountaining on 31 December 1969. High-fountaining episodes 1–12 are labeled on the top panel. Dates on figure in m/d/yyyy format. See also caption for figure 5.1.

Table 5.4. Kīlauea eruptions, intrusions, and earthquakes, Mauna Ulu IB (see figures 5.5, E22).

[In rows with multiple entries text applies down to the next entry; dates in m/d/yyyy format; do = ditto (same as above); data for eruptions and traditional intrusions are emphasized by grey shading]

Time and Date		Region[1]	Event Type[2]	No.[3]	Tilt[4]		Lag[5]	Comment[6]	Fig.[7]	Ref.[8]
Start	End				Mag	Az				
17:38 1/8/1970	05:32 1/15/1970	sf3kuer	SDI	40				Broken earthquake swarm; continuation of 12/1969 SDI	E23	
00:29 1/22/1970	03:41 1/22/1970	ms1	I	61	7.0	308		Tilt 1/17-24; I–A tilt 1/18 1/23	**5.6**; E24; *43.39*	8
00:29 2/4/1970	14:57 2/8/1970	ei4sswr	II	30	3.9	306		Tilt 2/4-9; broken earthquake swarm; paired rift intrusion	E25	
16:02 2/8/1970	03:04 2/9/1970	ei3uer	II	8			-52h 34m	I–A tilt 1/30-2/8 inflation; paired rift intrusion		
01:26 2/9/1970	01:55 2/11/1970	ei4swr	II	18	3.7	113		Tilt 2/9-11	*43.40*	8
09:12 2/9/1970	05:14 2/11/1970	ei3uer	II	20				Broken earthquake swarm	do	
15:45 3/17/1970	05:20 3/23/1970	ei4sswr	II	64				Tilt 3/19-22; broken earthquake swarm	E26; *43.41*	3, 8
22:38 3/20/1970	20:24 3/22/1970	ei3uer	II	13	7.3	296		Broken earthquake swarm; I–A tilt 3/9 -3/26: inflation		
21:29 4/4/1970	18:50 4/8/1970	ei4sswr	II	100	6.4	291		Tilt 4/2-8; I–A tilt 4/2 -4/10	E27; *43.42*	
02:21 4/7/1970	18:45 4/8/1970	ei3uer	I	44	10.0	102	-42h 00m	Tilt 4/8-10	E27	7, 8
02:53 5/15/1970	08:10 5/15/1970	sf2mer	EQS	10				South flank anticipation/accompaniment	E28	
03:38 5/15/1970	01:33 5/16/1970	sf3kuer	EQS	34				South flank accompaniment		
11:04 5/15/1970		sf3kuer	EQ					*M*3.8 with aftershocks embedded in earthquake swarm		
04:34 5/15/1970	05:59 5/16/1970	ms1	EQS	44						8
04:40 5/15/1970	13:25 5/18/1970	ei3uer	I	256	9.1	125	-0h 47m	Tilt 5/14-16; I–A tilt: inflation and deflation from 5/15 8:00-5/15 22:00; downrift/uprift migration	*43.43*	
04:43 5/15/1970	02:09 5/16/1970	koae	I	9						
04:48 5/15/1970	06:36 5/15/1970	ei2mer	I	6				Continuing south flank response		
06:08 5/19/1970	17:52 5/19/1970	sf2mer	SDI?	5				Broken earthquake swarm		
10:50 5/19/1970	10:56 5/21/1970	sf3kuer	SDI?	23				Heightened background seismicity		
11:05 7/15/1970	02:02 7/16/1970	sf2mer	EQS	10					E29	
01:27 9/21/1970		sf3kuer	EQ					*M*4.46; as pattern resembles SDI	E30	
13:30 10/26/1970	14:57 10/27/1970	ms1	I	19	5.1	297		Tilt 10/22-26; I–A tilt 10/22-10/27	E31	

Table 5.4. Kīlauea eruptions, intrusions, and earthquakes, Mauna Ulu IB (see figures 5.5, E22).—Continued

[In rows with multiple entries text applies down to the next entry; dates in m/d/yyyy format; do, ditto (same as above); data for eruptions and traditional intrusions are emphasized by grey shading]

Time and Date		Region[1]	Event Type[2]	No.[3]	Tilt[4]		Lag[5]	Comment[6]	Fig.[7]	Ref.[8]
Start	End				Mag	Az				
10:55 12/12/1970	13:19 12/14/1970	ms1	II	37	4.6	288		Tilt 12/10-13; I–A tilt 12/11-12/14		
21:19 12/12/1970	05:26 12/14/1970	er3	II	8				Broken earthquake swarm		8
23:03 12/12/1970	11:26 12/13/1970	sf3kuer	SDI?	6				SDI triggered by preceding II?	E32; *43.44*	8
16:0612/22/1970	08:23 12/23/1970	ms1	II	22	flat tilt			IA tilt 12/20 -12/31: inflation	E33*A*; *43.45*	
00:35 12/25/1970	21:33 12/26/1970	ms1	II	24	do			Summit/seismic southwest rift intrusion	E33*B*	
08:18 12/28/1970	06:57 12/29/1970	ei4sswr	II	22	do			do	E33*C*	
15:20 12/29/1970	01:55 1/01/1971	do	II	22	do			do	E33*D*	
21:06 1/03/1971	11:45 1/05/1971	do	II	24	do			do	E33*E*	
00:29 4/25/1971	04:59 4/27/1971	koae3	EQS	6				No swarm		
08:03 4/25/1971	00:55 4/30/1971	ms3	EQS	10				Scattered; includes *M*4.5 24 km 4/25/71	E34	
20:09 4/25/1971	04:40 4/26/1971	ms4/5	EQS	14					E35; *43.48*	8
17:06 4/26/1971	05:42 4/27/1971	ms2	EQS	5	5.0	121		Tilt 4/25-28; IA tilt flat ± 1μrad	E36; *43.49*	8
08:52 6/1/1971	11:40 6/2/1971	ms1	EQS	16				9 additional events to 06:39 6/1/71		
08:01 6/8/1971	14:54 6/10/1971	ms1	II	36				IA tilt 6/2 13:0-6/10 20:30: inflation		
07:01 6/11/1971	08:02 6/14/1971	ei4sswr	II	177				IA tilt 6/10 20:30-6/16 06:00: deflation		

[1]Earthquake classification abbreviations are given according to the classification in appendix A, table A3, and locations of regions are shown in appendix A, figure A4.

[2]E, Eruption; intrusion ("traditional" I; "inflationary" II; "suspected deep" SDI [see Wright and Klein, 2008, table 2]); south flank earthquake swarm EQS.

[3]Minimum number of events defining a swarm: 20 for south flank; 10 for all other regions.

[4]Magnitude in microradians and azimuth of daily tilt measurements from the water-tube tiltmeter in Uwēkahuna Vault.

[5]Lag times separating the onset of the earliest earthquake swarm (excluding south flank) for a given event and the beginning of deflation or inflation measured by the continuously recording Ideal-Arrowsmith tiltmeter in Uwēkahuna Vault. (+) tilt leads, (-) tilt lags.

[6]"I–A" in comment column refers to readings from the Ideal-Arrowsmith tiltmeter in Uwēkahuna Vault.

[7]Text figures **bold text**; appendix figures plain text; 43.xx = figures in Klein and others, 1987.

[8]References coded as follows: 1. Fiske and Kinoshita, 1969, 2. Kinoshita and others, 1969, 3. Wright and Klein, 2008, 4. Jackson and others, 1975, 5. Swanson and others, 1976b, 6. Swanson and others, 1979, 7. Duffield and others, 1974, 8. Klein and others, 1987, 9. Tilling and others, 1987, 10. Nielsen and others, 1977, 11. Lockwood and others, 1999; 12. Klein and others, 2006. Note: This is a master list for all tables in chapter 5. Only some references will be cited in this table.

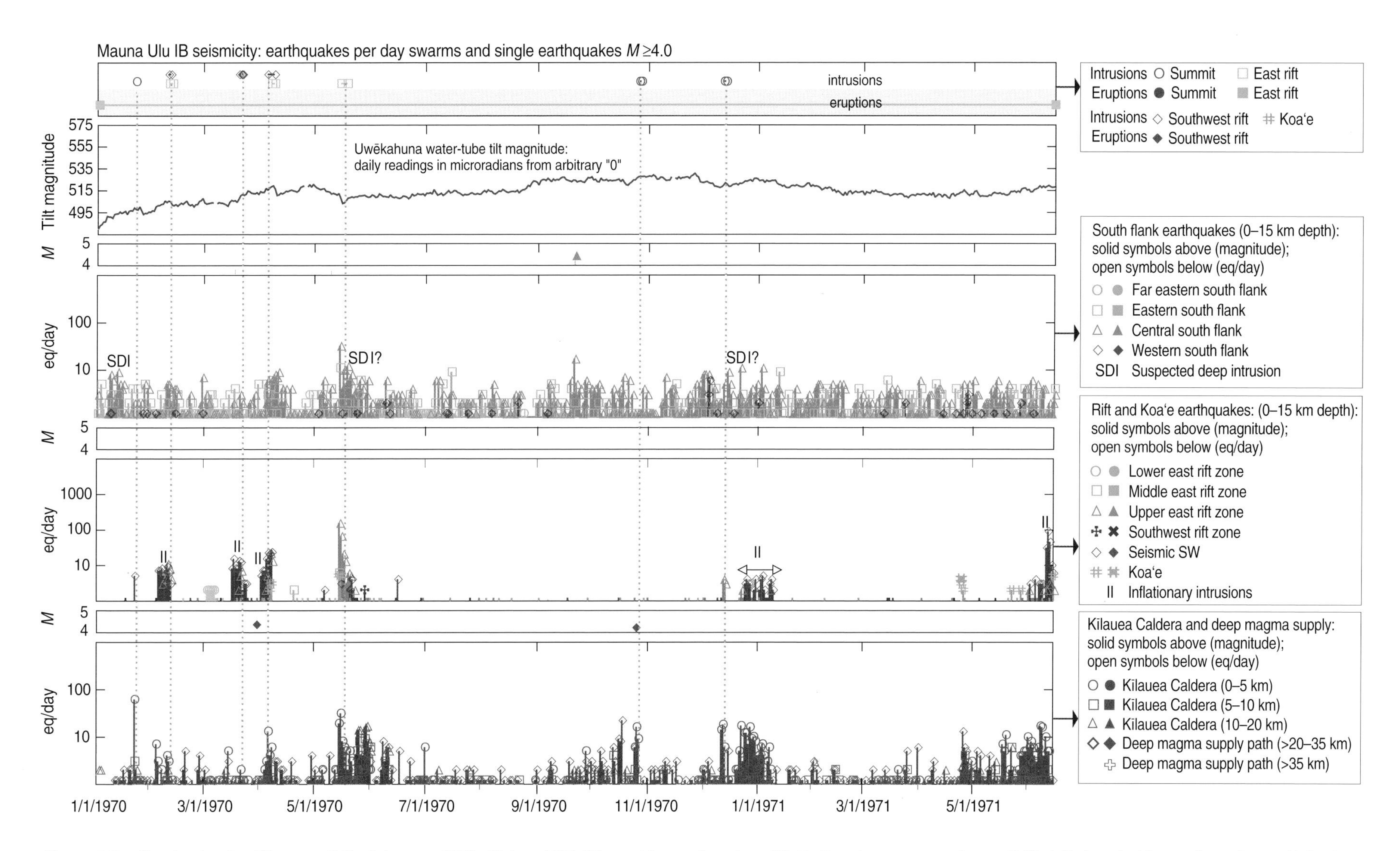

Figure 5.5. Graphs showing Kīlauea activity, 1 January 1970–15 June 1971 (Mauna Ulu eruption stage IB): Earthquake swarms and summit tilt plotted against times of eruption and intrusion. Dates on figure in m/d/yyyy format. See also caption for figure 5.1.

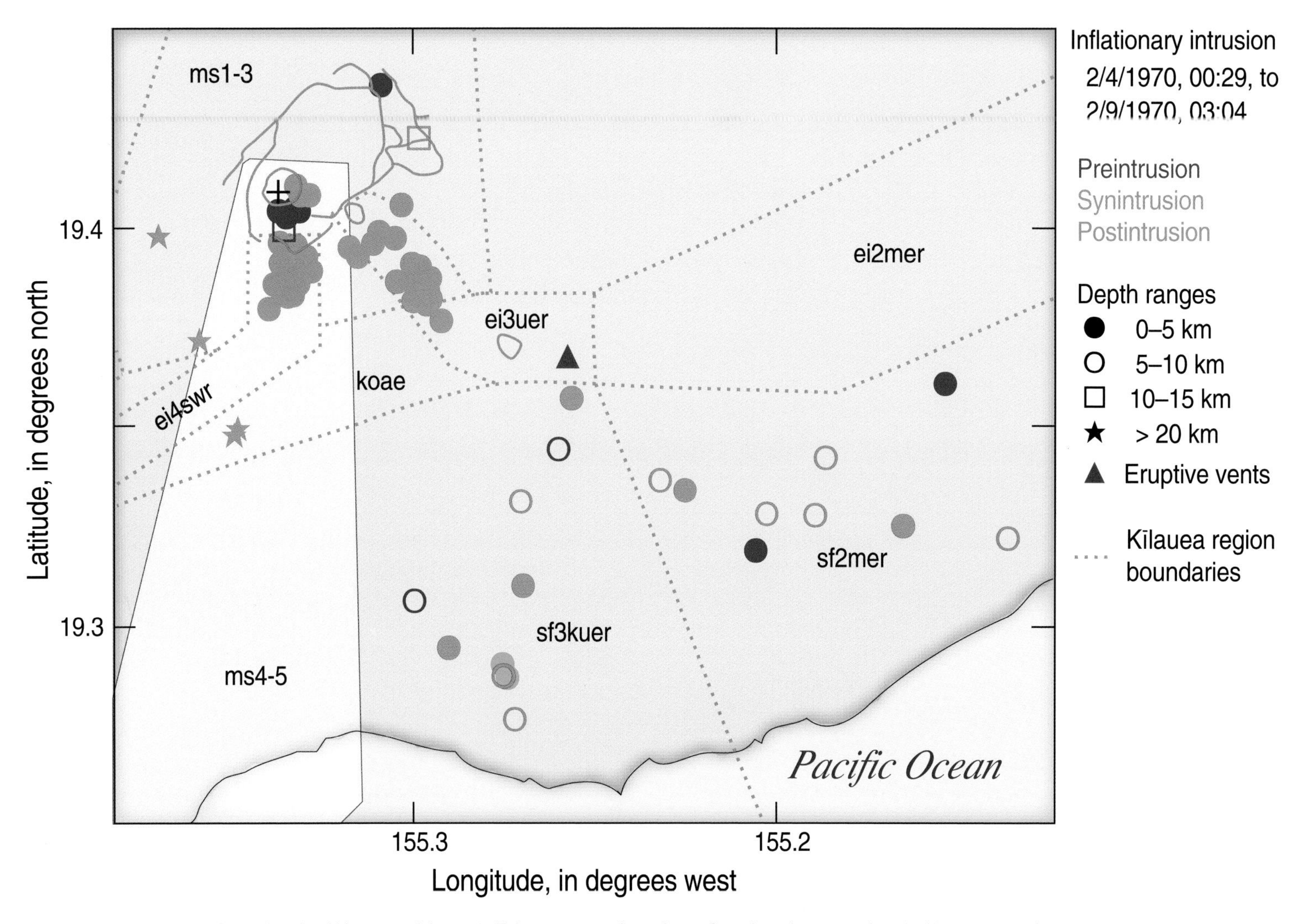

Figure 5.6. Map showing Kīlauea activity, 2–9 February 1970. Locations of earthquakes associated with a paired inflationary intrusion (4–9 February), with activity on the uppermost parts of the east rift zone (ei3uer) and seismic southwest rift zone (ei4sswr). Data are shown for periods before (blue), during (orange), and after (green) the event, and the Mauna Ulu vent is plotted as a red triangle. A small inflationary tilt vector of 3.9 microradians (table 5.3) and the shallow earthquake locations suggest that a small summit intrusion accompanied the inflationary intrusion. Dates on figure in m/d/yyyy format.

intrusions beneath the upper east rift zone and a south flank response that included continuation of the December 1971 sequence on the southwest side. The last of these on 26–28 January, one week before eruption resumed, was beneath the east rift uprift of the Mauna Ulu vent.

Mauna Ulu Period IIA: 4 February 1972–26 April 1973

Continuous eruption was renewed at Mauna Ulu on approximately 4 February 1972 (table 5.6) and continued until the summer of 1974 (Tilling and others, 1987). The immediate precursors included steady seismicity beneath both rift zones and a swarm of deep magma-supply earthquakes that occurred less than a day before eruption resumed (fig. 5.9). The two periods of continuous eruption are divided at the time of a deep *M*6.6 earthquake beneath the offshore east flank of Mauna Kea (table 5.6; fig. 5.9). That distant earthquake caused damage at Kīlauea's summit, and the character of the eruption changed significantly following the earthquake, as documented by Tilling and others (1987, p. 419–421).

The period from February 1972 to 26 April 1973 was the most stable period of the eruption. Summit tilt showed very slight deflation, and shallow seismicity beneath the summit and rift zones was low (fig. 5.9). Long-period (l-p) earthquakes were first catalogued by HVO seismic analysts in May 1972. Subsequently, l-p earthquakes become an important part of Kīlauea's seismic record. Magma-supply activity below 10-km depth and south flank seismicity remained elevated. An intrusion on 18 March 1972 was associated with the opening of a new fissure on the southwest flank of Mauna Ulu (Tilling and others, 1987, p. 413).

Mauna Ulu Period IIB: 26 April 1973–15 June 1974

The Mauna Kea (Honomū) earthquake of 26 April 1973 at 10:26 a.m. was followed immediately by an abrupt inflation of nearly 25 µrad measured at Uwēkahuna and, 7 hours later, by a small flurry of deep (20–35 km) magma-supply earthquakes that overlapped with increased seismicity beneath Kīlauea's east rift zone (table 5.6; fig. 5.9). The Uwēkahuna vector indicates possible inflation of the uppermost east rift zone, consistent with the increased seismicity.

On 5 May 5 1973 the first of two eruptions occurred in Pauahi Crater, coincident with the temporary cessation of activity at Mauna Ulu (fig.5.10). Following this eruption earthquakes extended into the Koa'e Fault Zone. South flank seismicity preceded this sequence, but was apparently diminished[22] during the eruption and the Koa'e intrusion. Mauna Ulu returned to active eruption on 8 May (table 5.6). Two intrusions followed. The first, on 8–10 June 1973, mimicked the rift/Koa'e seismicity associated with the May eruption/intrusion. The second, on 25 July, occurred downrift from Mauna Ulu, with all seismicity displaced to the east.

A second Pauahi eruption occurred on 10 November 1973, and this eruption continued at a low level through 9 December 1973. Mauna Ulu was quiet during this interval but returned to activity on 10 December. During the November Pauahi eruption, south flank seismic activity occurred widely distributed in time along a broad east-west swath and also within the SDI 2 region, suggesting that a suspected deep intrusion may have occurred during the continuing low-level activity at Pauahi. Another suspected deep intrusion occurred in January 1974. From late January to late March 1974 several low fountaining episodes occurred at Mauna Ulu (Tilling and others, 1987, table 16.3, p. 437–439), associated with a few microradians of summit deflation. During this time intrusions alternated with times of greater surface activity at Mauna Ulu.

The end of Mauna Ulu stage II was presaged by heightened seismic activity in several sectors of the volcano within a few months of the final decline of the Mauna Ulu lava lake. A summit intrusion occurred on 22 May 1974, followed on 24–30 May by increased seismicity beneath both rift zones and a suspected deep intrusion beneath site SDI 1 (fig. 5.9). Using seismicity as our basis, we have placed the end of the Mauna Ulu eruption on 15 June, after the last overflow on 2 June, but before lava disappeared from the Mauna Ulu vent on 22 July (Tilling and others, 1987, p. 442).

Post-Mauna Ulu to 1975 Earthquake: 18 June 1974–28 November 1975

The aftermath of the Mauna Ulu eruption (table 5.7; fig. 5.11) was remarkably similar to the period following the end of the Mauna Ulu stage I in the summer of 1971. Two summit eruptions occurred in July (fig. 5.12) and September, the vents extending toward the east and southwest rift zones, respectively (Lockwood and others, 1999). There were almost no located earthquakes during the two eruptions, possibly because of the high level of tremor. A series of inflationary intrusions beneath both the upper east rift zone and seismic southwest rift zone occurred between 6 October and 20 December 1974.

A third but much larger eruption/intrusion on a long segment of the seismic southwest rift zone occurred on 31 December 1974 (figs. 5.13–5.16). The eruption was preceded by a deep magma-supply swarm on 24 December, followed by an upper east rift inflationary

[22] It is possible that south flank activity near the rift was masked by the intense rift earthquake swarm.

Table 5.5. Kīlauea eruptions, intrusions, and earthquakes, Mauna Ulu pause (see figures 5.7, E37).

[In rows with multiple entries text applies down to the next entry; dates in m/d/yyyy format; do, ditto (same as above); data for eruptions and traditional intrusions are emphasized by grey shading]

Time and Date		Region[1]	Event Type[2]	No.[3]	Tilt[4]		Lag[5]	Comment[6]	Fig.[7]	Ref.[8]
Start	End				Mag	Az				
00:17 7/4/1971	23:45 7/5/1971	sf3kuer	SDI	25				Embedded *M*3.5 mainshock at 11:07 7/5	**5.8**	
02:10 7/12/1971	19:59 7/14/1971	ei4sswr	II	42	3.5	294		Tilt 7/12-16; broken swarm	E38	
09:33 7/18//1971	12:46 7/19/1971	ei4sswr	II	18	flat				E39; *43.50*	8
19:25 7/21/1971	14:37 7/24/1971	ei3uer	II	45	2.2	310		Tilt 7/21-25; broken swarm	E40	
18:32 7/27/1971	01:53 7/30/1971	ei3uer	II	35	6.1	295		Tilt 7/26-28; I–A: 4 µrad inflation 7/27-29	E41	
17:55 7/28/1971	06:06 7/30/1971	ms1	II	25						
22:11 7/28/1971	21:41 7/30/1971	ei4sswr	II	34				South flank response		
04:59 8/1/1971	23:37 8/1/1971	sf3kuer	EQS							
23:20 8/5/1971	15:36 8/6/1971	ei3uer	I	10	3.8	309		Tilt 8/5-8	E42	
01:04 8/8/1971	08:418/14/1971	ei4sswr	II	197				Broken swarm	E43; *43.51*	8
01:29 8/9/1971	10:00 8/13/1971	ms1	II	163	13.5	340		Tilt 8/9-13	do	
13:04 8/10/1971	08:52 8/14/1971	ei3uer	II	115					do	
21:41 8/10/1971	09:28 8/12/1971	ms2	EQS	28				8 additional events to 11:41 8/13	do	
20:51 8/13/1971	08:55 8/14/1971	ms1	EQS	29					do	
08:55 8/14/1971	19:00 8/14/1971	KC	E		14.3	118	+0h 55m	Kīlauea Caldera; tilt: 8/13-15	E44	7
15:36 8/15/1971		ms4.5	EQ					*M*4.9 mainshock with 11 aftershocks		
22:18 8/15/1971		sf3kuer	EQ					*M*4.1 mainshock with 13 aftershocks	*43.52*	8, 12
20:08 8/27/1971	00:06 8/29/1971	ms2	EQS	14	flat			Upward movement of new magma following deep magma supply earthquakes on 8/15	E45	
22:59 8/27/1971	05:55 8/28/1971	ms3	EQS	8						
19:23 8/28/1971	23:14 8/29/1971	ms1	II	10						
15:29 9/6/1971	17:42 9/8/1971	ei4sswr	II	25	4.0	339		Tilt 9/5-8	E46	
19:37 9/8/1971	06:21 9/9/1971	ei3uer	II	19					*43.53*	8
22:10 9/12/1971	02:41 9/14/1971	ei4sswr	II	16	8.3	314		Tilt 9/12-15	E47	
14:29 9/17/1971	03:28 9/19/1971	ei3uer	II	20	10.1	301		Tilt 9/15-24	E48*A-E*	
21:16 9/17/1971	06:27 9/18/1971	ms1	II	12				I–A: abrupt inflation offset	*43.54*	8
17:53 9/18/1971	07:30 9/19/1971	ei4sswr	II	36						
20:18 9/19/1971	10:01 9/21/1971	ei4sswr	II	24						
04:12 9/22/1971	06:01 9/22/1971	ms4/5	EQS	11				Deep magma supply anticipation	E48*F*	
13:16 9/23/1971	15:00 9/24/1971	ei3uer	I	15				Preeruption	E49*A*	
15:27 9/24/1971	02:27 9/26/1971	ei4sswr	I	21				Pre, during, and posteruption		
16:56 9/24/1971	07:40 9/25/1971	ms1	I	64				do	E49*B*	
19:20 9/24/1971	20:00 9/24/1971	KC/SWR	E		101.1	354	+1h 20m	Kīlauea Caldera/SW rift zone; tilt: 9/24-25; I–A 8/24: 20 µrad inflation	*43.54*	7, 8
23:53 9/24/1971	02:41 10/1/1971	sf3kuer	EQS	75				South flank response; includes *M*4.2 with aftershocks to 21:28 9/30/1971	E49*C*; *43.55*	
14:02 9/27/1971	17:46 9/30/1971	sf4swr	EQS	41						

Table 5.5. Kīlauea eruptions, intrusions, and earthquakes, Mauna Ulu pause (see figures 5.7, E37).—Continued

[In rows with multiple entries text applies down to the next entry; dates in m/d/yyyy format; do, ditto (same as above); data for eruptions and traditional intrusions are emphasized by grey shading]

Time and Date		Region[1]	Event Type[2]	No.[3]	Tilt[4]		Lag[5]	Comment[6]	Fig.[7]	Ref.[8]
Start	End				Mag	Az				
19:20 10/8/1971	17:48 10/09/1971	sf3kuer	EQS	10	10.6	296		Tilt 10/6-14; I–A 10/7-13: inflation	E50	
08:14 11/14/1971	04:55 11/15/1971	sf2mer	EQS	11	9.2	331.8		Tilt 11/11–17	E51	
19:33 11/15/1971	17:27 11/17/1971	sf3kuer	SDI	60						
02:15 12/9/1971		sf2mer	EQ					*M*4.3 with aftershocks		12
20:49 12/12/1971	08:40 12/13/1971	ei3uer	II	10	3.7	321		Tilt 12/10-13	E52	
09:05 12/22/1971	06:38 12/28/1971	sf3kuer	SDI	349				South flank anticipation	E53 *A-E*	
18:36 12/22/1971	14:45 12/23/1971	sf2mer	SDI	14	4.3	321		tilt: 12/23-26		
07:08 12/24/1971	23:04 12/24/1971	ei3uer	II	12						
12:54 12/24/1971	20:14 12/29/1971	sf4swr	SDI	325				South flank accompaniment/response	*43.56*	8
14:37 12/28/1971	08:08 12/29/1971	sf3kuer	SDI	46				Continued south flank response; I–A 12/27-29 3 µrad deflation		
19:52 12/31/1971	07:00 1/4/1972	sf2mer		10				No swarm	E54	
20:31 12/31/1971	02:18 1/4/1972	sf3kuer	SDI?	20				Broken swarm	do	
14:01 12/31/1971	23:40 1/3/1972	sf4swr	do	9				No swarm	do	
Precursors: return to eruption at Mauna Ulu										
15:32 1/9/1972	09:56 1/10/1972	ms1	II	10	4.6	299		Tilt 1/7-9		
21:27 1/10/1972	08:07 1/12/1972	ei3uer	II	14	2.1	124		Tilt 1/9-10; broken swarm	E55	
02:56 1/16/1972	03:01 1/17/1972	sf3.4	SDI	5, 3				Broken swarm; related to preceding and succeeding inflationary intrusions		
03:12 1/19/1972	13:13 1/19/1972	sf3	SDI	6				Related to succeeding inflationary intrusion		
19:06 1/19/1972	00:54 1/20/1972	ei3uer	II	16	flat			I–A 1/17-21 2.5 µrad inflation	E56*A*; *43.57*	8
10:42 1/20/1972	07:07 1/22/1972	ei3uer	II	71						
11:46 1/23/1972	02:13 1/24/1972	ms1	EQS	8						
22:30 1/23/1972	09:22 1/24/1972	ms4/5	EQS	8				Deep magma supply event		
15:52 1/26/1972	11:01 1/28/1972	ei3uer	II	91	2.2	293		Tilt 1/24-28	E56*B*; *43.57*	8

[1]Earthquake classification abbreviations are given according to the classification in appendix A, table A3, and locations of regions are shown in appendix A, figure A4. Eruption locations are designated in bold type as follows: KC, Kīlauea Caldera; LERZ, lower east rift zone; MERZ, middle east rift zone; UERZ, upper east rift zone; SWR, southwest rift zone.

[2]E, Eruption; intrusion ("traditional" I; "inflationary" II; "suspected deep intrusion" SDI [see Wright and Klein, 2008, table 2]); earthquake, EQ; earthquake swarm, EQS.

[3]Minimum number of events defining a swarm: 20 for south flank; 10 for all other regions.

[4]Magnitude in microradians and azimuth of daily tilt measurements from the water-tube tiltmeter in Uwēkahuna Vault.

[5]Lag times separating the onset of the earliest earthquake swarm (excluding south flank) for a given event and the beginning of deflation or inflation measured by the continuously recording Ideal-Arrowsmith tiltmeter in Uwēkahuna Vault. (+) tilt leads, (-) tilt lags.

[6]Abbreviations as follows: I–A, Ideal-Arrowsmith continuously recording tiltmeter in Uwēkahuna Vault.

[7]Text figures **bold text**; appendix figures plain text; 43.xx = figures in Klein and others, 1987.

[8]References coded as follows: 1. Fiske and Kinoshita, 1969, 2. Kinoshita and others, 1969, 3. Wright and Klein, 2008, 4. Jackson and others, 1975, 5. Swanson and others, 1976b, 6. Swanson and others, 1979, 7. Duffield and others, 1974, 8. Klein and others, 1987, 9. Tilling and others, 1987, 10. Nielsen and others, 1977, 11. Lockwood and others, 1999; 12. Klein and others, 2006.

Table 5.6. Kīlauea eruptions, intrusions, and earthquakes, Mauna Ulu II (see figure 5.9).

[In rows with multiple entries text applies down to the next entry; dates in m/d/yyyy format; do, ditto (same as above); data for eruptions and traditional intrusions are emphasized by grey shading]

Time and Date		Region[1]	Event Type[2]	No.[3]	Tilt[4]		Lag[5]	Comment[6]	Fig.[7]	Ref.[8]
Start	End				Mag	Az				
03:48 2/4/1972	16:14 2/4/1972	ms4/5	EQS	9	7.1	313		Deep magma supply anticipation; Tilt 1/29–2/5	E58; *43.58*	8, 9
16:28 2/4/1972	4/26/1973	MERZ	E					Mauna Ulu IIA		
2/26/1972	3/3/1972				9.6	104		Continuous deflation		
18:44 3/18/1972	20:07 3/18/1972	ei3uer	I	9	13.8	106	+0h 44m	Tilt: 3/18–22; I–A, 3/18–23 11 µrad deflation	E59	
09:52 5/1/1972	09:31 5/2/1972	sf3kuer	SDI	15					E60	
01:32 9/5/1972		sf3kuer	EQ					*M*5.2 mainshock with 24 aftershocks		
04:36 3/7/1973	02:09 3/8/1973	sf3kuer	SDI	9					E61	
00:01 4/15/1973		sf2mer	EQ					*M*4.46. aftershock pattern mimics SDI	E62	
10:26 4/26/1973			EQ		23.3	305		Mauna Kea *M*6.6 35 km; Tilt 4/26; I–A 17 µrad inflation; triggered Kīlauea earthquakes and changed eruption	**5.9**	3, 9, 10
4/26/1973	6/18/1974	MERZ	E					Mauna Ulu IIB		8, 9
07:14 5/5/1973	09:48 5/5/1973	sf3kuer	EQS	7				South flank anticipation/accompaniment		
07:20 5/5/1973	21:16 5/5/1973	ms1	EQS	15						
07:24 5/5/1973	15:46 5/5/1973	koae	I	75				Koaʻe/upper east rift intrusion; Tilt 5/4-9		
10:11 5/5/1973	07:31 5/6/1973	UERZ	I	164	26.0	108	+0h 24m	do	**5.10**; *43.59*	8, 9
10:25 5/5/1973	12:00 5/5/1973		E					Pauahi-Hiiaka[9]; eruption at Mauna Ulu resumed on 5/8		
04:43 6/4/1973	06:44 6/6/1973	ei3uer	II	11				Additional earthquakes on 6/1 (6) and 6/6 (4)		
23:57 6/8/1973	00:25 6/10/1973	ei3uer	I	82	7.1	102	-0h 3m	Tilt 6/8-9	E63	
01:10 6/9/1973	12:38 6/9/1973	koae	I	31					*43.60*	8, 9
02:26 7/25/1973	23:44 7/25/1973	ei2mer	I	7	2.5	111		Tilt: 7/24-26	E64*F*	
02:25 8/3/1973	18:37 8/3/1973	sf3kuer	SDI	11						
17:22 11/10/1973	08:08 11/11/1973	ei3uer	I	41				Tilt 11/10-13	E65	
21:47 11/10/1973	12:00 12/9/1973	UERZ	E		26.9	109	-0h 38m	Pauahi-Hiiaka[9]; eruption at Mauna Ulu resumed on 12/10	*43.61*	8, 9
08:04 12/25/1973	06:00 12/26/1973	sf3kuer	SDI	13					E66	
06:04 1/12/1974		sf2mer	EQ					*M*4.7 with aftershocks		12
17:49 2/23/1974	07:00 3/4/1974	ei3uer	II	14	5.9	313		Tilt 2/28-3/2; no earthquake swarm	E67	

Table 5.6. Kīlauea eruptions, intrusions, and earthquakes, Mauna Ulu II (see figures 5.9, E57).—Continued

[In rows with multiple entries text applies down to the next entry; dates in m/d/yyyy format; do, ditto (same as above); data for eruptions and traditional intrusions are emphasized by grey shading]

Time and Date		Region[1]	Event Type[2]	No.[3]	Tilt[4]		Lag[5]	Comment[6]	Fig.[7]	Ref.[8]
Start	End				Mag	Az				
00:46 3/11/1974	12:59 3/17/1974	ei3uer	II	28	flat			Broken eqs; I-A: 14:00 3/17 5.5 µrad deflation	E68	
23:23 3/23/1974	03:44 3/24/1974	sf3kue	EQS	7				South flank anticipation	*43.62*	8
03:13 3/24/1974	05:28 3/24/1974	ei3uer	I	11	5.4	106	+1h 13m	Tilt 3/24–24; IA; def from 3/23	E69	
12:51 3/24/1974	23:08 3/27/1974	sf3kuer	EQS	38				Broken eqs south flank response		
05:15 3/26/1974	12:42 3/26/1974	ms4/5	EQS	9				Deep magma supply		
10:03 5/12/1974	05:50 5/22/1974	ei3uer	II					No eq; I–A: inflation	E70*A*	
02:12 5/21/1974	10:45 5/23/1974	sf3kuer	SDI	51				South flank anticipation		
03:09 5/22/1974	11:37 5/22/1974	ms1	I	9	12.5	310		Tilt 5/21-22; I–A: <1 µrad inflation		
5/22/1974	5/23/1974				2.8	129		Tilt 5/22-23; deflation follows inflation	E70*B*	
17:49 5/24/1974	14:07 5/25/1974	sf3kuer	SDI	16				South flank response		
21:37 5/25/74	12:44 5/30/1974	ms3		9				Broken earthquake swarm	E70*C*	
04:24 5/28/1974	03:06 5/30/1974	ei3uer	II	13				Broken earthquake swarm		
6/18/1974		MERZ						End of Mauna Ulu eruption[10]		9

[1]Earthquake classification abbreviations are given according to the classification in appendix A, table A3, and locations of regions are shown in appendix A, figure A4. Eruption locations are designated in bold type as follows: KC, Kīlauea Caldera; LERZ, lower east rift zone; MERZ, middle east rift zone; UERZ, upper east rift zone; SWR, southwest rift zone.

[2]E, Eruption; intrusion ("traditional" I; "inflationary" II; "Suspected deep intrusions" SDI); EQ, earthquake; EQS, earthquake swarm.

[3]Minimum number of events defining a swarm: 20 for south flank; 10 for all other regions.

[4]Magnitude in microradians and azimuth of daily tilt measurements from the water-tube tiltmeter in Uwēkahuna Vault.

[5]Lag times separating the onset of the earliest earthquake swarm (excluding south flank) for a given event and the beginning of deflation or inflation measured by the continuously recording Ideal-Arrowsmith tiltmeter in Uwēkahuna Vault. (+) tilt leads, (-) tilt lags.

[6]Abbreviations as follows: I–A, Ideal-Arrowsmith continuously recording tiltmeter in Uwēkahuna Vault.

[7]Text figures **bold text**; appendix figures plain text; 43.xx = figures in Klein and others, 1987.

[8]References coded as follows: 1. Fiske and Kinoshita, 1969, 2. Kinoshita and others, 1969, 3. Wright and Klein, 2008, 4. Jackson and others, 1975, 5. Swanson and others, 1976b, 6. Swanson and others, 1979, 7. Duffield and others, 1974, 8. Klein and others, 1987, 9. Tilling and others, 1987, 10. Nielsen and others, 1977, 11. Lockwood and others, 1999; 12. Klein and others, 2006. Note: This is a master list for all tables in chapter 5. Only some references will be cited in this table.

[9]Mauna Ulu pauses during each eruption at Pauahi, 5/6- 5/8/1973; 11/9–12/12/1973.

[10]Ending time arbitrary, based on seismicity. Lava visible in Mauna Ulu to 7/22/1974.

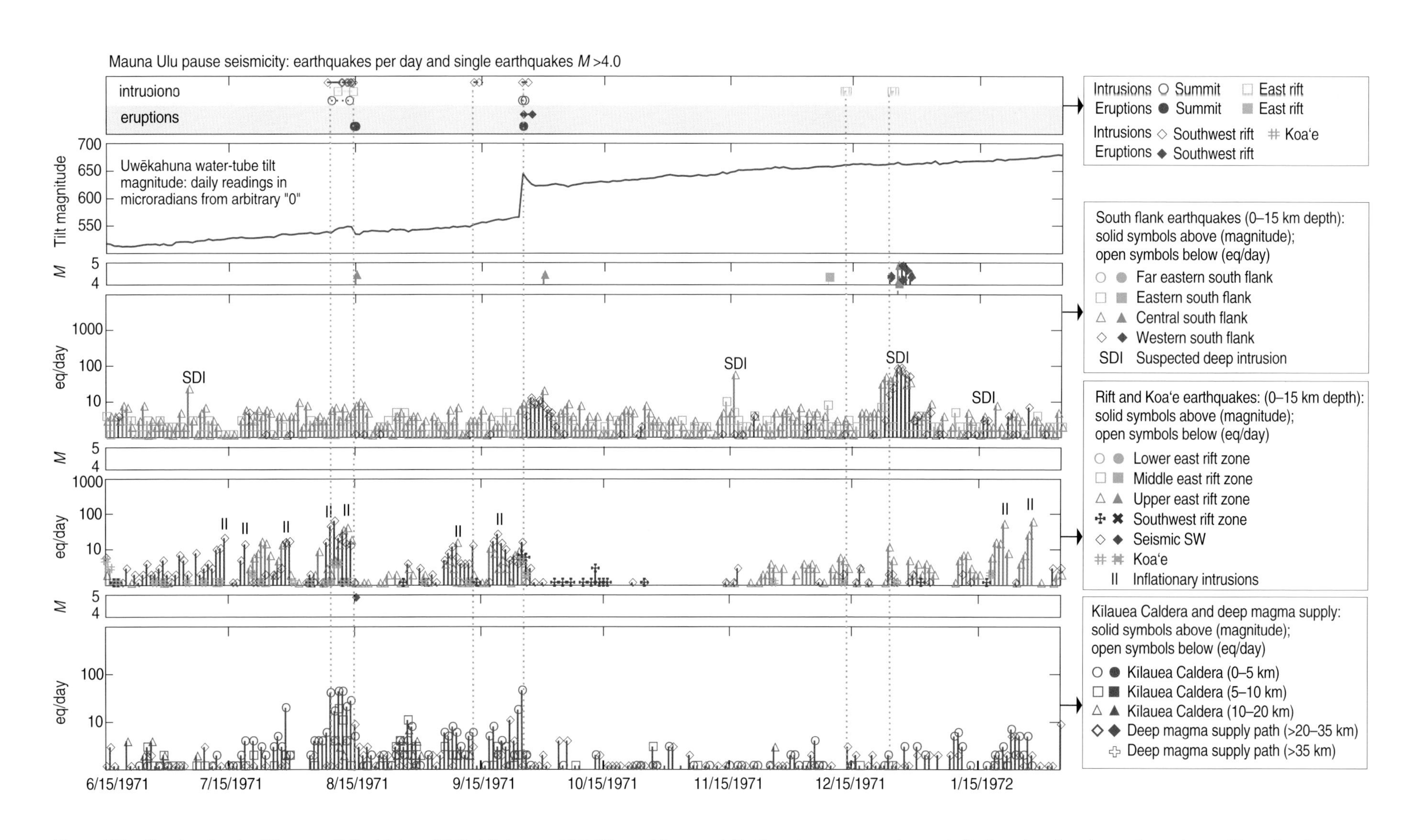

Figure 5.7. Graphs showing Kīlauea activity, 16 June 1971–4 February 1972 (Mauna Ulu pause): Earthquake swarms and summit tilt plotted against times of eruption and intrusion. Dates on figure in m/d/yyyy format. See also caption for figure 5.1.

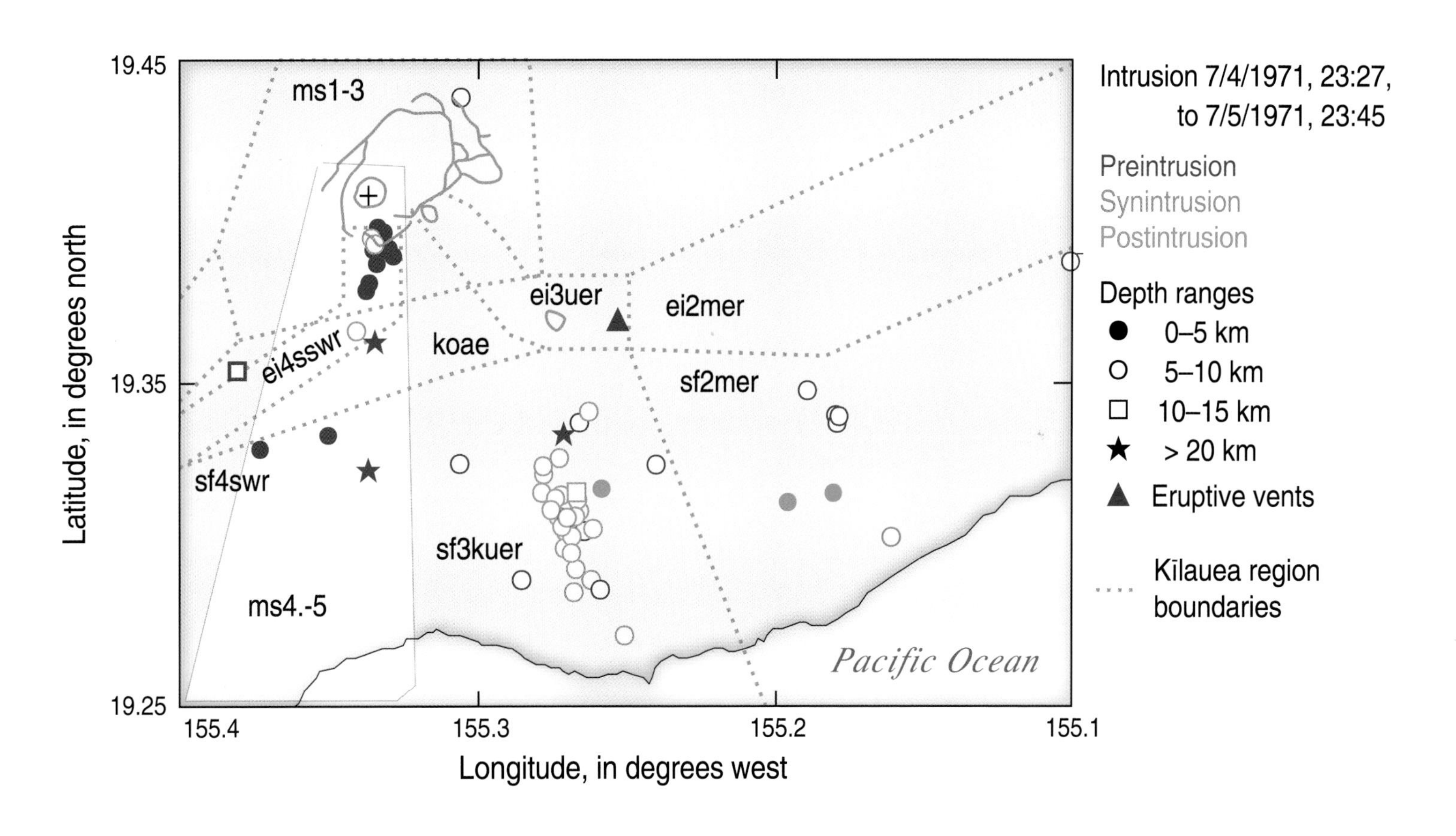

Figure 5.8. Map showing Kīlauea activity, 2–6 July 1971: Location of earthquakes associated with a suspected deep intrusion (4–5 July) at site SDI 2 (see fig. 5.2). Data are shown for periods before, during, and after the intrusion. The Mauna Ulu vent is plotted as a triangle. Dates on figure in m/d/yyyy format.

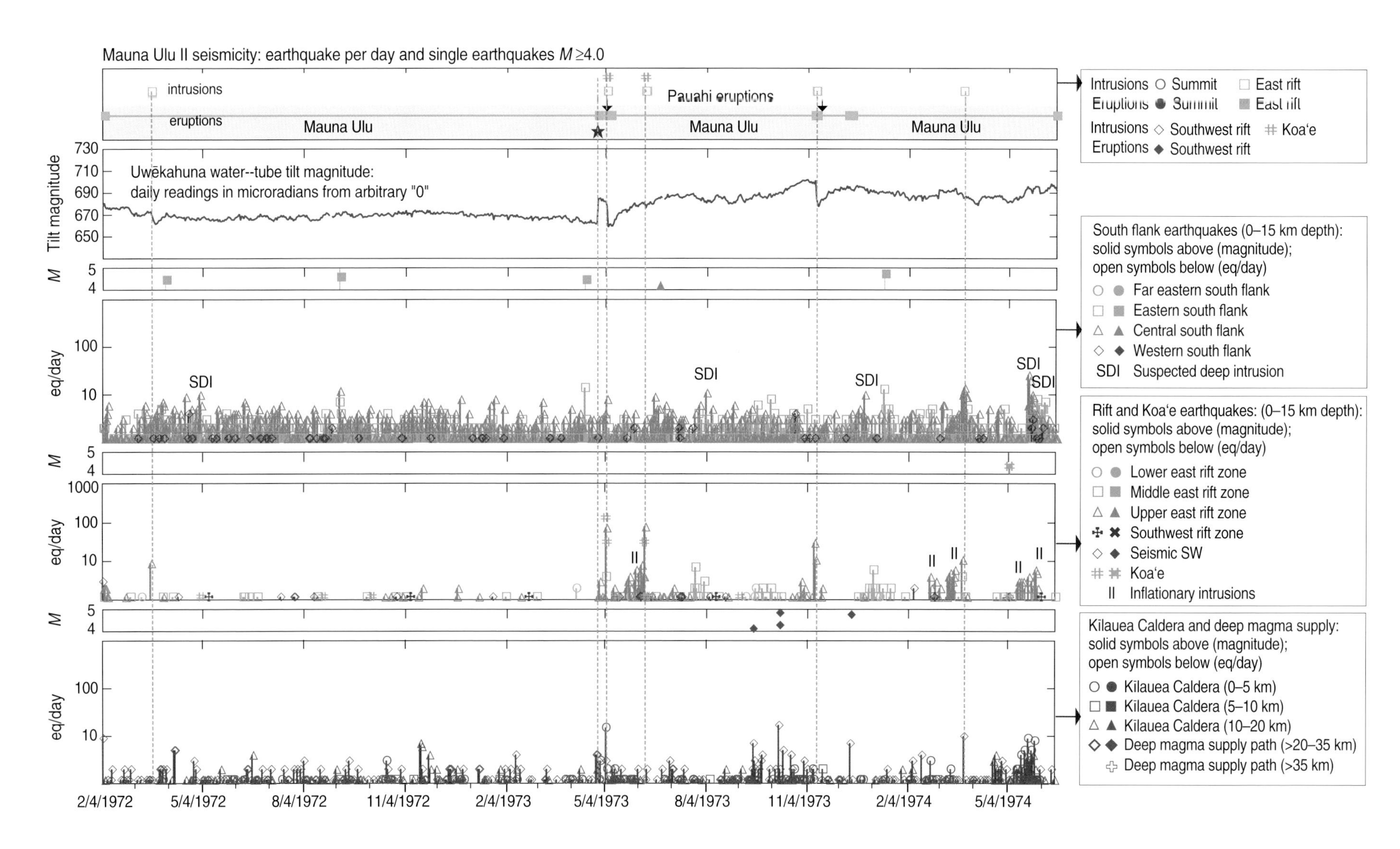

Figure 5.9. Graphs showing Kīlauea activity, 4 February 1972–14 June 1974 (Mauna Ulu stage II): Earthquake swarms and summit tilt plotted against times of eruption and intrusion. Dates on figure in m/d/yyyy format. See also caption for figure 5.1. The star (★) marks the *M*6.6 Honomū earthquake beneath Mauna Kea.

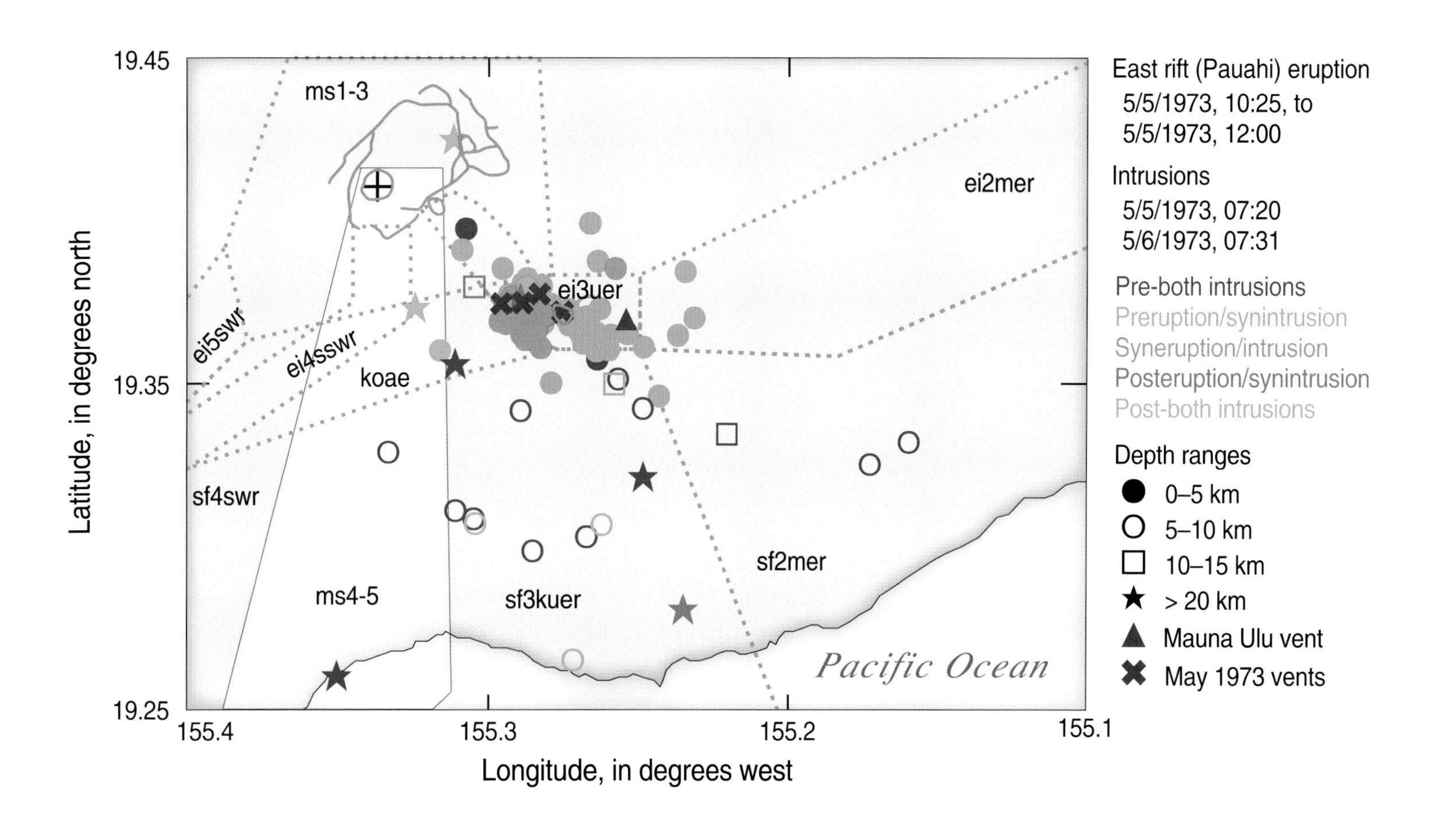

Figure 5.10. Map of Kīlauea activity, 2–9 May 1973: Location of earthquakes associated with the east rift eruption in and near Pauahi Crater (5 May) accompanied and followed by intrusion (5–6 May) extending into the Koaʻe Fault Zone, and preceded by enhanced south flank seismicity. Shown are deep earthquakes triggered by the Honomū earthquake and additional deep earthquakes within the period of eruption/intrusion. Data are shown for periods before, during, and after the events, and the Mauna Ulu vent is plotted as a red triangle. Dates on figure in m/d/yyyy format.

Table 5.7. Kīlauea eruptions, intrusions, and earthquakes, post-Mauna Ulu to 1975 earthquake (see figure 5.11, E71).

[In rows with multiple entries text applies down to the next entry; dates in m/d/yyyy format; do = ditto (same as above); data for eruptions and traditional intrusions are emphasized by grey shading]

Time and Date		Region[1]	Event Type[2]	No.[3]	Tilt[4]		Lag[5]	Comment[6]	Fig.[7]	Ref.[8]
Start	End				Mag	Az				
6/19/1974	11/29/1975							Post-Mauna Ulu		11
20:50 6/20/1974		sf3kuer	EQ					*M*4.4 ms; 1 aftershock		
22:14 6/21/1974	11:31 6/22/1974	sf3kuer	SDI?	11				Delayed aftershocks for previous event?	E72	
11:06 6/27/1974	16:59 6/27/1974	ms1	I	6	6.3	317		Tilt 6/29-30	E73	
06:21 7/17/1974	16:00 7/17/1974	sf3kuer	EQS	7				South flank anticipation	*43.63*	8
03:02 7/19/1974	7/19/1974 11:32	ei3uer	I	189	15.9	107	+1h 58m	Tilt 7/19-20; I–A: deflation		
04:11 7/19/1974	12:28 7/19/1974	ms1	EQS	7						
12:30 7/19/1974	06:00 7/22/1974	KC	E					Halema‘uma‘u and east caldera	**5.12**	
15:37 7/23/1974	21:16 7/24/1974	ms3	EQS	14				Broken earthquake swarm		
06:02 9/16/1974	19:48 9/17/1974	ei4sswr	I	11				Precursory seismicity	*43.64*	8
01:21 9/19/1974	15:00 9/19/1974	KC/SWR	E/I		28.7	331	-1h 21m	Southwest rift zone; summit intrusion; tilt 9/18-19	E74	
14:09 10/6/1974	04:02 10/9/1974	ei3uer	II	17	7.9	332		Tilt 10/5-11; broken earthquake swarm	E75	
16:41 10/7/1974	17:19 10/8/1974	ms1	II	8				Broken earthquake swarm	*43.65*	8
15:38 10/12/1974	07:50 10/15/1974	ei3uer	II	14				Broken earthquake swarm		
19:05 10/31/1974	11/2/1974 16:20	ei4sswr	II	21	7.7	326		Tilt 10/27-11/7; broken earthquake swarm	E76	
19:37 11/5/974	03:52 11/6/1974	ei3uer	II	8						
15:02 11/21/1974	23:40 11/25/1974	ei4sswr	II	18	15.3	314		No swarm; I-A: flat	E77	
08:08 12/1/1974	12:35 12/5/1974	ei4sswr	II	28				Tilt 11/29-12/5; broken earthquake swarm	E77	
02:04 12/6/1974	12:42 12/9/1974	ei3uer	II	27	flat			Broken earthquake swarm; I–A: inflation	E78	
13:36 12/10/1974	13:49 12/14/1974	ei3uer	II	47				Broken earthquake swarm; I–A: inflation		
21:55 12/17/1974	08:52 12/21/1974	ei3uer	II	23	11.0	311		Tilt 12/15-22; broken earthquake swarm	E79	
23:59 12/18/1974	22:30 12/20/1974	ei4sswr	II	11				Broken earthquake swarm		
20:26 12/25/1974	23:36 12/30/1974	ei3uer	II	81	42.7	107		Tilt 12/30-31; east rift zone anticipation of southwest rift eruption?	**5.13, 5.14**	
21:43 12/27/1974	08:02 1/6/1975	ei4sswr	I	602	86.5	135		Tilt 12/31/1974-1/1/1975	*43.66*	8
14:44 12/29/1974	23:55 12/29/1974	sf3kuer	EQS	12				South flank anticipation	**5.14**	
02:56 12/31/1974	08:50 12/31/1974	SWR	E		33.2	151	-2h 56m	Southwest rift zone; Tilt 1/1-2/19755; east rift earthquakes end	**5.15, 5.16,** *43.67*	
08:17 12/31/1974	19:24 1/7/1975	sf3kuer	EQS	306	9.1	169		South flank response; Tilt 1/3-4/1975; I–A deflation ends 1/3	E80*A–F*	
12:52 12/31/1974	05:01 1/8/1975	sf4swr	EQS	548	7.1	192		do		

Table 5.7. Kilauea eruptions, intrusions, and earthquakes, post-Mauna Ulu to 1975 earthquake (see figure 5.11, E71).—Continued

[In rows with multiple entries text applies down to the next entry; dates in m/d/yyyy format; do = ditto (same as above); data for eruptions and traditional intrusions are emphasized by grey shading]

Time and Date		Region[1]	Event Type[2]	No.[3]	Tilt[4]		Lag[5]	Comment[6]	Fig.[7]	Ref.[8]
Start	End				Mag	Az				
12/24/1974	2/25/1975	All reg.	I/E/I					Summary plot: eruption/intrusion	**5.16**	
11:11 3/15/1975	08:17 3/16/1975	sf3kuer	EQS	10	Flat			I–A: inflation	E81	
22:54 3/18/1975	09:14 3/20/1975	sf3kuer	do	13	do			Broken earthquake swarm		
21:06 3/20/1975	14:34 3/21/1975	lpms2	do	6	do					
23:42 3/21/1975	23:22 3/22/1975	lpms3	do	11	do					
19:07 3/24/1975	02:50 3/26/1975	sf3kuer	EQS	11	Flat				E82	
02:00 4/2/1975	10:14 4/3/1975	sf3kuer	SDI	11	flat			I–A: inflation	E83	
04:14 4/5/1975	18:34 4/5/1975	sf3kuer	SDI	12					E83	
14:47 4/16/1975	15:01 4/18/1975	sf3kuer	SDI	25	3.1	311		Tilt 4/16-18; I-A: flat tilt	E84	
07:36 8/5/1975	19:22 8/5/1975	sf4swr	SDI	25	flat			I–A: flat tilt	E85	
00:32 10/23/1975	18:42 10/23/1975	sf3kuer	SDI	11	flat			East rift earthquakes (ei3uer) reappear; I-A: flat tilt	E86	
17:04 11/12/1975	03:58 11/14/1975	ei4sswr	II	11	flat			Broken earthquake swarm; I–A: flat tilt	E87	
12:56 11/15/1975		sf3kuer	EQ					*M*4.5 mainshock with aftershocks		12
04:48 11/29/1975		sf2mer	EQ					*M*7.2 south flank earthquake (see Ando, 1979)		9
05:32 11/29/1975	22:00 11/29/1975	KC	E				-2h 0m	Halema‘uma‘u; triggered by earthquake		9
11/27/1975	11/281975				2.7	108		Tilt 11/27-28/1975; clockwise rotation		
11/281975	11/30/1975				36.8	134		Tilt 11/28-30/1975		
11/30/1975	12/10/1975				178.2	154		Tilt 11/30-12/10/1975		
12/10/1975	3/5/1976				39.2	198		Tilt 12/10/1975-3/5/1976		

[1]Earthquake classification abbreviations are given according to the classification in appendix A, table A3, and locations of regions are shown in appendix A, figure A4. Eruption locations are as follows: KC, Kīlauea Caldera; LERZ, lower east rift zone; MERZ, middle east rift zone; UERZ, upper east rift zone; SWR, southwest rift zone.

[2]E, Eruption; intrusion (“traditional” I; “inflationary” II; “suspected deep intrusion” SDI); EQ, earthquake; EQS, earthquake swarm.

[3]Minimum number of events defining a swarm: 20 for south flank; 10 for all other regions.

[4]Magnitude in microradians and azimuth of daily tilt measurements from the water-tube tiltmeter in Uwēkahuna vault.

[5]Lag times separating the onset of the earliest earthquake swarm (excluding south flank) for a given event and the beginning of deflation or inflation measured by the continuously recording Ideal-Arrowsmith tiltmeter in Uwēkahuna vault. (+) tilt leads, (-) tilt lags.

[6]I–A, Ideal-Arrowsmith continuously recording tiltmeter in Uwēkahuna vault.

[7]Text figures **bold text**; appendix figures plain text; 43.xx = figures in Klein and others, 1987.

[8]References coded as follows: (1) (Fiske and Kinoshita, 1969), (2) (Kinoshita and others, 1969), (3) (Wright and Klein, 2008), (4) (Jackson and others, 1975), (5) (Swanson and others, 1976b), (6) (Swanson and others, 1979), (7) (Duffield and others, 1974), (8) (Klein and others, 1987), (9) (Tilling and others, 1987), (10) (Nielsen and others, 1977), (11) (Lockwood and others, 1999); (12) (Klein and others, 2006). Note: This is a master list for all tables in chapter 5. Only some references will be cited in this table.

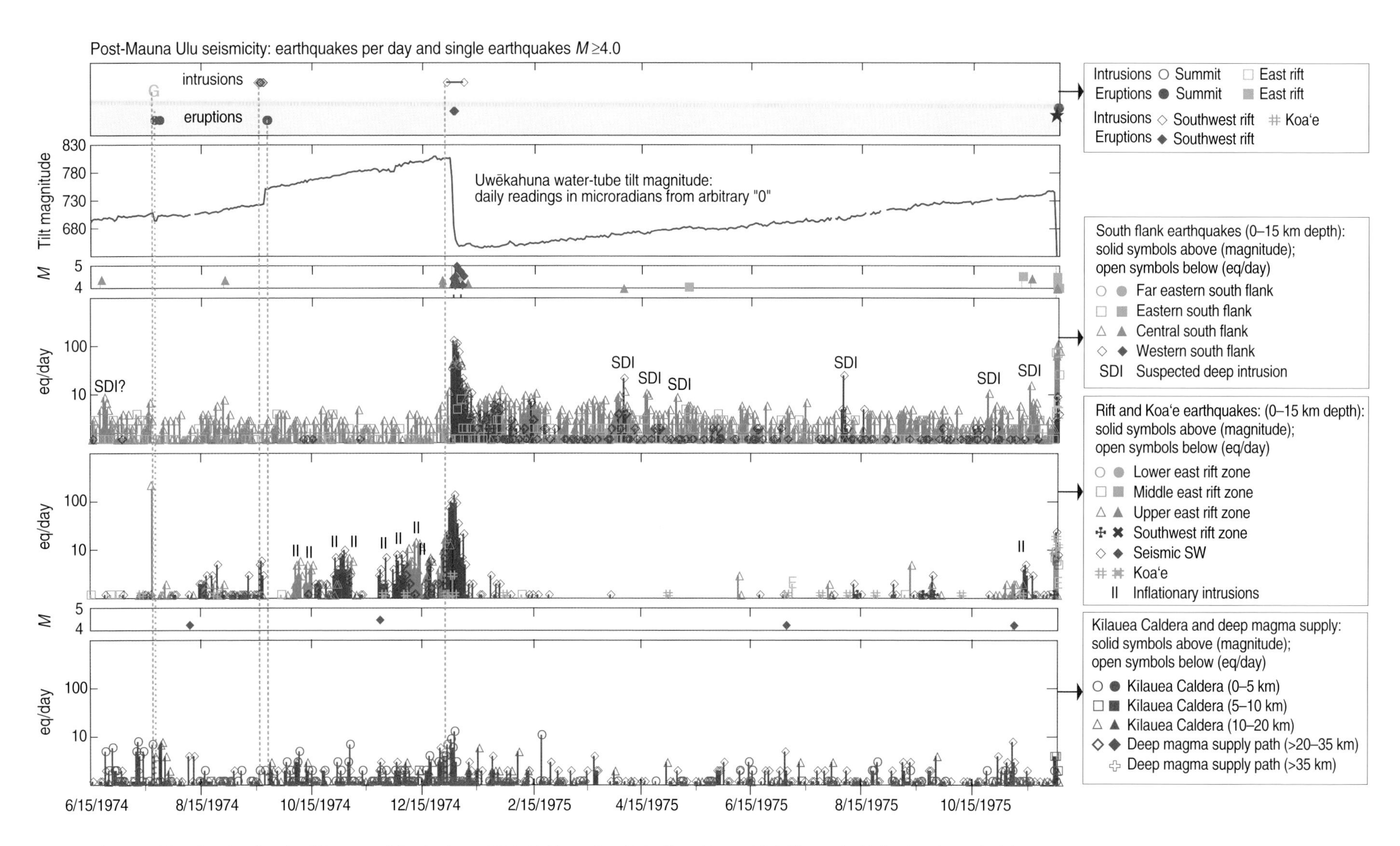

Figure 5.11. Graphs showing Kīlauea activity, 15 June 1974–29 November 1975 (Post Mauna Ulu). The period following the end of the Mauna Ulu eruption ends with the *M*7.2 south flank earthquake of 29 November 1975 (blue star). Earthquake swarms and summit tilt plotted against times of eruption and intrusion. Dates on figure in m/d/yyyy format. See also caption for figure 5.1.

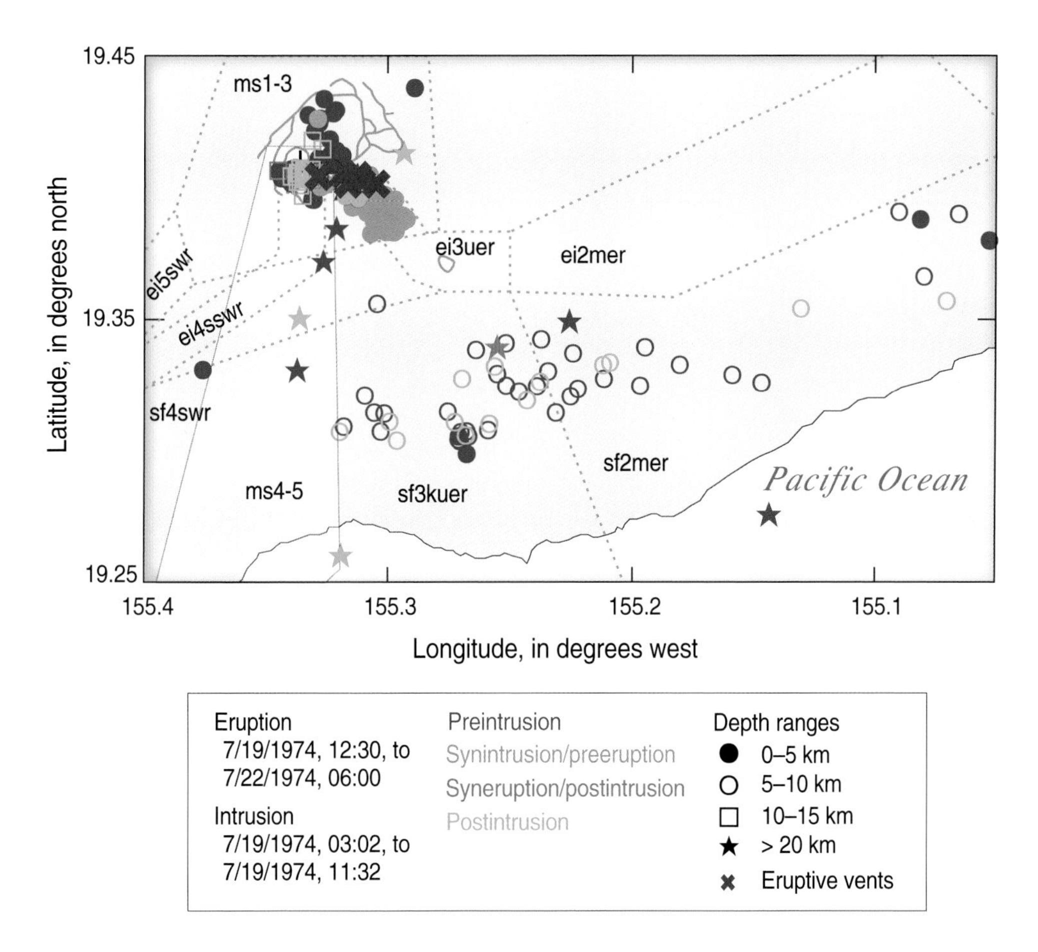

Figure 5.12. Map showing Kīlauea activity, 10–29 July 1974 : Location of earthquakes associated with eruption and intrusion on 19–22 July 1974. Data are shown for periods before, during, and after the event. Continuing intrusion may have been masked by eruption tremor. South flank seismicity precedes and follows eruption/intrusion. Dates on figure in m/d/yyyy format.

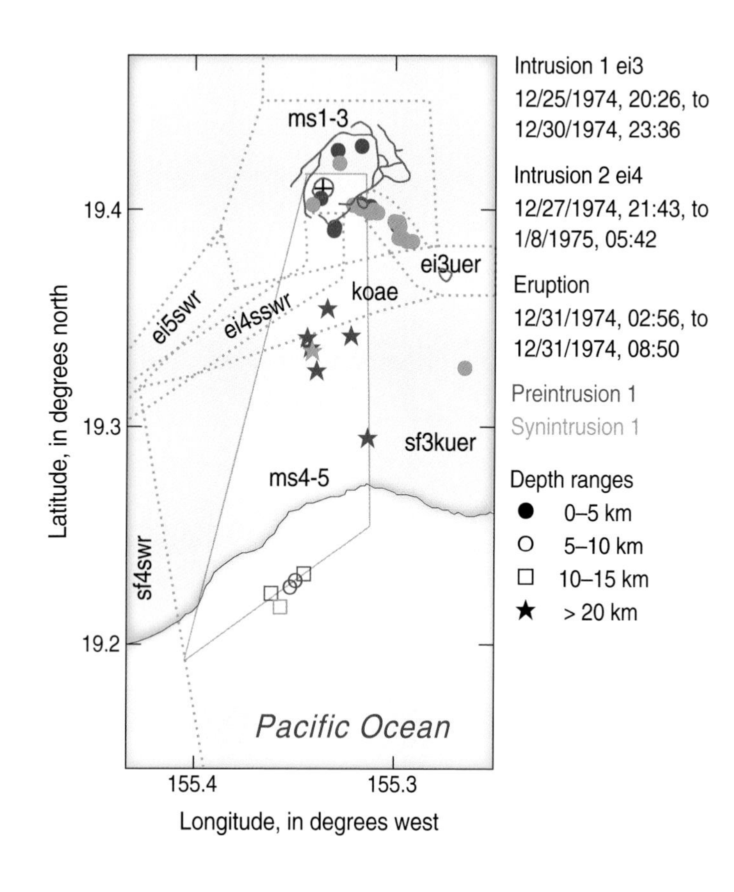

Figure 5.13. Map showing locations of earthquakes precursory to the 31 December 1974 southwest rift zone eruption and intrusion. This map covers the period 24–26 December 1974, during which there was a traditional intrusion beneath the upper east rift zone preceded by deep magma-supply seismicity. See also figures 5.14 and 5.15. Dates on figure in m/d/yyyy format.

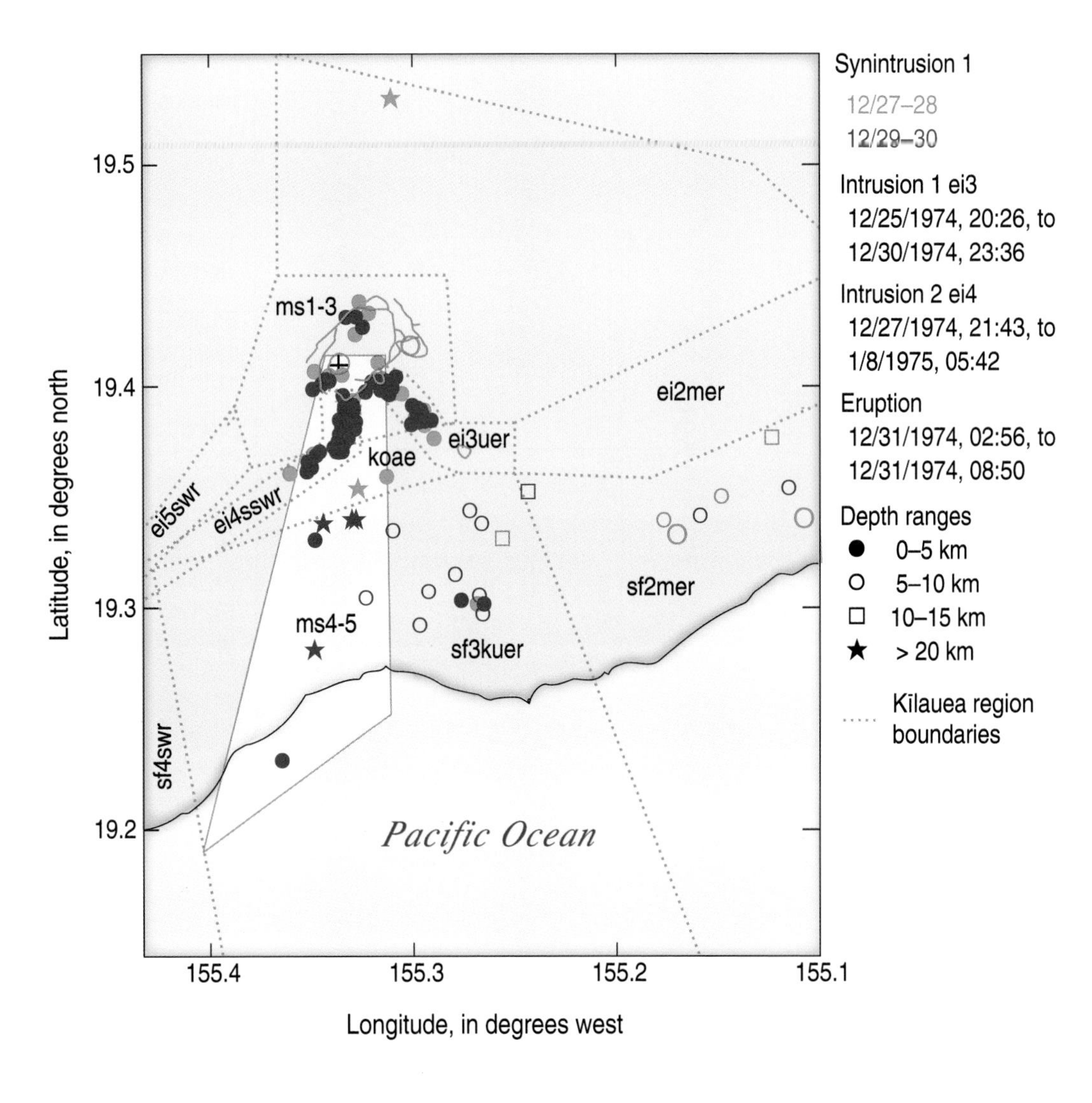

Figure 5.14. Map showing locations of earthquakes precursory to the 31 December 1974 southwest rift zone eruption and intrusion. This map covers the period 27–30 December 1974, during which there was inflationary intrusion beneath the upper part of both rift zones. See also figures 5.13 and 5.15. Dates on figure in m/d/yyyy format.

intrusion on 25 December (fig. 5.13). Inflationary intrusion expanded to the seismic southwest rift on 27–30 December (fig. 5.14). Coincident with and following the short eruption on 31 December seismicity moved down the seismic southwest rift zone and onto the western south flank (fig. 5.15). Intrusion continued for a week following the eruption (fig. 5.16), gradually tapering off during the following 6 weeks. Heightened seismicity beneath the south flank and in the magma-supply regions deeper than 10 km persisted up to the time of the 1975 earthquake.

The vents for the September 1971, September 1974, and December 1974 eruptions were located at increasing distance from the traditional southwest rift zone (1971) southeastward to the trace of the seismic southwest rift zone (December 1974).

Following the end of the large 1974 intrusion into the seismic southwest rift zone, there were several suspected deep intrusions (table 5.7) beneath both sites SDI 1 and SDI 2 in the central south flank. A final *M*4.5 south flank earthquake (table 5.6) beneath the central south flank occurred on 15 November, 2 weeks before the *M*7.2 earthquake beneath the eastern south flank on 29 November 1975.

Interpretations 1967–1975

We interpret this period to have one of steadily increasing magma supply. We also identify cycles in which the eruption efficiency, defined as the ratio of magma erupted to magma intruded, increased with continued use of the plumbing. A full discussion of both of these subjects is deferred to chapter 8.

The events in the period between the 1967–68 summit eruption in Halema‘uma‘u and the 1975 *M*7.2 south flank earthquake are a dramatic illustration of the internal adjustments in the Kīlauea plumbing needed to accommodate an increasing magma supply from the mantle (Wright and Klein, 2008). The period is also notable for revealing the variety of responses of Kīlauea's south flank to eruptions and the three different types of intrusions. Finally the period offers a clue to precursory signs that could indicate the impending occurrence of a large south flank earthquake.

Response of the Magma Plumbing to Increased Magma Supply

A reliable signal that continuing magma supply is stressing the capacity of the plumbing system is inflation of Kīlauea's summit. The types of accommodation of magma identified during the 1967–75 period are as follows:

1. *Traditional eruption and intrusion within the rift zones.* Magma pressure is relieved by eruption and intrusion, and space for intrusion is created by ongoing dilation of the rift zones during flank spreading and by uplift of the areas adjacent to the intrusion.

2. *Stepwise increase in spreading rate associated with suspected deep intrusions.* In our interpretation, suspected deep intrusions behave in a similar manner to the more recently defined "slow" or "silent" earthquakes, which have measured jumps in seaward spreading of Kīlauea's south flank (see discussion in chapter 8). Unfortunately it was not possible for deformation measurements made during 1967–75 to capture instantaneous movement of Kīlauea's south flank.

3. *Carrying capacity of the magma plumbing.* Repeated use of the magma delivery system leads to increasing eruption efficiency with time (Wright and Klein, 2008, figure 6; discussion in chapter 8).

4. *Change of location.* The east rift zone is the preferred site for eruption and intrusion. When the east rift plumbing is unable to accommodate the magma supply, eruption and intrusion may occur beneath Kīlauea's summit and (or) seismic southwest rift zone. Such shifts occurred during the Mauna Ulu pause in 1971 and following the end of the Mauna Ulu eruption in 1974.

Forecasting the 29 November 1975 *M*7.2 South Flank Earthquake

The earthquake of 29 November 1975 and its accompanying tsunami were widely studied (see, for example, Ando, 1979; Furumoto and Kovach, 1979; Lipman and others, 1985), those studies identifying, for the first time, a deep subhorizontal surface along which the south flank moved seaward. The 1975 earthquake was anticipated within a year of

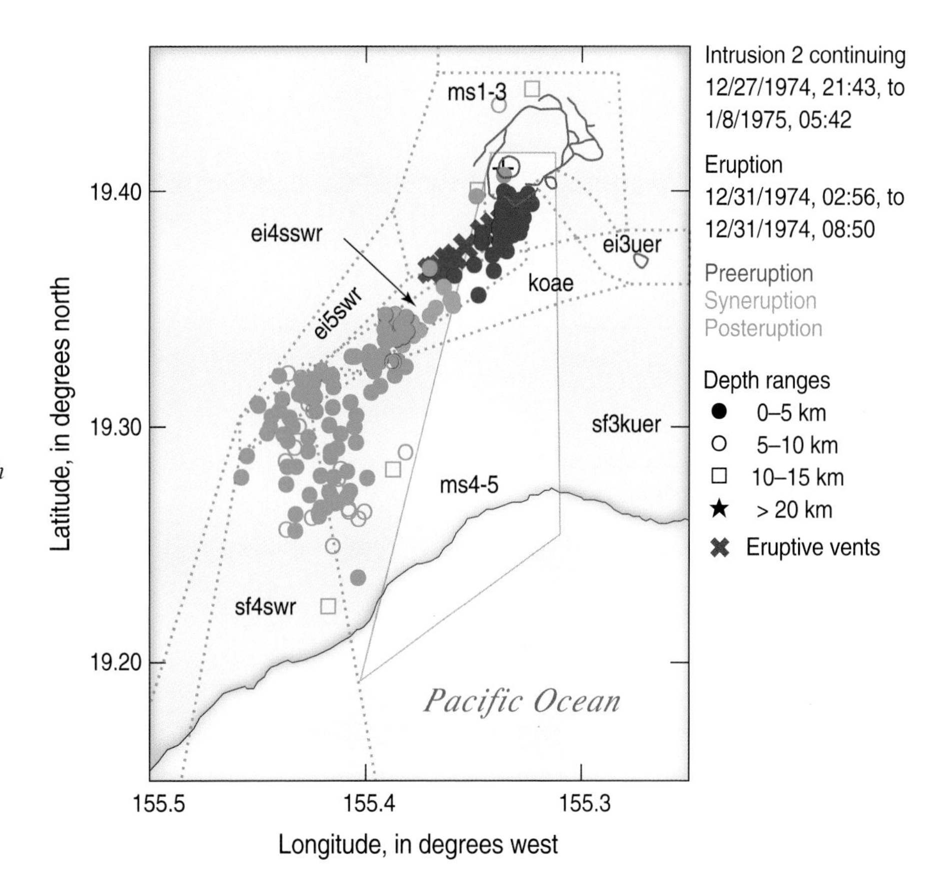

Figure 5.15. Map showing locations of earthquakes precursory to the 31 December 1974 southwest rift zone eruption and intrusion. This map covers the events of 31 December 1974 only, during which time there occurred a southwest rift zone eruption. The eruption is preceded by intrusion beneath the upper seismic southwest rift zone that then migrates downrift during and after the short eruption. A strong western south flank response accompanies the post-eruption intrusion. Note the complete absence of east rift and associated south flank activity. Dates on figure in m/d/yyyy format.

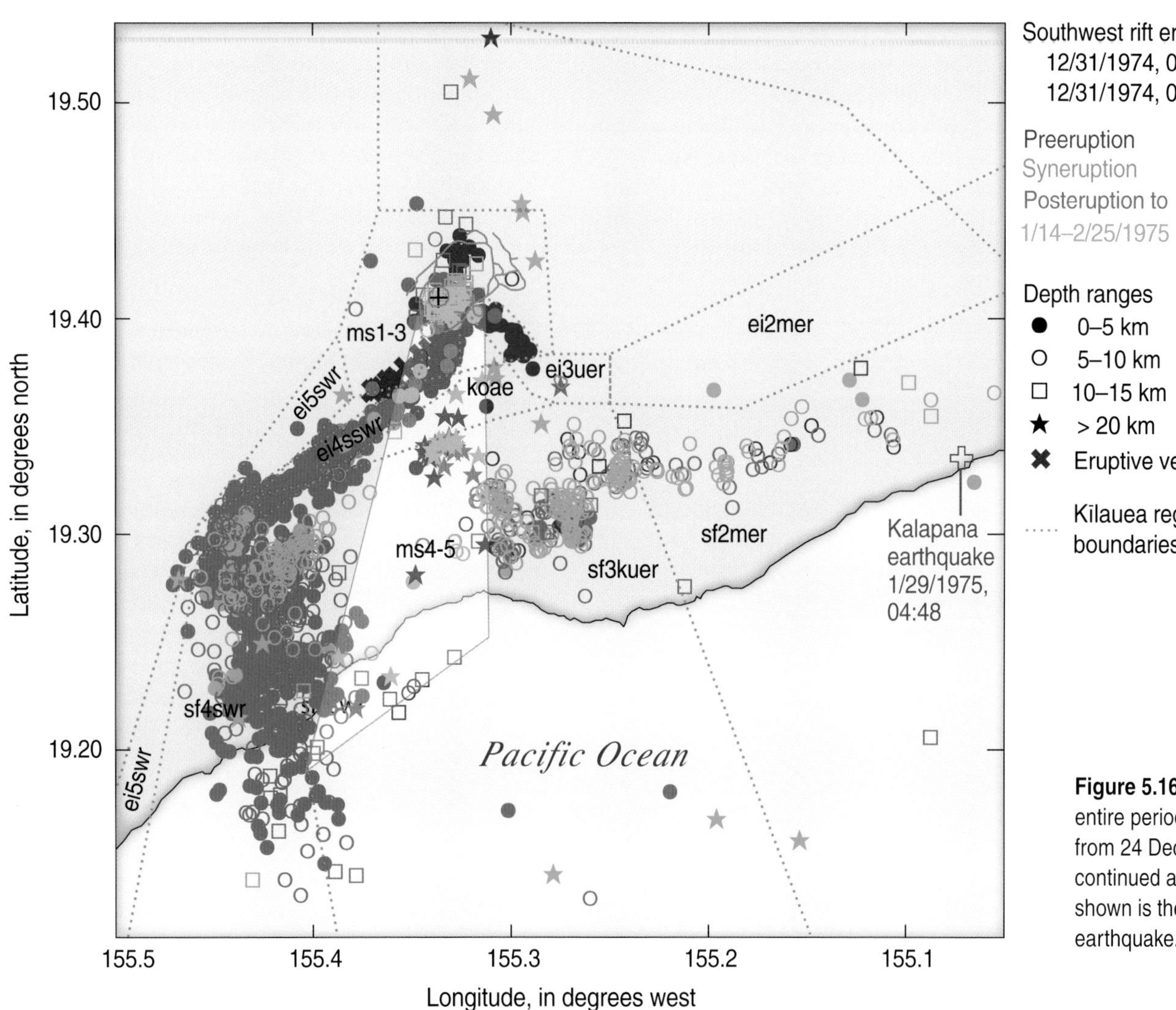

Figure 5.16. Map showing locations of earthquakes for the entire period surrounding the 31 December 1974 eruption. Data from 24 December 1974 through 25 February 1975. Intrusion continued at diminished intensity into mid-February 1975. Also shown is the epicenter of the 29 November 1975 Kalapana earthquake. Dates on figure in m/d/yyyy format.

its occurrence, using the following words (Swanson and others, 1976, p. 35), although the paper was published after the event: " . . . we anticipate a subsidence (strain-release) event of unknown magnitude in the not too far distant future."

A detailed analysis of the earthquake was published in 1981 with the following conclusion (Wyss and others, 1981, abstract): "We conclude that the Kalapana earthquake was preceded by a preparatory process which lasted 3.8 ± 0.3 years and which had dimensions of 45 × 10 km covering approximately the aftershock area."

The period of nearly 4 years mentioned extends from the beginning of stage II of the Mauna Ulu eruption. We conclude that the increase in magma supply, beginning even earlier, stressed the magma plumbing, which in turn stressed the south flank, culminating in the very large intrusion into the western part of Kīlauea's south flank in early 1975. We consider that this was the proximate event that readied the south flank for major failure. Following the intrusion, magma supply continued to increase, Kīlauea's summit was rapidly inflating, and seismicity within the south flank and along the deep magma-supply path remained high right up to the earthquake. A south flank earthquake of this magnitude may be one kind of major adjustment of an edifice within which the plumbing system cannot accommodate by any other means a relentlessly increasing magma supply.

In summary, this period represents the best demonstration of the complexities of magma transport beneath Kīlauea. Incoming magma is always seeking pathways in which its volume can be accommodated over and above space created by inflation of Kīlauea's summit during expansion of the 2–6-km-deep storage reservoir. Extra storage is divided between space found or created near sites of eruption and intrusion and deeper space created by rift dilation during seaward spreading of Kīlauea's south flank. One cannot forecast exactly how and where the accommodation will occur.

The south flank of Kīlauea acts as a useful monitor of the stresses involved in accommodating the pressure of incoming magma (compare Dieterich and others, 2000). Earthquake swarms beneath the south flank are initiated before many eruptions and traditional intrusions, and they continue during and after the volcanic event. Flank patches, including seismic zones nearly normal to the nearest rift zone, are activated during suspected deep intrusions as well as during some traditional intrusions, although most of the south flank seismicity associated with the latter events is distributed parallel to the rift zones.

Supplementary Material

Supplementary material for this chapter appears in appendix E, which is only available in the digital versions of this work—in the DVD that accompanies the printed volume and as a separate file accompanying this volume on the Web at http://pubs.usgs.gov/pp/1806/. Appendix E comprises the following:

Table E1 shows the calculation of magma supply and eruption efficiency during this period.

Figure E1*A*,*B* presents time series data covering the entire period

Figure E2*A*–*R* presents yearly data from 1967 through 1976 in plots similar to figures 5.1 and E1.

Figures E3–E8 present time series and earthquake locations for events during the pre-Mauna Ulu period listed in table 5.1 and figure 5.1.

Figures E9–E21 present time series and earthquake locations for events during period IA of the Mauna Ulu eruption listed in table 5.3 and figure 5.4.

Figures E22–E36 present time series and earthquake locations for events during period IB of the Mauna Ulu eruption listed in table 5.4 and figure 5.5.

Figures E37–E56 present time series and earthquake locations for events during the Mauna Ulu pause listed in table 5.5 and figure 5.7.

Figures E57–E70 present time series and earthquake locations for events during period II of the Mauna Ulu eruption listed in table 5.6 and figure 5.9.

Figures E71–E87 present time series and earthquake locations for events during the period between the end of Mauna Ulu II and the 1975 earthquake listed in table 5.7 and figure 5.11.

Geodimeter instrument setup. USGS photograph taken on 7 June 1971

Geodimeter mirror setup at Kīlauea. USGS photograph by R.T. Holcomb, 5 December 1972

Chapter 6

Eruptive and Intrusive Activity, 1975–1983

Recovery from the 1975 earthquake, in a series of intrusions on both sides of the volcano, culminates in the intrusion associated with episode 1 of the east-rift eruption that continues in the second decade of the 21st century. The spreading regime also changes from one driven by magma supply to one where spreading rates are decoupled from changes in magma supply rate.

A coconut grove at Halapē on Kīlauea's south coast, under water because of coastal subsidence following the 29 November 1975 earthquake. USGS photo by R.T. Holcomb, 4 December 1975.

Kīlauea eruptions and intrusions following the $M7.2$ earthquake on 29 November 1975 are summarized in table 6.1; seismic data from that earthquake through the end of 1979 are shown in figure 6.1, and earthquake locations related to the 1977 eruption are shown in figure 6.2. Seismic data covering the rest of the period, from 1 January 1980 to episode 1 of the Pu'u 'Ō'ō-Kupaianaha eruption, are shown in figure 6.3 These are supplemented by online figures that are cross-referenced in table 6.1. Additional seismic analysis for the entire period is included in Klein and others (1987), and figures from that paper are also cross-referenced in table 6.1. In the following paragraphs we briefly summarize the eruptive activity of this period and its interpretation.

The Kalapana Earthquake of 29 November 1975

The second largest earthquake in recorded Hawaiian history ruptured Kīlauea's south flank on 29 November 1975 (Klein and others, 2001). This earthquake had a surface-wave magnitude (M_S) of 7.2, and it has recently been assigned a moment magnitude (M_W) of 7.7. Motion in this Kalapana earthquake caused as much as 3.5 m subsidence at the south coast, as much as 1.5 m slip on the normal Hilina Pali faults, as much as 8 m of lateral motion of the coast, a tsunami with maximum run-up of 15 m along the south coast (causing two deaths), maximum shaking intensities (Modified Mercalli) of VIII in Puna and Hilo, including building damage, aftershocks that extended at least into late 1976, and a small eruption in Kīlauea Caldera (Tilling, 1976).

The 1975 Kalapana earthquake is unlike the more common strike-slip, thrust, and normal faulting seen in continental and plate-boundary environments. Fault slip in the Kalapana earthquake was at about 9-km depth on the decollement surface at the oceanic sediment layer between the oceanic crust and the overlying volcanic rocks, with a slight 5–10° northward dip of the fault plane (Ando, 1979). Lateral, compressive stress applied for decades by the east rift zone on the south flank enabled thrust motion on the decollement (Swanson and others, 1976a). Gravity aided slip on the Hilina Pali normal faults, subsidence at the coast, and slumping along the submarine toe of the Hilina slump block (Denlinger and Okubo, 1995; Morgan and others, 2000). The aftershocks defining the rupture extend for 40 km along most of the south flank at depths of 8–9 km and show focal mechanisms with slip on near-horizontal fault planes with slip vectors directed south-southeast and the upper block moving seaward (Crosson and Endo, 1982). An isoseismal map of shaking intensity also shows slip along the south flank, with areas of high intensity extending northward from the rupture zone to Hilo (Wyss and Koyanagi, 1992). Rupture during the mainshock was relatively slow and jerky, with secondary slip probably corresponding to high-stress fault asperities or stiff patches (Harvey and Wyss, 1986). Horizontal slip of benchmarks at the south coast, relative to the stable north flank of Kīlauea, was as much as 8 m south-southeast (Lipman and others, 1985). The surface deformation, teleseismic Love wave records, and tsunami recordings can be modeled by seaward slip on a subhorizontal fault of length 40 km and width of 20–40 km, extending offshore to give upward motion of the seafloor and an initial compression motion to the tsunami (Ando, 1979; Owen and Bürgmann, 2006). The interpretation of the tsunami mechanism is improved by modeling a rotating toe and slumping of the submarine extension of the south flank block, with subsidence at the coast and uplift 15–20 km offshore (Ma and others, 1999). Modeling of benchmark displacement during the earthquake involves both dilation of the east rift and subhorizontal slip in the buried decollement surface, similar to that modeled by Dvorak and others (1994) for 1976–88 and by Owen and Bürgmann (2006) for 1990–93.

The 29 November 1975 earthquake was anticipated because of the understanding and thoughtful analysis of the repeated dike injections in the east rift zone and geodetically measured compression of the south flank before the earthquake occurred (Swanson and others, 1976a). Several earthquake precursors were studied after the earthquake occurred, including seismic quiescence, foreshocks, earthquake clustering, lower P-wave velocities seen by teleseismic P residuals, and reversal of the compression of a geodetic line crossing a hypothesized asperity (Wyss and others, 1981), though the latter was not as strong an effect as first believed (Delaney and others, 1992).

Although magma pressure applied to the south flank over the preceding decades had set up conditions for the earthquake to occur, magma movement at the time of the earthquake was an effect of the earthquake, rather than a cause of it. The summit collapsed 1.2 m (Lipman and others, 1985), and about 0.1 km^3 of magma flowed into the east rift zone (Owen and Bürgmann, 2006). No east rift eruption took place, and most of the magma thus filled voids created by the earthquake. No direct measurements of the amount of rifting exist, but benchmarks separated by 3–6 km spanning the rift separated by 0.5 to 0.7 m (Lipman and others, 1985), and a geodetic inversion indicates that as much as 3–5 m of rifting took place (Owen and Bürgmann, 2006). There was a minor Kīlauea Caldera eruption that erupted about 250,000 m^3 of lava starting about 1 hr after the earthquake (Tilling, 1976). We interpret this summit eruption as magma moving upward from

Table 6.1. Earthquake swarms 11/29/1975 to 2/1/1983 (see figs. 6.1 and 6.3).

[In rows with multiple entries text applies down to the next entry; dates in m/d/yyyy format; do, ditto (same as above); data for eruptions and traditional intrusions are emphasized by gray bands]

Date and time		Class[1]	Event Type[2]	No.[3]	Tilt[4]		Lag[5]	Comment[6]	Fig.[7]
Start	End				Mag	Az			
11/29/1975 04:48		sf2mer	EQ					*M*7.2 with aftershocks	F4
11/29/1975 05:32	11/29/1975 22:00	Hm	E[8]					Halema‘uma‘u; induced by earthquake	
6/21/1976 09:15	6/22/1976 08:33	ei3uer	I	66	7.6	120	-4h45m	Tilt 6/20–22; uprift migration 1.3 km/hr	F5a; 43.68
7/14/1976 09:36	7/15/1976 08:54	ei3uer	I	49	6.6	108	-4h24m	Tilt 7/14–15	F5b; 43.69
1/22/1977 05:28	1/23/1977 02:39	ei3uer	II	17	5.3	299	-4h32m	Tilt 1/20–24; downrift migration 0.7 km/hr	F6; 43.70
2/7/1977 20:28	2/9/1977 00:44	ei3uer	I	65	6.6	108	-22h33m	Tilt 2/7–11; downrift migration 0.37 km/hr	F7; 43.71
2/8/1977 19:04	2/8/1977 22:43	koae	I	6				Koa‘e events aligned along east rift boundary	
9/12/1977 19:01	9/20/1977 05:25	sf2mer	EQS	302				South flank anticipation/response	
9/12/1977 21:33	9/17/1977 05:04	ei2mer	I	84	43.7	107	-0h27m	Tilt 9/11-13; downrift migration 1.23 Ð0.23 km/hr	43.72
9/13/1977 00:38	9/16/1977 09:41	lpms1	EQS	277	26.3	123		Tilt 9/13-14; magma recharge?	**6.2**
9/13/1977 19:13	9/17/1977 20:45	MERZ	E[9]		30.8	130		Stage 1; Tilt 9/14-16	**6.2**
9/14/1977 00:20	9/14/1977 10:08	koae	EQS	8					
9/18/1977 10:15	9/20/1977 12:00	MERZ	E[9]		12.9	163		Stage 2; Tilt 9/16–22	**6.2**; 43.73
9/25/1977 23:50	10/1/1977 15:30	MERZ	E		14.4	145		Stage 3; Tilt 9/26–10/9	
9/27/1977 10:35	10/8/1977 06:45	lpms1	EQS	130	14.4	145	-7h25m	Tilt 9/26–10/9; lp response; earthquake swarms of 47, 14, 23, 23 and 23 events; downrift migration 0.01 km/hr	**6.2 inset**; F8
5/29/1979 16:21	5/29/1979 21:06	sf3kue	EQS	22				South flank anticipation/response	F9; 43.75
5/29/1979 17:15	5/30/1979 04:36	ei3uer	I	33	4.3	133	+1h15m	Tilt 5/28–30; uprift migration 0.77 km/hr; downrift migration 0.65 km/hr	
5/29/1979 18:59	5/30/1979 00:34	ei2mer	I	27					
8/12/1979 06:02	8/12/1979 22:18	ei3uer	I	54	3.5	100	-3h58m	Tilt 8/11–13; downrift migration 1.6 km/hr	F10; 43.76
9/21/1979 21:59			EQ					*M*5.7; 3 foreshocks? from 17:16; >100 aftershocks	F11; 43.77
9/22/1979 13:33	9/22/1979 16:01	lpms1	II	22			+15h33m	I–A: 3.9 µrad inflation over 6 h	
10/11/1979 13:04	10/12/1979 19:03	ei3uer	II	21	nd		na	I–A: gradual inflation	F12
11/15/1979 00:06	11/15/1979 05:32	ei3uer	I	11					F13
11/15/1979 06:10	11/16/1979 21:51	sf3kuer		25				South flank anticipation/response	
11/15/1979 20:27	11/16/1979 21:51	ei3uer	I	138	8.2	101	-1h33m	Tilt 11/15–16; complex uprift migration/downrift migration	43.78
11/15/1979 22:21	11/17/1979 01:38	koae	I	33				I–A: deflation to 11/17, 03:00	F13
11/16/1979 08:18	11/17/1979 07:30	UERZ	E[10]					Pauahi Crater	F13
11/19/1979 23:30	11/21/1979 14:06	ei3uer	I	16					F13
1/16/1980 15:35	1/17/1980 04:06	ei3uer	II	11	nd		-6h25m	I–A: small deflation	F14
2/2/1980 00:57	2/2/1980 15:41	ei3uer	II	11	nd			Gradual inflation from 2/1, 22:00	F15
3/1/1980 00:31	3/1/1980 23:58	sf2mer	EQS	9				South flank anticipation	
3/2/1980 07:51	3/2/1980 20:55	ei3uer	I	114	2.7	122	+0h 51m	Tilt 3/1-3	F16; 43.79
3/2/1980 08:51	3/3/1980 05:50	sf3kuer	EQS	9				South flank accompaniment/response	
3/2/1980 14:12	3/3/1980 07:27	sf2mer	EQS	10				South flank accompaniment/response	

Table 6.1. Earthquake swarms 11/29/1975 to 2/1/1983 (see figs. 6.1 and 6.3).—Continued

[In rows with multiple entries text applies down to the next entry; dates in m/d/yyyy format; do, ditto (same as above); data for eruptions and traditional intrusions are emphasized by gray bands]

Date and time		Class[1]	Event Type[2]	No.[3]	Tilt[4]		Lag[5]	Comment[6]	Fig.[7]
Start	End				Mag	Az			
3/10/1980 07:48	3/12/1980 23:05	sf3kuer	EQS	88	13.0	112	-3h 39m	South flank anticipation/response, tilt 3/10–11; uprift migration 0.25 km/hr	F17; 43:80
3/10/1980 18:21	3/11/1980 22:38	er2mer	I	43					
3/10/1980 21:55	3/12/1980 03:46	er3uer	I	130					
3/11/1980	3/12/1980	UERZ	E[11]					Near Devil's Throat	
3/11/1980 17:41	3/12/1980 07:58	koae		8					
7/30/1980 07:24	7/30/1980 08:44	ei3uer	I	36	1.6	158		Tilt 7/29–30; on I-A: gradual inflation	F18; 43.81
8/27/1980 14:30	8/28/1980 09:49	ei3uer	I	172	9.7	113	-0h 30m	Tilt 8/26–27; I-A: 8.6 µrad deflation; uprift and downrift migration	F19; 43.82
8/27/1980 15:01	8/28/1980 03:31	ms1	EQS	24					
10/21/1980 19:01	10/21/1980 22:04	ei2mer	I	12	9.0	135	-0h 04m	Tilt 10/20–22	F20; 43.83
10/22/1980 18:41	10/22/1980 21:36	ei3uer	I	34					
11/2/1980 14:11	11/3/1980 01:32	ei3uer	I	142	8.5	123	-0h 19m	Tilt 11/2–3; uprift migration and downrift migration	F21; 43.84
1/20/1981 03:10	1/21/1981 06:18	ei4sswr	I	94	2.7	147	32h 23m	Tilt 1/20–21; uprift migration 0.04 km/hr I-A: 1/18–20 slow deflation;	F22A;
1/24/1981 20:23	1/25/1981 18:45	ei4sswr	I	24	flat			I–A: slight inflation	43.86
1/26/1981 02:31	1/28/1981 01:05	ei4sswr	I	29	flat		22h	I–A: deflation from 1/25–1/26	F22B; 43.87
2/10/1981 13:19	2/13/1981 21:03	sf4swr	EQS	60				South flank anticipation/accompaniment	F23
2/13/1981 07:59	2/14/1981 06:24	ei4sswr	I	14	2.9	127		Tilt 2/9-10	F23; 43.88
2/14/1981 09:55	2/17/1981 01:17	sf4swr	EQS	40				South flank accompaniment/response	
2/14/1981 14:40	2/16/1981 02:20	ei4sswr	I	17	13.4	104		Tilt 2/12–18; I-A: ~15 µrad deflation 2/8–17	
2/16/1981 09:18	2/16/1981 22:59	ei4sswr	I	10					
3/19/1981 00:18	3/20/1981 4:14	sf2mer	SDI?	8				Scattered earthquakes	
3/19/1981 1:21	3/20/1981 3:58	sf3kuer	SDI?	5				Scattered earthquakes	
3/19/1981 4:32	3/20/1981 5:28	sf4swr	SDI	15				Concentrated earthquake swarm	
4/25/1981 01:02	4/27/1981 11:29	sf3kuer	SDI?	21	flat			I-A: gradual inflation	F24
6/25/1981 15:17	6/25/1981 17:15	ms1	I	5	13.8	301		Tilt 6/23-29; summit intrusion	F25
6/26/1981 18:48	6/27/1981 02:55	ms1	I	5			+1h 17m	I-A: sharp inflation on 6/25, gradual inflation on 6/26	
7/20/1981 19:56	7/21/1981 13:13	ei4sswr	II	10	flat		+8h 56m	I-A: ~1 µrad deflation	F26
8/1/1981 16:33	8/4/1981 05:26	ei4sswr	I	40			+2h 33m	I-A : deflation 8/1-8/4	F27
8/4/1981 19:39	8/7/1981 03:01	ei4sswr	I	22	7.0	289		Tilt 8/1-7	
8/9/1981 17:20	8/15/1981 04:53	ei4sswr	I	277	28.9	86	-11h 10m	Tilt 8/10; uprift migration 3.5 km/hr	F28; 43.89
8/10/1981 07:23	8/13/1981 08:32	sf3kuer	EQS	50	22.2	118		Tilt 8/10 07:40-09:30; south flank response	
8/10/1981 17:36	8/13/1981 11:36	sf4swr	EQS	76	46.6	128		Tilt 8/10; downrift migration 2.6, 0.28 km/hr	F28; 43.90
8/13/1981 23:30	8/14/1981 20:58	lpms1	I	13	20.3	157		Tilt 8/11–14; summit intrusion?	
11/10/1981 03:03		sf3kuer	EQ					*M*4.5 ms with >10 aftershocks	

Table 6.1. Earthquake swarms 11/29/1975 to 2/1/1983 (see figs. 6.1 and 6.3).—Continued

[In rows with multiple entries text applies down to the next entry; dates in m/d/yyyy format; do, ditto (same as above); Data for eruptions and traditional intrusions are emphasized by gray bands]

Date and time		Class[1]	Event Type[2]	No.[3]	Tilt[4]		Lag[5]	Comment[6]	Fig.[7]
Start	End				Mag	Az			
1/15/1982 18:44	1/16/1982 16:16	ei4sswr	II	10	flat		na	I-A : < 1 µrad inflation	F29[12]
2/27/1982 22:44	2/28/1982 07:56	ei4sswr	II	11	flat		-1h	I–A: ~ 1 µrad inflation	F30
3/3/1982 07:35	3/3/1982 17:29	ei4sswr	II	11	flat		+30m	I–A: ~ 1 µrad inflation	F31
3/9/1982 01:24	3/9/1982 15:08	ei4sswr	II	11	flat		+6h 24m	I–A: ~ 1 µrad r inflation	F32
3/23/1982 09:05	3/23/1982 16:10	ei4sswr	II	22	8.3	314		Tilt 3/22–25; I-A: ~ 2 µrad gradual inflation following earthquake swarm	F33[12]; 43.91
4/30/1982 08:55	4/30/1982 11:41	ei3uer	I	10	10.1	323		Tilt 4/30-5/2; uprift migration 14 km/hr	F34; 43.92
4/30/1982 08:55	4/30/1982 12:46	ms1	I	30			+17m	I-A: 5.6 µrad inf	
4/30/1982 11:37	5/1/1982 06:30	KC	E[13]/I					Kīlauea Caldera	
6/8/1982 15:30	6/9/1982 07:45	ei4sswr	II	13	8.6	340		Tilt 6/7-12; I-A shows <1 µrad inf/def	F35
6/22/1982 02:57	6/27/1982 15:38	sf3kuer	EQS	447	21.0	120		Tilt 6/21-23; South flank anticipation/response	F36
6/22/1982 07:04	6/24/1982 19:25	ei4sswr	I	131	32.8	141	+27h	Tilt 6/23-24; I-A: 39 µrad gradual deflation 6/21–25; downrift migration 0.37 km/hr	43.93
6/23/1982 17:55	6/24/1982 20:21	sf4swr	EQS	31	9.6	134		Tilt 6/24-30; continuing south flank response	
6/28/1982 17:04	7/1/1982 13:35	sf3kuer	EQS	37				Continued south flank response?	
8/10/1982 1:21	8/10/1982 1:37	sf3kuer	EQ					Double mainshock *M*3.8, 3.6 with aftershocks	
9/19/1982 2:21	9/19/1982 14:20	sf3kuer	EQS	11				South flank ant?	
9/25/1982 16:51	9/25/1982 20:24	ei4sswr	I	54	23.3	330	+0h 26m	Tilt 9/23-25 17:40; I-A: ~15 µrad sharp inflation	F37
9/25/1982 16:52	9/26/1982 0:12	ms1	I	30				Tilt 9/25; downrift migration 1.4 km/hr	43.94
9/25/1982 17:19	9/27/1982 13:51	ei3uer	I	180	10.1	320			
9/25/1982 18:45	9/26/1982 08:30	KC	E/I[13]					Kīlauea Caldera	
10/1/1982 14:46	10/3/1982 08:34	ei3uer	II	24	6.0	335		Tilt 9/2710/3; a series of small inflationary intrusions; I-A shows cyclic inflation-deflation of 0.21.4 µrad that does not matching intrusion times	F38
10/5/1982 01:50	10/6/1982 13:12	ei3uer	II	24	flat				
10/7/1982 20:15	10/8/1982 21:27	ei3uer	II	20	flat				
10/10/1982 10:54	10/11/1982 10:36	ei3uer	II	27	flat				
12/9/1982 17:46	12/11/1982 10:54	ei3uer	I	185	4.4	148	+0h 27m	Tilt 12/9 08:4521:00; IA: 3 µrad sharp deflation; downrift migration 6.4 km/hr	**6.4***A*; 43.96
12/9/1982 17:49	12/10/1982 01:13	ms1	EQS	10					
12/21/1982 02:15	12/21/1982 14:44	Sf2.3	EQS	9		flat		South flank ant?	
12/21/1982 14:06	12/21/1982 23:42	ei3uer	II	10				I-A; no change	**6.4***A*
12/30/1982 01:36	1/3/1983 00:31				flat			Precursory sequence to ep. 1	**6.4***A-B*
12/30/1982 01:36	12/30/1982 20:25	ei2mer	II	10	flat		-2h 12m	I-A: ~1.2 µrad gradual deflation through both intrusions	**6.4***A*
12/29/1982 20:08	12/31/1982 04:56	ei3uer	II	10					43.97
12/30/1982 17:21	12/31/1982 23:37	sf3kuer	EQS	12				South flank anticipation/response	
1/1/1983 04:52	1/6/1983 11:32	sf2mer	EQS	95					

Table 6.1. Earthquake swarms 11/29/1975 to 2/1/1983 (see figs. 6.1 and 6.3).—Continued

[In rows with multiple entries text applies down to the next entry; dates in m/d/yyyy format; do, ditto (same as above); Data for eruptions and traditional intrusions are emphasized by gray bands]

Date and time		Class[1]	Event Type[2]	No.[3]	Tilt[4]		Lag[5]	Comment[6]	Fig.[7]
Start	End				Mag	Az			
1/1/1983 22:36	1/6/1983 13:59	ei2mer	I	324	45.6	119	-5h 24m	Tilt 1/2–1/3; downrift migration 0.6 km/hr followed by uprift migration 0.06 km/hr	**6.5**
1/2/1983 15:32	1/3/1983 09:10	ms1	EQS	11	86.3	134		Tilt 1/3 13:50-1/8 07:15	
1/3/1983 00:31	1/3/1983 15:21	MERZ	E/I					Episode 1 Pu'u 'Ō'ō eruption	
1/4/1983 11:01	1/5/1983 11:45	sf3kuer	EQS	19				South flank response	
1/5/1983 11:44	1/5/1983 23:17	ms1	EQS	14					
1/5/1983 11:23	1/6/1983 09:55	MERZ	E				+8h 44m	Reactivate episode 1	**6.5**; 43:97
1/6/1983 10:11	1/6/1983 20:49	MERZ	E					do	
1/6/1983 18:19	1/6/1983 20:07	sf2mer	EQS	58				South flank response	
1/7/1983 00:07	1/7/1983 21:33	ei2mer	I	81				Downrift migration 0.61 km/hr	**6.5**; 43.97
1/7/1983 01:19	1/8/1983 01:05	ms1	EQS	18					
1/7/1983 09:57 1/8/1983 19:57	1/8/1983 15:04 1/8/1993 23:22	MERZ MERZ	E E					Reactivate episode 1; I–A: flat do	
1/10/1983 05:02 1/11/1983 01:30 1/15/1983 03:12	1/10/1983 14:50 1/11/1983 12:30 1/15/1983 18:35	MERZ MERZ MERZ	E E E		8.8	161		Do; Tilt 1/9-14; I–A: flat do do	**6.5**
1/18/1983 12:38	1/19/1983 23:32	sf2mer	EQS	18				South flank response	
1/23/1983 10:11	1/23/1983 23:04	sf2mer	EQS	14				South flank anticipation/response	**6.5**
1/23/1983 18:30	1/23/1983 19:30	MERZ	E					Episode 1 end; I–A: flat	

[1]Earthquake classification abbreviations are given according to the classification in appendix table A3, and locations are shown on appendix figure A4.

[2]Event types defined in chapter 1 are abbreviated as follows: Eruption (E); intrusion ("traditional" I; "inflationary" II; "suspected deep" SDI); earthquake swarms EQS; Earthquake ≥ $M4$ (EQ).

[3]Minimum number of events defining a swarm: 20 for south flank; 10 for all other regions.

[4]Magnitude in microradians and azimuth of daily tilt measurements from the water-tube tiltmeter in Uwēkahuna Vault; nd, record not available.

[5]Lag times separating the onset of the earliest earthquake swarm (excluding south flank) for a given event and the beginning of deflation or inflation measured by the continuously recording Ideal-Arrowsmith tiltmeter in Uwēkahuna Vault. (+) tilt leads, (-) tilt lags.

[6]Abbreviations as follows: eq, earthquake; eqs, earthquake swarm; fs, foreshock; as, aftershock; ms, mainshock; sf, south flank; inf , inflation; def, deflation; ant, anticipation (preceding event); acc, accompaniment (during event); resp, response (following event); during an earthquake swarm: drm, downrift migration of epicenters; urm, uprift migration of epicenters; I–A, Ideal-Arrowsmith continuously recording tiltmeter in Uwēkahuna Vault.

[7]Text figures **bold text**; appendix figures plain text; 43.xx = figures in Klein and others, 1987.

[8]See Tilling (1976) for a description of the eruption.

[9]See Moore and others (1980) for a description of the eruption.

[10]Eruption information in unpublished HVO monthly reports.

[11]Eruption discovered in March 1982 as documented in unpublished HVO monthly reports.

[12]See also figure 43.85 in Klein and others (1987).

[13]Eruption information in unpublished HVO monthly reports and in Nancy Baker's unpublished Honor's thesis at the University of Hawai'i at Manoa (Baker, 1987).

the reservoir a short distance through cracks shaken open by the earthquake and the movement of magma into the rift as a passive magma flow resulting from breaking of a magma barrier separating the reservoir and the rift zone.

Effects of the Kalapana Earthquake

The Kalapana earthquake profoundly altered stress on the south flank and "softened" it to the stress imposed by magma traversing the east rift. Aftershocks initially followed a normal t^{-p} time decay with p = 0.82 for the first 100 days (Klein and others, 2006), but did not return to pre-1975 levels until stage II of the Pu'u 'Ō'ō eruption (see chap. 7; Dvorak, 1994; Klein and Wright, 2008). The high rate of aftershocks in the 17 years following 1975 means the stress released by the mainshock was large, but the relatively high rates of seismicity in the south flank before the earthquake also means that the rate of stress accumulation on the flank actually decreased at the time of the earthquake (Klein and others, 2006). This means that rift eruptions and intrusions stressed the flank more before the Kalapana earthquake than they did after it. Thus the flank was stiffer and offered more resistance before November 1975, but became softer and offered less resistance afterward.

When magma intrudes into the rift zone, it may stop before it erupts if the rift can accommodate all of the magma. But if conditions are right, the intrusion can turn into an eruption, in which case the magma pressure is partly relieved. After the 1975 earthquake, intrusions outnumbered eruptions (Klein, 1982), and movement of magma in the rift was often at a slower rate (Decker, 1987). This geologic change is another effect of the flank becoming more pliable and able to absorb magmatic stress without offering resistance to a dike, making magma less likely to be forced up to the surface. The "softening" of the flank as a result of the earthquake can also be seen in the change of the rate of extension or compression measured in the south flank: The line lengths, measured within the flank before the earthquake, showed extension of about 1 cm/yr during 1970–75 (Swanson and others, 1976a), but this changed to compression of about 6 cm/yr during 1976–80 (Delaney and others, 1998; Dvorak and others, 1994). Thus, after the flank suddenly expanded during the earthquake, intrusions could more easily squeeze the "soft" flank and compress it.

Following the earthquake, Kīlauea has been unable to sustain the high levels of magma in its summit reservoir as it did in 1971–75 (Lipman and others, 1985). Kīlauea's summit inflated almost 2 m between 1965 and 1975, but it deflated by almost 2.5 m between 1976 and 1997 (Delaney and others, 1998, figure 2a). Except for the collapse associated with the 1977 eruption, Kīlauea neither inflated nor deflated from 1976 to 1983, but it steadily deflated at 7.8 cm/yr after the Pu'u 'Ō'ō-Kupaianaha eruption began (see chap. 7).

Intrusions in the 1½ Years Following the 1975 Earthquake

A series of four east rift intrusions in the 18 months after the Kalapana earthquake followed typical patterns for Kīlauea intrusions (table 6.1; fig. 6.1). The east rift intrusions of 21 June and 14 July 1976 were similar in that the intense earthquake swarms began near Pauahi Crater and earthquakes then migrated about 5 km uprift. Earthquakes also migrated upward from the seismically defined conduit at a depth of 3 km, but the dike did not reach the surface. Minimum intruded magma volumes were 3.4×10^6 m^3 and 3.0×10^6 m^3. In both intrusions the summit reservoir began deflating about 4 hours after an intense earthquake swarm began beneath the rift zone.

We interpret the 1976 intruded magma, and hence the new or rejuvenated dike, as having originated at the rift zone magma reservoir under Pauahi Crater (fig. 6.5) that has also fed many other uprift and downrift migrating east rift intrusions (Klein and others, 1987). Intruded magma was ultimately resupplied from the summit reservoir. Dzurisin and others (1980) used Kīlauea's approximate magma supply rate of 9×10^6 m^3/month and the 40–90×10^6 m^3 void space created during the 1975 earthquake to infer that rift voids were filled in the 7 months before the June 1976 intrusion. We believe it took much longer than 7 months to fill all the 1975 void space because there was no summit inflation prior to the 1977 east rift eruption and no net inflation from the bottom of the 1975 collapse until after 1981 (figs. 6.1, 6.3).

Aftershocks of the Kalapana earthquake continued during 1976, but there were some small changes in south flank seismicity resulting from the intrusions that year. There was minor triggered seismicity in the flank immediately after the 21 June 1976 intrusion, but the amount is typical of day-to-day fluctuations and may not be related to the intrusion. A brief but noticeable spike in south flank activity occurred on the day of the 14 July 1976 intrusion, indicating that the intrusion incrementally added to south flank stress. The level of south flank seismicity was noticeably reduced during the 2–3 weeks after the 21 June 1976 intrusion (fig. 6.1), raising the possibility that redistribution of magma within the rift during the intrusion may have lessened stress on the flank.

Two intrusions in early 1977 are related to each other. Earthquakes in the primary east rift zone conduit during the 22 January 1977 inflationary intrusion (table 6.1) extended downrift to the vicinity of Kōko'olau Crater, and depths were confined between

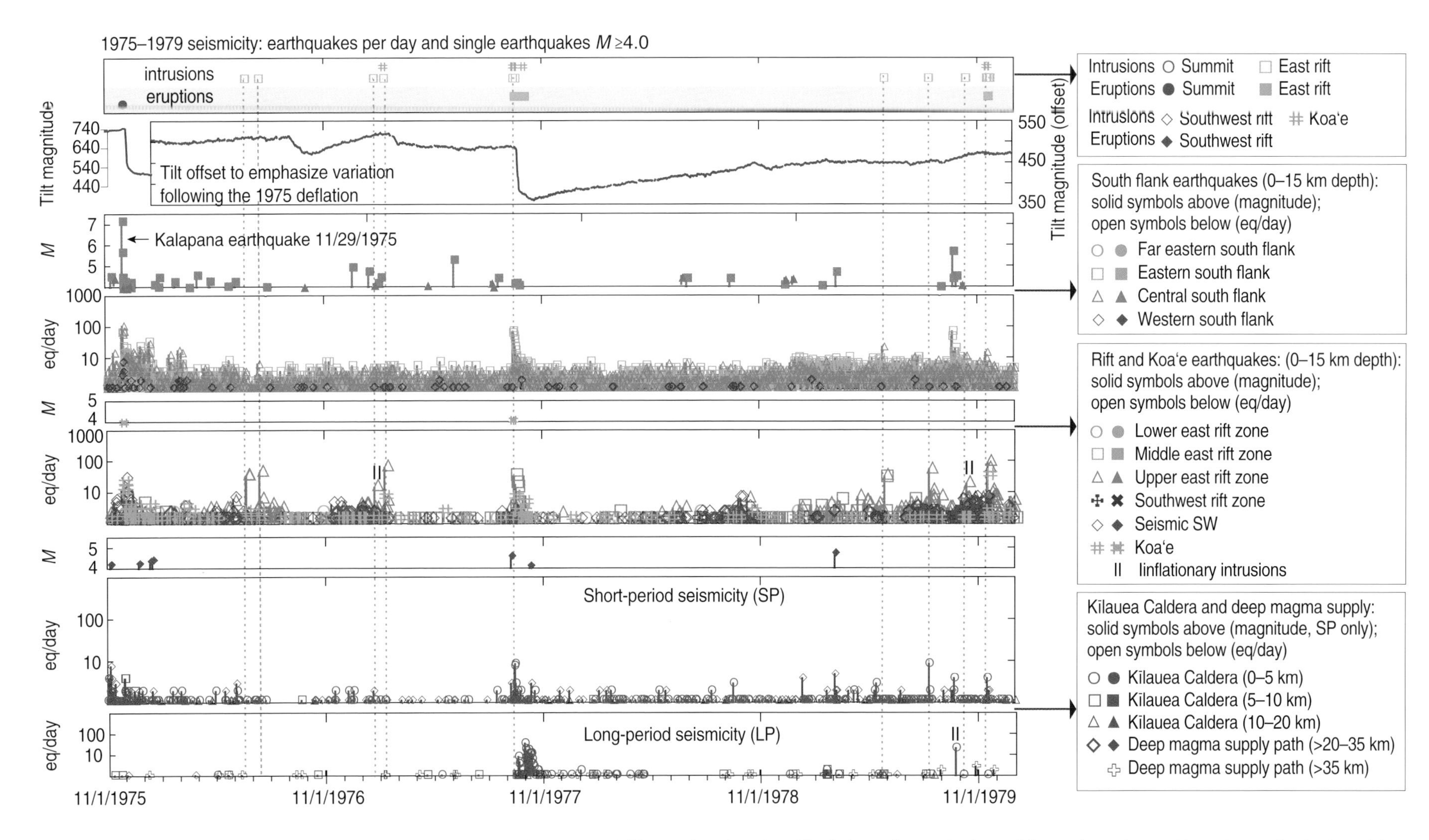

Figure 6.1. Graphs showing Kīlauea activity, 1 November 1975–1 January 1980. Time series plots show Uwēkahuna tilt magnitude, and times of eruption and intrusion, and seismic activity. Symbols are given on the plots. Top panel: Times of eruption and traditional intrusion. Second panel from top: Uwēkahuna tilt magnitude related to times of eruption and intrusion emphasized by vertical dotted lines. Tilt magnitudes are given in microradians. Bottom seven panels: Seismicity is plotted, from bottom to top, for the magma supply path, rift zones and Koa‘e, and south flank. Earthquakes per day (eq/day) and magnitudes (*M*) greater than or equal to 4.0 are given for each region. Three types of intrusions are defined in chapter 1 and the following information applies to all time series figures in this chapter. Traditional intrusions are shown as intrusion symbols. Inflationary intrusions are labeled (II), and suspected deep intrusions are labeled (SDI) but not plotted as intrusions. Dates on figure in m/d/yyyy format.

2 and 4 km. The 8 February 1977 intrusion (table 6.1) began just uprift of Kōkoʻolau Crater and extended the newly intruded magma of 22 January downrift to Pauahi. Rift earthquakes associated with the February intrusion occurred between 4-km depth and the surface, a much larger cross sectional area than during the January intrusion. Both intrusions showed downrift earthquake migration (Klein and others, 1987, figure 43.70D and 43.71D). Neither of the early 1977 intrusions caused a south flank earthquake response.

We interpret the initial January 1977 inflationary intrusion as low volume and low pressure because the summit actually inflated while an earthquake swarm migrated away from the summit. In this and similar instances magma can leak into the rift zone at rates equal to or less than the rate of resupply. Earthquakes were confined to the main conduit near 3-km depth without expanding to form a larger dike. Earthquakes and magma stopped near Kōkoʻolau Crater, perhaps at a barrier within the rift or because of insufficient magma pressure. An opening in the reservoir complex on 8 February 1977 permitted a larger and more pressurized volume of magma to extend the intrusion downrift to the next rift barrier below Pauahi (fig. 6.5).

East Rift Eruption of 13 September–1 October 1977

What was then the largest Kīlauea eruption since the 1969–74 Mauna Ulu eruption began on 13 September 1977 (table 6.1, figs. 6.1, 6.2). An eruption had not occurred this far down the middle east rift since September 1961. The erupted lava had a fractionated chemical composition indicating that it had been stored in the rift (Moore and others, 1980). In chapter 4 (table 4.2) we suggested that the erupted magma had been emplaced in 1955 and cooled and fractionated in the intervening 22 years.

The eruption in September 1977 was accompanied by a large summit deflation that continued for 4 days, indicating that an uprift intrusion from the summit had forced the stored magma to the surface. Earthquakes produced by the intrusion immediately before the September 1977 eruption occurred in four main zones (fig. 6.2). Three zones, the southern Kīlauea Caldera, the Koaʻe Fault Zone, and the Makaopuhi section of the middle east rift zone, started producing earthquakes at essentially the same time late on 12 September 27 minutes before the summit began rapid deflation (table 6.1). The simultaneity of seismicity in these separate zones and the lack of earthquakes along the rift between them suggest that the part of the rift that is uprift of the eventual intrusion and vent area was open, fluid, and saw magma pressure and flow transmitted easily (Klein and others, 1987, figure 43.72). From Makaopuhi Crater to the eruption site, earthquake hypocenters migrated downrift at about 1.2 km/hr along the main conduit at 2–3-km depth. Earthquakes then slowed to 0.23 km/hr downrift, as the earthquakes traced the upward growth of the dike to the surface vents.

Earthquakes were triggered immediately in the south flank adjacent to the east rift zone, especially in the sections that saw no shallow rift earthquakes (fig. 6.2). This demonstrates that the growth of the dike in this section was by widening of an already fluid-filled dike: had there been voids in the rift or a destressed flank, magma filling would have occurred for some time before the flank accumulated enough stress to make earthquakes. The large number of triggered flank earthquakes implies that there was a great increase in stress rate in the south flank adjacent to the vents, suggesting that the stress applied by the November 1975 Kalapana earthquake was not fully released in this area[23]. The rate of earthquakes beneath the central and eastern flank during the 4 months following the September 1977 eruption was about twice that before and after the September–December period (fig. 6.1). The flank seismicity suggests that flank spreading was temporarily accelerated by the intrusion of new summit magma, although there were no geodetic measurements to confirm this.

Unlike other recent east rift zone eruptions, the September 1977 eruption was accompanied by a large earthquake swarm in and under Kīlauea Caldera. Many of these were long-period earthquakes but with seismogram onsets impulsive enough to time and locate the earthquakes in the upper 3 km of brittle rock above the summit magma reservoir (fig. 6.1). The swarms of long-period earthquakes continued through the deflation (fig. 6.2 inset) and, in decreasing numbers, beyond the end of the eruption (fig. 6.1). This was the first eruption to be accompanied by a large swarm of shallow long-period earthquakes beneath the caldera since HVO began counting them.

Earthquakes in the Koaʻe Fault Zone accompanied both the initial east rift zone intrusion on 13–19 September 1977 and the later pulse on 27 September–1 October 1. We do not believe that new magma entered the Koaʻe from the east rift zone because (1) the earthquakes were located in the central Koaʻe south of the caldera and not near the east rift zone from which new magma would be fed as it was in May 1973 (chap. 5, fig. 5.10) and (2) the earthquakes during 13–19 September 1977 started in the central Koaʻe and migrated toward instead of away from the east rift zone. This suggests that the Koaʻe can respond to major south flank movements by tectonic rifting unaccompanied by magma intrusion.

[23] This may explain why the 1977 magma was not erupted immediately following the 1975 earthquake.

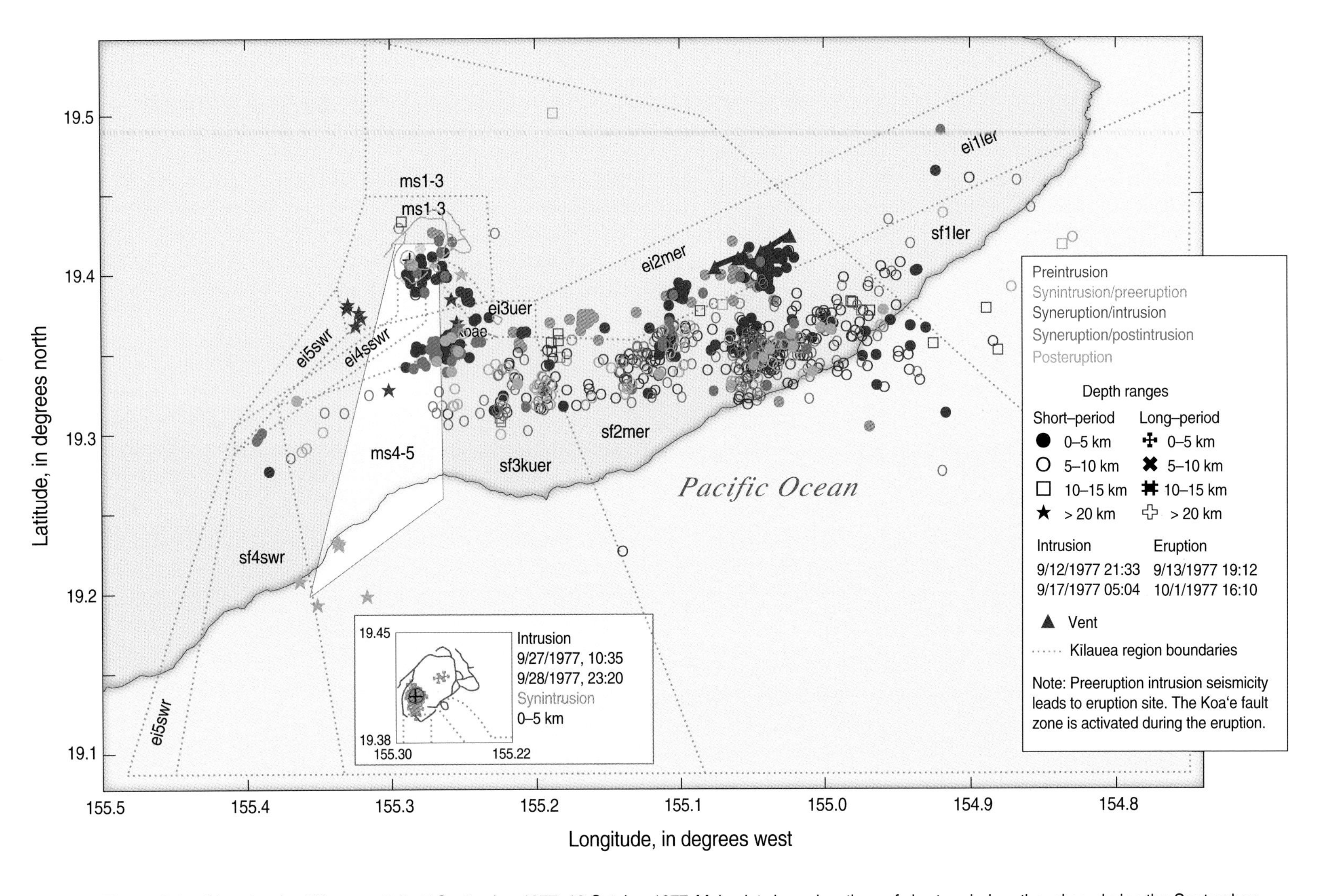

Figure 6.2. Map showing Kīlauea activity, 5 September 1977–13 October 1977. Main plot shows locations of short-period earthquakes during the September 1977 eruption and the surrounding time period. Eruption was preceded by nonswarm earthquakes beneath the rift zone to the west of the vents and was accompanied by intense south flank seismicity and intrusion beneath the Koaʻe Fault Zone and the uppermost seismic southwest rift zone. Intrusion occurred in two pulses separated by seven hours and in the same location. Data are shown for periods before (blue), during (orange), and after (green) the intrusion and eruption, and eruptive vents are plotted as red triangles. Inset shows a long-period earthquake swarm on 27–28 September. The unusually tight cluster of epicenters combined with inflationary tilt indicates a small zone of magma movement beneath Kīlauea's summit. Dates on figure in m/d/yyyy format.

East Rift Intrusions and Eruption, 1979

The summit reservoir very slowly but steadily inflated in the 2 years after the major deflation of the September 1977 eruption and before the intrusions of 1980. The relative slowness of the inflation suggests that magma was still filling voids in the east rift zone left by the major south flank movement during the November 1975 Kalapana earthquake. The deflation of September 1977 was so large, and subsequent reinflation so slow, that there was not a single recognized intrusion until 29 May 1979.

The 29 May 1979 intrusion began 3 km below Mauna Ulu on the east rift zone (table 6.1), and earthquakes migrated uprift, downrift, upward, and downward during the 6-hour-long swarm. This intrusion fed a dike, 2 by 5 km in extent, initially supplied with magma from a reservoir located in the rift below Mauna Ulu and about 1 km uprift of it. This apparent magma reservoir was located near the bend in the east rift zone and its junction with the Koa‘e Fault Zone, and it is the starting point for many other intrusions (fig. 6.5; Klein and others, 1987).

Kīlauea inflated during September and October 1979, accompanied by inflationary earthquake swarms (1) in the south caldera area on 22 September and (2) in the uppermost east and southwest rift zones in the first half of October 1979 (table 6.1), culminating in intense seismicity on 11–12 October. The high sustained inflation rate of September was comparable to the inflation of the eruption recovery after September 1977, suggesting that more intrusions were possible and that the voids in the east rift zone left by the 1975 earthquake were mostly filled.

The largest intrusive event in 1979 started on 15 November and culminated with a small eruption in and near Pauahi Crater (table 6.1). The November 1979 intrusion was the first sizable one to occur after the installation of an Eclipse computer at HVO, which resulted in more precise timing of earthquake arrivals within a dense part of the seismic network. This improvement in earthquake location resolved individual horizontal magma conduits at 1-km and 3-km depths, defined the blade of the dike that pushed upward toward the surface vent, and provided details of the earthquake migration and dike propagation in time (Klein and others, 1987, fig. 43.78, discussion, p. 1133). The 15 November eruption did not trigger earthquakes in the adjacent south flank, suggesting that the intrusion may have lessened stress on the flank. We interpret the November 1979 event dominantly as a redistribution of magma within the rift, with minimal contribution from the summit.

East Rift Intrusions of 1980

The seismic network and accurate earthquake locations enabled intrusions of the early 1980s to be studied in great detail. Kīlauea inflated steadily after the 15 November 1979 eruption, at a very low rate of about 2 µrad per month. Inflationary seismicity was in the caldera and uppermost east rift zone, with representative swarm peaks on 16–17 January 1980 and 2 February 1980 (table 6.1, fig. 6.3).

The east rift intrusions of 2 March and 10–12 March 1980 (table 6.1) are probably linked. The first, and smaller, intrusion in the upper east rift on 2–3 March 1980 was followed a week later by intrusion in the middle east rift on 10–12 March. The latter intrusion was accompanied by significant summit deflation and south flank earthquake response. An accompanying eruption (table 6.1) was inferred from discovery of a small pad of lava in March 1982 near the site of the earthquake swarm. The 10–12 March 1980 earthquake swarm and magma migration preceded summit deflation by nearly 4 hours, suggesting that stored east rift magma fed the eruption and intrusions, to be later resupplied from the summit via a different, parallel, and probably deeper conduit. This resupply conduit produced few if any earthquakes and was probably an open and hot conduit in direct connection to the summit reservoir.

East rift activity was temporarily suspended following four more intrusions fom August to November 1980 (table 6.1). A pair of linked east rift intrusions occurred on 30 July and 27 August 1980. The first, smaller intrusion (no measurable deflation) began uprift of Puhimau Crater on 30 July 1980. On 27 August a larger intrusion started at Puhimau where the previous intrusion terminated. We interpret the first intrusion as stopping at a rift conduit barrier, where magma initially pooled but then broke through the barrier 28 days later and enlarged a dike in both uprift and downrift directions. The final two intrusions were on 21–22 October and 2 November 1980.

Southwest Rift Intrusions of 1981–1982

1981 marked a shift in magmatic activity from the east rift zone to the seismic southwest rift zone (sswr). The first intrusion in this period was in January–February 1981, followed by almost continuous sswr earthquake activity from February through August, which culminated in a large sswr intrusion on 10 August 1981. Intrusions extended into 1982, ending with a large event on 22 June (table 6.1; fig. 6.3). Each of these three intrusion sequences had different earthquake and time-development characteristics, implying different sswr pathways.

Earthquake swarms between January and August 1981 show an erratic progression down the sswr. During the first intrusion on 20 January 1981,

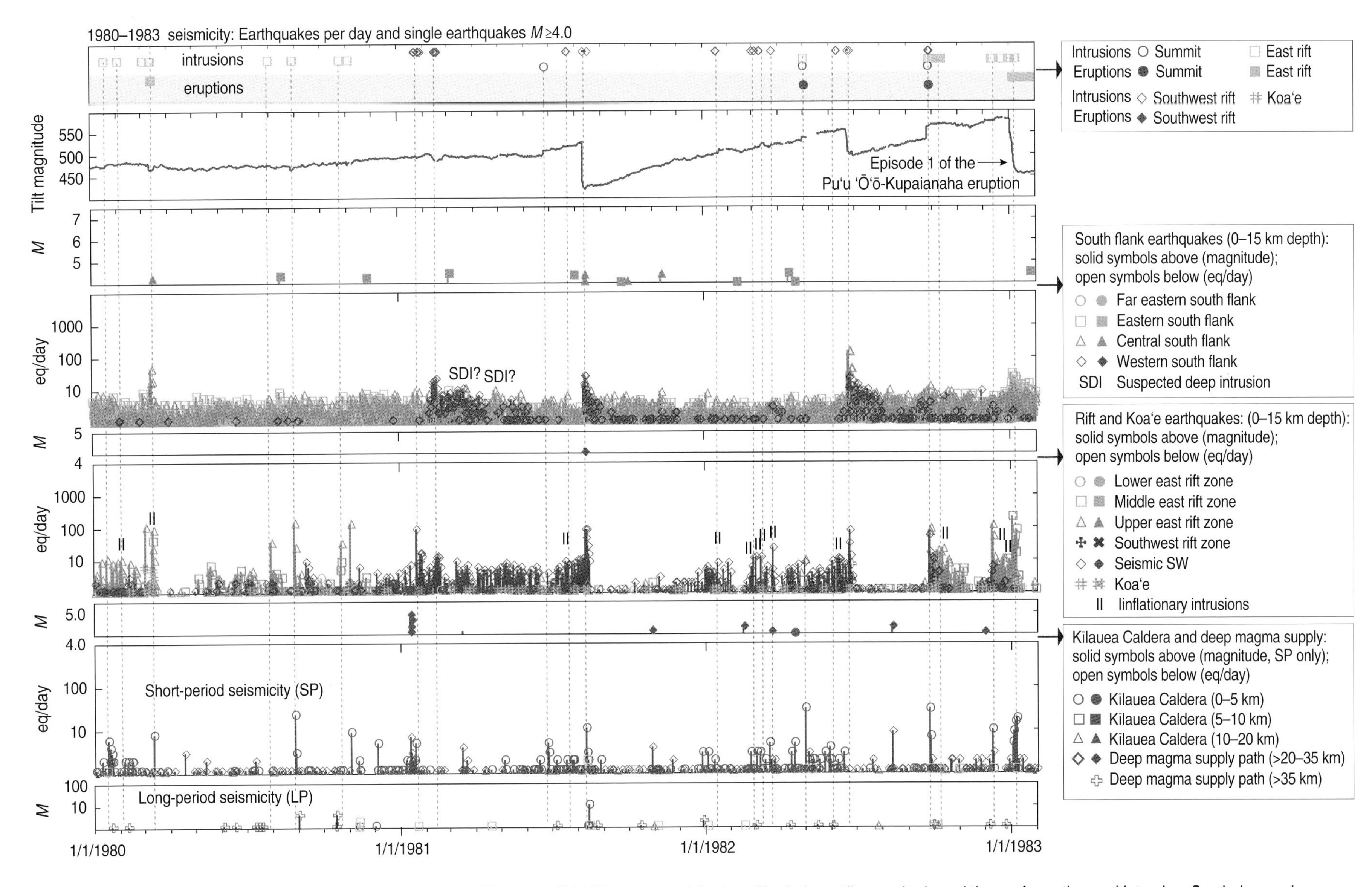

Figure 6.3. Graphs showing Kīlauea activity, 1 January 1980–1 February 1983. Time series plots show Uwēkahuna tilt magnitude and times of eruption and intrusion. Symbols are given on the plots. Three types of intrusions are defined in chapter 1. Traditional intrusions and inflationary intrusions are shown as intrusion symbols and inflationary intrusions are labeled (II). Suspected deep intrusions are labeled (SDI) but not plotted as intrusions. See also caption of figure 6.1. Dates on figure in m/d/yyyy format.

earthquakes ventured no more than 2 km south of the south caldera boundary into the sswr. On 24 January magma and earthquakes progressed 5 km further down the rift zone. During 9–16 February, the intrusion progressed to and activated a patch near Pu‘ukou at a bend and possible blockage in the rift zone, about 20 km from the caldera. During the next intrusions earthquake swarms were again concentrated in the uppermost sswr.

Unlike the rapid deflation characteristic of east rift zone intrusions, during which the beginning of deflation followed by minutes to hours the onset of earthquake swarms, deflation during these sswr intrusions was both gradual and began a day or more before the rift earthquake swarms. This suggests that the sswr offered resistance to magma flow in a conduit that was not open and liquid filled, and that the adjacent flank offered some resistance to spreading.

August 1981 Intrusion

A much larger sswr intrusion began on 10 August 1981 and followed a pattern very similar to the intrusion associated with the December 1974 eruption (chap. 5, figs. 5.15 and 5.16). The August 1981 intrusion was preceded by nearly continuous elevated sswr seismicity from March through July 1981, including a possible suspected deep intrusion in April (fig. 6.3). Summit inflation was low through mid-June, then accelerated in the 2 months following the minor summit intrusion of 25–27 June 1981. That intrusion marks a time when magma largely ceased flowing down the sswr and inflated the summit reservoir instead. Clearly the sswr and summit were building for a possible event. On 2 August 1981, the gradual buildup of earthquakes in the upper sswr increased to more than 10 per day, and earthquakes suddenly advanced down the sswr to a point 5 km from the caldera.

The August intrusion started from a point just south of the caldera and moved along the sswr at a speed of 2.6 km/hr, twice the speed of the similar December 1974 intrusion, perhaps because the rift had been opened during the 1974–75 intrusion, making it easier for further intrusion in 1981. A dike cross section of 1 m width by 1.5 km height is consistent with the earthquake migration speed, deflation rate, earthquake cross section, and geodetic observations (Klein and others, 1987, page 1153). Shallow sswr earthquakes stopped on 15 August, but the triggered flank earthquakes continued above background for about a month (fig. 6.3).

Early 1982 Intrusions

Kīlauea reinflated steadily in the 10 months following the 10 August 1981 sswr intrusion (fig. 6.3). Inflationary intrusions increased in frequency as the caldera reservoir swelled with magma. Peaks of intrusive activity occurred in 1982 on 15 January, 27 February, 3, 9, and 23 March, and 8 June. Because earthquakes did not migrate down the rift, the magma buildup was confined to the near-summit region.

A small caldera eruption occurred on 30 April 1982 during this inflationary period. Earthquakes were confined to the vent area within the caldera (Klein and others, 1987, fig 43.92). The Uwēkahuna tiltmeter showed an “inflationary” jump of 10 μrad (table 6.1; figure 6.3), suggesting that magma moved rapidly upward to a shallower depth as it would for a new dike under the caldera vents, creating an “inflationary” bulge from a shallow source. On 22 June 1982, another intrusion started—the last sswr intrusion before the Pu‘u ‘Ō‘ō-Kupaianaha eruption. This was a major intrusion accompanied by a downrift seismic progression similar to that shown by the August 1981 and 1974–75 intrusions.

Latter 1982 Intrusions and Return to the East Rift Zone

A caldera eruption and an upper east rift zone intrusion in the latter half of 1982 signaled a return of volcanism from the sswr to the east rift zone. The 25 September 1982 eruption formed a small, short-lived fissure in the southern caldera and was accompanied by an intrusion beneath the fissure (table 6.1). Modeling the eruption as inflation of a shallow source and deflation of a deeper source yields outward tilt at nearby stations. The outward tilt of nearly 35 mrad (table 6.1) did not recover (fig. 6.3), which indicated a permanent September 1982 dike emplacement rather than reservoir inflation.

Following the summit eruption/intrusion of 25 September 1982, inflationary intrusions into the upper east rift zone occurred as the summit continued to inflate. Four small intrusions between 1 and 11 October repeat the seismic patterns seen in September. A series of three intrusions occurred on 9, 21, and 29–30 December (table 6.1; fig. 6.4). The first intrusion was an intense upper east rift zone intrusion with downrift earthquake migration between the caldera and Hiiaka Crater. The latter two were slow moving and less intense inflationary intrusions, which did not interrupt the inflation (fig. 6.3).

Episode 1 of the Pu‘u ‘Ō‘ō-Kupaianaha Eruption

The earthquake swarm of 29–30 December 1982 continued into 1983 and was the immediate precursor to the long-lived Pu‘u ‘Ō‘ō-Kupaianaha eruption that began on 3 January 1983 and is still continuing as of this writing (2014). We consider episode 1 to be the culminating event in the recovery from the 1975 earthquake,

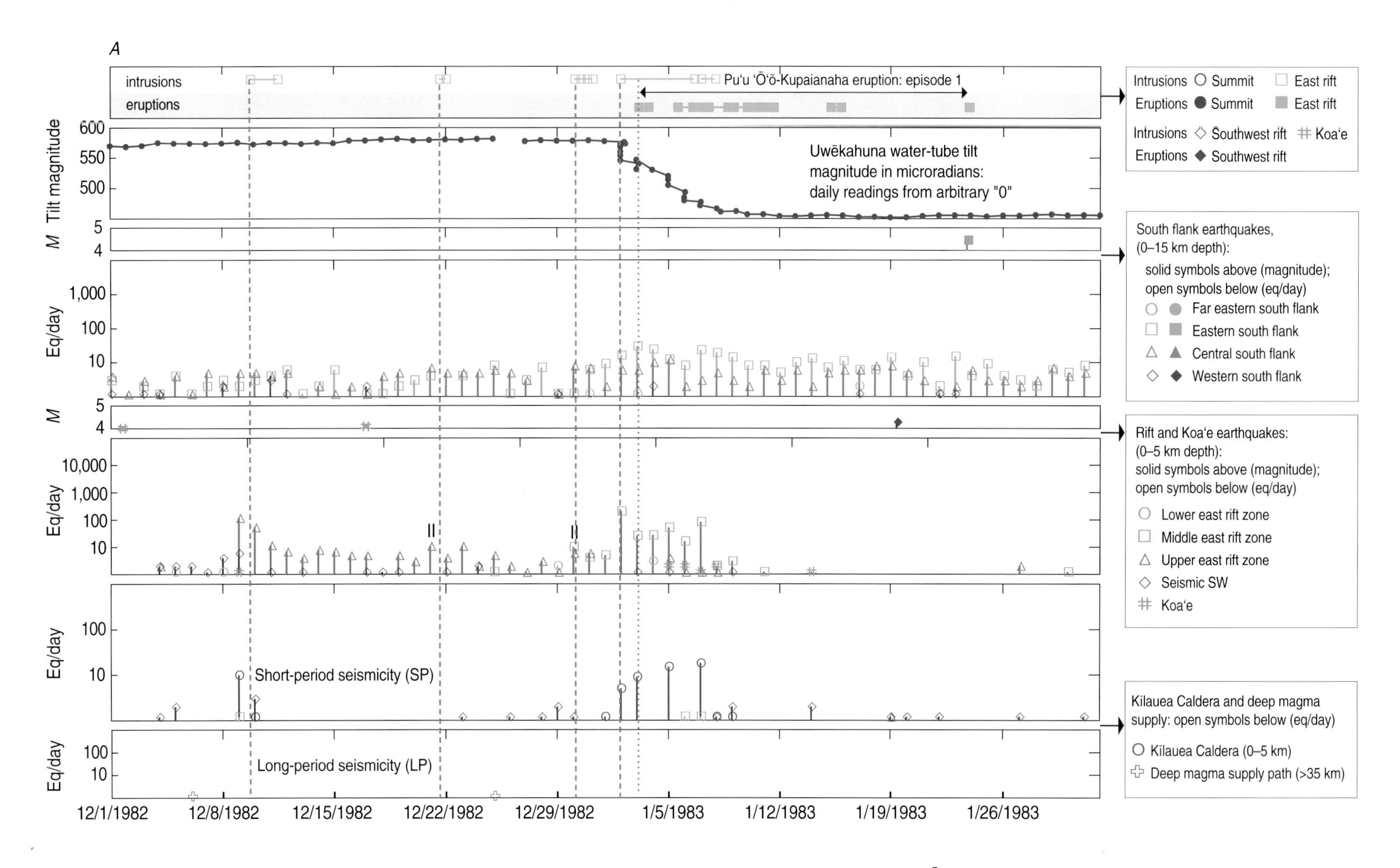

Figure 6.4. Graphs and map showing Kīlauea activity, 1 December 1982–1 February 1983. Precursors to episode 1 of the Pu'u 'Ō'ō-Kupaianaha eruption. Dates on figure in m/d/yyyy format. ***A***, Time series plot for 1 December 1982–1 February 1983. Pu'u 'Ō'ō-Kupaianaha eruption episode 1 and precursory intrusions. The last intrusion precedes eruption by about 1 day, and the summit begins to deflate after the beginning of the intrusion earthquake swarm (table 6.1). See also caption to figure 6.1. ***B***, Map showing earthquake locations for the period 7–31 December 1982. Each intrusion moves progressively downrift toward the episode 1 vents.

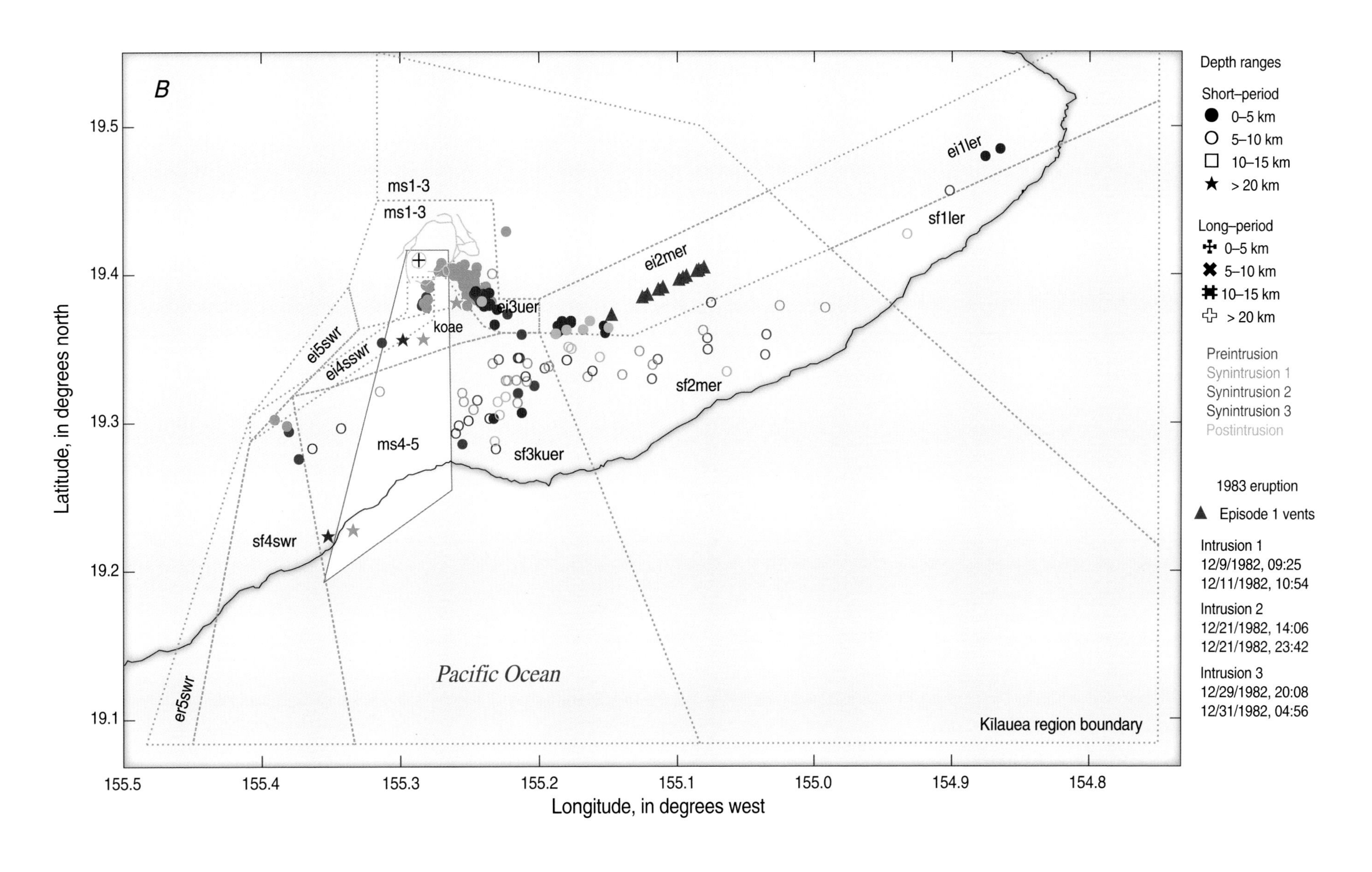

Figure 6.4.—Continued

because the character of seismic activity in all regions of the volcano as well as deformation patterns changed dramatically following episode 1. Further discussion of episode 1 leads off the description of the entire Pu'u 'Ō'ō-Kupaianaha eruption in the next chapter.

Interpretations 1975–1982

Stages of recovery from the 29 November 1975 earthquake can be summarized as follows:

November 1975–September 1977: no inflation, east rift intrusions refilling the volume formed by dilation of the rift zones during the earthquake.

September 1977–May 1979: rapid reinflation without further shallow intrusion following the 1977 eruption and deflation of Kīlauea's summit.

May 1979–December 1980: continued rapid inflation accompanied by frequent east rift zone intrusions.

January 1981–June 1982: continued rapid inflation accompanied by seismic southwest rift zone intrusions.

June 1982–January 1983: rapid inflation accompanied by east rift zone intrusions precursory to deflation of Kīlauea's summit during episode 1 of the Pu'u 'Ō'ō-Kupaianaha eruption.

East Rift Intrusions and Uprift/Downrift Migration of Associated Earthquake Swarms

Zones of identified magma storage are shown in fig. 6.5. The summit caldera reservoir is principally defined by inflation and deflation seen at the surface and as the source of intrusions revealed by earthquakes, as well as numerous geophysical and geological studies. The Pauahi magma reservoir was identified by Tilling and others (1987, fig. 16.41, 16.43) from localized inflation prior to the Pauahi eruptions of 1973, in the pause of the 1969–74 Mauna Ulu eruption. Klein and others (1987) identified the Pauahi reservoir as an origin point of intrusions in the 1960s, 1970s, and 1980s, evidenced by earthquakes migrating uprift (west) and downrift (east) initially fed from the Pauahi reservoir. The Mauna Ulu reservoir was suggested by Tilling and others (1987, fig. 16.43) as the site of deflation during 1973–74 within the Mauna Ulu eruption, but it is less well expressed from other evidence and may only be short lived. The Makaopuhi reservoir was identified by Jackson and others (1975, figs. 32, 35) and Swanson and others (1976, fig. 20) from local magma differentiation and fractionation before and after the October 1968 eruption and from uplift before the February 1969 eruption seen by leveling and seismicity. The intense seismicity and tremor under and just before the eruption cited by these authors do not necessarily require a local magma reservoir as a source, but are certainly the result of dike propagation from the conduit to the surface. Other unrecognized or temporary magma reservoirs undoubtedly exist, but the summit, Pauahi, and Makaopuhi reservoirs are the most persistent and best documented. The shape of the reservoirs is approximate. The reservoirs within the rift have unknown shape and may be only zones where the conduit widens and stores liquid magma for a period of time. An approximate minimum magma volume of 0.001–0.003 km^3 is consistent with the temporary magma storage behavior, which could be any shape or combination of liquid and solid having a minimum dimension of 50–100 m, necessary to minimize cooling between times of magma replenishment. Wyss and others (2001, fig. 6) found indirect evidence for persistent Pauahi and Makaopuhi reservoirs because b-value anomalies (a relative excess of small magnitude earthquakes suggesting heterogeneous fractures and stress or high pore pressure) are in the south flank adjacent to both reservoirs.

During the 1977–82 inflation period, the summit reservoir was in magmatic communication with the upper east rift zone down to a conduit barrier located near Mauna Ulu (fig. 6.5). This barrier was active during 1976–82 as a stopping point of earthquake swarms progressing downrift. In addition, there was aseismic passage of magma along upper rift pathways to the central and lower rift preceding the eruption of 13 September 1977. Earthquake migration with time shows that more than half of the intrusions during the period from June 1976 to early 1982 began near Mauna Ulu and migrated both uprift and downrift simultaneously, with resupply from Kīlauea's summit. Because the summit was deflating at the same time as magma was actually moving uprift from Mauna Ulu, it is likely that intrusions were forming dikes within a shallow pathway originating beneath Mauna Ulu, while magma was being resupplied aseismically from Kīlauea's summit via another, deeper east rift zone pathway .

Magma pathways in the upper east rift between the caldera and Mauna Ulu are seismically defined by earthquake bands at depths of 1.5 and 3 km (fig 6.5*B*). Many larger intrusions, such as that of 15 November 1979, produce earthquakes in both depth zones (Klein and others, 1987, figure 43.78). Typical intrusions, especially those associated with eruption, first show earthquakes in the 3-km-deep zone that migrate upward through the 1.5-km-deep zone to the surface, but there is no simple pattern of the seismic zones at the two depths that would indicate that they operate as independent conduits (Klein and others, 1987). Multiplet relocation of shallow background (nonintrusion) earthquakes in the upper east rift (Gillard and others, 1996) and the January 1983 intrusion in the middle east rift (Rubin

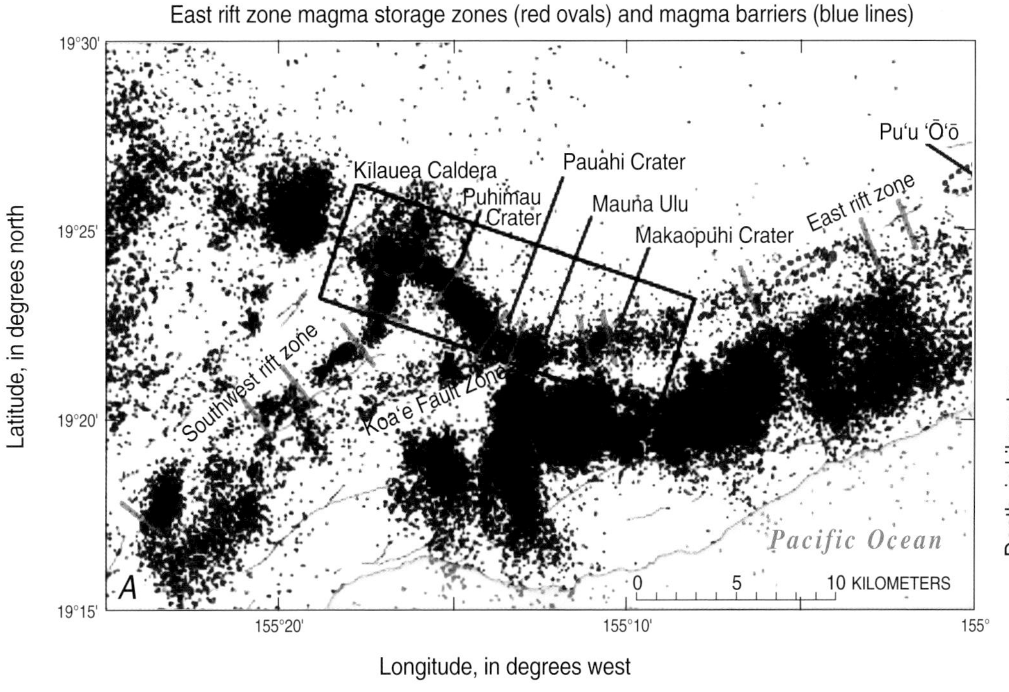

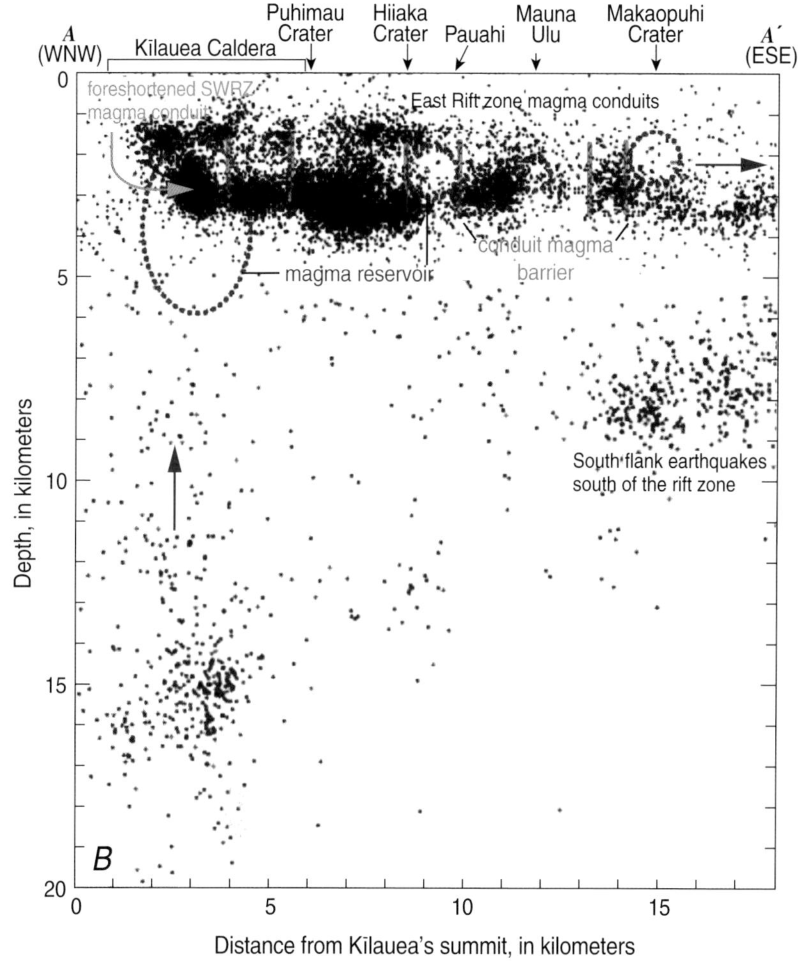

Figure 6.5. Map and cross-section showing location of east rift magma storage zones and magma barriers. ***A***, Map of Kīlauea earthquakes from 1976 through 2007. Earthquakes shown are short period (brittle fracture), and those designated as long period are excluded. ***B***, Cross-section of earthquakes of the east rift zone from Kīlauea Caldera to below Makaopuhi Crater. Rectangle on *A* is the cross-section A-A' shown in *B*. Red dashed circles in both *A* and *B* mark the locations of magma storage areas within the rift zone and the summit magma reservoir, defined by reduced seismicity, magnetic anomalies (Kauahikaua, 1993), self potential (Kauahikaua, 1993; Zablocki and Koyanagi, 1979; Zablocki, 1976) or by other factors. Blue lines on both *A* and *B* indicate inferred magma barriers, which are often points where intrusions start and stop. Barriers and the magma reservoir at long 155°5′W. outside the section zone are from Klein and others (1987). The reservoir near longitude 155°0'W. (Pu‘u ‘Ō‘ō-Heiheiahulu region) is inferred from accumulated magma before the Pu‘u ‘Ō‘ō eruption (Wolfe and others, 1988) and the eruption of differentiated lava in the 1977 eruption (Moore, 1983). The bands of earthquakes centered near 1.5 and 3 km depth on *B* show the magma conduits that produced earthquakes by brittle fracture through dike propagation along the subhorizontal magma conduit. Red arrows on *B* show direction of magma flow to replenish the summit reservoir from depth and feed eruptions and intrusions in the east rift zone.

and others, 1998) show the deeper zone as a narrow ribbon, with earthquakes at depths of about 3 km ± 50 meters. They interpret this 3-km-deep seismicity as the region of high stress concentration above the deep rift magma body. It is possible that the 1.5-km and 3-km depths are the brittle, dike-forming zones above and below a fluid filled, aseismic conduit, but this cannot be proven because the relative absence of earthquakes near 2-km-depth indicates the absence of brittle rock under stress—that rock may or may not be fluid. The 3-km-deep earthquake zone is narrow and responds to stresses from intrusions propagating uprift and downrift as well as from rift growth between intrusions, and the earthquakes reveal dikes that propagate upward from to 2–3-km depth to the surface.

Another barrier closer to Kīlauea's summit is under Hiiaka Crater and appears to have stopped intrusions migrating downrift on 12 August 1979 and 27 August 1980. The Hiiaka barrier may help contain the Mauna Ulu magma storage reservoir just downrift of it. The Pauahi magma reservoir, just uprift of Mauna Ulu, is visible as a gap in seismicity and may feed intrusions up and down rift (fig 6.5).

Some barriers, such as that under Mauna Ulu, are structurally controlled by a junction with the Koa'e Fault Zone and the eastward bend of the rift and may persist for long periods of time. Other barriers may have a short lifetime as the rift grows and evolves. The relation of intrusions to rift barriers is not always straightforward. In some cases an intrusion may stop at a barrier and be followed by an intrusion that breaches the barrier as magma moves downrift in discrete pulses. An example is the 22 January 1977 inflationary intrusion that paused uprift of the Hiiaka barrier, then continued downrift in the 8 February 1977 intrusion. A general discussion of rift zone barriers and magma storage reservoirs is given by Klein and others (1987).

Intrusion Pathway, Speed, and Flank Resistance on the Seismic Southwest Rift Zone

The June 1982 sswr intrusion was different in depth and time compared with other rift intrusions. The main intrusive seismic zone started on 22 June and was between 7 and 16 km distant from the caldera. Earthquakes defining the connecting magma pathway adjacent to the caldera occurred 2–4 days earlier on 18–20 June, rather than being part of the larger earthquake swarm defining the intrusion. Deflation began at a slow rate on 21 June, about 1 day before the downrift earthquakes started. The deflation rate accelerated about 18 hours later, but still more slowly than in previous intrusions. This suggests that magma moved slowly and aseismically through the upper 7 km of rift, probably through the still-molten dike formed in August 1981.

Unlike most other rift intrusions, the June intrusion's seismic zone was mostly at depths between 6 and 9 km, with few earthquakes defining the shallow sswr magma pathway near 3–4 km depth revealed in the August 1981 intrusion. The June 1982 intrusion was thus similar to but smaller than the December 1974 intrusion and did not produce earthquakes in the shallow rift conduit as in the December 1974 and August 1981 intrusions. The 22–24 June 1982 intrusion was relatively slow, taking 3 days to deflate compared to the 1 day of the December 1974 and August 1981 intrusions. This slow deflation and deeper earthquake location suggests a deeper magma pathway. The deeper conduit of 1982 offered more flow resistance than the shallower conduits of December 1974 and August 1981, as evidenced by (1) the slowness of deflation (taking 3 times as long) and (2) slower earthquake migration speed, 0.2 km/hr for 1982 vs. 1.3 km/hr for 1974 (table 6.1). The last and farthest part of the August 1981 intrusion slowed from 2.6 km/hr to 0.28 km/hr (table 6.1; Klein and others, 1987, figure 43.90), suggesting that it too encountered the higher flow resistance of an intrusion that must push against the deeper parts of the adjacent south flank.

Restoration of the Magmatic System Following the Kalapana Earthquake

During the earthquake of 29 November 1975 the flank moved several meters seaward and the summit deflated by 225 µrad, sending about 0.08–0.12 km^3 of magma into the east rift zone[24]. This was at least 10 percent of the rift dilation volume of 0.8–0.9 km^3 created during the earthquake (Owen and Bürgmann, 2006). After the earthquake, the tilt record shows no net inflation or deflation until the eruption of September 1977. We interpret the lack of summit magma accumulation to indicate that all of Kīlauea's magma supply during this 1975–77 period went to fill voids or volumes in the east rift left by the flank slip to the south. The four east rift zone intrusions during this period were of small magma volume compared to the 1975 collapse, and they encountered a rift and flank that offered little resistance to intrusion and easy ability to accommodate subsurface magma without eruption.

The 1977 eruption served to further delay the recovery from the 1975 earthquake. The 91 µrad Uwēkahuna deflation in 1977 means approximately

[24] The lower value represents deflation below the central caldera. The higher value represents the sum of volumes calculated as the deflationary tilt varied in azimuth. Our tilt to volume factors may not apply to the entire range of azimuths. The lower value falls within the range of 0.04–0.09 km^3 estimated from gravity measurement made after the earthquake (Dzurisin and others, 1980). Our values are higher than a model estimate of 0.04 km^3 used by Owen and Bürgmann (2006).

0.04 km^3 left the summit reservoir, leaving that volume to be filled. Reinflation of the summit took 2 years, up to the eruption in November 1979, to recover the magma lost from the summit reservoir in the September 1977 eruption. The slow reinflation suggests that magma was also still refilling the volume loss produced during the earthquake as well as the added volume of rift dilation from continued spreading (Delaney and others, 1998). After this 1978–79 recovery and during 1980, tilt was approximately level, indicating that the magma supply was still filling the remaining rift volume lost during the earthquake. Magma did not leak into the east rift zone steadily but in pulses, as tilt oscillated once or twice per month with amplitudes of 3 to 5 µrad (Klein and others, 1987, fig 43.74). Earthquakes also showed this oscillation, with caldera earthquakes peaking at times of peak inflation as the caldera extended and rift earthquakes swarming at times of deflation. This was a time of stable oscillation of filling and draining when no volcanic event was large enough to alter the equilibrium or change the level or pressure in the magma column. There were no large, forceful intrusions during this 1978–79 time period, and Kīlauea was in a steady reinflation mode.

The beginning of summit reinflation above the level of the bottom of the 1975 collapse (including recovery from the 1977 collapse) signified the end of passive intrusion and the beginning of a different filling mode. The new filling mode found intrusions moving into the volume produced during ongoing spreading and also occasionally breaking through barriers to complete the filling of volume lost during the 1975 earthquake. Our interpretation of the recovery from the 1975 earthquake is similar to our interpretation of the recovery from the 1924 collapse as described in chapter 3.

The shift to the southwest rift zone in 1981 and early 1982 signified that the east rift volume deficit was effectively made up, and the remaining volume deficit was made up by intrusion beneath the seismic southwest rift zone (sswr). We view the sswr as a region of overflow that is only activated when the east rift zone can no longer accept magma and the summit is not fully inflated. A similar sequence occurred at the end of both stages of the Mauna Ulu eruption in 1971 and 1974 as described in chapter 5.

The year 1982 marked the end of the post-1975 earthquake recovery era. The summit caldera underwent intrusion and eruption in April 1982, and the last significant southwest rift intrusion was in June 1982. It was not until the summit eruption of September 1982, and simultaneous intrusions into both rifts, that magmatic intrusions switched back to the east rift zone, which had undergone further dilation during spreading over the previous 1.5 years of 1981–82 southwest rift and summit intrusions. East rift intrusions in the latter part of 1982 prepared the way for release of magma to the surface of the east rift in January 1983. The large and continuing deflation and the long-lived 1983 and later eruption indicated that the recovery from the 1975 earthquake was complete and an equilibrium was established between the magma supply rate and eruption.

Supplementary Material

Supplementary material for this chapter appears in appendix F, which is only available in the digital versions of this work—in the DVD that accompanies the printed volume and as a separate file accompanying this volume on the Web at http://pubs.usgs.gov/pp/1806/. Appendix F comprises the following:

Table F1 presents tilt volume, eruption efficiency, and magma supply rate for the period 1975–83.

Figure F1 shows short-period and long-period earthquake swarms for all regions.

Figure F2A–H. shows short-period earthquake counts and earthquakes of $M \geq 4$ at 1-year intervals from 1 Feburary 1975 to 1 February 1983.

Figure F2I–P shows long-period earthquake counts and earthquakes of $M \geq 4$ at 1-year intervals from 2/1 February 1975 to 1 Feburary 1983

Figure F3 shows time series plots at 1-year intervals for 1975–1983.

Figure F4 presents the M7.2 south flank earthquake of 29 November 1975. Aftershocks are shown through 5 December 1975.

Figures F5–F38 show locations of earthquakes for eruptions and intrusions between June 1976 and December 1982.

High fountaining at Pu'u 'Ō'ō. USGS photo by C.C. Heliker, 19 September 1984.

Chapter 7

Eruptive and Intrusive Activity, 1983–2008

Documents the continuing decades of the east-rift eruption, beginning with a balance between magma supply and erupted lava and ending with increased intrusion. With more sophisticated instrumental monitoring, eruption pauses and small-scale deformation events not recorded in previous eruptions are described and interpreted. The first documentation of "silent" earthquakes is equated with our longer term identification of "suspected deep" intrusions through similarity in their patterns of south flank seismicity. Documents the changes driven by continuing increase in magma supply rate, which culminates in March 2008 with a new eruption in Halema'uma'u with no dimunition of the east rift activity, a unique occurrence in Kīlauea's recorded history of such simultaneous eruptions and the event that closes our study

Aerial view of the summit and east rift zone of Kīlauea. Kīlauea Caldera and its pit crater, Halema'uma'u, are in the foreground, with HVO buildings back from the caldera rim to the lower left. The steaming area at the top of the photo marks activity at the Pu'u 'Ō'ō vent active since January 1983. USGS photo by J.D. Griggs, 10 January 1985.

Early on 2 January 1983, an earthquake swarm and summit subsidence heralded the beginning of what was to be long-term activity in the middle east rift zone—activity that continues to the present time (2014). The eruption took place at several vents over a 3-week time period (table 7.1; figs. 7.1, 7.2). An intense south flank earthquake response occurred adjacent to the erupting vents and also south of the upper east rift zone and for several hundred meters downrift from the erupting vents. The complexity in rate of both downrift and uprift migration of earthquakes from different points beneath the rift indicates that this eruption was fed from a series of separate intrusions (Klein and others, 1987, figure 43.97, discussion on p. 1167). In order to understand this period we build on the recently published summary of the first 20 years of this eruption (Heliker and others, 2003).

On 28 March 2008 a small explosive eruption occurred at Halema'uma'u, followed within weeks by identification of an active magma body just below the Halema'uma'u floor (Orr and others, 2008; Poland and Sutton, 2008; Wilson and others, 2008). The ongoing eruption on Kīlauea's east rift zone continued through this event, yielding the first instance in Kīlauea's historical period of simultaneous eruption at two different sites. The Halema'uma'u eruption provides a convenient closing date for our study and also provides a context for this chapter—that is, we seek to answer the question of how Kīlauea's plumbing evolved between 1983 and 2008 to allow simultaneous eruption at Kīlauea's summit and east rift zone.

Episode 1 of the ongoing Pu'u 'Ō'ō-Kupaianaha eruption is interpreted as the culminating event of recovery from the 1975 earthquake, as indicated in chapter 6. For convenience we have divided the eruption into three stages. Stage I comprises 9 years: the 4 years of episodic eruption at Pu'u 'Ō'ō beginning with episode 2 (stage IA), followed by more than 5 years of continuous eruption from a lava lake developed at Kupaianaha (stage IB). Stage II begins on 8 November 1991, during the dying of the Kupaianaha vent and just before the return of activity to Pu'u 'Ō'ō in episode 49. The early events of stage II feature several short episodes at different places on the Pu'u 'Ō'ō cone (episodes 49–51), and the period ends just before the beginning of reinflation of Kīlauea's summit in December 2003. Stage II is arbitrarily divided into stages IIA and IIB at 1 January 1997, before the intense eruption and intrusion of January–February 1997 and the beginning of the long-lived episode 55 at Pu'u 'Ō'ō. Stage III begins on 1 December 2003, a date preceding the beginning both of reinflation and of a dramatic increase in CO_2 emission at Kīlauea's summit. Stage III ends with the explosive eruption in Halema'uma'u on 19 March 2008. In order to emphasize the year of intense activity precursory to the Halema'uma'u eruption, we divide stage III into stages IIIA and IIIB at 18 May 2007, a date which precedes the east rift intrusion of 24 May 24, the following Father's Day eruption and intrusion of 17 June, and the renewal of east rift eruption downrift from Pu'u 'Ō'ō on 21 July.

The terminology used here to describe eruptive and intrusive events is as follows:

Episode: An eruption event that forms part of a series repeated at the same site. Stage IA comprised 47 episodes of high-fountaining (figure 7.1) that built the 260-m-high cone Pu'u 'Ō'ō (Heliker and Mattox, 2003; Wolfe and others, 1987; Wolfe and others, 1988). The Hawaiian Volcano Observatory (HVO) continues to use the term episode to apply to every new eruption. We emphasize the change in location of the eruption with a name in addition to the episode designation for eruptions within the Pu'u 'Ō'ō edifice and use other naming terminology to refer to eruptions outside of Pu'u 'Ō'ō. Thus episode 48 refers to the fissure eruption that led to continuous eruption at Kupaianaha (Heliker and Mattox, 2003; Heliker and Wright, 1991), thereafter designated in this paper as the "Kupaianaha" eruption. Eruption terminology for this chapter is listed in the following text table.

Stage	Begin	End	Location	Eruption name*
Episode 1	1/3/1983	1/23/1983	Various, including Pu'u 'Ō'ō	Pu'u 'Ō'ō; episode 1
IA	2/13/1983	7/18/1986	Pu'u 'Ō'ō	Pu'u 'Ō'ō; episodes 2–47
IB	7/18/1986	11/8/1991		*Kupaianaha* (episode 48)
	7/18/1986	2/7/1992	Kupaianaha	
IIA	11/8/1991	1/1/1997		
	11/8/1991	1/29/1997	Pu'u 'Ō'ō	Pu'u 'Ō'ō episodes 49–53
IIB	1/1/1997	12/1/2003		
	1/30/1997	1/31/1997	Nāpau	*1997 Nāpau* (episode 54)
	2/24/1997	6/17/2007	Pu'u 'Ō'ō	Pu'u 'Ō'ō episode 55
IIIA	12/1/2003	5/18/2008	Pu'u 'Ō'ō	Pu'u 'Ō'ō episode 55
IIIB	5/18/2007	3/19/2008		
	6//19/2007	6/19/2007	Kane nui o Hamo	*Father's Day* (episode 56)
	7/2/2007	7/21/2007	Pu'u 'Ō'ō	Pu'u 'Ō'ō episode 57
	7/21/2007	present	East rift	*2007 fissure* (episode 58)

* Proposed name in italic; current name in parentheses.

Table 7.1. Pu‘u ‘Ō‘ō-Kupaianaha eruption[1]: Stage IA—2/1/1983–7/20/1986 (see fig. 7.1).

[In rows with multiple entries text applies down to the next entry; dates in m/d/yyyy format at; do, ditto (same as above); data for eruptions and traditional intrusions are emphasized by grey shading]

Date and time		Reg.[2]	Event Type[3]	No.[4]	Tilt[5]		Lag[6]	Comment[7]	Fig.[8]
Start	End				Mag	Az			
2/10/1983 10:30	2/25/1983 09:00	MERZ	E					Episode 02—low-level fountaining	
2/25/1983 09:00	3/4/1983 14:51	do	do		13.9	117	+6h	do—vigorous fountaining; tilt 2/24–3/3	
2/12/1983 07:21	2/12/1983 12:37	sf4swr	EQS	9				Possibly a suspected deep intrusion (SDI)?	
	3/21/1983 13:12	sf2mer	EQS	10				South flank anticipation	
3/21/1983 6:00	3/28/1983 1:00	MERZ	E		14.9	115	+0h	Episode 03—low-level fountaining	
3/28/1983 1:00	4/9/1983 02:57	do	do		19.2	133		do—vigorous fountaining; tilt 3/27–4/2; Tilt 4/2–9-clockwise rotation	
6/13/1983 10:25	6/17/1983 14:13	MERZ	E		16.2	113	-4h 35m	Episode 04; tilt 6/12–17	
6/29/1983 12:51	7/3/1983 7:15	MERZ	E		9.3	120	-10h 9m	Episode 05; tilt 6/30–7/3	
7/22/1983 15:30	7/25/1983 16:30	MERZ	E		16.7	117	-8h 30m	Episode 06; tilt 7/22–25	
8/15/1983 7:41	8/17/1983 16:00	MERZ	E		19.7	117	-4h 19m	Episode 07; tilt 8/15–18	
8/15/1983 22:51	8/16/1983 22:49	sf3kuer	EQS	18				South flank response—*M*3.7 with aftershocks	
9/6/1983 5:11	9/7/1983 5:26	MERZ	E		14.8	117	-5h 49m	Episode 08; tilt 9/5–7	
9/9/1983 06:30 9/9/1983 06:37	9/9/1983 23:02 9/10/1983 23:35	sf2mer sf2.3	EQ					*M*5.7; 3 foreshocks? from 9/8 20:26 26 aftershocks extending to the west from mainshock in regions sf2mer and sf3kuer	
9/15/1983 15:41	9/17/1983 19:20	MERZ	E		16.2	115	-3h 19m	Episode 09; tilt 9/15–17	
10/5/1983 1:06	10/7/1983 16:50	MERZ	E		16.9	108	-4h 54m	Episode 10; tilt 10/5–7	
11/5/1983 23:50	11/7/1983 18:45	MERZ	E		21.1	110	-2h 10m	Episode 11; tilt 11/5–8	
11/16/1983 06:13 11/16/1983 10:02	 11/18/1983 10:52	 all	EQ					Ka‘ōiki mainshock *M*6.7 Triggered Kīlauea aftershock sequence	 **7.3**
11/30/1983 4:47	12/1/1983 15:45	MERZ	E		24.6	111	-1h 13m	Episode 12; tilt 11/30–12/2	
12/18/1983 11:39	12/18/1983 18:27	ms2	EQS	9	flat			Nāmakanipaio[9]	
1/20/1984 17:24	1/22/1984 11:23	MERZ	E		10.0	111	-3h 36m	Episode 13; tilt 1/19–23	
1/30/1984 17:45	1/31/1984 13:18	MERZ	E		12.1	112	-2h 15m	Episode 14; tilt 1/29–2/1	
2/14/1984 19:40	2/15/1984 15:01	MERZ	E		15.7	110	-1h 20m	Episode 15; tilt 2/12–17	
3/3/1984 14:50	3/4/1984 22:31	MERZ	E		18.2	105	-2h 10m	Episode 16; tilt 3/3–5	
3/30/1984 4:48	3/31/1984 3:24	MERZ	E		14.6	120	-4h 12m	Episode 17; tilt 3/30–31	

Table 7.1. Pu'u 'Ō'ō-Kupaianaha eruption[1]: Stage IA—2/1/1983–7/20/1986 (see fig. 7.1).—Continued

[In rows with multiple entries text applies down to the next entry; dates in m/d/yyyy format at; do, ditto (same as above); data for eruptions and traditional intrusions are emphasized by grey shading]

Date and time		Reg.[2]	Event Type[3]	No.[4]	Tilt[5]		Lag[6]	Comment[7]	Fig.[8]
Start	End				Mag	Az			
4/18/1984 18:00	4/21/1984 5:33	MERZ	E		30.1	112	-4h 0m	Episode 18; tilt 4/18-22	
4/21/1984 21:15	4/22/1984 09:00	sf3kuer	EQS	7				south flank response; *M*3.8 (2) with aftershocks	
5/16/1984 5:00	5/18/1984 0:50	MERZ	E		8.6	127	-5h 0m	Episode 19; tilt 5/14–18	
6/7/1984 21:04	6/8/1984 6:25	MERZ	E		21.8	124	-0h 56m	Episode 20; tilt 6/6–11	
6/30/1984 10:28	6/30/1984 18:27	MERZ	E		12.7	107	-0h 32m	Episode 21; tilt 6/30–7/1	
7/8/1984 19:30	7/9/1984 10:17	MERZ	E		11.3	117	-1h 30m	Episode 22; tilt 7/7–10	
7/28/1984 12:00	7/29/1984 5:40	MERZ	E		16.2	113	-1h 0m	Episode 23; tilt 7/28–29	**7.4**
8/19/1984 21:52	8/20/1984 17:25	MERZ	E		12.0	117	-1h 8m	Episode 24; tilt 8/19–21	**7.4**
9/19/1984 16:04	9/20/1984 5:32	MERZ	E		20.9	114	-0h 56m	Episode 25; tilt 9/19–20	
11/2/1984 11:40	11/2/1984 16:36	MERZ	E		9.9	112	-0h 20m	Episode 26; tilt11/2–4	
11/20/1984 0:05	11/20/1984 10:06	MERZ	E		10.5	106	-0h 55m	Episode 27; tilt 11/19–20	
12/3/1984 19:05	12/4/1984 9:41	MERZ	E		10.4	124	+0h 5m	Episode 28; tilt 12/3–4	
12/23/1984 06:14	12/23/1984 13:37	sf4swr	SDI	24	flat			*M*3.2 and aftershocks during inter-Episode inflation	G5
1/3/1985 13:15	1/4/1985 5:04	MERZ	E		16.5	113	-0h 45m	Episode 29; tilt 1/2v7	
2/4/1985 5:46	2/5/1985 2:46	MERZ	E		31.1	108	-2h 14m	Episode 30; tilt 2/3–5	
2/21/1985 19:48		sf3kuer	EQ					*M*4.8[10] with aftershocks	
3/13/1985 6:00	3/14/1985 4:55	MERZ	E		30.9	117	-3h	Episode 31; tilt 3/11–14	
4/21/1985 15:16	4/22/1985 9:06	MERZ	E		21.4	114	-4h 44m	Episode 32; tilt 4/21–23	
5/20/1985 06:16	5/20/1985 08:10	ms2	EQS	6	flat			Nāmakanipaio[9]	
6/11/1985 05:19	6/11/1985 19:12	sf3kuer	EQS	9				South flank anticipation	
6/12/1985 23:06	6/13/1985 4:53	MERZ	E		7.7	110	-0h 54m	Episode 33; tilt 6/12–14; 2-step def	
6/13/1985 11:13	6/13/1985 16:58	sf3kuer	EQS	9				South flank response	
7/6/1985 19:03	7/7/1985 8:50	MERZ	E		18.3	118	-3h 57m	Episode 34; tilt 7/6–8	
7/26/1985 2:52	7/26/1985 9:52	MERZ	E		10.5	119	+0h 52m	Episode 35; tilt 7/25–26	
7/27/1985 4:14	8/12/1985 4:30	MERZ	E					Episode 35a—fissure continuation	
9/2/1985 14:00	9/2/1985 23:35	MERZ	E		14.4	110	-3h	Episode 36; tilt 9/2–3	
9/24/1985 18:08	9/25/1985 6:19	MERZ	E		14.3	104	-5h 52m	Episode 37; tilt 9/24–25	

Table 7.1. Pu'u 'Ō'ō-Kupaianaha eruption[1]: Stage IA—2/1/1983–7/20/1986 (see fig. 7.1).—Continued

[In rows with multiple entries text applies down to the next entry; dates in m/d/yyyy format at; do, ditto (same as above); data for eruptions and traditional intrusions are emphasized by grey shading]

Date and time		Reg.[2]	Event Type[3]	No.[4]	Tilt[5]		Lag[6]	Comment[7]	Fig.[8]
Start	End				Mag	Az			
10/5/1985 21:12	10/6/1985 20:44	sf3.2	SDI??	8, 4	flat			Possibly a suspected deep intrusion	G6
10/21/1985 3:00	10/21/1985 11:24	MERZ	E		17.2	113	+1h	Episode 38; tilt10/20–22	
11/13/1985 15:34	11/14/1985 1:24	MERZ	E		16.0	110	-1h 26m	Episode 39; tilt 11/13–15	
12/31/1985 13:35	12/31/1985 17:55	ms2	EQS	8				Nāmakanipaio[9]	
1/1/1986 13:09	1/2/1986 2:38	MERZ	E		14.5	114	-1h 51m	Episode 40; tilt 1/1–2	
1/27/1986 20:35	1/28/1986 7:57	MERZ	E		14.0	112	-0h 25m	Episode 41; tilt 1/27–28	
1/27/1986 13:36	1/30/1986 06:54	sf3kuer	EQS	11				Weak south flank anticipation/response	
2/22/1986 15:15	2/23/1986 4:20	MERZ	E		6.4	111	-1h 45m	Episode 42; tilt 2/22–23	
3/22/1986 4:50	3/22/1986 15:56	MERZ	E		12.7	107	-0h 45m	Episode 43; tilt 3/21–23	
4/13/1986 20:54	4/14/1986 7:56	MERZ	E		24.5	102	-3h 6m	Episode 44; tilt 4/11–14	
5/7/1986 15:45	5/7/1986 21:07	sf3kuer	EQS	8				Weak south flank anticipation	
5/7/1986 22:41	5/8/1986 11:06	MERZ	E		13.3	123	no data	Episode 45; tilt 5/7–8	
6/2/1986 2:29	6/2/1986 13:20	MERZ	E		14.6	105	no data	Episode 46; tilt 6/1–2	
6/26/1986 4:19	6/26/1986 16:35	MERZ	E		23.5	116	no data	Episode 47; tilt 6/25–29	
7/18/1986 12:05	7/19/1986 09:30	MERZ	E		17.9	106	no data	Episode 48 fissure eruption	G7

[1]References as follows: Wolfe and others (1987, 1988) cover the first 20 episodes. Heliker and others (2003) summarize the first 20 years of eruption. Other references to the eruption are covered in the Hawai'i bibliographic database (Wright and Takahashi, 1998) from which all references pertaining to the eruption may be obtained by searching on the keyword "kl.erz.1983".

[2]Earthquake classification abbreviations are given according to the classification in appendix table A3, and locations are shown on appendix figure A4.

[3]Event types defined in chapter 1 are abbreviated as follows: **E**, Eruption; intrusion ("traditional" **I;** "inflationary" **II;** "suspected deep" **SDI**); **EQS**, earthquake swarms; EQ, earthquake $M \geq 4$; "surge"—abrupt inflation at Kīlauea's summit followed by an increase in magma output at Pu'u 'Ō'ō. These later become "d-i-d" events in which there is a preliminary deflation at Kīlauea's summit followed by sharp inflation and then deflation. This pattern is matched by the electronic tilt at Pu'u 'Ō'ō with a slight time delay.

[4]Minimum number of events defining a swarm: 20 for south flank; 10 for all other regions.

[5]Magnitude in microradians (μrad) and azimuth of daily tilt measurements from the water-tube tiltmeter in Uwēkahuna vault.

[6]Lag times separating the onset of the earliest earthquake swarm (excluding south flank) or eruption (in the absence of a precursory swarm) for a given event and the beginning of deflation or inflation measured by the continuously recording Ideal-Arrowsmith tiltmeter in Uwēkahuna vault. (+) tilt leads, (-) tilt lags. For the later eruption surges (Deflation-Inflation-Deflation events) lag times (+) are given for the delayed response of the tiltmeter at Pu'u 'Ō'ō (POC) time lag in minutes, and the increase in eruptive activity at or near the Pu'u 'Ō'ō vent (E) time lag in hours.

[7]Abbreviations as follows: ftn, fountaining; eq, earthquake; eqs, earthquake swarm; fs, foreshock; as, aftershock; ms, mainshock; sf, south flank; inf, inflation; def, deflation; ant, anticipation (preceding event); acc, accompaniment (during event); resp, response (following event); I-A, Ideal-Arrowsmith continuously recording tiltmeter in Uwēkahuna vault; drm, downrift migration of earthquakes; urm, uprift migration of earthquakes; abs, missing data.

[8]Text figures **bold text**; appendix figures plain text.

[9]Continuation of earthquakes triggered by the 1983 Ka'ōiki earthquake. Locations are beneath the Nāmakanipaio campground in Hawai'i Volcanoes National Park.

[10]Klein and others, 2006.

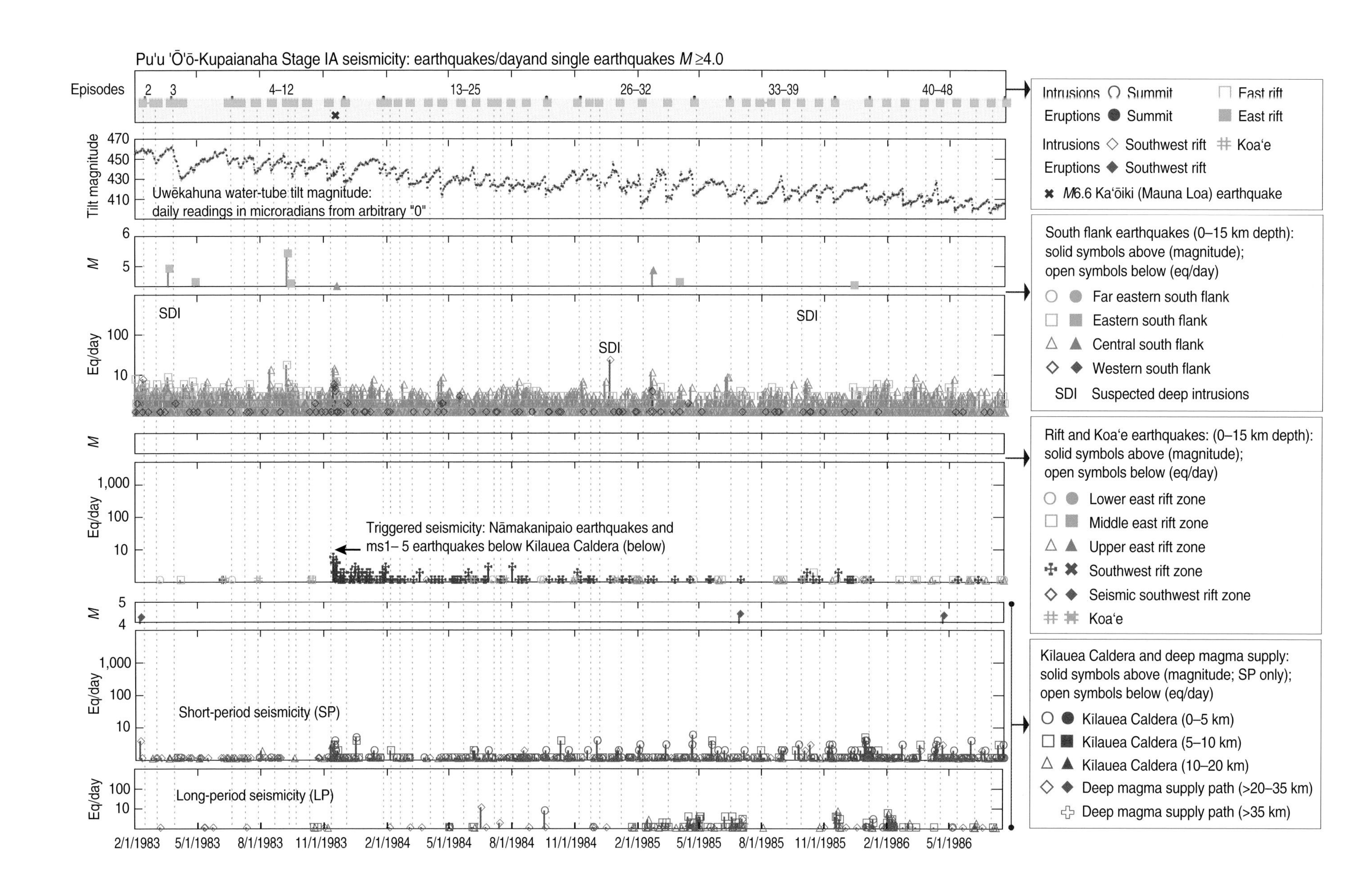

Pu'u 'Ō'ō-Kupaianaha Stage IA seismicity: earthquakes/dayand single earthquakes M ≥4.0
Episodes
2
3
4–12
13–25
26–32
33–39
40–48
Tilt magnitude
470
450
430
410
Uwēkahuna water-tube tilt magnitude:
daily readings in microradians from arbitrary "0"
M
6
5
Eq/day
SDI
SDI
SDI
100
10
M
Eq/day
1,000
100
10
Triggered seismicity: Nāmakanipaio earthquakes and
ms1– 5 earthquakes below Kīlauea Caldera (below)
M
5
4
Eq/day
1,000
100
10
Short-period seismicity (SP)
Eq/day
100
10
Long-period seismicity (LP)
2/1/1983
5/1/1983
8/1/1983
11/1/1983
2/1/1984
5/1/1984
8/1/1984
11/1/1984
2/1/1985
5/1/1985
8/1/1985
11/1/1985
2/1/1986
5/1/1986
Intrusions Summit
East rift
Eruptions Summit
East rift
Intrusions Southwest rift
Koa'e
Eruptions Southwest rift
M6.6 Ka'ōiki (Mauna Loa) earthquake
South flank earthquakes (0–15 km depth):
solid symbols above (magnitude);
open symbols below (eq/day)
Far eastern south flank
Eastern south flank
Central south flank
Western south flank
SDI Suspected deep intrusions
Rift and Koa'e earthquakes: (0–15 km depth):
solid symbols above (magnitude);
open symbols below (eq/day)
Lower east rift zone
Middle east rift zone
Upper east rift zone
Southwest rift zone
Seismic southwest rift zone
Koa'e
Kīlauea Caldera and deep magma supply:
solid symbols above (magnitude; SP only);
open symbols below (eq/day)
Kīlauea Caldera (0–5 km)
Kīlauea Caldera (5–10 km)
Kīlauea Caldera (10–20 km)
Deep magma supply path (>20–35 km)
Deep magma supply path (>35 km)

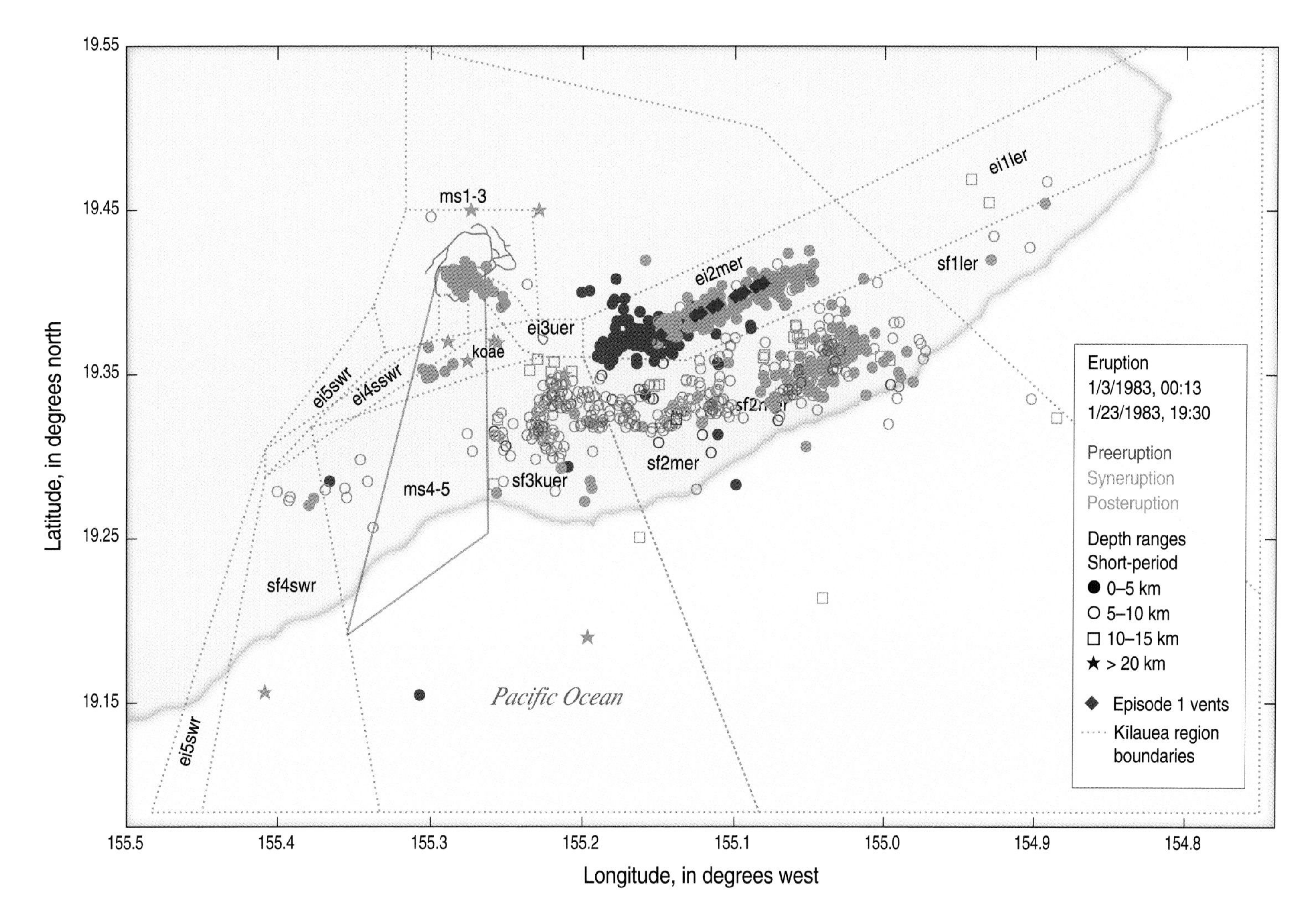

Figure 7.2. Map showing Kīlauea activity, January 1 through January 31, 1983: Pu'u 'Ō'ō-eruption episode 1. Preeruption intrusion is uprift of the vents, and intrusion during the eruption occurs beneath the erupting vents on the upper east rift zone and extends somewhat farther east. The zone due south of the upper east rift zone, which is the locus of many suspected deep intrusions, was apparently activated by a seismic passage of magma on the way to being erupted. Dates on figure in m/d/yyyy format.

Figure 7.1. Graphs showing Kīlauea activity, 1 February 1983–20 July 1986: Stage IA of the Pu'u 'Ō'ō-Kupaianaha eruption. Time-series plots show the number of earthquakes per day and magnitudes (*M*) ≥4 plotted against times of eruption and intrusion and the Uwēkahuna tilt magnitude. Symbols are given on the plots. Seismic regions are shown in figure A4 of appendix A. Top panel: Times of eruption and traditional intrusion. Numbered episodes of the Pu'u 'Ō'ō-Kupaianaha eruption are shown. Vertical dashed lines connect seismicity associated with eruptions and associated traditional intrusions. Second panel from top: Uwēkahuna tilt magnitude related to times of eruption and intrusion emphasized by vertical dotted lines. Tilt magnitudes are given in microradians. Bottom seven panels: seismicity is plotted, from bottom to top, for the magma supply path (divided into short-period and long-period earthquakes), rift zones and Koa'e, and south flank. Earthquakes per day (eq/day) and magnitudes (*M*) greater than or equal to 4.0 are given for each region. Three types of intrusions are defined in chapter 1. Traditional intrusions are shown as intrusion symbols. Inflationary intrusions are labeled (II), and suspected deep intrusions are labeled (SDI) but are not plotted as intrusions. Dates on figure in m/d/yyyy format.

Pause: A temporary cessation in eruptive activity. The earliest pauses were accompanied by significant tilt and earthquake signals, whereas later pauses gradually came to lack such association.

Surge: A visible increase in eruptive activity at Pu‘u ‘Ō‘ō. The earliest surges followed a pattern of correlated gradual deflation at Kīlauea’s summit and Pu‘u ‘Ō‘ō, followed by rapid inflation of several microradians and a subsequent less rapid deflation, as measured by continuously recording tiltmeters. Once the tilt correlation was recognized, these become known as “deflation-inflation-deflation” (D-I-D) events. Eventually the D-I-D events were no longer associated with visible changes in eruptive activity and the tilt signal shifted to a smaller paired and nearly simultaneous rapid deflation and inflation at Kīlauea’s summit and at Pu‘u ‘Ō‘ō.

Earthquake swarm (eqs): Located earthquakes are from the HVO catalog and consist of both brittle-failure (short-period) and long-period events. A continuous sequence consisting of at least 10 earthquakes of one type within a single region separated by times of less than 6 hours is designated as an earthquake swarm. Most earthquake swarms consist of more than 10 events occurring at rates exceeding one earthquake per hour. Swarms are caused by a sustained increase in stressing rate and are almost always associated with magma movement of some kind.

Silent earthquakes: “Silent” or “slow” earthquakes (see, for example, Cervelli and others, 2002b; Segall and others, 2006) represent an abrupt increase in the rate of south flank spreading and are accompanied by a sequence of south flank earthquakes not related to a rift intrusion.

Intrusion: We recognize three types of intrusions, “traditional,” ”inflationary.” and “suspected deep,” as defined in chapter 1. Traditional and inflationary intrusions are associated with earthquake swarms at depths of 2 km or less. Traditional intrusions have a sharp tilt drop at Kīlauea’s summit and may occur alone or as part of an eruptive sequence. Inflationary intrusions occur during stable tilt or continuing inflation and are confined to the near-summit parts of the rift zones. The suspected deep intrusions were originally named “slow” (Klein and others, 1987; Wright and Klein, 2008). The name change is made to emphasize their association with deeper magmatic processes and to avoid confusion with traditional intrusions that evolve slowly.

Tilt changes: Tilt changes are measured daily by the short-base water-tube tiltmeter located in the Uwēkahuna Vault and are expressed as an azimuth in degrees measured clockwise from north and a magnitude in microradians. The water-tube results are compared with readings from a single component (east-west) continuously recording Ideal-Aerosmith (“I-A”) tiltmeter co-located in the Uwēkahuna Vault from June 1965 to May 1992. Beginning in January 1999, a two-component continuously recording borehole tiltmeter (“UWE”) was installed near the Uwēkahuna Vault site.

“Lag” times (in tables): The difference in time, given in hours (h) and minutes (m) between (1) the onset of an earthquake swarm near the site of eruption and a deflection on the continuously recording tiltmeters I-A and UWE or (2) the difference in time between the onset of sharp inflation or deflation at Kīlauea’s summit and observed increase in eruptive activity. Most intrusions have a negative value for (1), which means that the onset of the earthquake swarm began before the change in tilt. All values for (2) are positive because tilt changes begin at Kīlauea’s summit before changes on the rift zone.

Stage IA: Episodic Eruption at Pu‘u ‘Ō‘ō, 1 February 1983–20 July 1986

Table 7.1 and figure 7.1 summarize all eruptions, intrusions, earthquake swarms, and other significant events for stage IA. Figures showing earthquake locations for selected events are referenced in the right-hand column of the table 7.1. Some figures may show the near-simultaneous occurrence of more than one type of event.

Following the end of episode 1, the eruption devolved into a series of 46 high-fountaining events, each lasting for 6–60 hours, separated by quiet intervals of about a month. The tilt and seismicity patterns of individual Pu‘u ‘Ō‘ō episodes during stage IA mimic the behavior of previous small rift eruptions and of early episodes of Mauna Ulu in 1969. Each episode was immediately preceded by a sharp drop in summit tilt, with partial recovery in between. Each episode shows a remarkably similar deflation azimuth, ranging from 102 to 127 degrees, mostly between 110 and 120 degrees, closely matching the initial deflation azimuth of episode 1. By the end of stage IA the net deflation at Kīlauea’s summit was more than 190 µrad at an azimuth of 150 degrees.

Mauna Loa's Ka'ōiki Earthquake of 16 November 1983: Triggered Seismicity at Kīlauea

On 16 November 1983 a *M*6.7 earthquake occurred on Mauna Loa's Ka'ōiki Fault Zone[25] (see Buchanan-Banks, 1987; Jackson and others, 1992). The earthquake triggered seismicity beneath Kīlauea, shown well by the earthquake counts (fig. 7.1). Counts of earthquakes deeper than 20 km increased dramatically, as did located events immediately west of Kīlauea Caldera, referred to as "Nāmakanipaio" for their proximity to a Kīlauea campground of that name (Okubo and Nakata, 2003, p. 178)[26]. The latter occur in two regions (fig. 7.3), which became active because they are part of the aftershock zone (not shown). The southern region may correspond to a Ka'ōiki fault plane, now concealed beneath lava erupted from Kīlauea's southwest rift zone. The northern region may correspond to either the northern continuation of the same Ka'ōiki fault or a concealed former Kīlauea Caldera boundary fault now covered by lava erupted from Kīlauea's summit (A. Miklius, oral. commun., 2005). Both of these regions are active over the first 4 days following the 1983 earthquake. The southern region was not active again, but the northern region has shown periodic swarms during the entire Pu'u 'Ō'ō eruption (events labeled Nāmakanipaio in table 7.1) and nonswarm activity at many other times. The counts of events deeper than 20 km decrease and reach background levels by February 1984 (fig.7.1).

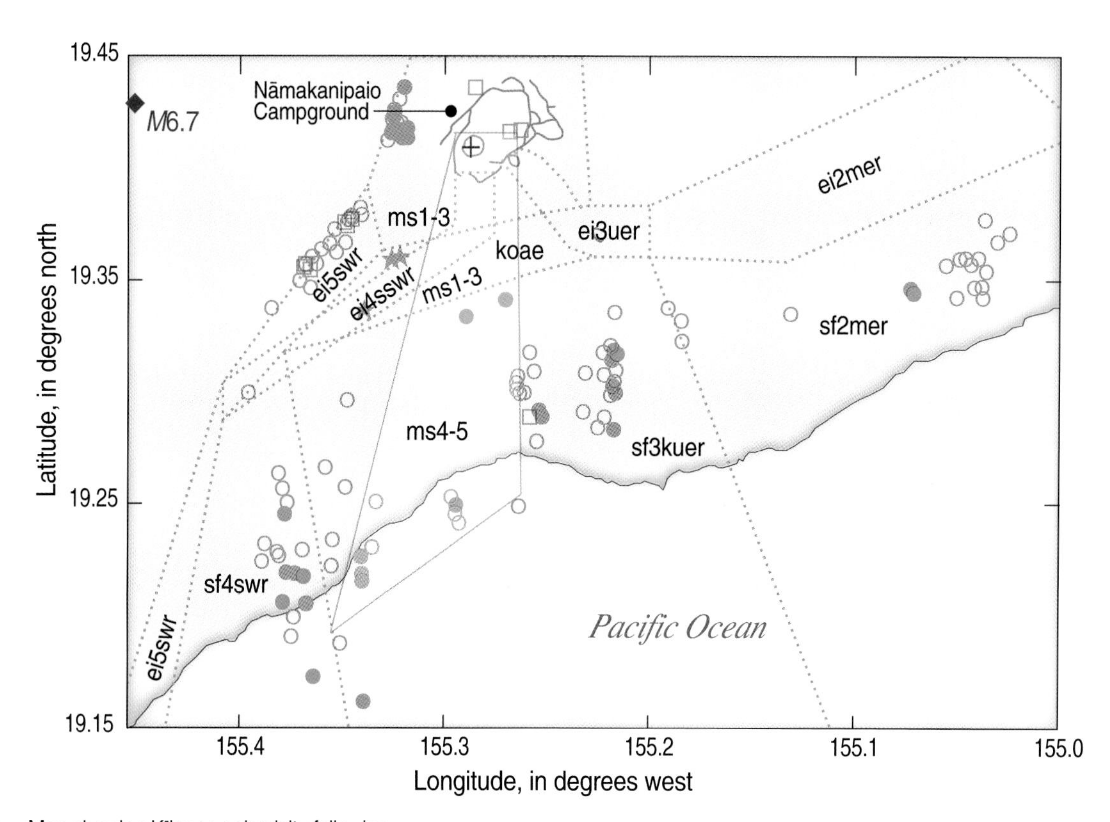

Figure 7.3. Map showing Kīlauea seismicity following the 16 November 1983 Ka'ōiki earthquake (*M*6.7). Other Mauna Loa earthquakes are not shown. A sequence of Nāmakanipaio earthquakes was triggered about 4 hours after the mainshock, with locations both beneath Nāmakanipaio and the upper southwest rift zone. Map shows events through 21 November 1983. Subsequent Nāmakanipaio swarms did not affect the southwest rift zone. Also shown are earthquake swarms beneath Kīlauea's south flank contemporaneous with the Nāmakanipaio swarm, which have the characteristics of suspected deep intrusions at four different places from the western to the eastern south flank. Dates on figure in m/d/yyyy format.

[25] Located on the lower slopes of Mauna Loa's south flank adjacent to Kīlauea.

[26] Nāmakanipaio can be regarded as an extension of the aftershock zone.

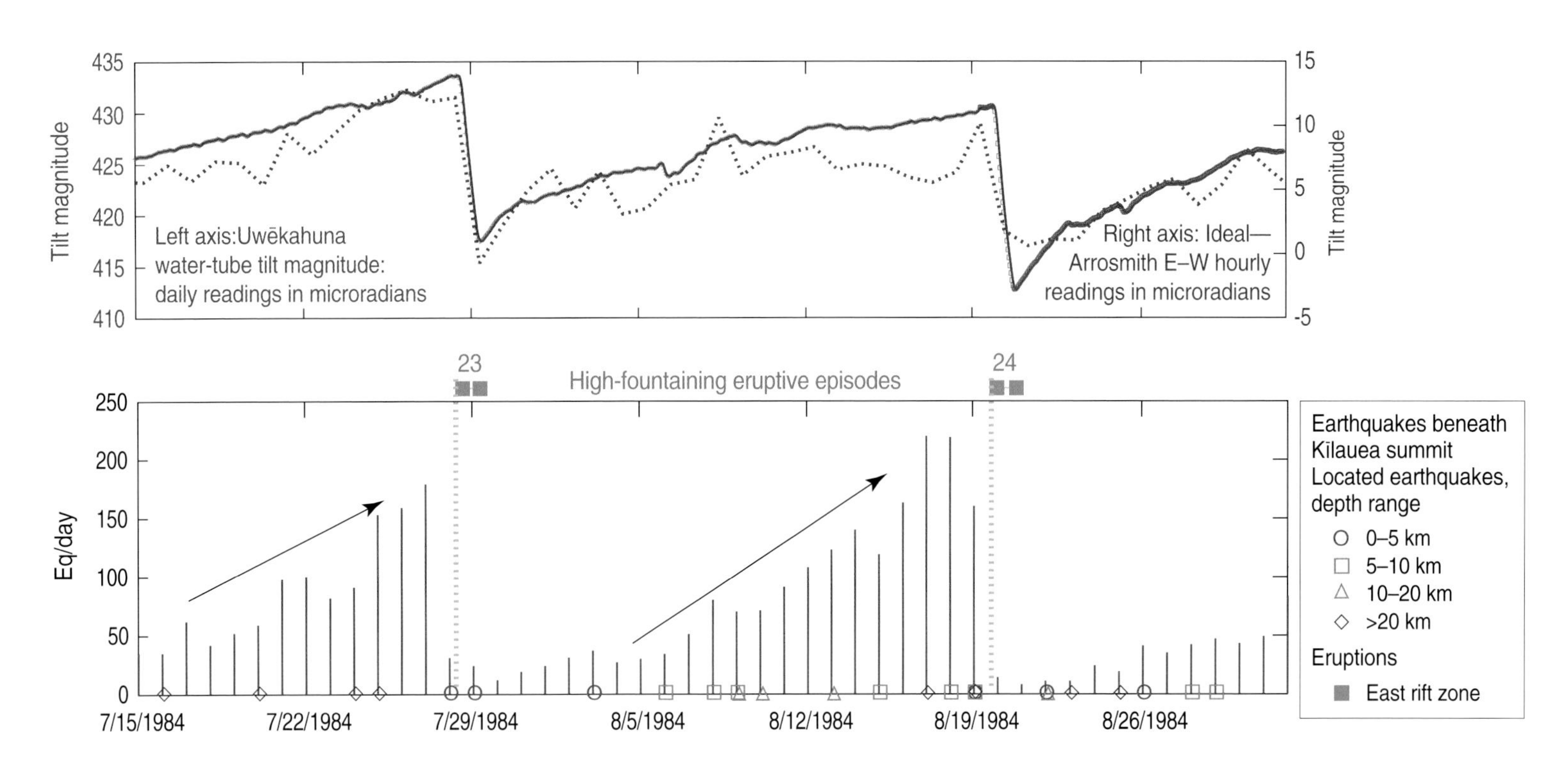

Figure 7.4. Plots showing cyclic variation of summit tilt (top panel) and numbers of short-period earthquakes per day (eq/day) too small to locate (red bars, bottom panel) at 0–5 km beneath Kīlauea's summit during stage IA of the Puʻu ʻŌʻō-Kupaianaha eruption. Shown here are patterns for episodes 23 and 24; data from 15 July to 1 September 1984. Earthquake counts increase before each episode, then decrease during the eruptive fountaining. The cyclic variation of summit tilt is shown by both water-tube tilt measured daily and Ideal-Aerosmith electronic tilt measured continuously but recorded hourly. Dates on figure in m/d/yyyy format.

Located earthquakes at 10–20 km depth beneath Kīlauea's summit (ms3, defined in appendix A, table A3) increase following the Ka'ōiki earthquake, but then die off quickly.

Several rhythmic variations in earthquake counts are associated with eruptive episodes. Short-period daily counts of summit earthquakes above 5-km depth increase before eruptive episodes (fig. 7.4). Long-period counts at the same depth show a less steady rhythmic pattern. The rhythmic increase in shallow short-period counts before eruptive episodes is less obvious in the period preceding episodes 12 and 13, immediately following the Ka'ōiki earthquake.

There were no traditional intrusions, as revealed by rift earthquake swarms, associated with any of the high-fountaining episodes following episode 1. Two periods of increased south flank seismicity between episodes, which are classified as potential suspected deep intrusions, are listed in table 7.1. The lack of continuously recorded deformation data precludes the identification of spreading steps associated with these earthquake swarm events.

Stage IB: Continuous Eruption at Kupaianaha, 20 July 1986–8 November 1991

A fissure eruption extending downrift from Pu'u 'Ō'ō on 18 and 19 July 1986 (HVO episode 48) marks the transition from episodic eruption at Pu'u 'Ō'ō to continuous eruption at Kupaianaha, which is at the eastern or lower end of the fissure system (Heliker and Mattox, 2003). There is no obvious precursory seismicity to indicate that emplacement of a new dike caused a change of eruption style and location. A small shield surmounted by an active lava lake was built over the Kupaianaha vent, and Kupaianaha grew both by endogenous processes and by periodic overflow. Activity at Kupaianaha gradually declined toward the end of stage IB. Its demise on 7 February 1992 was predicted several months before that by extrapolation of its declining flow rate (Kauahikaua and others, 1996). Events of this period are summarized in table 7.2 and figure 7.5.

Kīlauea seismic activity continued at near-normal levels during stage IB as the summit continued to deflate. The average daily Uwēkahuna tilt azimuth for the entire Kupaianaha period (1986–92) is 198 degrees, near the westernmost Fiske-Kinoshita inflation-deflation centers defined in chapter 4, figure 4.2. This is consistent with the shift of summit subsidence to include Kīlauea's southwest rift zone (see, for example, HVO monthly report for February 1990; (Delaney and others, 1998)). The overall rate of seismicity beneath the shallow summit, rift zones, and Koa'e in stage IB increases relative to stage IA, but the south flank seismicity decreases and magma supply seismicity deeper than 5 km remains about the same.

The following events (table 7.2) occurred within stage IB: (1) the onset of located long-period earthquake swarms deeper than 5 km beneath Kīlauea's summit (lpc-c), beginning in May 1987[27] and continuing throughout stage IB (fig. 7.5), (2) a suspected deep intrusion on 22–24 March 1990 (fig. 7.6), (3) pauses in the continuous eruption beginning in April 1988 (fig. 7.4), (4) a magnitude 6.2 south flank earthquake on 26 June 1989 with aftershocks, and (5) four intrusions into the upper east rift zone.

Pauses in the supply of magma to the eruption have a seismic as well as a tilt signature. The first eruption pause occurred at the end of April 1988. Subsequently there was a series of pauses in 1990 that showed variation in summit tilt and seismic signature within a specific time sequence. All pauses were preceded by an increase in lpc-c (5–13 km) earthquake counts, followed shortly (except for the 1988 pause) by a small deflation measured on the continuously recording Ideal-Aerosmith tiltmeter. The pause in eruption began following the end of deflation and ended sometime after the summit began to reinflate. Tilt and seismicity associated with two typical pauses are shown in figure 7.7. This pattern gradually disappeared in the pauses of stage II.

Swarms of long-period earthquakes (lpc-c) greatly increased after May 1987 and became continuous between March 1989 and March 1990 (fig. 7.5). They were unusual in that all previous long-period swarms had been shallow, as documented for the 1977 eruption in chapter 6. The deep long-period earthquake swarms decreased to background levels by the end of 1991.

The earthquake of 26 June 1989 (Arnadottir and others, 1991; Delaney and others, 1993; Dvorak, 1994) occurred within a normal seismic background (fig. 7.5). The earthquake was followed by an aftershock sequence extending offshore and produced a south flank displacement of more than 25 cm and an almost equal amount of subsidence south of the lower east rift zone (Delaney and others, 1993; Dvorak, 1994). Focal mechanisms indicate that offset occurred on planes shallower than the decollement (Arnadottir and others, 1991; Bryan, 1992). Although not recognized at the time, the declining occurrence rate of such $M \geq 5$ south flank earthquakes may have been influenced by the declining residual stress field of the 1868 *M*8 event (Klein and Wright, 2008) that occurred more than a century earlier. The 1989 earthquake had no apparent effect on the eruption.

The intrusion of 4 December 1990 (table 7.2; fig. 7.5) was the first large intrusion to occur in the

[27] Previous to 1987, long-period earthquakes deeper than 5 km had occurred individually, but not in noticeable swarms.

Table 7.2. Pu‘u ‘Ō‘ō-Kupaianaha eruption[1]: Stage IB—7/18/1986–11/8/1991 (see fig. 7.5).

[In rows with multiple entries text applies down to the next entry; dates in m/d/yyyy format; do, ditto (same as above); data for eruptions and traditional intrusions are emphasized by grey shading; "no data" in Lag column refers to gaps in the record of the Ideal-Arrowsmith tiltmeter]

Date and Time		Reg.[2]	Event Type[3]	No.[4]	Tilt[5]		Lag[6]	Comment[7]	Fig.[8]
Start	End				Mag	Az			
7/18/1986 12:05	7/19/1986 9:30	MERZ	E[9]		23.5	106	no data	Episode 48[9] tilt 7/18–19	G8
7/20/1986 8:30	2/7/1992 0:00	MERZ	E[10]		5.4	122	no data	Episode 48[10]; tilt 7/25–27	
9/13/1986 12:15	9/13/1986 19:58	ms2.1	EQS	13, 1	4.8	303		Nāmakanipaio[11]; tilt 9/12–15 inflation	G9
5/19/1987 3:13	5/19/1987 7:30	sf3kuer	SDI	8				Weak swarm	
6/7/1987 20:26	6/8/1987 08:55	lpms3	EQS	22					
6/11/1987 13:12	6/13/1987 09:03	lpms3	EQS	27					
8/8/1987 21:23	8/9/1987 02:37	sf3.2	SDI	9, 3					
2/17/1988 00:02	2/18/1988 21:52	lpms3	EQS	37					
3/11/1988 03:13	3/14/1988 07:44	lpms3	EQS	44					
7/11/1988 17:43	7/12/1988 01:19	ms1-2	EQS	16				Nāmakanipaio[11]	G10
8/16/1988 16:29	8/17/1988 03:57	lpms3	EQS	23					
9/17/1988 12:56	9/17/1988 15:41	ei3uer	I	9	3.4	160		Tilt 9/18–20	G11
6/25/1989 17:27			EQ					*M*6.21 mainshock[12]	G12
6/25/1989 17:32	6/29/1989 01:23			152				Aftershocks continue to 7/10	
7/25/1989 20:04	7/30/1989 17:45	lpms2.3	EQS	104					
8/5/1989 00:23	8/6/1989 03:59	lpms2	EQS	11					
8/6/1989 18:05	8/7/1989 17:08	do	do	17					
8/8/1989 04:07	8/9/1989 08:04	do	do	11					
8/29/1989 03:14	8/31/1989 12:22	lpms2	EQS	27					
9/10/1989 17:36	9/11/1989 15:30	lpms2	EQS	24					
10/1/1989 17:54	10/6/1989 09:45	lpms2	EQS	61					
10/10/1989 08:50	10/12/1989 06:10	lpms2	EQS	78					
10/10/1989 10:59	10/12/1989 09:45	lpms3	EQS	62					
10/18/1989 03:27	10/20/1989 03:13	lpms3	EQS	92					
10/19/1989 05:25	10/20/1989 03:04	lpms2	EQS	79					
11/21/1989 14:13	11/25/1989 01:41	lpms3	EQS	44					
11/24/1989 09:35	11/27/1989 11:49	lpms2	EQS	45					
11/27/1989 02:22	11/29/1989 09:14	lpms3	EQS	30					
12/6/1989 19:42	12/9/1989 16:36	lpms2	EQS	46					
12/28/1989 20:50	12/30/1989 04:35	lpms2	EQS	33					
12/28/1989 21:12	12/30/1989 07:10	lpms3	EQS	29					

Table 7.2. Pu'u 'Ō'ō-Kupaianaha eruption[1]: Stage IB—7/18/1986–11/8/1991 (see fig. 7.5).—Continued

[In rows with multiple entries text applies down to the next entry; dates in m/d/yyyy format; do, ditto (same as above); data for eruptions and traditional intrusions are emphasized by grey shading; "no data" in Lag column refers to gaps in the record of the Ideal-Arrowsmith tiltmeter]

Date and Time		Reg.[2]	Event Type[3]	No.[4]	Tilt[5]		Lag[6]	Comment[7]	Fig.[8]
Start	End				Mag	Az			
1/17/1990 22:06	1/19/1990 10:47	lpms2	EQS	27					
2/4/1990 07:14	2/5/1990 22:04	lpms3	EQS	45					
3/22/1990 22:54	3/24/1990 06:15	sf3kuer	SDI	22					**7.6**
5/4/1990 23:09	5/5/1990 12:49	lpms3	EQS	27					
8/8/1990 16:06	8/9/1990 09:27	sf2mer	EQ					*M*4.7	
11/11/1990 18:20	11/11/1990 23:10	ms1.2	EQS	2, 4				Precursory Nāmakanipaio[11, 13]	G13
12/4/1990 16:29	12/4/1990 19:29	ms1	I	11	9.2	302	no data	Tilt 12/3–4	G13
12/4/1990 17:13	12/4/1990 22:24	er3.2	I	29, 4				Downrift migration 6.3 km/hr	
1/7/1991 17:04	1/15/1991 06:29	lpms3	EQS	204	8.8	101		Tilt 1/5–19	
3/13/1991 10:27	3/13/1991 11:33	ms1.2	EQS	7, 6				Precursory Nāmakanipaio[11, 13]	G14
3/26/1991 05:33	3/26/1991 15:27	ei3uer	I	97	17.7	107	+1h 33m	Tilt 3/21–26; downrift migration 1.3 km/hr; uprift migration 0.53 km/hr	G14
3/26/1991 05:39	3/26/1991 11:55	sf3kuer	EQS	13				South flank accompaniment—6 eq on 3/28	
8/10/1991 19:08	8/10/1991 23:05	ms1.2		1, 6				Precursory Nāmakanipaio[11, 13]	G15
8/21/1991 11:05	8/21/1991 15:23	ei3uer	I	58	5.0	291	+0h 05m	Tilt 8/20–24—inflation; I-A: < 1 µrad deflation 8/21 11:00–12:00	G15

[1]References as follows: Wolfe and others (1987, 1988) cover the first 20 episodes. Heliker and others (2003) summarize the first 20 years of eruption. Other references to the eruption are covered in the Hawai'i bibliographic database (Wright and Takahashi, 1998) from which all references pertaining to the eruption may be obtained by searching on the keyword "kl.erz.1983".

[2]Earthquake classification abbreviations are given according to the classification in appendix table A3 and locations are shown on appendix figure A4.

[3]Event types defined in chapter 1 are abbreviated as follows: **E**, Eruption; intrusion ("traditional" **I;** "inflationary" **II;** "suspected deep" **SDI**); **EQS**; earthquake swarms; EQ, Earthquake $M \geq 4$; "surge"—abrupt inflation at Kīlauea's summit followed by an increase in magma output at Pu'u 'Ō'ō. These later become "d-i-d" events in which there is a preliminary deflation at Kīlauea's summit followed by sharp inflation and then deflation. This pattern is matched by the electronic tilt at Pu'u 'Ō'ō with a slight time delay.

[4]Minimum number of events defining a swarm: 20 for south flank; 10 for all other regions.

[5]Magnitude in microradians (µrad) and azimuth of daily tilt measurements from the water-tube tiltmeter in Uwēkahuna Vault.

[6]Lag times separating the onset of the earliest earthquake swarm (excluding south flank) or eruption (in the absence of a precursory swarm) for a given event and the beginning of deflation or inflation measured by the continuously recording Ideal-Arrowsmith tiltmeter in Uwēkahuna Vault. (+) tilt leads, (-) tilt lags. For the later eruption surges (Deflation-Inflation-Deflation events) lag times (+) are given for the delayed response of the tiltmeter at Pu'u 'Ō'ō (POC) time lag in minutes, and the increase in eruptive activity at or near the Pu'u 'Ō'ō vent (E) time lag in hours.

[7]Abbreviations as follows: ftn, fountaining; eq, earthquake; eqs, earthquake swarm; fs, foreshock; as, aftershock; ms, mainshock; sf, south flank; inf, inflation; def, deflation; ant, anticipation (preceding event); acc, accompaniment (during event); resp, response (following event); I-A, Ideal-Arrowsmith continuously recording tiltmeter in Uwēkahuna Vault; drm, downrift migration of earthquakes; urm, uprift migration of earthquakes; abs, missing data.

[8]Text figures **bold text**; appendix figures plain text.

[9]Fissure eruption.

[10]Continuous eruption.

[11]Continuation of earthquakes triggered by the 1983 Ka'ōiki earthquake. Locations are beneath the Nāmakanipaio campground in Hawai'i Volcanoes National Park.

[12]Klein and others, 2006.

[13]Identified as precursor to following east rift intrusion (Okubo and Nakata, 2003).

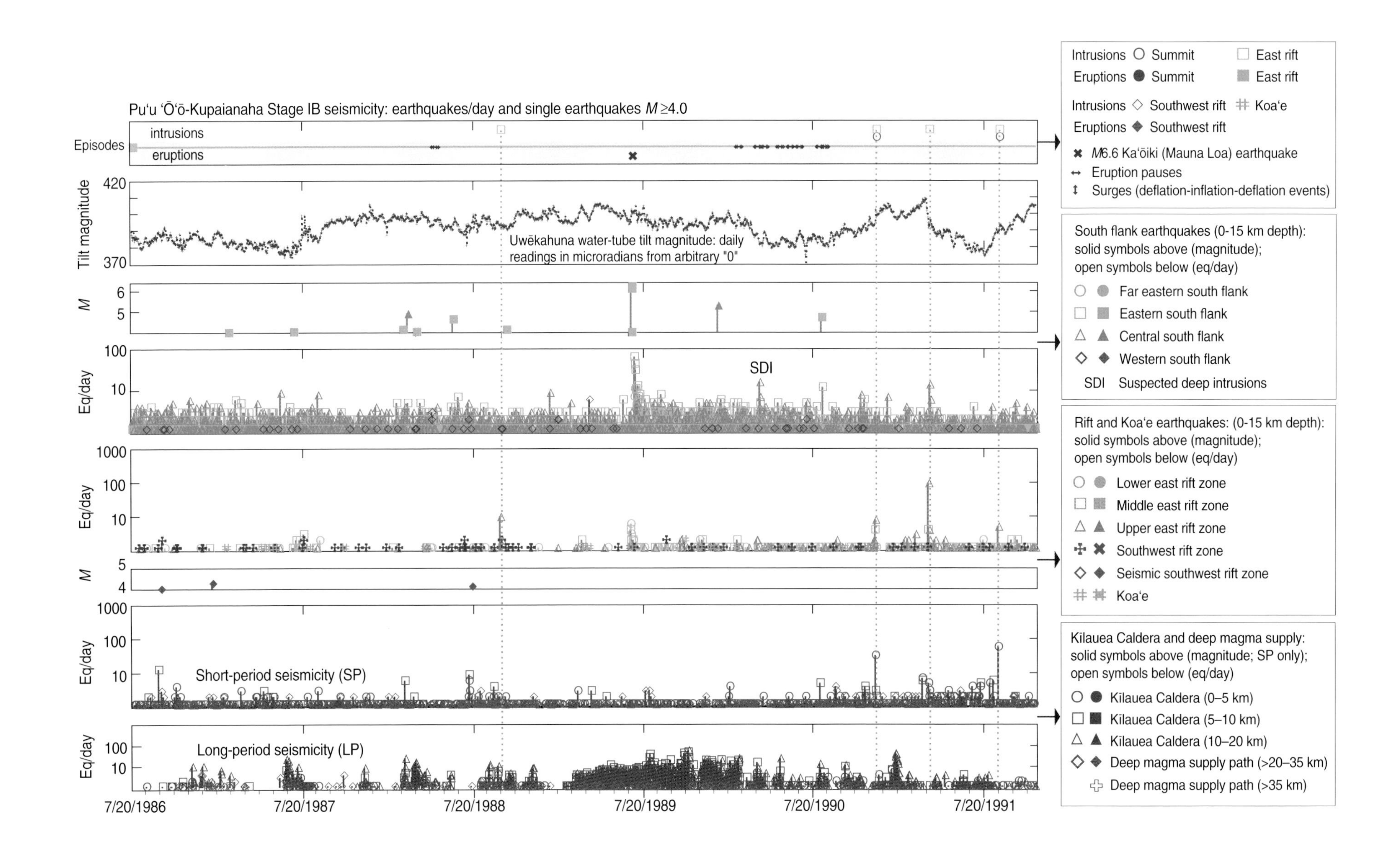

Pu‘u ‘Ō‘ō-Kupaianaha Stage IB seismicity: earthquakes/day and single earthquakes M ≥4.0
Episodes
intrusions
eruptions
Tilt magnitude
420
370
Uwēkahuna water-tube tilt magnitude: daily readings in microradians from arbitrary "0"
M
6
5
Eq/day
100
10
SDI
1000
4
Short-period seismicity (SP)
Long-period seismicity (LP)
7/20/1986
7/20/1987
7/20/1988
7/20/1989
7/20/1990
7/20/1991
Intrusions Summit East rift
Eruptions Summit East rift
Intrusions Southwest rift Koa‘e
Eruptions Southwest rift
M6.6 Ka‘ōiki (Mauna Loa) earthquake
Eruption pauses
Surges (deflation-inflation-deflation events)
South flank earthquakes (0-15 km depth): solid symbols above (magnitude); open symbols below (eq/day)
Far eastern south flank
Eastern south flank
Central south flank
Western south flank
SDI Suspected deep intrusions
Rift and Koa‘e earthquakes: (0-15 km depth): solid symbols above (magnitude); open symbols below (eq/day)
Lower east rift zone
Middle east rift zone
Upper east rift zone
Southwest rift zone
Seismic southwest rift zone
Koa‘e
Kīlauea Caldera and deep magma supply: solid symbols above (magnitude; SP only); open symbols below (eq/day)
Kīlauea Caldera (0–5 km)
Kīlauea Caldera (5–10 km)
Kīlauea Caldera (10–20 km)
Deep magma supply path (>20–35 km)
Deep magma supply path (>35 km)

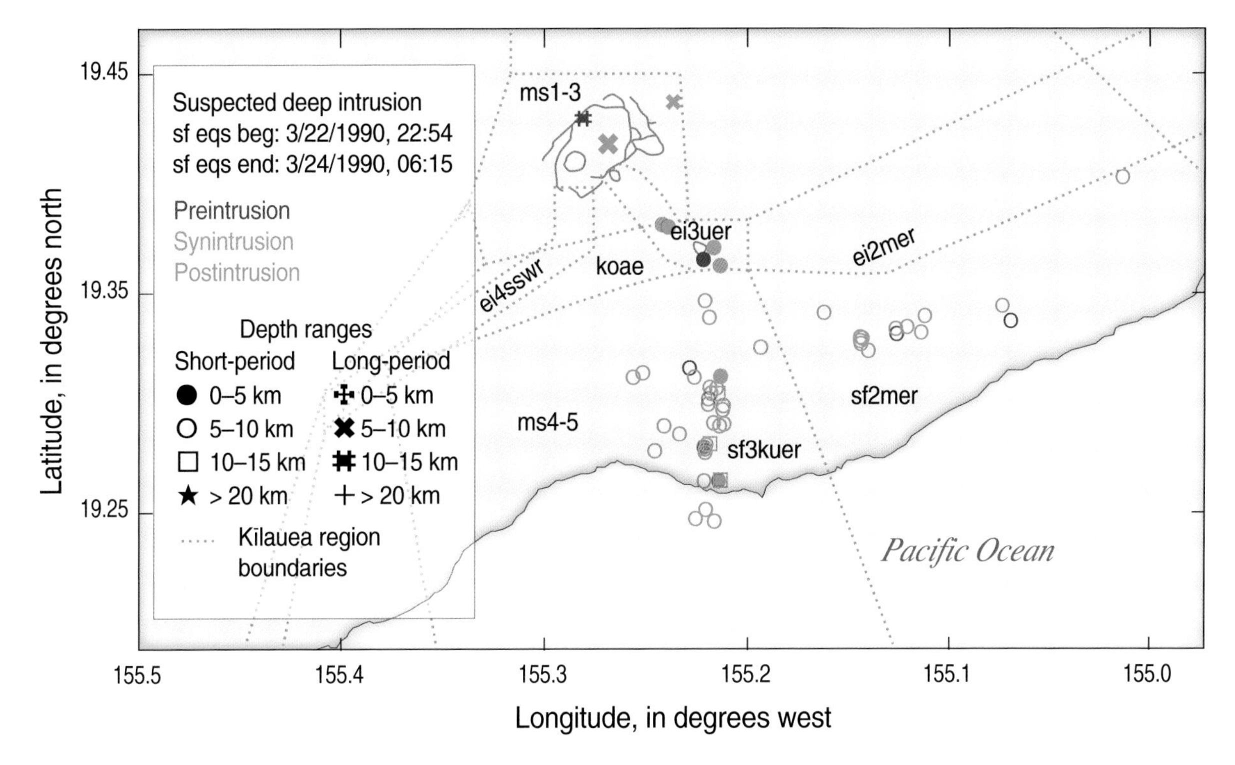

Figure 7.6. Map showing seismicity on Kīlauea, 22–28 March 1990 associated with a strong suspected deep intrusion beneath the central south flank from 22 to 24 March, possibly associated with weak intrusive activity beneath the upper east rift. Dates on figure in m/d/yyyy format.

Figure 7.5. Graphs showing Kīlauea activity, 20 July 1986–8 November 1991: Stage IB of the Pu'u 'Ō'ō-Kupaianaha eruption. Time-series plots show the number of earthquakes per day (eq/day) and magnitudes (M) ≥4 plotted against times of eruption and intrusion and the Uwēkahuna tilt magnitude. Symbols are given on the plots. Seismic regions are shown in figure A4 of appendix A. Three types of intrusions are defined in chapter 1. Traditional intrusions are shown as intrusion symbols. Inflationary intrusions are labeled (II), and suspected deep intrusions are labeled (SDI) but not plotted as intrusions. Panels from top to bottom show the following: (a), Times of eruption and intrusion. Vertical dotted lines connect seismicity associated with eruptions and associated traditional intrusions. Vertical dashed lines connect traditional and inflationary intrusions not associated with an eruption. (b), Uwēkahuna tilt magnitudes in microradians. (c), South flank earthquake magnitudes. (d), South flank earthquakes per day. (e), Rift and Koa'e earthquake magnitudes. (f), Rift and Koa'e earthquakes per day. (g), Kīlauea Caldera and deep magma-supply earthquake magnitudes. (h), Kīlauea Caldera and deep magma-supply earthquakes per day—short-period data. (i), Kīlauea Caldera and deep magma-supply earthquakes per day—long-period data. Dates on figure in m/d/yyyy format.

more than 7 years since the initial intrusion associated with the beginning of eruption in January 1983, and it was the second of four intrusions during stage IB. Less than a month before the December 1990 intrusion, a swarm of Nāmakanipaio earthquakes occurred west of Kīlauea Caldera (Okubo and Nakata, 2003; table 7.2). During the intrusion of 4 December 1990 the daily tilt at Uwēkahuna recorded a sudden inflation east of Halema'uma'u between Fiske-Kinoshita centers 1 and 2. Earthquakes began beneath the summit and extended southeast to the upper east rift zone. A few south flank earthquakes occurred both before and after this intrusion.

The intrusion of 26 March 1991 also was preceded by a Nāmakanipaio earthquake swarm west of Kīlauea Caldera (table 7.2). The summit began deflating several days before the intrusion, and the rate of deflation accelerated about 1.5 hours before the east rift earthquake swarm. The extent of uplift and horizontal displacement are not known, because ground deformation data were not available close to the time of intrusion. The earthquake swarm was concentrated in the upper east rift zone near the eastern termination of the previous intrusion. South flank seismicity during the intrusion was focused parallel to and south of the rift zone.

A fourth intrusion in the same upper east rift region occurred on 21 August 1991, preceded by Nāmakanipaio seismicity on 10 August 1991 (table 7.2). The intrusion was preceded and followed by very few south flank earthquakes.

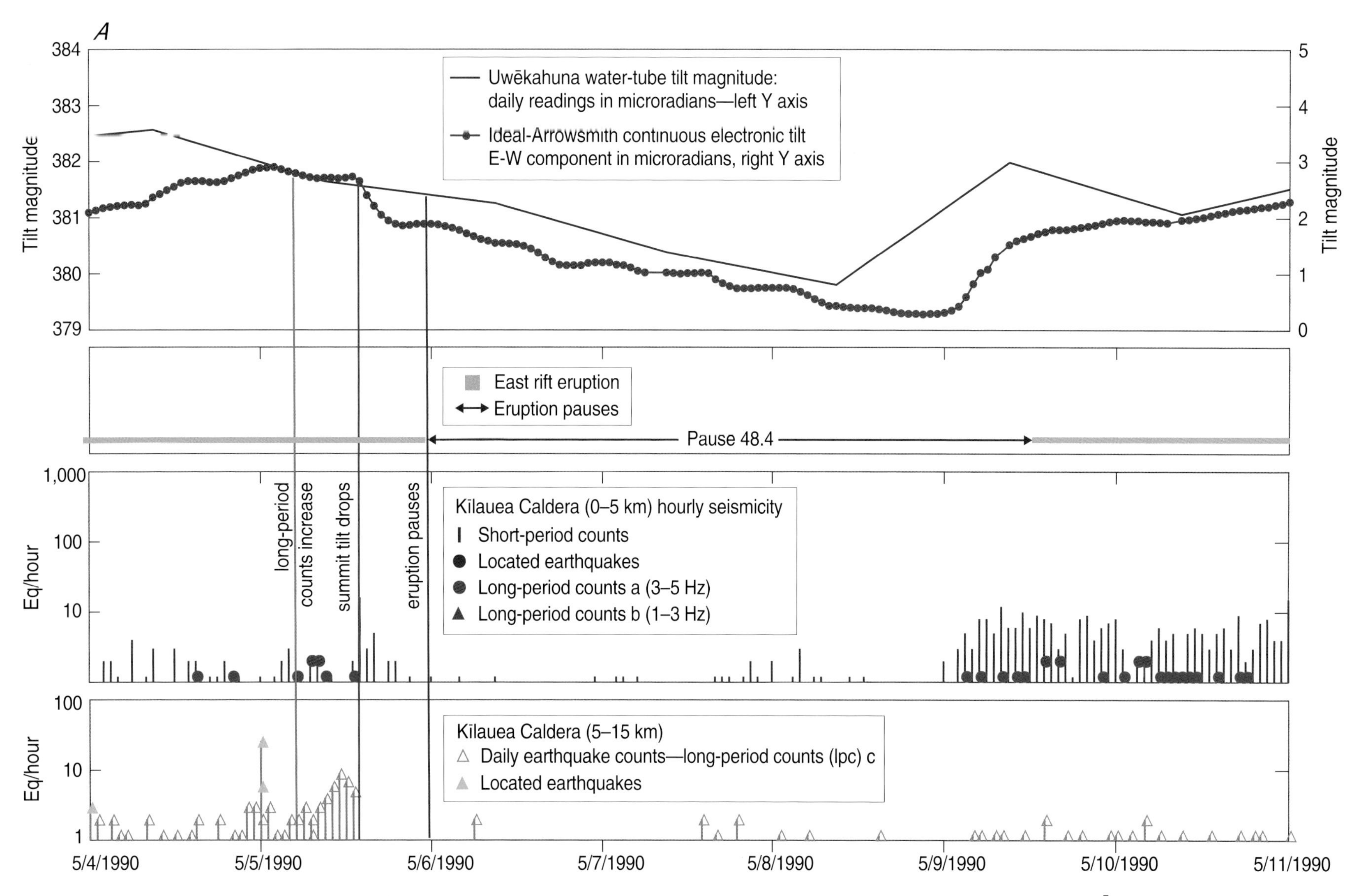

Figure 7.7. Plots of long-period earthquakes per hour (eq/hour) beneath Kīlauea's summit associated with pauses in the eruption during stage IB of the Pu'u 'Ō'ō-Kupaianaha eruption. Pauses began in 1988 (pause 48.1), and eight more occurred in 1990 (pauses 48.2–48.9). See Heliker and Mattox (2003). Each pause was preceded by an increase in lpc-c earthquakes beneath Kīlauea's summit, followed by a small deflation at Kīlauea's summit. lpc-c earthquakes are mostly absent during the pause, and shallow short-period earthquake counts pick up before the eruption resumes. See text for further explanation. Dates on figure in m/d/yyyy format. ***A***, Seismic sequences for pause 48.4, 6–9 May 1990. ***B***, Seismic sequences for pause 48.7, 31 July–2 August 1990.

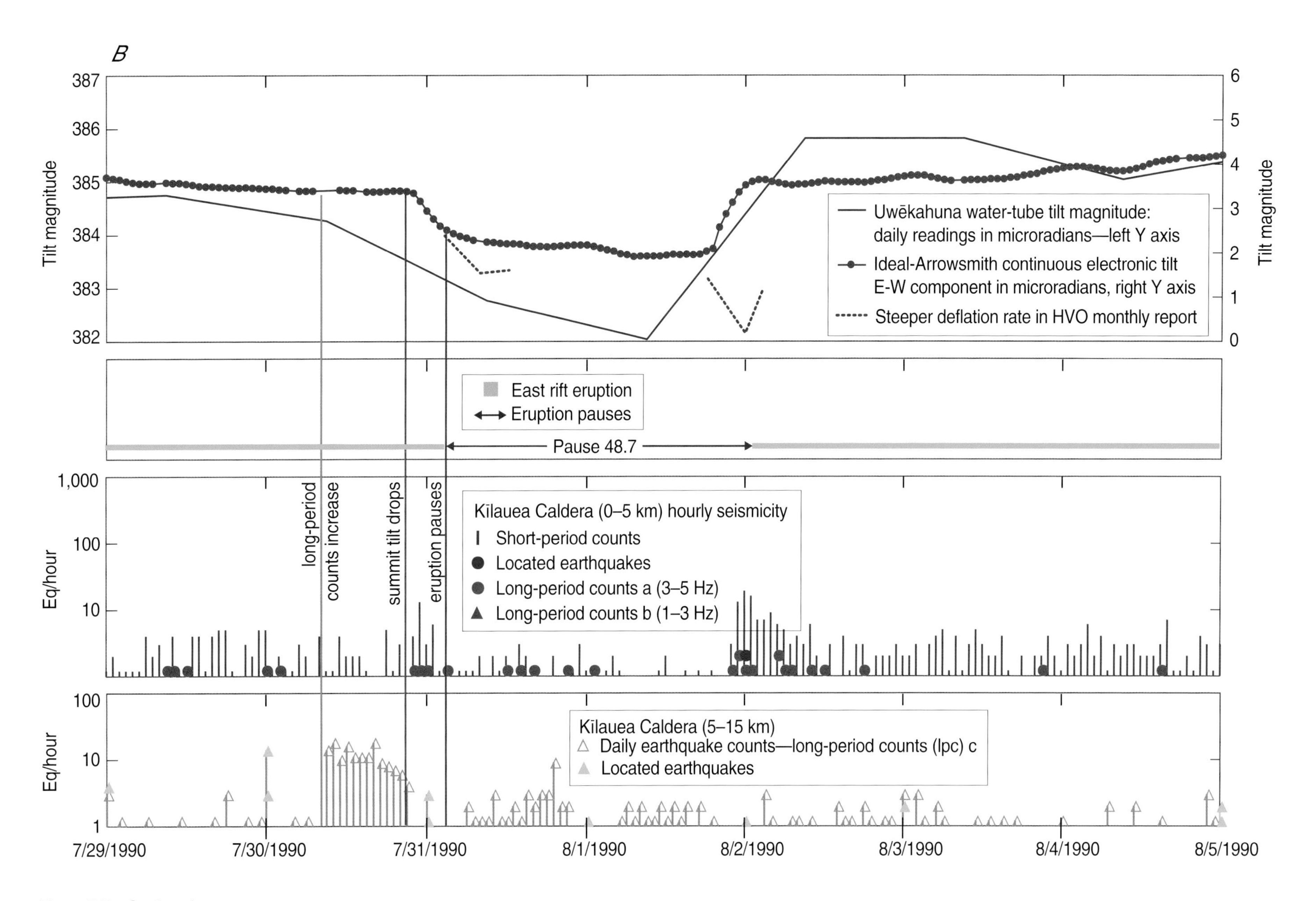

Figure 7.7.—Continued

Stage IIA: Return of Eruption to Pu‘u ‘Ō‘ō, 8 November 1991–1 January 1997

Lava returned to the flanks of Pu‘u ‘Ō‘ō on 8 November 1991 in episode 49 (Heliker and others, 1998; Mangan and others, 1995). Episode 49 was preceded by about 3.5 months of summit inflation, with no increase in seismicity anywhere on the volcano (figs. 7.5, 7.8), and was accompanied by a sharp summit deflation (table 7.3, fig. 7.8) and by a few long-period earthquakes down to 10 km beneath Kīlauea’s summit. Episode 50 began on 17 February 1992, 10 days after the cessation of activity at Kupaianaha. Episode 50 was also accompanied by a sharp summit deflation and was preceded by steady inflation and increased earthquake activity beneath the upper and middle east rift during the week before. Episode 51 began on 7 March 1992, preceded on 3–4 March by a traditional intrusion. The eruption a few days later took place quietly, accompanied by no tilt change and only a few earthquakes beneath the upper east rift zone. The onset of episode 52 on 3 October 1992 (Heliker and Mattox, 2003, table 2) was accompanied by a moderate summit deflation and scattered south flank seismicity.

The intrusion of 7–9 February 1993 was accompanied by a sharp deflation of Kīlauea’s summit, but it differed from the preceding two intrusions in combining an earthquake swarm beneath Kīlauea’s summit with activity on the upper east rift and an intense swarm of earthquakes beneath the middle east rift (figs. 7.8, 7.9). A very large sequence of Nāmakanipaio earthquakes on 24–26 January preceded the intrusion (table 7.3). South flank seismicity during the February intrusion was focused beneath the eastern south flank. Following the intrusion, earthquakes were again focused in the uppermost east rift, accompanied by an earthquake swarm beneath the central south flank. The east rift seismicity extended to the site of later eruption and intrusion in January 1997. The February 1993 intrusion led to the end of episode 52. Following an eruptive pause of less than 2 weeks duration, episode 53 began on 20 February 1993 from yet another fissure system on the flanks of Pu‘u ‘Ō‘ō.

The remainder of 1993 through the end of 1995 was seismically quiet with the exception of a major deep-magma supply earthquake on 1 February 1994 and two possible suspected deep intrusions in September 1993 and October 1994 (table 7.3). Most of the eruptive pauses during stage IIA (fig. 7.8) lacked the definitive seismic signatures that characterized the pauses of 1990. All but the first pause in March 1992 lacked rapid precursory deflation and only a few pauses were preceded by lpc-c seismic activity. Unlike the pauses of 1990, the 1993–95 pauses that were preceded by lpc-c seismic activity are not distinguished from periods of enhanced lpc-c activity that did not result in an eruptive pause.

Installation of a Broadband Seismic Array

A broadband seismic array was installed at Kīlauea’s summit in 1994 to study tremor and long-period seismicity in more detail than had been previously possible (Dawson and others, 1998). Early studies with this array identified a shallow upward extension of Kīlauea’s summit magma reservoir close to the northeast margin of Halema‘uma‘u at a depth of less than 1 km (Almendros and others, 2002a,b; Dawson and others, 2004). This shallow reservoir was not detected by earlier ground deformation studies, at least through the end of episodic high-fountaining and summit deflation characterizing stage IA. Subsequently the existence of this shallow reservoir was confirmed by the installation of continuously recording tiltmeters around Kīlauea’s summit (see below).

Figure 7.8. Graphs showing Kīlauea activity, 8 November 1991–1 January 1997: Stage II A of the Pu‘u ‘Ō‘ō-Kupaianaha eruption. Time-series plots show the number of earthquakes per day (eq/day) and magnitudes (*M*) ≥4 plotted against times of eruption and intrusion and the Uwēkahuna tilt magnitude. Symbols are given on the plots. Seismic regions are shown in figure A4 of appendix A. Three types of intrusions are defined in chapter 1. Traditional intrusions are shown as intrusion symbols, and inflationary intrusions are labeled (II). Suspected deep intrusions are labeled (SDI) but not plotted as intrusions. Dates on figure in m/d/yyyy format. Panels from top to bottom show the following: *(a)*, Times of eruption and intrusion. Vertical dotted lines connect seismicity associated with eruptions and associated traditional intrusions. Vertical dashed lines connect traditional intrusions not associated with an eruption. *(b)*, Uwēkahuna tilt magnitudes in microradians. *(c)*, South flank earthquake magnitudes. *(d)*, South flank earthquakes per day. *(e)*, Rift and Koa‘e earthquakes per day. *(f)*, Kīlauea Caldera and deep magma supply earthquake magnitudes. *(g)*, Kīlauea Caldera and deep magma-supply earthquakes per day—short-period data. *(h)*, Kīlauea Caldera and deep magma-supply earthquakes per day—long-period data.

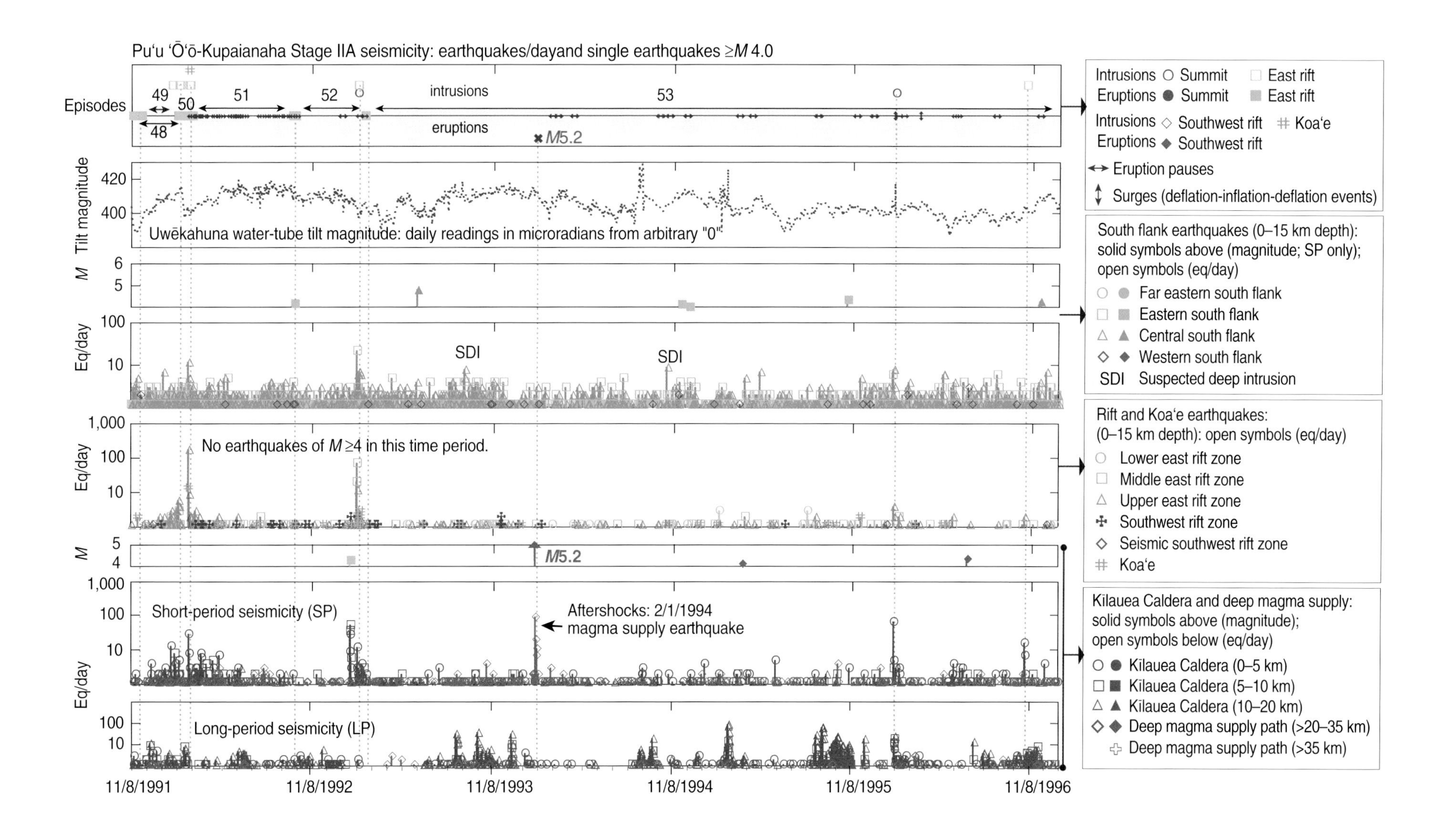

Pu'u 'Ō'ō-Kupaianaha Stage IIA seismicity: earthquakes/dayand single earthquakes ≥M 4.0
Episodes
49
50
51
52
53
48
intrusions
eruptions
M5.2
Tilt magnitude
Uwēkahuna water-tube tilt magnitude: daily readings in microradians from arbitrary "0"
M
Eq/day
SDI
SDI
No earthquakes of M ≥4 in this time period.
M5.2
Short-period seismicity (SP)
Aftershocks: 2/1/1994
magma supply earthquake
Long-period seismicity (LP)
11/8/1991
11/8/1992
11/8/1993
11/8/1994
11/8/1995
11/8/1996
Intrusions Summit East rift
Eruptions Summit East rift
Intrusions Southwest rift Koa'e
Eruptions Southwest rift
Eruption pauses
Surges (deflation-inflation-deflation events)
South flank earthquakes (0–15 km depth):
solid symbols above (magnitude; SP only);
open symbols (eq/day)
Far eastern south flank
Eastern south flank
Central south flank
Western south flank
SDI Suspected deep intrusion
Rift and Koa'e earthquakes:
(0–15 km depth): open symbols (eq/day)
Lower east rift zone
Middle east rift zone
Upper east rift zone
Southwest rift zone
Seismic southwest rift zone
Koa'e
Kilauea Caldera and deep magma supply:
solid symbols above (magnitude);
open symbols below (eq/day)
Kilauea Caldera (0–5 km)
Kilauea Caldera (5–10 km)
Kilauea Caldera (10–20 km)
Deep magma supply path (>20–35 km)
Deep magma supply path (>35 km)

Table 7.3. Pu‘u ‘Ō‘ō-Kupaianaha eruption[1]: Stage IIA—11/8/1991–1/1/1997 (see fig. 7.8).

[In rows with multiple entries text applies down to the next entry; dates in m/d/yyyy format; do, ditto (same as above); data for eruptions and traditional intrusions are emphasized by grey shading; "no data" in Lag column refers to gaps in the record of the Ideal-Arrowsmith tiltmeter]

Date and Time		Reg.[2]	Event Type[3]	No.[4]	Tilt[5]		Lag[6]	Comment[7]	Fig.[8]
Start	End				Mag	Az			
11/8/1991 4:45	11/26/1991 0:00	MERZ	E		15.5	117	+1h 45m	Episode 49[9]—Tilt 11/7–15	G17
1/28/1992 07:18	1/29/1992 03:15	ei3uer	II	13	0.8	135	+4h 18m	Tilt 1/28–29; I-A: 1/28 8.4 µrad deflation	G18
2/4/1992 08:36	2/4/1992 17:09	ms1.2	EQS	1, 7				Precursory Nāmakanipaio[10, 11]	G19
2/11/1992 00:30	2/13/1992 13:22	ei2.3	I	18	flat			Broken swarm	
2/17/1992 19:30	3/3/1992 1:30	MERZ	E		18.0	103	+2h 30m	Episode 50[12]—Tilt 2/17–22	G19
3/3/1992 00:45	3/4/1992 22:20	ei3uer	I	225	3.9	132	+0h 45m	Tilt 3/2–3	G20
3/3/1992 02:28	3/3/1992 07:41	koae	I	12					
3/3/1992 06:08	3/4/1992 15:04	sf3kuer	EQS	16				South flank response similar to SDI[3]	
3/7/1992 12:45	9/27/1992 6:00	MERZ	E		flat		+ 5h 45m	Episode 51[12]	
10/3/1992 3:30	2/20/1993 14:50	MERZ	E		7.1	122	no data	Episode 52[12]—Tilt 10/4-6	
1/24/1993 19:22	1/24/1993 22:13	ms1.2	EQ	41,10				Foreshocks for earthquake below	
1/24/1993 22:14		ms 2	EQ		flat			Precursory Nāmakanipaio[10]; *M*4.3	**7.9**
1/24/1993 19:48	1/26/1993 07:34	ms 2	EQ	83				Aftershocks for earthquake above	
2/7/1993 23:28	2/9/1993 02:23	ei2mer	I	96	5.1	128	no data	Tilt 2/7–8 deflation; 2/9–10 missing	**7.9**
2/7/1993 23:28	2/8/1993 13:23	ei3uer	I	13					
2/7/1993 23:35	2/8/1993 20:35	ms1	EQS	14					
2/7/1993 23:35	2/8/1993 16:46	sf2mer	EQS	23				South flank response similar to SDI[3]	
2/10/1993 18:01	2/14/1993 06:59	sf3kuer	EQS	22				Broken swarm	
2/20/1993 14:50	1/29/1997 18:52	MERZ	E		7.6	127		Episode 53[12]—Tilt 2/23–3/2	
6/8/1993 02:57			EQ					Ka‘ōiki *M*5.3 with aftershocks[13]	
8/22/1993 22:51	8/23/1993 14:14	lpms3	EQS	34					
8/24/1993 13:17	8/27/1993 00:59	lpms3	EQS	39					
9/11/1993 12:27	9/12/1993 12:13	sf3.2	SDI?	13, 5					G21
10/5/1993 14:53	10/8/1993 12:48	lpms3	EQS	68					
12/12/1993 23:40	12/15/1993 00:11	lpms3	EQS	43					
1/31/1994 21:29	1/31/1994 23:00	ms4.5	EQ	2				Foreshocks for earthquake below	
2/1/1994 01:55		do	EQ					*M*5.2, 35 km; magma supply path	G22
2/1/1994 04:03	2/3/1994 06:01	do	EQ	113				Aftershocks for earthquake above	
2/3/1994 17:57	2/4/1994 18:37	do	EQ	13				do	
9/25/1994 02:18	9/26/1994 05:20	lpms3	EQS	24					
10/25/1994 23:20	10/26/1994 14:22	sf3.2	SDI?	9, 2					G23
2/24/1995 12:54	2/28/1995 11:00	lpms3	EQS	233				Mixed deflation/inflation	
2/25/1995 17:40	2/27/1995 05:17	lpms2	EQS	21					
5/3/1995 12:28		sf3kuer	EQ					*M*3.9 mainshock; 2 foreshocks, 4 aftershocks	

Table 7.3. Pu'u 'Ō'ō-Kupaianaha eruption[1]: Stage IIA—11/8/1991–1/1/1997 (see fig. 7.8).—Continued

[In rows with multiple entries text applies down to the next entry; dates in m/d/yyyy format; do, ditto (same as above); data for eruptions and traditional intrusions are emphasized by grey shading; "no data" in Lag column refers to gaps in the record of the Ideal-Arrowsmith tiltmeter]

Date and Time		Reg.[2]	Event Type[3]	No.[4]	Tilt[5]		Lag[6]	Comment[7]	Fig.[8]
Start	End				Mag	Az			
8/24/1995 16:33	8/26/1995 05:53	lpms3	EQS	52					
9/5/1995 11:42	9/9/1995 18:58	lpms3	EQS	207				Tilt—9/5–8 missing	
9/27/1995 05:47	9/29/1995 16:04	lpms3	EQS	28					
10/2/1995 18:42	10/10/1995 17:20	lpms3	EQS	129	flat				
10/29/1995 10:25	10/31/1995 22:40	lpms3	EQS	22					
12/1/1995 23:29	12/3/1995 12:06	lpms3	EQS	26					
2/1/1996 08:09	2/1/1996 12:10		surge				E: 3h	I-A: 15 µrad 08:24–11:45⊠deflation	G23 A
2/1/1996 08:53	2/2/1996 05:41	ms1	I	70	18.5	307		Tilt 1/31-2/1	G23 A
2/1/1996 13:17	2/2/1996 05:43	sf3kuer	EQS	10					
2/1/1996 14:48	2/2/1996 23:45	lpms3	EQS	33					G23B
2/1/1996 16:16	2/2/1996 06:51	lpms1	EQS	11					G23B
2/1/1996 17:28	2/2/1996 14:17	lpms2	EQS	84					G23B
3/24/1996 10:36	3/24/1996 11:51		surge				E: ~3h	I-A: 4.3 µrad inflation 10:36–11:33⊠deflation	
5/13/1996 11:00	5/13/1996 20:37	sf3kuer	SDI	7				Weak swarm	
10/25/1996 21:41	10/26/1996 12:46	ei3uer	I	22	0.93	117		Tilt 10/25–28	G24

[1]References as follows: Wolfe and others (1987, 1988) cover the first 20 episodes. Heliker and others (2003) summarize the first 20 years of eruption. Other references to the eruption are covered in the Hawai'i bibliographic database (Wright and Takahashi, 1998) from which all references pertaining to the eruption may be obtained by searching on the keyword "kl.erz.1983".

[2]Earthquake classification abbreviations are given according to the classification in appendix table A3, and locations are shown on appendix figure A4.

[3]Event types defined in Chapter 1 are abbreviated as follows: **E**, Eruption; intrusion ("traditional" **I;** "inflationary" **II;** "suspected deep" **SDI**); **EQS**, earthquake swarms; EQ, Earthquake $M \geq 4$; "surge"—abrupt inflation at Kīlauea's summit followed by an increase in magma output at Pu'u 'Ō'ō. These later become "d-i-d" events in which there is a preliminary deflation at Kīlauea's summit followed by sharp inflation and then deflation. This pattern is matched by the electronic tilt at Pu'u 'Ō'ō with a slight time delay.

[4]Minimum number of events defining a swarm: 20 for south flank; 10 for all other regions.

[5]Magnitude in microradians (µrad) and azimuth of daily tilt measurements from the water-tube tiltmeter in Uwēkahuna Vault.

[6]Lag times separating the onset of the earliest earthquake swarm (excluding south flank) or eruption (in the absence of a precursory swarm) for a given event and the beginning of deflation or inflation measured by the continuously recording Ideal-Arrowsmith tiltmeter in Uwēkahuna Vault. (+) tilt leads, (-) tilt lags. For the later eruption surges (Deflation-Inflation-Deflation events) lag times (+) are given for the delayed response of the tiltmeter at Pu'u 'Ō'ō (POC) time lag in minutes, and the increase in eruptive activity at or near the Pu'u 'Ō'ō vent (E) time lag in hours.

[7]Abbreviations as follows: sf, south flank; ftn, fountaining; eq, earthquake; eqs, earthquake swarm; fs, foreshock; as, aftershock; ms, mainshock; sf, south flank; inf, inflation; def, deflation; ant, anticipation (preceding event); acc, accompaniment (during event); resp, response (following event); I-A, Ideal-Arrowsmith continuously recording tiltmeter in Uwēkahuna Vault, drm, downrift migration of earthquakes; urm, uprift migration of earthquakes; abs, missing data.

[8]Text figures **bold text**; appendix figures plain text.

[9]Eruption described by Mangan and others, 1995.

[10]Continuation of earthquakes triggered by the 1983 Ka'ōiki earthquake. Locations are beneath the Nāmakanipaio campground in Hawai'i Volcanoes National Park.

[11]Identified as precursor to following east rift intrusion (Okubo and Nakata, 2003).

[12]Eruption described by Heliker and others, 1998.

[13]Klein and others, 2006.

Table 7.4. Pu'u 'Ō'ō-Kupaianaha eruption[1]: Stage IIB—1/1/1997–12/1/2003 (see fig. 7.10).

[In rows with multiple entries text applies down to the next entry; dates in m/d/yyyy format; do, ditto (same as above); data for eruptions and traditional intrusions are emphasized by grey shading; "no data" in Lag column refers to gaps in the record of the Ideal-Arrowsmith tiltmeter]

Date and Time		Reg.[2]	Event Type[3]	No.[4]	Tilt[5]		Lag[6]	Comment[7]	Fig.[8]
Start	End				Mag	Az			
1/29/1997 18:52	1/31/1997 12:14	ei2mer	I	48	26.8	118		Tilt 1/27-30; I-A no data	**7.11**
1/29/1997 19:34	2/16/1997 13:00	lpms1	EQS	2318				2 7–8 hour pauses in swarm	**7.11**
1/30/1997 2:40	1/31/1997 0:33	MERZ	E					Episode 54: 1997 Nāpau	**7.11**
2/9/1997 08:48	2/10/1997 20:17	lpms2	EQS	21					
2/24/1997 7:00	6/17/2007 9:30	MERZ	E		flat			Episode 55	
6/2/1997 03:35	6/3/1997 19:45	ms1-2	EQS	34				Nāmakanipaio sequence[9]	G26
6/30/1997 05:47		sf2mer	EQ		flat			*M*5.5 with aftershocks[10]	
7/28/1997 16:19	7/29/1997 13:11	lpms2	EQS	45					
7/28/1997 19:28	7/30/1997 01:49	lpms1	EQS	15					
8/14/1997 15:54		sf2mer	EQ		flat			*M*5.0 with aftershocks[10]	
9/16/1997 22:59	9/18/1997 10:31	lpms2	EQS	63					
9/17/1997 01:41	9/18/1997 07:14	lpms3	EQS	47					
9/17/1997 03:27	9/17/1997 14:41	lpms1	EQS	10					
9/20/1997 14:47	9/21/1997 10:27	lpms2	EQS	28					
10/1/1997 10:18	10/2/1997 07:28	lpms3	EQS	50					
10/1/1997 11:42	10/2/1997 08:13	lpms2	EQS	56					
1/14/1998 18:20	1/14/1998 20:35		Surge[6]				E 4h	I-A:7.2 µrad inflation 18:20-20:35⊠deflation	G27*A*
1/14/1998 18:57	1/14/1998 23:00	ei3/kc	I	17	8.0	129		Tilt 1/13-15	G27*A*
1/15/1998 00:30	1/15/1998 14:26	lpms3	EQS	21					G27*B*
1/27/1998 16:55	1/28/1998 08:33	ms1.2	EQS	9, 36				Nāmakanipaio sequence[9]—2 *M*4, many *M*3+	G28
2/20/1998			SE?[11]					2 sf with no located eq 2 days before and after	
3/8/1998 08:30	3/9/1988 22:40	lpms3	EQS	20					
9/19/1998 06:08	9/20/1998 03:12	sf3kuer	SE[11]	17	2.4	166		Tilt 9/19-21; 9/20 missing; SE on 9/19; also classified as SDI	G29
9/27/1998 21:56		ms1	EQ		2.7	148		Glenwood *M*4.6[10]; Tilt 9/27–29	G30
9/28/1998 20:39	9/29/1998 22:29	sf2mer	EQ					*M*4.8 with aftershocks[10]	
1/10/1999			SE?[11]					2 widely separated south flank eq	
3/18/1999 11:42	3/18/1999 18:36	sf3kuer	SDI/EQ	9				*M*3.3 mainshock at 11:42 with 8 aftershocks?	
4/16/1999 14:46			EQ					Hīlea *M*5.6[10]	G31
4/16/1999 15:48	4/17/1999 05:23	ei5swr	EQ	15				Triggered Kīlauea aftershocks	
5/26/1999 06:01		ms2	EQ					Nāmakanipaio[9] mainshock *M*4.3	G32
5/26/1999 06:04	5/27/1999 01:27	ms2	EQ	34				Aftershocks	
9/12/1999 01:31	9/12/1999 15:17	ei3uer	I	61	7.1	159	0h 03m	Tilt 9/11–13	G33

Table 7.4. Pu‘u ‘Ō‘ō-Kupaianaha eruption[1]: Stage IIB—1/1/1997–12/1/2003 (see fig. 7.10).—Continued

[In rows with multiple entries text applies down to the next entry; dates in m/d/yyyy format; do, ditto (same as above); data for eruptions and traditional intrusions are emphasized by grey shading; "no data" in Lag column refers to gaps in the record of the Ideal-Arrowsmith tiltmeter]

Date and Time		Reg.[2]	Event Type[3]	No.[4]	Tilt[5]		Lag[6]	Comment[7]	Fig.[8]
Start	End				Mag	Az			
9/12/1999 01:40	9/12/1999 05:01	sf3kuer	EQS	11				South flank accompaniment; sf seismicity rate increases following this intrusion	
11/21/1999 11:38			SE?[11]					No located sf eq on this date	
2/23/2000 13:41	2/24/2000 19:45	ei3uer	I	107	8.5	147		Tilt 2/20–22 deflation; uprift migration 8.6 km/hr	G34 a
2/23/2000 15:09	2/24/2000 01:47	sf3kuer	EQS	7				South flank accompaniment	
2/23/2000 15:58	2/24/2000 17:58	lpms2.3	EQS	7				Broken swarm 2/17–2/29	G34 b
4/1/2000 20:18			EQ					*M*5.0 mainshock with at least 6 aftershocks	
4/24/2000 13:59	4/25/2000 03:16	lpms2	EQS	20	7.6	115		Tilt 4/19–28	
5/4/2000 21:38	5/7/2000 18:27	lpms2	EQS	45	4.2	124		Tilt 5/3–9	
5/5/2000 22:07	5/8/2000 06:09	lpms3	EQS	39					
5/8/2000 21:14	5/11/2000 03:56	lpms2	EQS	51	5.7	319		Tilt 5/14–19 inflation?	
5/29/2000 16:10	5/30/2000 15:00	sf3kuer	SE[11]	23	7.2	124		Tilt 5/29–6/2; *M*3.9 with aftershocks; SE at 5/29 11:23?; SE at 6/2 12:00?; also classified as SDI	**7.12**
6/1/2000 17:17	6/2/2000 21:47	do	do	11					
9/24/2000 10:30	9/25/2000 17:38		Surge[6]				POC 22 m E 3h	I-A: >6 µrad inflation 9/24 21:00-21:47⊠deflation[11]!	G35
11/8/2000 22:24	11/9/2000 10:53	sf3kuer	SE[11]	10	3.0	119		Tilt 11/5-7 SE on 11/9; also classified as SDI	G36
1/26/2001 10:44	1/27/2001 21:04	lpms3	EQS	20					
4/25/2001 17:34		ms2						*M*4.4[10], depth 6.34 km	
5/20/2001 04:02	5/21/2001 16:40		Surge[6]				POC 18m E 1.5h	I-A: 11 µrad inflation 5/20 1629-1743⊠deflation	G37
5/20/2001 20:43	5/20/2001 23:42	lpms3	EQS	11					
6/23/2001 21:57	6/24/2001 06:10	sf3.2	SDI	10, 3					G38
9/19/2001 04:03	9/19/2001 22:24	sf2	SE?[11]	4				SE at 11:40–9/19	G39
9/17/2001 01:14	9/22/2001 19:15	sf2.3		18				Scattered seismicity	
12/8/2001 10:40	12/10/2001 09:00		D-I-D		6.3	294	POC 41m	Inflation 12/9 08:54–09:50; no effect on eruption	G40
12/8/2001 12:00	12/8/2001 16:45	lpms2.3	EQS	5, 3					
12/9/2001 09:31	12/9/2001 09:54	ei3uer	I	5				Summit intrusion	
12/9/2001 20:54	12/10/2001 11:03	ms1-3	I	11					
12/19/2001 08:49	12/19/2001 11:24	ei3uer	I	6	4.1	129		Tilt 12/20–21; no rift deformation	G41
2/12/2002 06:47	2/12/2002 20:41	sf3kuer	SDI	7					G42
4/4/2002 20:50	4/6/2002 12:00		D-I-D				POC 11m E 94m	Inflation 4/5 15:56–16:47; no eqs but increased eruption ~4/5 17:30[13]	G43
8/21/2002 15:55	8/21/2002 20:00	sf3kuer	SDI	6				2 suspected deep intrusions	G44
8/28/2002 12:02	8/29/2002 03:21	sf3kuer	SDI	8					G44

Table 7.4. Pu'u 'Ō'ō-Kupaianaha eruption[1]: Stage IIB—1/1/1997–12/1/2003 (see fig. 7.10).—Continued

[In rows with multiple entries text applies down to the next entry; dates in m/d/yyyy format; do, ditto (same as above); data for eruptions and traditional intrusions are emphasized by grey shading; "no data" in Lag column refers to gaps in the record of the Ideal-Arrowsmith tiltmeter]

Date and Time		Reg.[2]	Event Type[3]	No.[4]	Tilt[5]		Lag[6]	Comment[7]	Fig.[8]
Start	End				Mag	Az			
12/15/2002 18:54	12/16/2002 15:19	sf3.2	SE[11]	7, 3	5.2	171		Tilt 12/14–16; 12/15 missing; SE 12/17; also classified as SDI	G45
1/20/2003 17:00	1/22/2003 06:50		D-I-D				POC 26m E 4h 25m	Inflation 1/21 16:16–16:52	G46
4/6/2003 21:27	4/7/2003 03:23	ms1	EQS	13				Nāmakanipaio[9]	G47
6/23/2003 16:30	6/23/2006 23:07	lpms2	EQS	10					
6/23/2003 19:04	6/25/2003 02:16	lpms3	EQS	16					
7/1/2003 16:56	7/2/2003 00:00	sf2mer	SE[11]	10	4.8	115		Tilt 6/30–7/1; SE 7/1; also classified as SDI	G48
7/11/2003 20:09	7/12/2003 16:35	lpms3	EQS	19					
7/15/2003 12:07	7/15/2003 18:37	lpms3	EQS	11					
8/4/2003 07:24	8/4/2003 18:03	lpms2	EQS	11					
8/7/2003 22:36	8/10/2003 00:00		d-i-d[6]				POC 69m E 19h	Inflation 8/8 19:32–22:48; 4 ms1 eq	G49
11/5/2003 06:54	11/5/2003 23:16	lpms2	EQS	14					
11/18/2003 09:34	11/21/2003 02:45	lpms2	EQS	36	6.4	117		Tilt 11/17–18	

[1]References as follows: Wolfe and others (1987, 1988) cover the first 20 episodes. Heliker and others (2003) summarize the first 20 years of eruption. Other references to the eruption are covered in the Hawai'i bibliographic database (Wright and Takahashi, 1998) from which all references pertaining to the eruption may be obtained by searching on the keyword "kl.erz.1983".

[2]Earthquake classification abbreviations are given according to the classification in appendix table A3, and locations are shown on appendix figure A4.

[3]Event types defined in chapter 1 are abbreviated as follows: **E**, Eruption; intrusion ("traditional" **I;** "inflationary" **II;** "suspected deep" **SDI**); **EQS**, earthquake swarms; EQ, Earthquake $M \geq 4$; "surge"—abrupt inflation at Kīlauea's summit followed by an increase in magma output at Pu'u 'Ō'ō. These later become "d-i-d" events in which there is a preliminary deflation at Kīlauea's summit followed by sharp inflation and then deflation. This pattern is matched by the electronic tilt at Pu'u 'Ō'ō with a slight time delay.

[4]Minimum number of events defining a swarm: 20 for south flank; 10 for all other regions.

[5]Magnitude in microradians (μrad) and azimuth of daily tilt measurements from the water-tube tiltmeter in Uwēkahuna Vault.

[6]Lag times separating the onset of the earliest earthquake swarm (excluding south flank) or eruption (in the absence of a precursory swarm) for a given event and the beginning of deflation or inflation measured by the continuously recording Ideal-Arrowsmith tiltmeter in Uwēkahuna Vault. (+) tilt leads, (-) tilt lags. For the later eruption surges (Deflation-Inflation-Deflation events) lag times (+) are given for the delayed response of the tiltmeter at Pu'u 'Ō'ō (POC) time lag in minutes, and the increase in eruptive activity at or near the Pu'u 'Ō'ō vent (E) time lag in hours.

[7]Abbreviations as follows: sf, south flank; ftn, fountaining; eq, earthquake; eqs, earthquake swarm; fs, foreshock; as, aftershock; ms, mainshock; sf, south flank; inf, inflation; def, deflation; ant, anticipation (preceding event); acc, accompaniment (during event); resp, response (following event); I-A, Ideal-Arrowsmith continuously recording tiltmeter in Uwēkahuna Vault, drm, downrift migration of earthquakes; urm, uprift migration of earthquakes; abs, missing data.

[8]Text figures **bold text**; appendix figures plain text.

[9]Continuation of earthquakes triggered by the 1983 Ka'ōiki earthquake. Locations are beneath the Nāmakanipaio campground in Hawai'i Volcanoes National Park.

[10]Klein and others, 2006.

[11]Silent earthquakes documented in 4 papers—summary in (Montgomery-Brown and others, 2009, table 1), including earthquakes with a small deformation step and scant seismic expression designated "SE?": (Brooks and others, 2006; Cervelli and others, 2002b; Montgomery-Brown and others, 2009; Segall and others, 2006).

[12]Recorded on strainmeter on the West side of Mauna Loa.

[13]Tilt source was determined ESE of Halema'uma'u Crater near Fiske-Kinoshita site 2 (see fig. 4.2) at a depth of ~0.45 km.

Narrative of Later Intrusion and Eruption During Stage IIA

Seismic activity increased dramatically in 1996, with a major intrusion beneath Kīlauea's summit on 1 February defined by a swarm of short-period earthquakes (table 7.3). This swarm was accompanied by a small swarm of long-period earthquakes shallower than 5 km, and by a very large swarm of long-period earthquakes deeper than 5 km. This date also marks the first of several "surges" in eruptive activity (Thornber and others, 1996) that were later defined as deflation-inflation-deflation (D-I-D) events from the pattern seen on continuously recorded tiltmeters in the Uwēkahuna vault and at the eruption site (Cervelli and Miklius, 2003). The 1996 surges were defined by a sharp inflation at the summit followed about 3 hours later by a noticeable increase in eruptive activity at the Pu'u 'Ō'ō vent and its associated tube system delivering lava to the ocean. The location of a magma source shallower than 1 km and southeast of Halema'uma'u was modeled by three continuously recording tiltmeters operating during a surge on 24 March 1996, and this may be the source of other surges as well. The location of the shallow reservoir agreed with that determined by the broadband seismic array (Dawson and others, 1999).

Ten eruptive pauses took place over a period of 6 weeks in early 1997. These pauses show some positive correlation with the lpc-c seismicity, but there are times of elevated lpc-c seismicity both before and after this time during which the eruption didn't pause.

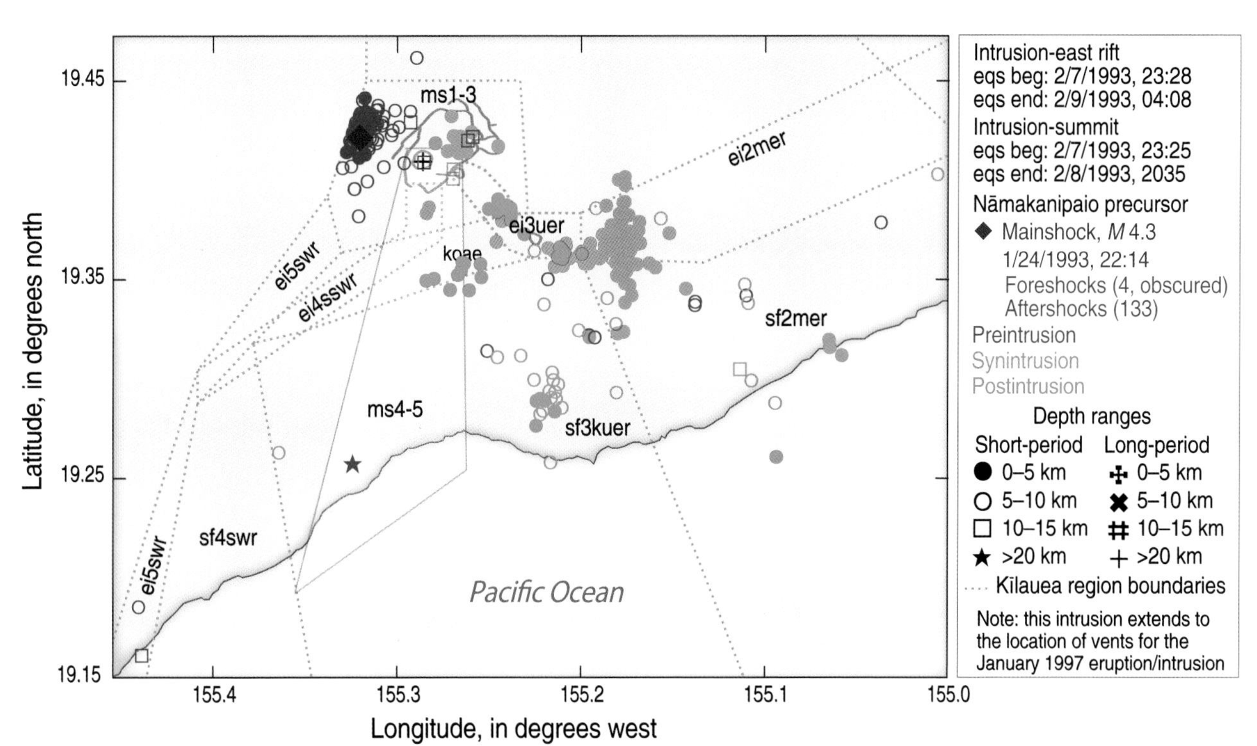

Figure 7.9. Map showing seismicity on Kilauea, 4–15 February 1993, associated with east rift and summit intrusions during 7–9 February 1993. A large middle east rift intrusion was accompanied by a small summit intrusion. These seismic swarms were preceded by a Nāmakanipaio swarm and followed by seismicity beneath the upper east rift and Koa'e Fault Zone and a possible suspected deep intrusion beneath the central south flank. Dates on figure in m/d/yyyy format.

Stage IIB: 1 January 1997–1 December 2003

Events in this period are summarized in table 7.4 and figure 7.10. Stage IIB marks the beginning of an increase in seismicity beneath the southwest rift zone and western south flank that is accentuated in stage III. Stage IIB began with the east rift eruption/intrusion near Nāpau Crater (fig. 7.11) that took place over the period 29–31 January 1997 (Desmarais and Segall, 2007; Owen and others, 2000b). The activity at Pu'u 'Ō'ō (episode 53) ended with the onset of an earthquake swarm beneath and slightly uprift of the eventual eruption site. Seismicity and Uwēkahuna daily tilt are tabulated in table 7.4 and shown in figure 7.10. Although HVO refers to this event as episode 54 of the eruption that began in 1983, we prefer to refer to it as the 1997 Nāpau eruption, because it was distant from Pu'u 'Ō'ō.

The 1997 Nāpau eruption/intrusion is noteworthy for the occurrence of a small swarm of Koa'e earthquakes that followed the eruption (fig. 7.10), the

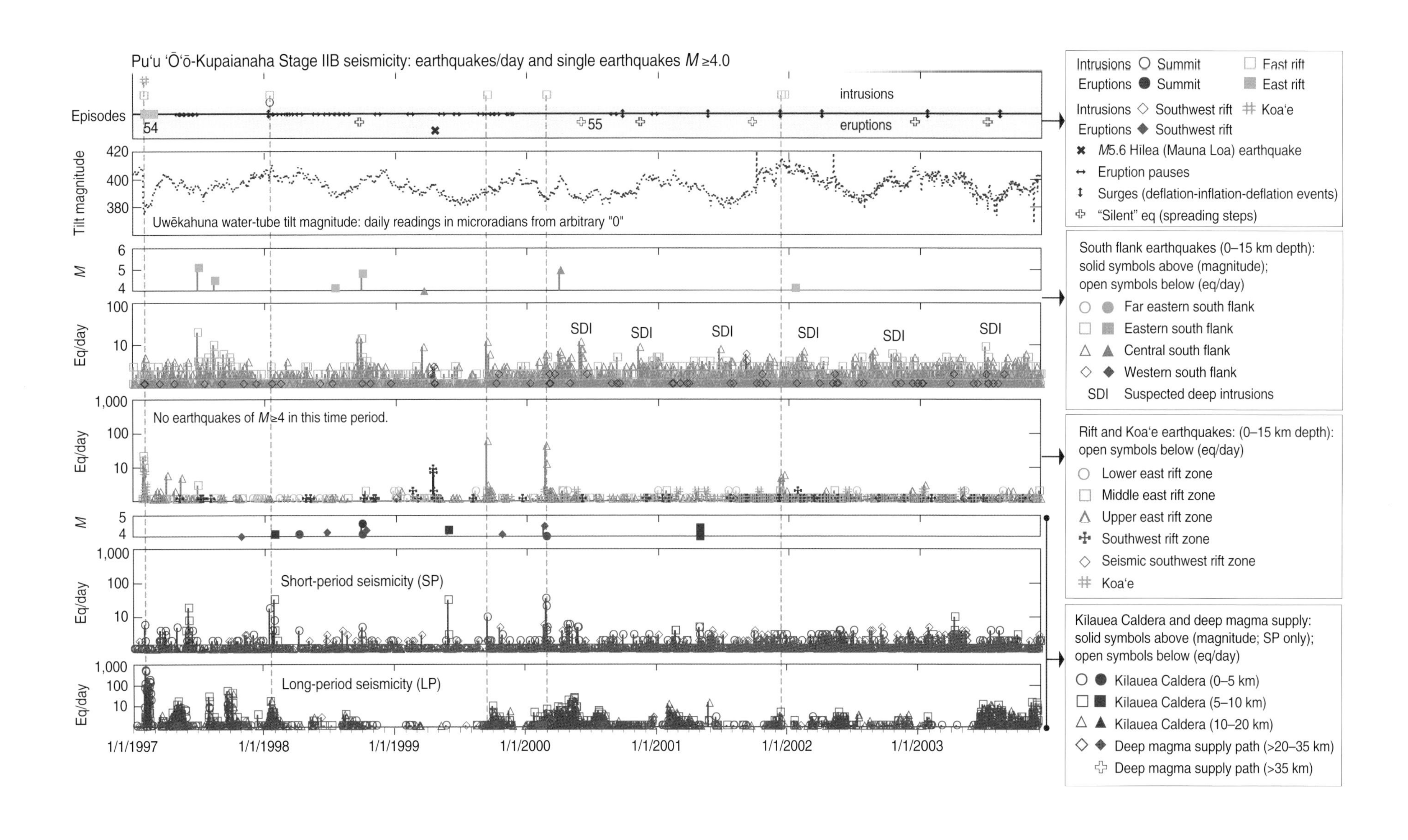

Pu‘u ‘Ō‘ō-Kupaianaha Stage IIB seismicity: earthquakes/day and single earthquakes M ≥4.0
Episodes
54
55
intrusions
eruptions
Tilt magnitude
420
400
380
Uwēkahuna water-tube tilt magnitude: daily readings in microradians from arbitrary "0"
M
6
5
4
Eq/day
100
10
SDI
No earthquakes of M≥4 in this time period.
1,000
Short-period seismicity (SP)
Long-period seismicity (LP)
1/1/1997
1/1/1998
1/1/1999
1/1/2000
1/1/2001
1/1/2002
1/1/2003
Intrusions Summit East rift
Eruptions Summit East rift
Intrusions Southwest rift Koa‘e
Eruptions Southwest rift
M5.6 Hilea (Mauna Loa) earthquake
Eruption pauses
Surges (deflation-inflation-deflation events)
“Silent” eq (spreading steps)
South flank earthquakes (0–15 km depth): solid symbols above (magnitude); open symbols below (eq/day)
Far eastern south flank
Eastern south flank
Central south flank
Western south flank
SDI Suspected deep intrusions
Rift and Koa‘e earthquakes: (0–15 km depth): open symbols below (eq/day)
Lower east rift zone
Middle east rift zone
Upper east rift zone
Southwest rift zone
Seismic southwest rift zone
Koa‘e
Kilauea Caldera and deep magma supply: solid symbols above (magnitude; SP only); open symbols below (eq/day)
Kilauea Caldera (0–5 km)
Kilauea Caldera (5–10 km)
Kilauea Caldera (10–20 km)
Deep magma supply path (>20–35 km)
Deep magma supply path (>35 km)

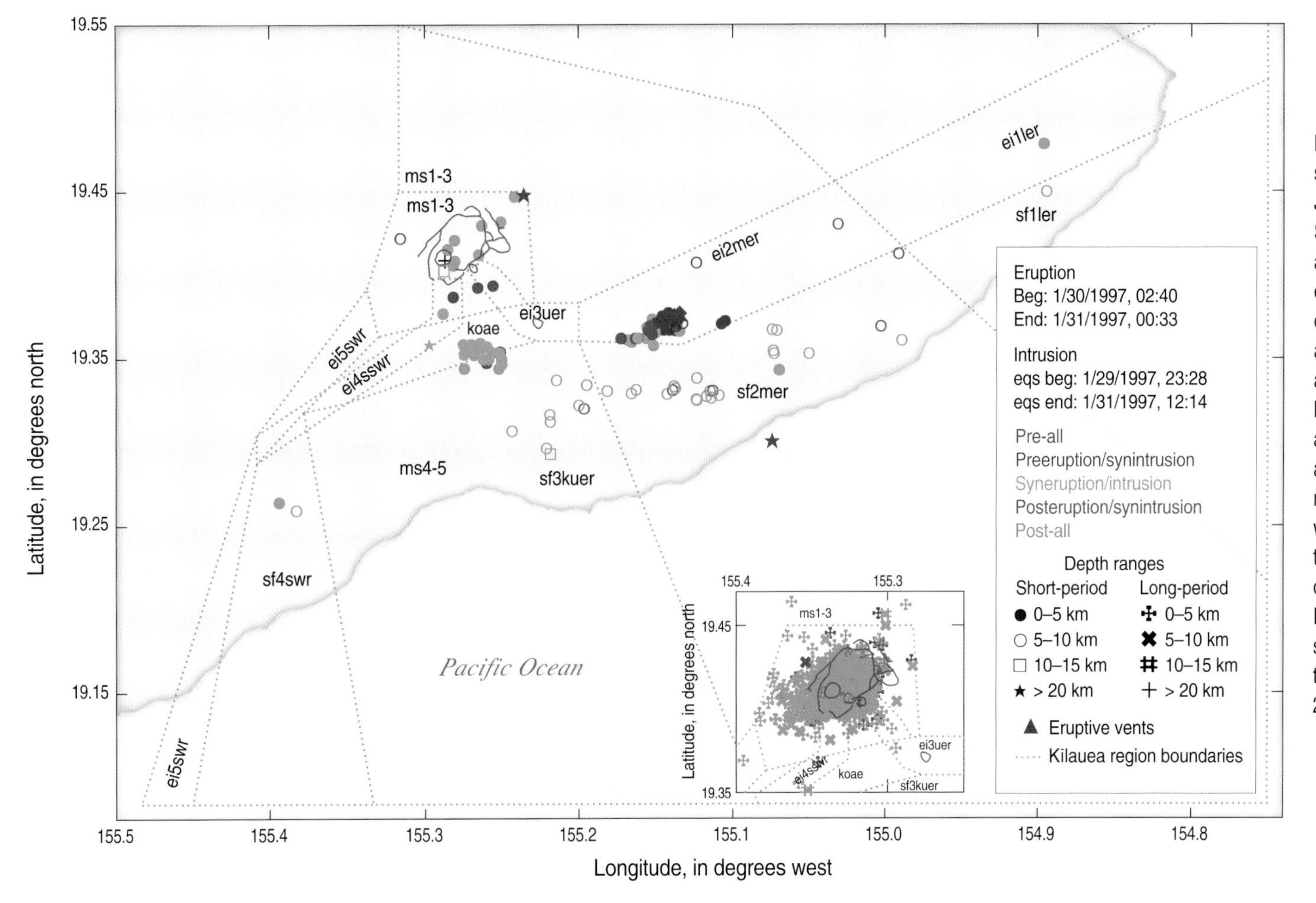

Figure 7.11. Map showing seismicity on Kīlauea, 28 January-16 February 1997. Short-period seismicity associated with the Nāpau eruption and intrusion of–29–31 January (Owen and others, 2000b; Thornber and others, 2003). An intense long-period earthquake swarm accompanied the intrusion and continued for another 3 months. The Koaʻe Fault Zone was briefly activated following the end of eruption. Dates on figure in m/d/yyyy format. Inset shows long-period seismicity associated with the eruption and intrusion of 29–31 January 1997.

Figure 7.10. Graphs showing Kīlauea activity, 1 January 1997–1 December 2003: Stage II B of the Puʻu ʻŌʻō-Kupaianaha eruption. Time-series plots show the number of earthquakes per day (eq/day) and magnitudes(M) ≥4 plotted against times of eruption and intrusion and the Uwēkahuna tilt magnitude. Symbols are given on the plots. Seismic regions are shown in figure A4 of appendix A. Three types of intrusions are defined in chapter 1. Traditional intrusions are shown as intrusion symbols, and suspected deep intrusions (SDI) are not plotted as intrusions. Dates on figure in m/d/yyyy format. Panels from top to bottom show the following: *(a)*, Times of eruption and intrusion. Vertical dotted lines in panel below connect seismicity associated with eruptions and associated traditional intrusions. Vertical dashed lines connect traditional and inflationary intrusions not associated with an eruption. *(b)*, Uwēkahuna tilt magnitudes in microradians. *(c)*, South flank earthquake magnitudes. *(d)*, South flank earthquakes per day. *(e)*, Rift and Koaʻe earthquakes per day. *(f)*, Kīlauea Caldera and deep magma-supply earthquake magnitudes. *(g)*, Kīlauea Caldera and deep magma-supply earthquakes per day—short-period data. *(h)*, Kīlauea Caldera and deep magma supply earthquakes per day—long-period data.

lack of preeruption summit inflation, and the absence of seismicity beneath the rift zone between Kīlauea's summit and the eruption site. It is tempting to interpret the lack of preeruption seismicity to opening of the rift zone during the February 1993 intrusion (see above), but apparently no extension occurred across the east rift zone at the time of the earlier intrusion (Heliker and others, 1998, p. 391).

The event was accompanied by a large summit deflation and by a long-period earthquake swarm above 5-km depth that consisted of more than 2,000 located and thousands of additional shallow lpc-a events spanning a period of nearly 3 weeks. A pause in seismicity occurred on 16 February, after which the lpc-a earthquake counts continued to decay, but they still exceeded preeruption levels until about 15 April 1997.

The 29 January 1997 eruption marked only a temporary shift in the eruption site. Lava returned to Pu'u 'Ō'ō on 24 February 1997 (episode 55) before the end of seismic activity associated with the 1997 Nāpau eruption/intrusion (fig. 7.10).

Long-period seismicity, principally deeper than 5 km, continued during stage IIB, as did intervals of pauses, surges, and Nāmakanipaio seismicity. Pauses during stage IIB have the same character as those during stage IIA (see above). The numbers of eruption pauses decreased in 2000 and 2001. Swarms of Nāmakanipaio earthquakes, mostly at depths of 5–10 km, occurred in June 1997, January 1998, and May 1999.

An intrusion occurred on the uppermost east rift on 14 January 1998, associated with a surge in eruptive activity and long-period seismicity at depths of 10–15 km (table 7.4). Large intrusions on 12 September 1999 (Cervelli and others, 2002a) and 23–24 January 2000 extended somewhat farther down the upper east rift (table 7.4). In April 1999, a Mauna Loa earthquake of magnitude 5.6 occurred beneath the Hīlea region and triggered seismicity adjacent to Kīlauea's lower southwest rift zone.

Additional surges occurred on 24–25 September 2000 and 20–21 May 2001. With the installation of continuous, digital recording tiltmeters at Kīlauea's summit (UWE tiltmeter) and at Pu'u 'Ō'ō (POC tiltmeter), the surges could now be more precisely defined by their character of gradual deflation (D) followed by inflation over about one hour (I) followed by decay to the value before the event (D). Later D-I-D events occurred on 9 December 2001, 5 April 2002, and 21 January and 8 August 2003. The time lag between the onset of the event at Kīlauea's summit and Pu'u 'Ō'ō was shown to be usually less than 30 minutes (table 7.4), and a noticeable increase in eruptive activity began a few hours after the beginning of inflation. Little seismic activity is associated with these events.

With the advent of a continuously recorded and telemetered GPS network, events of a new type—"slow" or "silent" earthquakes (Brooks and others, 2006; Cervelli and others, 2002b; Montgomery-Brown and others, 2009; Segall and others, 2006)—were identified on 19–20 and 28–29 September 1998, 29 May 2000 (fig. 7.12), 8–9 November 2000, 15–16 December 2002, and 1–2 July 2003 ("SE" in the "type" column of table 7.4, plotted as Greek crosses in figure 7.10). The largest of these events are associated with seaward movement of Kīlauea's south flank by several centimeters, with geodetic moments calculated as equivalent to an *M*5 or greater earthquake. The silent earthquakes are accompanied by swarms of low-magnitude south flank earthquakes of variable location and number of events.

Stage IIIA-B: Reinflation and Acceleration of Eruptive/Intrusive Activity, 1 Decembcr 2003–19 March 2008

Events of this period are given in tables 7.5 and 7.6 and figure 7.13. Stage III traces the evolution of Kīlauea from the end of continuous deflation to the explosive eruption in Halema'uma'u. We have arbitrarily divided the period at 1 May 2007 to isolate the buildup to the Father's Day eruption in June 2007.

Long-period earthquake swarms at the summit continued through 2004 (table 7.5). Nāmakanipaio events continued from late 2004. D-I-D events died out by September 2004 and the character changed to typical rapid deflation instead of inflation. Intrusions continued, increasing dramatically in 2006 (table 7.5).

An important change in the seismicity patterns associated with intrusions begins in 2005. Paired intrusions on the east rift zone and seismic southwest rift zone are seen in January 2005, the first time since before the beginning of the eruption in 1983 (table 7.5; fig. 7.13). This dual intrusion is accompanied by an apparent silent earthquake with an earthquake pattern identified with suspected deep intrusions and is followed within 3 days by a Nāmakanipaio sequence. Seismic southwest rift seismicity increases in 2006, particularly during a period of almost continuous unrest from 1 February through 2 March 2006 (fig. 7.13), and continues through March 2007. An additional suspected deep intrusion occurs on 3–5 April 2007, with no spreading step documented (Montgomery-Brown and others, 2009, table 1). A silent earthquake associated with a small spreading step is documented on 16 April, but there is very little seismic accompaniment and the few located earthquakes are scattered.

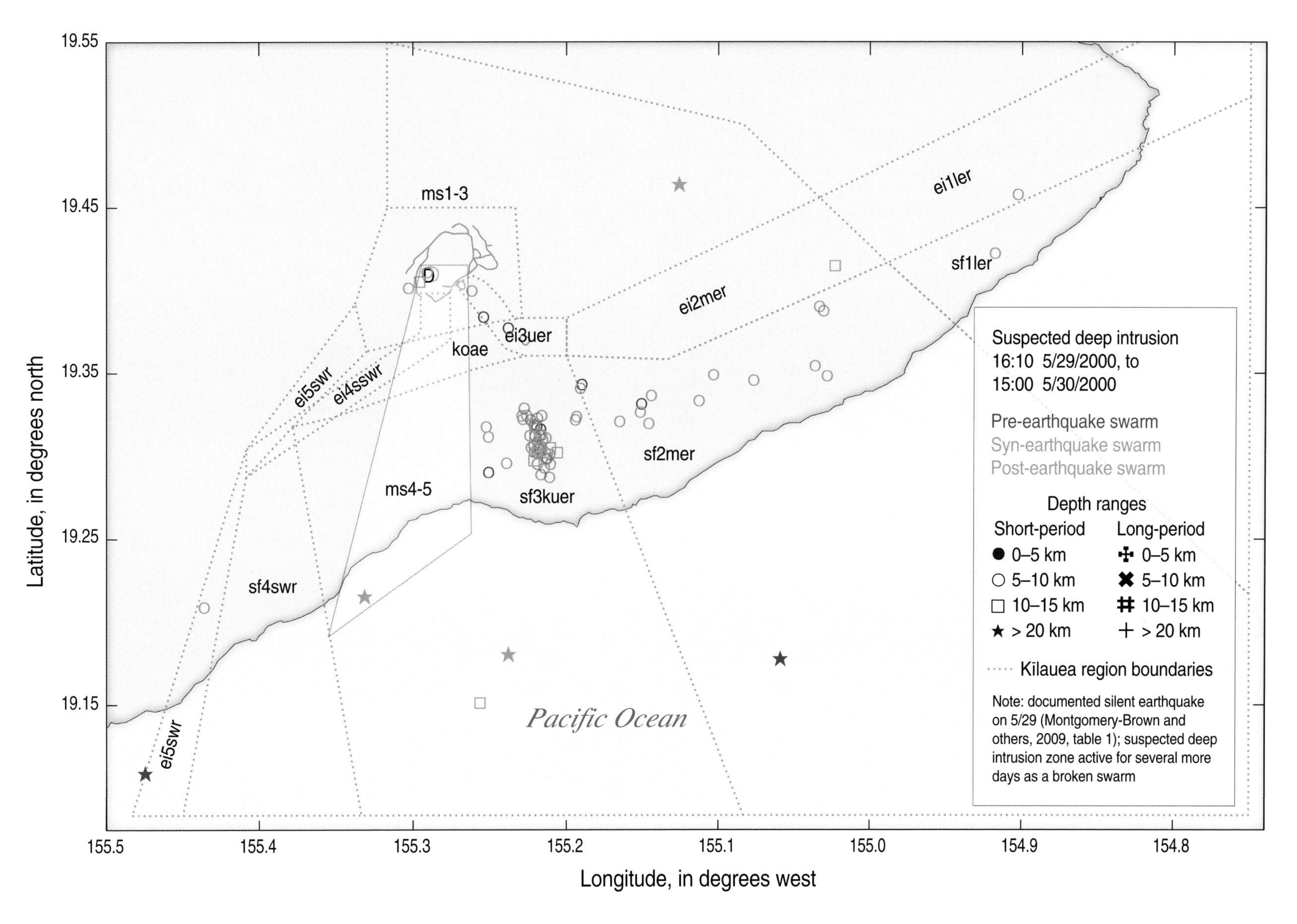

Figure 7.12. Map showing seismicity on Kīlauea, 26 May–6 June 2000. 5/29/2000: A suspected deep intrusion on 29–30 May beneath the central south flank was independently documented as a silent earthquake accompanied by a spreading step (Montgomery-Brown and others, 2009). Dates on figure in m/d/yyyy format.

Table 7.5. Pu‘u ‘Ō‘ō-Kupaianaha eruption[1]: Stage IIIA—12/1/2003–5/18/2007 (see fig. 7.13).

[In rows with multiple entries text applies down to the next entry; dates in m/d/yyyy format; do, ditto (same as above); data for eruptions and traditional intrusions are emphasized by grey shading; "no data" in Lag column refers to gaps in the record of the Ideal-Arrowsmith]

Date and Time		Reg.[2]	Event Type[3]	No.[4]	Tilt[5]		Lag[6]	Comment[7]	Fig.[8]
Start	End				Mag	Az			
12/5/2003 03:29	12/5/2003 18:25	lpms2	EQS	14	5.9	141		Tilt 12/4–5; 12/7–8 missing	
12/6/2003 07:56	12/6/2003 20:25	do	do	18					
12/10/2003 23:31	12/12/2003 20:13	lpms3	EQS	20				Break 12/11–12	
12/10/2003 23:32	12/13/2006 04:15	lpms2	EQS	61					
12/22/2003 11:16	12/22/2003 16:09	sf3kuer	SDI?	7					G51
12/27/2003 01:10	12/27/2003 11:33	lpms2	I?	12					
1/12/2004 19:58	1/12/2004 23:16	lpms2	I?	12					
1/23/2004 08:37	1/23/2004 15:20	lpms2	I?	10					
3/3/2004 03:00	3/5/2004 12:00		D-I-D				POC 30m E <3h	I-A: 17.5 µrad inflation 3/4 02:52–04:14 Tilt 3/2–3	
3/4/2004 03:39	3/4/2004 07:08	ei3uer	I	10	4.0	111			G52
3/20/2004 02:00	3/21/2004 15:50		D-I-D				POC 36m E 4h	I-A: 5.4 µrad inflation 3/20 13:49–15:41; no swarm	G53
5/6/2004 11:45			SE?[9]					2 widely separated sf eq on 5/5–6; probable SDI	
5/12/2004 08:05	5/13/2004 01:28	ei3uer	I	10	6.4	117	POC 26m	Tilt 5/12–14	G54
5/15/2004 06:06	5/17/2004 01:16		D-I-D				E ~8h	I-A: 6 µrad inflation 5/15 16:11–17:22	
6/4/2004 00:06	6/4/2004 17:18	ei3uer	I	8	8.0	116		Tilt 6/3-9 deflation	G55
7/9/2004 03:13	79/2004 18:24	ei3uer	I	8	6.9	135		Tilt 7/8–11; downrift migration of 0.19 km/hr[10]	G56
7/10/2004 04:00	7/10/2004 15:12	ei3uer	I	8				South flank eq scattered in time	
7/26/2004 18:00	7/28/2004 13:44		D-I-D				POC 56m	I-A: 6 µrad inflation 7/27 7:32–09:34	G57
7/27/2004 04:17	7/27/2004 17:27	ei3uer	I	10	9.8	130	E 8h	Tilt 7/27–30; downrift migration 8.2 km/hr[10]	G57
9/14/2004 03:13	9/16/2004 12:00		D-I				POC 71m	~2.5 µrad inflation 9/15 follows 5 lpms3 by 6 hr; d-i-d surges end	
11/2/2004 05:33	11/2/2004 09:32	ms4.5	EQS	5	flat			Tilt 11/2-3	G58
11/2/2004 12:45	11/2/2004 16:33	ms2	EQS	6					
11/2/2004 13:02	11/2/2004 16:56	ms1	EQS	17				Nāmakanipaio[11]	G58
11/24/2004 14:35	11/25/2004 12:45	lpms2	EQS	12					
12/1/2004 01:04	12/1/2004 03:22	lpms3	EQS	10					
12/13/2004 20:46	12/14/2004 11:38	lpms2	EQS	11	3.9	312		Tilt 12/13-14 inflation	
12/14/2004 02:33	12/14/2004 12:24	lpms3	EQS	15					
12/22/2004 07:07	12/23/2004 10:25	ei3uer	I	23	3.1	131		Tilt 12/23-27; downrift migration 1.2 km/hr[10]	G59
1/24/2005 18:35	1/25/2005 04:00	ei4swr	II	7	2.4	135		Tilt 1/24-26; inflation-deflation event	G60
1/24/2005 22:42	1/25/2005 04:00	ei3uer	II	13				UWE: 1/25 01:16 5.3 µrad 322.7	G60
1/25/2005 01:19	1/25/2005 11:17	ei3/kc	II	10					G60

Table 7.5. Pu‘u ‘Ō‘ō-Kupaianaha eruption[1]: Stage IIIA—12/1/2003–5/18/2007 (see fig. 7.13).—Continued

[In rows with multiple entries text applies down to the next entry; dates in m/d/yyyy format; do, ditto (same as above); data for eruptions and traditional intrusions are emphasized by grey shading; "no data" in Lag column refers to gaps in the record of the Ideal-Arrowsmith]

Date and Time		Reg.[2]	Event Type[3]	No.[4]	Tilt[5]		Lag[6]	Comment[7]	Fig.[8]
Start	End				Mag	Az			
1/25/2005 17:54	1/27/2005 02:37	sf3kuer	SDI?	57				SE 1/26; broken swarm	
1/26/2005 16:34	1/30/2005 09:28	sf2mer	EQS	18				Continuation of SE?	G60
1/27/2005 10:23	1/28/2005 05:18	ms1	EQS	19				Nāmakanipaio[11]	
1/29/2005 00:32	1/29/2005 16:35	ms2		14					
12/29/2005 15:54	12/29/2005 23:46	ei3.2	I	7, 4	flat				G61
1/15/2006 05:39	1/16/2006 01:50	ei3uer	II	10	1.7	149		Tilt 1/20–23 deflation; 1/21–22 missing	G62
1/22/2006 23:07	1/23/2006 11:40	ms1	EQS	16				Nāmakanipaio[11]	G63
1/25/2006 06:23	1/26/2006 22:36	ei3uer	II	15	4.7	308		Tilt 1/25–27 inflation	G64*A*
1/29/2006 11:31	1/30/2006 11:22	ei3uer	II	15	5.5	129		Tilt 1/27–2/3 deflation	G64*B*
1/30/2006 21:24	1/31/2006 18:13	ei3uer	II	12				UWE: flat	G64*B*
2/1/2006 04:47	2/8/2006 22:50	ei3uer	II	191				No rift deformation	G64*C*
2/9/2006 15:36	2/18/2006 07:02	ei3uer	II	113	flat			40, 38, 10, 12 and 48 events	G64*D*
						301		UWE: 2/10-17 2.3 µrad inflation	
2/22/2006 16:41	2/23/2006 09:47	ms2	EQS	23				Nāmakanipaio[11]	G65
2/23/2006 07:30	2/24/2006 17:29	ei4sswr	II	12				Tilt 2/23–28 inflation	G65
2/23/2006 21:08	2/24/2006 08:25	er3uer	II	12				No rift deformation	G65
2/26/2006 18:05	3/2/2006 20:05	ei4swr	II	17				Broken swarm ; Tilt 2/25–27 missing	G66
2/27/2006 22:42	3/1/2006 17:03	ei3uer	II	22				UWE: flat; no rift deformation	G66
3/1/2006 01:55	3/4/2006 20:25	ms1-2	EQS	65	5.1	297		Nāmakanipaio[11]	G67
3/1/2006 03:40	3/1/2006 17:03	er3.4	I	7, 3				Tilt 3/1-3; inflation	G67
3/6/2006 05:56	3/7/2006 02:07	ms2	EQS	17	8.8	128		Nāmakanipaio[11]	G68
3/6/2006 20:27	3/7/2006 03:49	ei4sswr	II	9				Tilt 3/3–8; deflation	G68
3/11/2006 02:00	3/16/2006 01:08	ms1-2	EQS	24	9.3	112		Tilt 3/10–13 deflation broken swarm	G69
3/16/2006 02:02	3/16/2006 22:51	sf3kuer	SDI	18	3.9.	132		Tilt 3/15–17 deflation; no spreading step recorded	G70
3/22/2006 18:37	3/22/2006 21:38	ei3uer	I	13	5.9	129		Tilt 3/20–22; UWE flat	G71
5/4/2006 12:18	5/5/2006 12:44	sf2mer	EQS	9	3.4	290			
5/9/2006 14:27	5/10/2006 00:10	ei3uer	I	12				Tilt 5/9–10	G72
7/29/2006 15:30	7/30/2006 10:21	ei4sswr[12]	II	13	9.5	290		Tilt 7/27–8/1; 7/29–31 missing	G73
8/3/2006 08:07	8/4/2006 02:30	ei3uer	I	11	flat			Tilt 8/3–7; 8/4–6 missing; UWE flat	G74
8/6/2006 10:18	8/6/2006 18:35	do	do	7					
4/3/2007 13:40	4/5/2007 07:20	sf3kuer	SDI	28	flat			No spreading step recorded	G75
4/16/2007			SE?[9]					Very few eq—resembles south flank response to intrusion; also classified as SDI	G76

Table 7.5. Pu'u 'Ō'ō-Kupaianaha eruption[1]: Stage IIIA—12/1/2003–5/18/2007 (see fig. 7.13).—Continued (footnotes)

[In rows with multiple entries text applies down to the next entry; dates in m/d/yyyy format; do, ditto (same as above); data for eruptions and traditional intrusions are emphasized by grey shading; "no data" in Lag column refers to gaps in the record of the Ideal-Arrowsmith]

[1]References as follows. Wolfe and others (1987, 1988) cover the first 20 episodes. Heliker and others (2003) summarize the first 20 years of eruption. Other references to the eruption are covered in the Hawai'i bibliographic database (Wright and Takahashi, 1998) from which all references pertaining to the eruption may be obtained by searching on the keyword "kl.erz.1983".

[2]Earthquake classification abbreviations are given according to the classification in appendix table A3, and locations are shown on appendix figure A4.

[3]Event types defined in chapter 1 are abbreviated as follows: **E**, Eruption; intrusion ("traditional" **I**; "inflationary" **II**;"suspected deep" **SDI**); **EQS**, earthquake swarms; EQ, Earthquake $M \geq 4$; "surge"—abrupt inflation at Kīlauea's summit followed by an increase in magma output at Pu'u 'Ō'ō. These later become "d-i-d" events in which there is a preliminary deflation at Kīlauea's summit followed by sharp inflation and then deflation. This pattern is matched by the electronic tilt at Pu'u 'Ō'ō with a slight time delay.

[4]Minimum number of events defining a swarm: 20 for south flank; 10 for all other regions.

[5]Magnitude in microradians (µrad) and azimuth of daily tilt measurements from the water-tube tiltmeter in Uwēkahuna vault.

[6]Lag times separating the onset of the earliest earthquake swarm (excluding south flank) or eruption (in the absence of a precursory swarm) for a given event and the beginning of deflation or inflation measured by the continuously recording Ideal-Arrowsmith tiltmeter in Uwēkahuna Vault. (+) tilt leads, (-) tilt lags. For the later eruption surges (Deflation-Inflation-Deflation events) lag times (+) are given for the delayed response of the tiltmeter at Pu'u 'Ō'ō (POC) time lag in minutes, and the increase in eruptive activity at or near the Pu'u 'Ō'ō vent (E) time lag in hours.

[7]Abbreviations as follows: sf, south flank; ftn, fountaining; eq, earthquake; eqs, earthquake swarm; fs, foreshock; as, aftershock; ms, mainshock; sf, south flank; inf, inflation; def, deflation; ant, anticipation (preceding event); acc, accompaniment (during event); resp, response (following event); Continuously recording tiltmeters in Uwēkahuna Vault: I-A, Ideal-Arrowsmith, UWE, Uwēkahuna; drm, downrift migration of earthquakes; urm, uprift migration of earthquakes; abs, missing data.

[8]Text figures **bold text**; appendix G figures plain text.

[9]Silent earthquakes documented in 4 papers—summary in (Montgomery-Brown and others, 2009, table 1), including earthquakes with a small deformation step and scant seismic expression designated "SE?": (Brooks and others, 2006; Cervelli and others, 2002b; Montgomery-Brown and others, 2009; Segall and others, 2006).

[10]After January 1983 earthquake migration rates are measured from plots similar to those in Klein and others (1987) that show migration rates of earthquakes between 1968 and 1983.

[11]Continuation of earthquakes triggered by the 1983 Ka'ōiki earthquake. Locations are beneath the Nāmakanipaio campground in Hawai'i Volcanoes National Park.

[12]er4 seismicity continues until 10/2006.

Table 7.6. Pu'u 'Ō'ō-Kupaianaha eruption[1]: Stage IIIB—5/18/2007–3/19/2008 (see fig. 7.13).

[In rows with multiple entries text applies down to the next entry; dates in m/d/yyyy format; do, ditto (same as above); data for eruptions and traditional intrusions are emphasized by grey shading]

Date and Time		Reg.[2]	Event Type[3]	No.[4]	Tilt[5]		Lag[6]	Comment[7]	Fig.[8]
Start	End				Mag	Az			
5/24/2007 07:40	5/25/2007 14:07	ei3uer	I	30	1.56	131	+ 1h 36m	Tilt 5/22–29; 5/26–28 missing	G77
5/25/2007 23:01	5/27/2007 11:25	ei3uer	I	19					
5/31/2007 21:29	6/1/2007 04:08	ei4sswr	II	8					G78
6/17/2007 02:16	6/18/2007 12:46	ei3uer	I	121	7.18	123	+ 0h 21m	Tilt 6/15–20 6/16–17 missing	
6/17/2007 02:31	6/17/2007 13:00	lpms1	EQS	27				Location beneath upper east rift—first ever long-period earthquake swarm in this region!	
6/17/2007 02:54	6/17/2007 12:22	sf3.2	EQS	14, 8				South flank response	
6/17/2007 03:01	6/17/2007 08:11	ms1	EQS	7				9 more ei2mer to 6/21/2007 19:24; 9 ei3uer from 6/19 11:58 to 6/22 07:56	
6/17/2007 03:03	6/18/2007 23:24	ei2mer	I	88				Silent earthquake; one 10-hr break	

Table 7.6. Pu‘u ‘Ō‘ō-Kupaianaha eruption [1]: Stage IIIB—5/18/2007–3/19/2008 (see fig. 7.13).—Continued

[In rows with multiple entries text applies down to the next entry; dates in m/d/yyyy format; do, ditto (same as above); data for eruptions and traditional intrusions are emphasized by grey shading]

Date and Time		Reg.[2]	Event Type[3]	No.[4]	Tilt[5]		Lag[6]	Comment[7]	Fig.[8]
Start	End				Mag	Az			
6/17/2007 20:52	6/18/2007 17:47	sf3kuer	SE[9]	30				South flank anticipation	G79
6/18/2007 14:12	6/18/2007 15:35	ms1	EQS	4					
6/20/2007 12:37	6/21/2007 12:27	sf3, sf2	EQS	5/5				South flank anticipation	
6/22/2007 18:35	6/22/2007 23:23	MERZ	E					Fathers day eruption (Episode 56)	**7.14**
6/22/2007 18:35	6/23/2007 02:56	sf3, sf2	EQS	6,1				South flank accompaniment/response	
7/2/2007 07:50	7/20/2007 22:23	MERZ	E					Ep 57; renew activity at Pu‘u ‘Ō‘ō	
7/3/2007 23:16	7/4/2007 12:37	ei3uer	I	28	0.97	133		Tilt 7/6-16; 58 er3 events 7/4–16	G80
7/5/2007 06:11	7/6/2007 01:38	ei3uer	I	13					G81
7/16/2007 21:07	7/17/2007 13:23	ei3uer	I	16				UWE 7/15–17 5.3 µrad deflation	G82
7/21/2007 00:39	6/1/2009	MERZ	E		flat			Ep 58; fissure East of Pu‘u ‘Ō‘ō	**7.15**
7/26/2007 12:29	7/26/2007 22:52	ms4.5	EQS	8					
8/13/2007 19:38	8/19/2007 08:26	sf2mer	EQS	74				Several swarm breaks after first day	
9/23/2007 16:05	9/24/2007 13:13	sf2mer	EQS	10					
11/4/2007 02:01	11/4/2007 14:59	sf3kuer	SDI	14				*M*3.3, 3.2 double mainshock? followed by aftershocks	

[1]References as follows: Wolfe and others (1987, 1988) cover the first 20 episodes. Heliker and others (2003) summarize the first 20 years of eruption. Other references to the eruption are covered in the Hawai‘i bibliographic database (Wright and Takahashi, 1998) from which all references pertaining to the eruption may be obtained by searching on the keyword “kl.erz.1983”.

[2]Earthquake classification abbreviations are given according to the classification in appendix table A3, and locations are shown on appendix figure A4.

[3]Event types defined in chapter 1 are abbreviated as follows: **E**, Eruption; intrusion (“traditional” **I**; “inflationary” **II**;“suspected deep”**SDI**); **EQS**, earthquake swarms; EQ, Earthquake $M \geq 4$; “surge”—abrupt inflation at Kīlauea’s summit followed by an increase in magma output at Pu‘u ‘Ō‘ō. These later become “d-i-d” events in which there is a preliminary deflation at Kīlauea’s summit followed by sharp inflation and then deflation. This pattern is matched by the electronic tilt at Pu‘u ‘Ō‘ō with a slight time delay.

[4]Minimum number of events defining a swarm: 20 for south flank; 10 for all other regions.

[5]Magnitude in microradians (µrad) and azimuth of daily tilt measurements from the water-tube tiltmeter in Uwēkahuna vault.

[6]Lag times separating the onset of the earliest earthquake swarm (excluding south flank) or eruption (in the absence of a precursory swarm) for a given event and the beginning of deflation or inflation measured by the continuously recording Ideal-Arrowsmith tiltmeter in Uwēkahuna Vault. (+) tilt leads, (-) tilt lags. For the later eruption surges (Deflation-Inflation-Deflation events) lag times (+) are given for the delayed response of the tiltmeter at Pu‘u ‘Ō‘ō (POC) time lag in minutes, and the increase in eruptive activity at or near the Pu‘u ‘Ō‘ō vent (E) time lag in hours.

[7]Abbreviations as follows: sf, south flank; ftn, fountaining; eq, earthquake; eqs, earthquake swarm; fs, foreshock; as, aftershock; ms, mainshock; sf, south flank; inf, inflation; def, deflation; ant, anticipation (preceding event); acc, accompaniment (during event); resp, response (following event); Continuously recording tiltmeters in Uwēkahuna Vault: I-A, Ideal-Arrowsmith, UWE, Uwēkahuna; drm, downrift migration of earthquakes; urm, uprift migration of earthquakes; abs, missing data.

[8]Text figures **bold text**; appendix G figures plain text.

[9]Silent earthquakes documented in 4 papers—summary in (Montgomery-Brown and others, 2009, table 1), including earthquakes with a small deformation step and scant seismic expression designated “SE?”: (Brooks and others, 2006; Cervelli and others, 2002b; Montgomery-Brown and others, 2009; Segall and others, 2006).

A major paired intrusion occurred in May 2007 (table 7.6), marking the beginning of stage IIIB. One month later the most dramatic event of this stage occurred, the Father's Day east rift eruption and traditional intrusion of 19 June 2007 (Poland and others, 2008). The volume of lava erupted was very small compared to the volume of intrusion. A silent earthquake accompanied the precursory intrusion (fig. 7.14; Montgomery-Brown and others, 2010). Eruption stopped at Pu'u 'Ō'ō on the day of the intrusion, ending episode 55.

Pu'u 'Ō'ō activity resumed on 2 July (episode 57) and ended on 20 July, one day before the fissure eruption (episode 58) located downrift of Pu'u 'Ō'ō (Poland and others, 2008)) Two intrusions occurred between the Father's Day eruption and episode 58 (table 7.6). South flank seismicity increased as rift seismicity moved eastward toward the site of fissure eruption. Rift seismicity virtually ended before the fissure eruption began. South flank seismicity continued through the time of fissure eruption and was focused both south of the eruption site and south of the uppermost east rift zone (fig. 7.15).

Volatile Release During Stage IIIB

East rift sulfur dioxide (SO_2) emissions at Pu'u 'Ō'ō increased dramatically following the 1997 Nāpau eruption and again in 2005 and 2006 (Elias and Sutton, 2002, 2007; Elias and others, 1998). Sulfur dioxide emissions at Kīlauea's summit showed only small changes through 2006, but emissions increased briefly during the Father's Day eruption and increased steadily before and during the eruption at Halema'uma'u in March 2008 (Wilson and others, 2008). Carbon dioxide (CO_2) emissions at Kīlauea's summit increased in 2004, peaking in 2005, and remained high through 2007 (J. Sutton, written commun., 2009).

Interpretations, 1983–2008

The onset of the Pu'u 'Ō'ō eruption in January 1983 (episode 1) is interpreted as the culminating event in the series of eruptions and intrusions marking the recovery from the 1975 earthquake. The succeeding eruption history is governed by the adjustment of the rift plumbing to accommodate an increasing magma supply (see chap. 8, figure 8.5). The current activity is unique in Kīlauea's historical period in that increased magma supply eventually resulted in two concurrent eruptions.

A broad comparison between seismicity in the year preceding and including episode 1 and the remainder of stage I (episodes 2–47) is shown in table 7.7, which counts all located earthquakes in the HVO catalog sorted into different regions. As a consequence of the 1975 earthquake and its aftershocks, the rate of south flank seismicity increased greatly and the rate of deep magma-supply seismicity was reduced (Klein and others, 1987, p. 1034, 1038, figure 43.8H, p. 1054). The release of magma to the surface and accompanying summit deflation in January 1983 resulted in a release of the stress on Kīlauea's south flank engendered during magma refilling under the east rift during the recovery from the 1975 earthquake (Delaney and others, 1993, 1998; Dieterich and others, 2003). The average number of earthquakes per day above 5-km depth beneath the summit (ms1), rift zones, and Koa'e at depths extending to the decollement is also greatly reduced following the end of episode 1. In contrast, the average number of earthquakes per day along the magma supply path at depths greater than 5 km, which was low during recovery from the 1975 earthquake, again increases in the period following episode 2.

The entire seismic and eruptive history from 1982 through 2007 is summarized in figures 7.16 and 7.17. The daily radial tilt component at Uwēkahuna vault is plotted with yearly averages of located short-period and long-period earthquakes, number of intrusions, SO_2 and CO_2 emission rates, and south flank spreading rates. Kīlauea's summit continued to deflate until the end of stage II (fig. 7.17*A*).

Figure 7.13. Graphs showing Kīlauea activity, 1 December 2003–19 March 2008: Stage III of the Pu'u 'Ō'ō-Kupaianaha eruption. Stage III is arbitrarily divided into IIIA and IIIB at 18 May 2007, preceding the east rift intrusion of 24 May and the Father's Day eruption in June. These events begin the final precursory period before the Halema'uma'u eruption of 19 March 2008. Time-series plots show the number of earthquakes per day (eq/day) and magnitudes (*M*) ≥4 plotted against times of eruption and intrusion and the Uwēkahuna tilt magnitude. Symbols are given on the plots. Seismic regions are shown in figure A4 of appendix A. Three types of intrusions are defined in chapter 1. Traditional intrusions are shown as intrusion symbols. Inflationary intrusions are labeled (II), and suspected deep intrusions are labeled (SDI) but are not plotted as intrusions. Dates on figure in m/d/yyyy format. Panels from top to bottom show the following: (a), Times of eruption and intrusion. Vertical dotted lines connect seismicity associated with eruptions and associated traditional intrusions. Vertical dashed lines connect traditional intrusions not associated with an eruption. (b), Uwēkahuna tilt magnitudes in microradians. (c), South flank earthquake magnitudes. (d), South flank earthquakes per day, Rift and Koa'e earthquake magnitudes. (e), Rift and Koa'e earthquakes per day. (f), Kīlauea Caldera and deep magma-supply earthquake magnitudes. (g), Kīlauea Caldera and deep magma-supply earthquakes per day—short-period data. (h), Kīlauea Caldera and deep magma-supply earthquakes per day—long-period data.

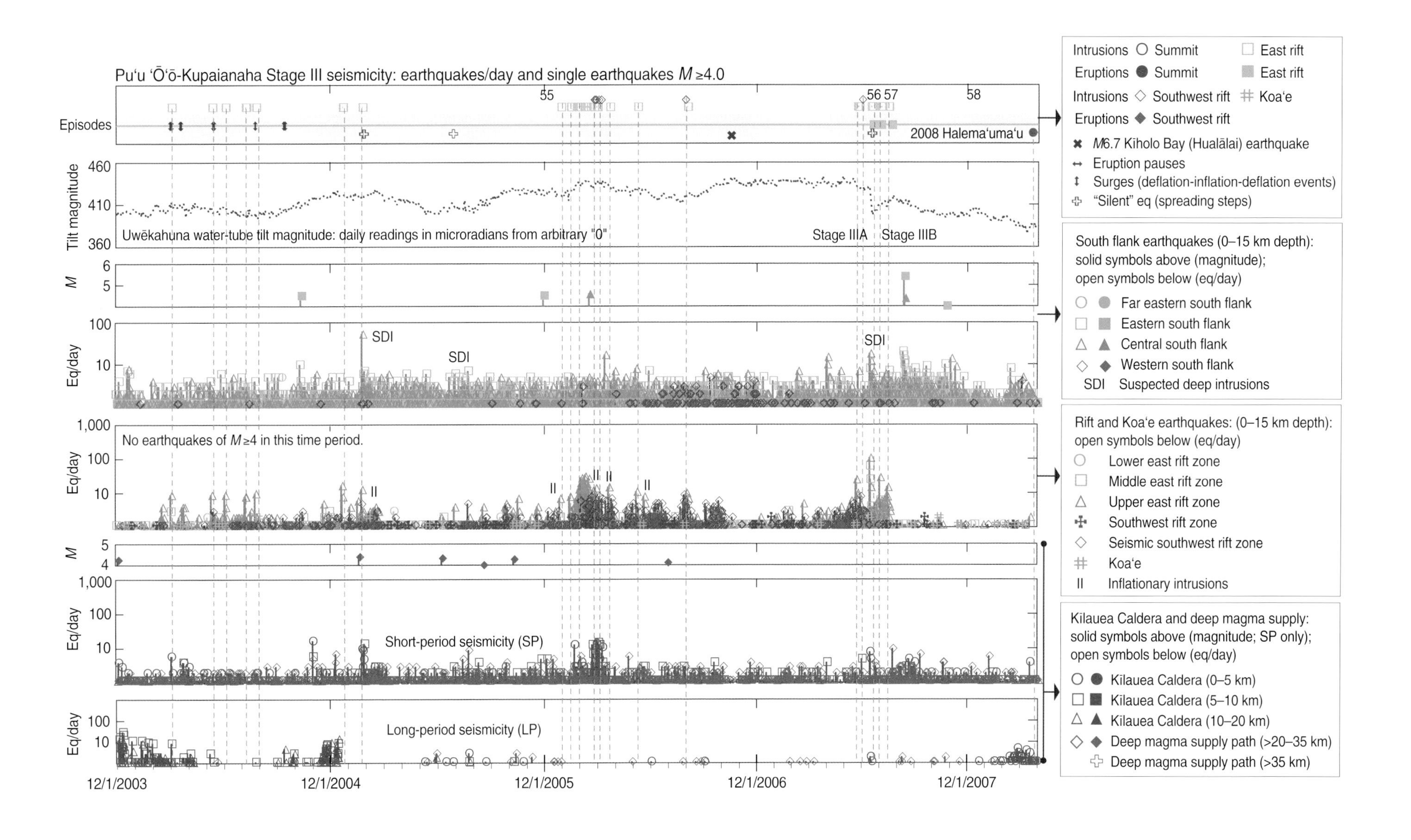
Pu‘u ‘Ō‘ō-Kupaianaha Stage III seismicity: earthquakes/day and single earthquakes M ≥4.0
Episodes
55
56 57
58
2008 Halema‘uma‘u
Tilt magnitude
460
410
360
Uwēkahuna water-tube tilt magnitude: daily readings in microradians from arbitrary "0"
Stage IIIA
Stage IIIB
M
Eq/day
SDI
No earthquakes of M ≥4 in this time period.
II
Short-period seismicity (SP)
Long-period seismicity (LP)
12/1/2003
12/1/2004
12/1/2005
12/1/2006
12/1/2007
Intrusions Summit East rift
Eruptions Summit East rift
Intrusions Southwest rift Koa‘e
Eruptions Southwest rift
M6.7 Kīholo Bay (Hualālai) earthquake
Eruption pauses
Surges (deflation-inflation-deflation events)
"Silent" eq (spreading steps)
South flank earthquakes (0–15 km depth): solid symbols above (magnitude); open symbols below (eq/day)
Far eastern south flank
Eastern south flank
Central south flank
Western south flank
SDI Suspected deep intrusions
Rift and Koa‘e earthquakes: (0–15 km depth): open symbols below (eq/day)
Lower east rift zone
Middle east rift zone
Upper east rift zone
Southwest rift zone
Seismic southwest rift zone
Koa‘e
II Inflationary intrusions
Kīlauea Caldera and deep magma supply: solid symbols above (magnitude; SP only); open symbols below (eq/day)
Kīlauea Caldera (0–5 km)
Kīlauea Caldera (5–10 km)
Kīlauea Caldera (10–20 km)
Deep magma supply path (>20–35 km)
Deep magma supply path (>35 km)

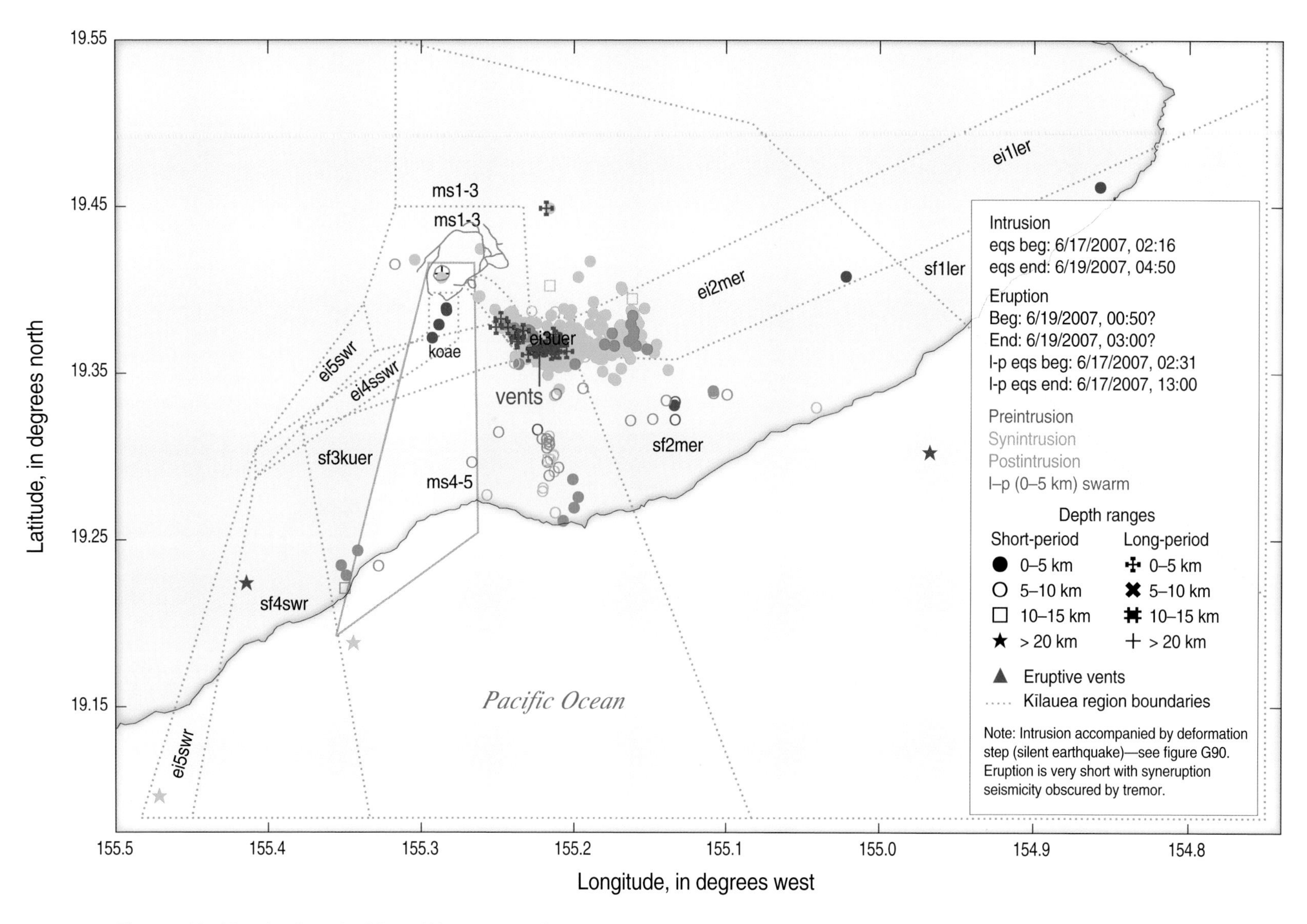

Figure 7.14. Map showing seismicity on Kīlauea, 13–23 June 2007: before, during, and after the "Father's Day eruption." That small eruption on 19 June was accompanied by a major intrusion on 17–19 June beneath the upper and middle east rift, accompanied and followed by a suspected deep intrusion beneath the central south flank. Other than that accompanying the suspected deep intrusion, the south flank seismicity is very sparse and distributed over a wide area. Dates on figure in m/d/yyyy format.

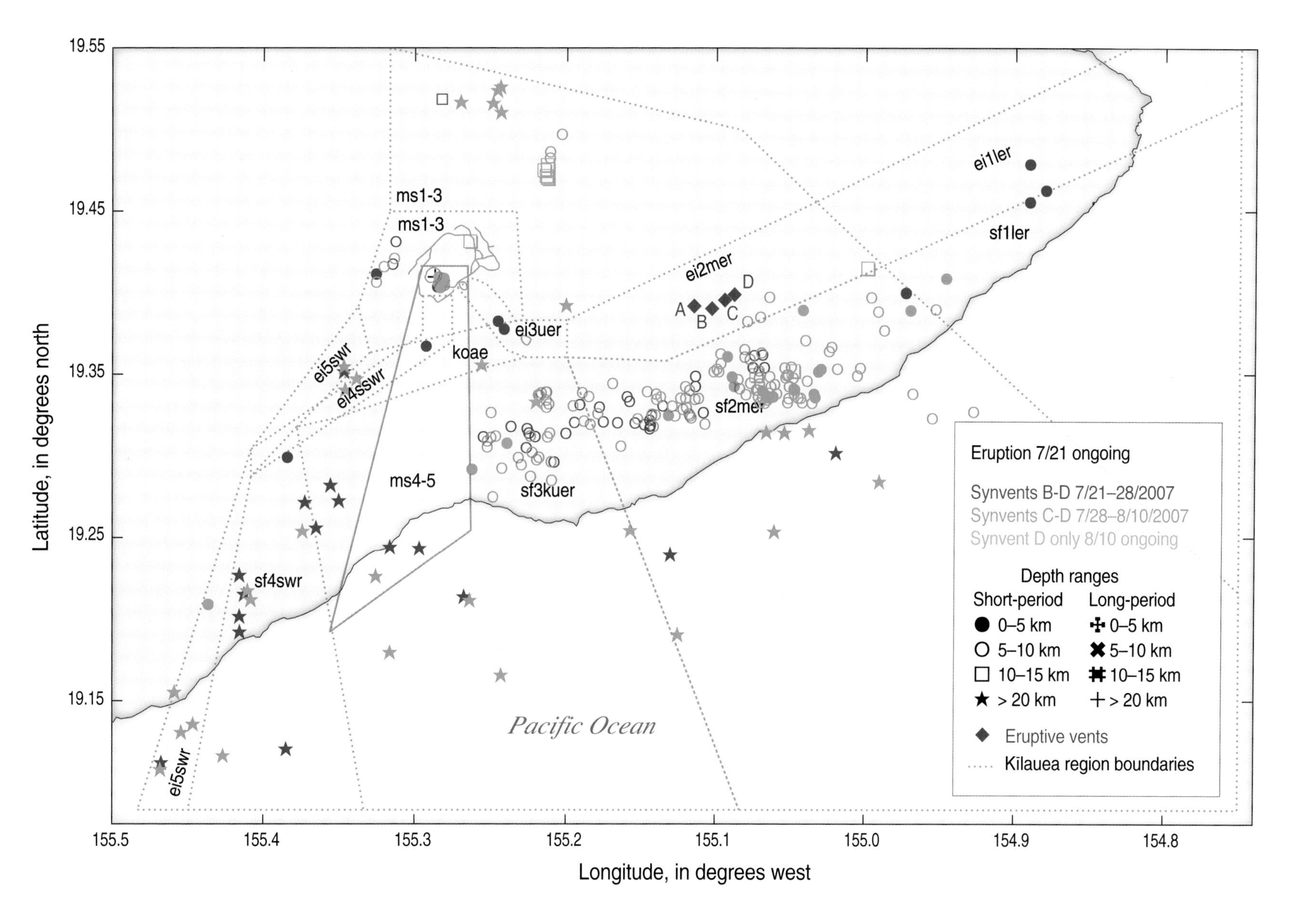

Figure 7.15. Map showing seismicity on Kīlauea, 22 July–19 August 2007. A fissure eruption on 21 July was accompanied and followed by very little rift seismicity, but it was followed by a very strong swarm of south flank earthquakes. Dates on figure in m/d/yyyy format.

Table 7.7. Number of short-period earthquakes per day in different regions through stages I–III of the Pu‘u ‘Ō‘ō–Kupaianaha eruption.

[Earthquake region abbreviations are given according to the classification in appendix A, table A3, and locations of regions are shown in appendix A, figure A4]

Regions Depth (km)	ms1[1] 0-5	ms1 all[2]	ms2.3[1] 5–15	ms2.3 all[2]	ms4.5 > 20	ei2.3 0-15	ei4 0-15	ei2.3 4 0-15	koae 0–15	cf2.3 0–15	sf4 0–15	sf all
Stages												
Pre: 2/1/1982-2/1/1983	0.57	0.57	0.04	0.04	0.04	3.28	1.6	4.90	0.09	6.97	0.86	7.83
IA: 2/1/1983-7/18/1986	0.13	0.22	0.06	0.16	0.12	0.03	0.01	0.03	0.01	2.78	0.11	2.89
IB: 7/18/1986-11/8/1991	0.26	0.28	0.04	0.14	0.15	0.19	0.00	0.19	0.01	1.90	0.04	1.94
IIA: 11/8/1991-1/1/1997	0.26	0.29	0.35	0.15	0.18	0.30	0.002	0.30	0.03	1.15	0.02	1.17
IIB: 1/1/1997-8/1/2003	0.32	0.34	0.09	0.22	0.32	0.20	0.01	0.31	0.09	1.17	0.04	1.54
III: 8/1/2003-3/18/2008	0.27	0.35	0.07	0.34	0.52	1.21	0.49	1.70	0.02	2.62	0.18	2.62

[1]Excluding seismicity triggered by the earthquake of 11/16/1983 and subsequent seismicity beneath the Nāmakanipaio region.

[2]Including seismicity beneath the Nāmakanipaio region.

Kīlauea's south flank continued to relax through the end of stage IIA, as indicated by decreasing numbers of earthquakes, but those numbers then increased through the beginning of eruption in Halema‘uma‘u in 2008 (figure 7.17*B, C*). The deep magma-supply seismicity matches the later south flank seismicity, and both began to increase in 1998–2000, several years before the beginning of inflation in 2004. The number of earthquakes beneath the rift zones and the number of intrusions, particularly on the southwest side of the volcano, increased coincident with the beginning of inflation. CO_2 emissions at Kīlauea's summit were low and constant until a spike in 2003–2004, also coincident with the beginning of inflation. SO_2 emissions measured on the east rift zone show a large increase associated with the 1997 eruption/intrusion and a smaller increase during stage III of the eruption. SO_2 emissions at the summit show little change until the advent of the March 2008 Halema‘uma‘u explosions, and the rate of SO_2 release has remained high since that time. The south flank spreading rate has varied little except for a spike associated with the January 1997 intrusion.

Magma Supply and the Evolution of Kīlauea's Plumbing, 1983–2008

Within the period from 1983 to 2008, Pu‘u ‘Ō‘ō-Kupaianaha stage I represents an approximate equilibrium between magma supply, spreading, and eruption rates at an estimated magma-supply rate of 0.18 km^3/yr (Cayol and others, 2000), which is higher than magma-supply estimates prior to the 1975 earthquake (Wright and Klein, 2008). During stage IA there were no intrusions, and eruptive episodes were fed from the traditional summit reservoir at 2–6-km depth depicted in chapter 2, figure 2.5. Intrusions began in 1988 and have been frequent since (fig. 7.17*D*).

Magma supply rates during stages I and II have been estimated using a variety of methods. Eruption rates during this time were estimated by mapping flow distribution, by measurement of SO_2 emission, by electromagnetic profiling over active lava tubes (Heliker and Mattox, 2003, table 1; Sutton and others, 2003, table 1, figures 2–4), and by estimating volume changes during D-I-D events (Cervelli and Miklius, 2003, table 1). Magma supply rates were calculated for the first 20 years of eruption and show good agreement among the various methods. When rift dilation is taken into account, magma supply rates agree with the 0.18 km^3/yr magma supply rate estimated for stage IA (Cayol and others, 2000), which may have increased to more than 0.2 km^3/yr by 2001(Cervelli and Miklius, 2003, table 1). A microgravity study indicates that mass was added to Kīlauea's summit reservoir in 1983–85 and 1991–93 and subtracted from it during1985–91 (Kauahikaua and Miklius, 2003, figure 2). Mass was also added following the December 1990 intrusion through the transition from stage I to stage II. A full discussion of the long-term history of magma supply and the evolution of Kīlauea's plumbing is given in the next chapter (see fig. 8.5).

A new magma source just southeast of Halema‘uma‘u at a depth of less than 1 km was identified in 1996 by the broadband seismic array and by continuous recording tiltmeters located around Kīlauea's summit (see section above on

"Installation of a Broadband Seismic Array"). This magma source supplied Puʻu ʻŌʻō during stages II and III (Cervelli and Miklius, 2003). Earthquakes associated with the summit eruption in April 1982 and the intrusion of December 1990 lie very close to the new magma center and may bracket the time when this source first became active. Subsequent eruptions at Puʻu ʻŌʻō may have been fed from this new shallow magma chamber. On the other hand, the large intrusion of 26 March 1991 appears to be more similar to earlier intrusions fed from the traditional reservoir at depths of 2–6 km. The March 1991 intrusion and others accompanied by a sizable summit deflation observed on the Uwēkahuna tiltmeter must be fed from the 2–6-km deep reservoir to produce the observed summit collapse. We suggest that subsequent intrusions continued to be fed from the deeper reservoir, whereas magma erupted from Puʻu ʻŌʻō followed the shallower path.

We suggest that a small increase in magma supply may have caused the transition from episodic (stage IA) to continuous (stage IB) eruption, in agreement with Parfitt and Wilson (1994). The supply increase caused a nonlinear transition from an oscillator to a continuous flow mechanism. A change in conduit geometry could also cause a transition from oscillator to continuous, but we prefer magma-supply increase as the cause because other evidence supports a supply increase. Another increase in magma supply triggered the 1997 eruption and intrusion. In the latter case the summit was not pressurized before the intrusion, because there was no prior inflation, indicating that pressure was applied from the deeper magma system beneath the rift zone. The application of magma pressure originating in the deep rift resulted in dilation of the rift and a temporary increase in spreading rate (fig. 7.17*B*; Owen and others, 2000b), as well as a temporary increase in SO_2 emission measured at the east rift zone (fig. 7.17*C*). The steady increase in short-period earthquakes deeper than 20 km from 1996 through 2007 (figure 7.16*D*) suggests that the magma supply from depth was increasing continuously during this period but that conditions within the crust caused the increase in magma supply to be manifested in a discontinuous fashion.

A further large increase in magma supply occurred in late 2003, evidenced by summit inflation (fig. 7.16*A*), an increase in CO_2 emission at Kīlauea's summit to double or triple the average values measured earlier in the eruption (fig. 7.17*C*; Poland and others, 2007, 2008; Sutton and others, 2009), and a great increase in the frequency of traditional and inflationary intrusions (figure 7.17*D*). Additional evidence for a magma-supply increase is provided by gravity measurements that show addition of mass beneath Kīlauea's summit through periods of both inflation and deflation (Eggers and Johnson, 2006; Tikku and others, 2006). Finally, the activation of Kīlauea's seismic southwest rift zone to accommodate increased magma supply, manifested as increased seismicity (table 7.7; fig. 7.16*B*), uplift (Poland and others, 2007), and increased intrusion beneath both rift zones (fig. 7.17*D*), mimics a pattern seen earlier in the Mauna Ulu period that preceded major changes in the eruptive regime (this paper, chap. 5; (Wright and Klein, 2008)).

The final events concluding stage III and leading to the 2008 Halemaʻumaʻu explosions begin with the Father's Day eruption. The great increase of sulfur dioxide emission at Kīlauea's summit during and following that eruption (Poland and others, 2009a) is a function of the greatly increased magma supply. In our interpretation, the ever-increasing magma supply resulted in the upward migration of the magma system to the surface through stoping of the weakened and formerly resistant rocks above the magma reservoir roof. At some time before the Halemaʻumaʻu explosions of March 2008, the shallow reservoir broke upward to yield a single magma source extending from the surface to 1 km depth (see chap. 8, fig. 8.1). In chapter 8 we interpret the monitored data in terms of this three-stage magma chamber evolution.

Long-Period Seismicity

Long-period earthquakes at different depths offer a clue to changes in plumbing during the Puʻu ʻŌʻō eruption. Long-period earthquakes at depths of 0–5 km represent excitation of the hydrothermal system around Kīlauea's summit reservoir (Almendros and others, 2001; Chouet, 1996). Before the current eruption, shallow long-period seismicity was seen below the caldera several days after the beginning of eruption and intrusion in September 1977 and persisted for many days after the eruption ended. Shallow long-period seismicity again dominated the response to the 1997 intrusion, this time persisting for many weeks after the end of eruption. Notably, shallow long-period seismicity was absent from every other eruption and intrusion since 1972, including episode 1 of the current eruption.

Long-period earthquakes below the magma chamber must reflect the flow regime along the vertical magma path connecting the magma chamber to its mantle source. Long-period earthquake swarms (at 5–15 km depth) become important beginning in 1985, reaching a maximum in 1989 and 1990, during stage IB (figure 7.16*D*). The seismicity most likely reflects a response to increased and consistent use of a single magma path in this depth range. As basalt is a good insulator, it is not likely

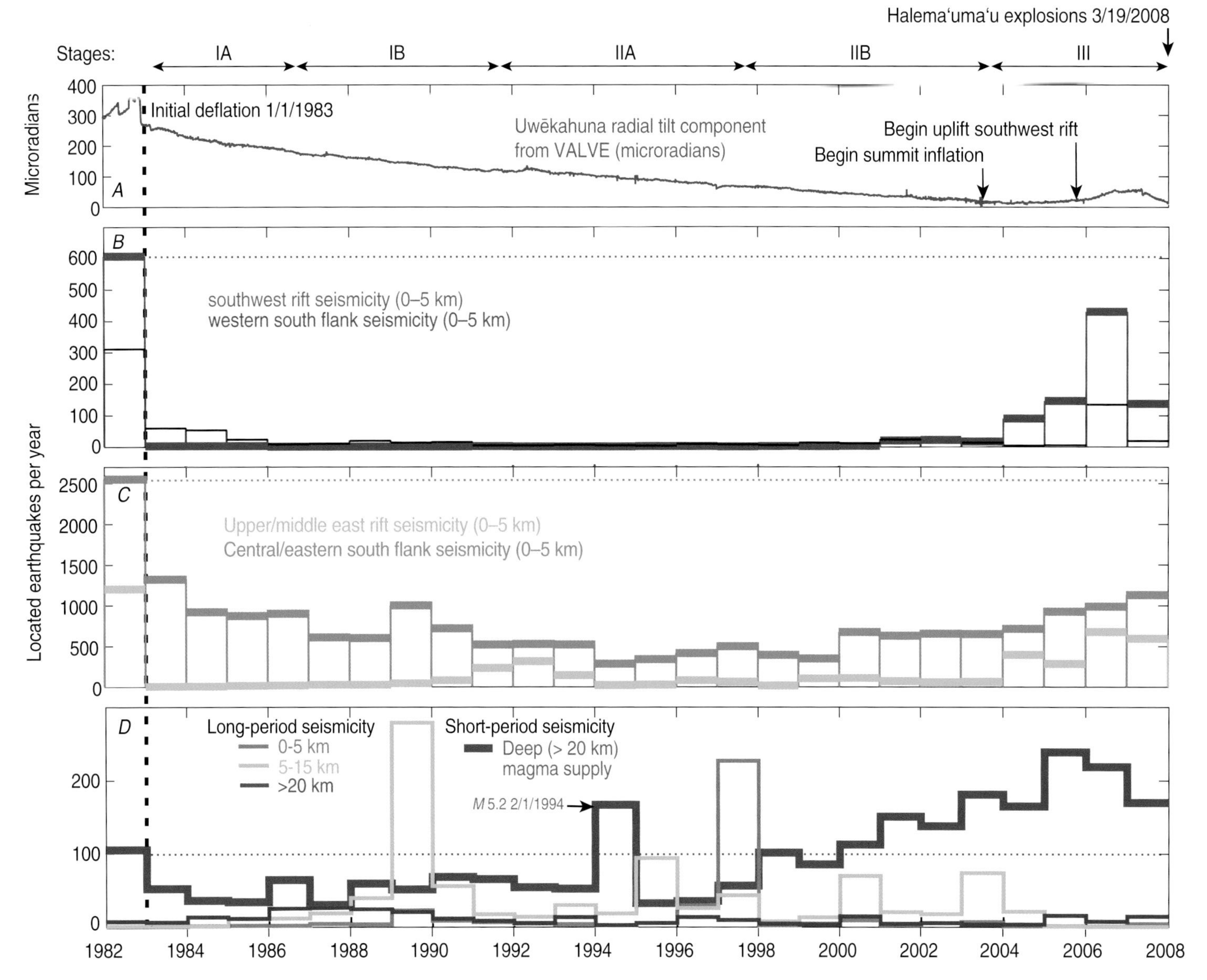

Figure 7.16. Graphs showing Kīlauea seismicity and tilt data, 1982–2008. The heavy dashed line separates data on the left covering a 1-year precursory period through episode 1 of the ongoing east rift eruption (1 February 1982–1 February 1983) from data on the right covering yearly data from 1 February 1983 to 1 February 2008. Arbitrarily defined stages of the Pu'u 'Ō'ō-Kupaianaha eruption are given across the top. Horizontal dotted lines represent the level of activity in the precursory period for comparison with the later eruption stages. Dates on figure in m/d/yyyy format. ***A***, Uwēkahuna water-tube tilt, radial component, from the Hawaiian Volcano Observatory's computer database (VALVE). ***B***, Southwest rift zone and western south flank earthquakes per year. ***C***, Upper/middle east rift and adjacent south flank earthquakes per year. ***D***, Long-period seismicity (thin bars) at 0–5, 5–15, and >20 km depth compared with short-period seismicity deeper than 20 km (thick bars). Earthquakes from 0 to 15 km depth are located beneath Kīlauea's summit. Earthquakes deeper than 20 km extend south and southwest from Kīlauea's summit along Kīlauea's magma supply path (Wright and Klein, 2006).

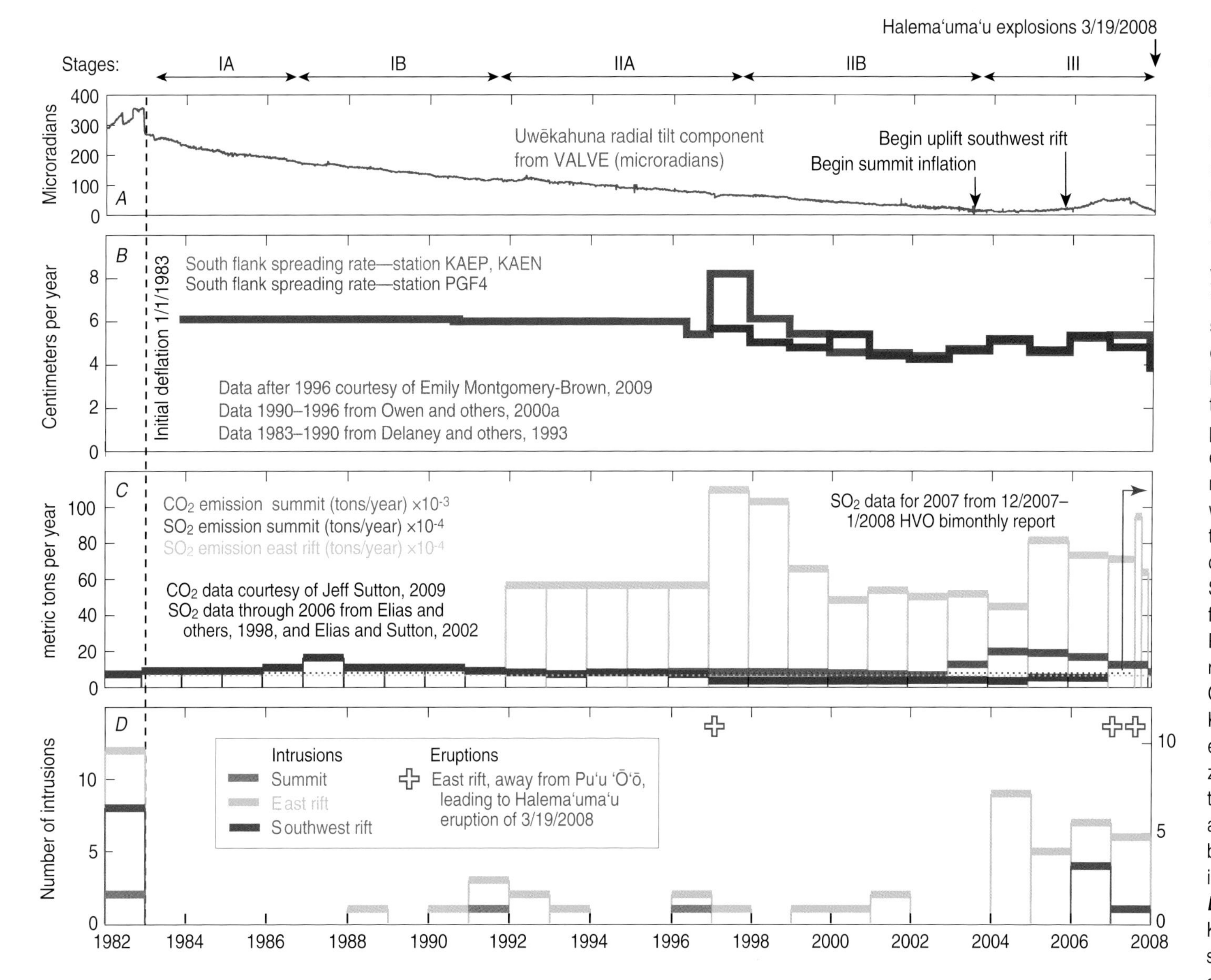

Figure 7.17. Graphs showing Kīlauea south flank spreading, gas emissions, intrusions, eruptions away from Puʻu ʻŌʻō, and tilt data, 1982–2008. The heavy dashed line separates data on the left covering a 1-year precursory period through episode 1 of the ongoing east rift eruption (1 February 1982–1 February 1983) from data on the right covering yearly data from 1 February 1983 to 1 February 2008. Arbitrarily defined stages of the Puʻu ʻŌʻō-Kupaianaha eruption are given across the top. Horizontal dotted lines represent the level of activity in the precursory period for comparison with the later eruption stages. Dates on figure in m/d/yyyy format. ***A***, Uwēkahuna water-tube tilt, radial component, from the Hawaiian Volcano Observatory's computer database (VALVE). ***B***, South flank spreading rates estimated from episodic and continuous Global Positioning System measurements modeled at two locations. ***C***, Yearly CO_2 and SO_2 emissions from Kīlauea's summit and yearly SO_2 emissions from Kīlauea's east rift zone. The blue arrow points to the very high rate of SO_2 release at Kīlauea's summit following the beginning of the March 2008 eruption in Halemaʻumaʻu that ends our study. ***D***, Number of intrusions beneath Kīlauea's summit, east rift zone, and seismic southwest rift zone and timing of eruptions away from Puʻu ʻŌʻō.

that softening of the rocks surrounding the magma conduit through heating could yield a broad distribution of long-period seismicity (Bruce Julian, oral commun., 2009). Mechanical disruption of the rocks surrounding the conduit is considered more likely (P. Dawson, oral commun., 2009), perhaps accompanying conduit widening as a response to long and continued use of the conduit at an ever–increasing magma supply rate. In any case, the physical or thermal environment of the conduit dramatically changed in the years after 1985.

More difficult to explain is the great decrease of long-period earthquakes deeper than 5 km in 2005 and their complete absence thereafter. This requires a process that would restore a stable physical and thermal structure of a possibly wider conduit noted above. It is also possible that the decrease is an artifact related to changes in seismic staffing at HVO, because the notation for long-period earthquakes is done through visual inspection by the analyst as part of normal earthquake processing.

Nāmakanipaio Earthquake Sequences

The ongoing occurrence of swarms of earthquakes a few kilometers west and north of Kīlauea Caldera in the vicinity of the Nāmakanipaio Campground is an interesting aspect of the eruption that began in 1983. Such earthquakes were first identified as Kīlauea seismicity triggered by the 16 November 1983 Ka‘ōiki earthquake. Five swarms in 1990–92 were identified as precursors to several intrusions on the east rift zone (Okubo and Nakata, 2003). However, such swarms have occurred both earlier and later and sometimes accompany east rift intrusion (tables 7.1–7.5). Their origin remains uncertain. One possibility is that they are triggered by increase in magma pressure beneath Kīlauea's summit on an older caldera ring fracture (A. Miklius, oral commun., 2009), which was weakened from the effects of the 1983 Ka‘ōiki earthquake.

Supplementary Material

Supplementary material for this chapter appears in appendix G, which is only available in the digital versions of this work—in the DVD that accompanies the printed volume and as a separate file accompanying this volume on the Web at http://pubs.usgs.gov/pp/1806/. Appendix G comprises the following:

Table G1. Tilt volume, eruption efficiency, and magma supply rate from 1983 to 2008.

Figures G1–G3. One-year time series plots between 1 February 1982 and 1 February 2008, showing details of numbers of both short-period and long-period earthquakes shown in text figures.

Figures G4–G7. Time series plot of earthquake swarms and plots of earthquake locations for the events of stage IA.

Figures G8–G15. Time series plot of earthquake swarms and plots of earthquake locations for the events of stage IB.

Figures G16–G24. Time series plot of earthquake swarms and plots of earthquake locations for the events of stage IIA.

Figures G25–G49. Time series plot of earthquake swarms and plots of earthquake locations for the events of stage IIB.

Figures G50–G82. Time series plot of earthquake swarms and plots of earthquake locations for the events of stage III.

Webcam setup on the rim of Pu‘o ‘Ō‘ō, part of the monitoring improvements following the period covered in this paper. Images from the camera are continuously transmitted to HVO to allow staff to remotely assess changes in the eruption. USGS photo by B. Gaddis, 18 November 2010.

Chapter 8

Summary

Presents summary interpretations of topics that cover multiple time periods in the recorded history of Kīlauea, including long-term magma supply, south flank anticipation of eruptions and intrusions, history of south flank spreading, and the relation between the plumbing systems of Kīlauea and Mauna Loa Volcanoes.

Continuous GPS station HOLE, located on Holei Pali on the south flank of Kīlauea Volcano. The GPS antenna is the white disk on the monument at the far right, while the antenna on the left side of the image transmits data back to the Hawaiian Volcano Observatory. The station is powered by solar panels (far left) that charge a bank of batteries, which reside, with the GPS receiver, in the hardened instrument box at the center of the photo. Photograph by M. Poland.

Our 200-year history of Kīlauea emphasizes the interpretation of magma transport and storage below the volcano surface in order to connect the long sequence of eruptive and intrusive activity as coherently as possible. For our interpretation we use the continuous record provided first by earthquakes and eruption chronology and later, following the founding of the Hawaiian Volcano Observatory, by tilting of the ground measured daily at Kīlauea's summit. We seek a long-term view of Kīlauea within which individual events and observations can be placed. Details of individual eruptions are provided by references cited in the text; our aim is to understand how each event in Kīlauea's history is affected by prior events and influences future events.

Magma Transport from Melting Source to Storage

Geometry of Kīlauea's Magma Plumbing

Magma destined for eruption at Kīlauea is generated by partial melting at depths of 80–100 km within a source region in the mantle designated as the Hawaiian hot spot. Evidence from long-period earthquakes indicates that most of Kīlauea's magma travels from initial melting at 80–100 km depth within the source region through a vertical (100–35 km depth) to horizontal (30±5 km depth) to vertical (above ~25 km depth) conduit to reach Kīlauea's summit magma chamber (Wright and Klein, 2006). Vertical transport from the mantle is aseismic to about 60-km depth as magma coalesces in a ductile region near the base of the lithosphere. Between 60-km depth and the base of the Kīlauea edifice at ~11-km depth, only about 10 percent of the energy associated with magma transport is released as a combination of long-period and short-period earthquakes (Aki and Koyanagi, 1981). In this realm the coalescing magma must maintain a hot core within a cooler shell in order to allow upward movement, much as magma transport in pāhoehoe lava is facilitated by the formation of lava tubes.

A shallow-dipping decollement at depths of 10–12 km separates the Kīlauea edifice from the older ocean floor and sediment layer present at Kīlauea's inception (Delaney and others, 1990). Magma transport through the edifice is accompanied by a much higher seismic response. A storage chamber has been identified at depths of 2–6 km beneath Kīlauea's summit caldera through modern seismic and geodetic study. The chamber consists of a small liquid core (see below) surrounded by magma-permeated rock, cool enough to result in rapid solidification of small packets of magma such that the carapace can continue to fracture during epsisodes of inflation and deflation. All magma passes through this chamber before eruption or intrusion at Kīlauea's summit or rift zones. We presume that such a chamber has existed throughout the shield-building stage of Kīlauea's history (Clague, 1987; Clague and Dixon, 2000). At different times in Kīlauea's history this magma chamber at 2–6 km depth has been augmented by a smaller magma storage chamber at a depth of 1 km or less, also beneath Kīlauea Caldera (fig. 8.1).

Below Kīlauea's rift zones, from 5-km depth down to the decollement, modern studies have postulated magma-permeated rock that is able to transmit pressure to the adjacent solid south flank. All of our historical research supports the existence of such a "deep [rift] magma system" (Delaney and others, 1990). Some-upward moving magma from the hot spot may be diverted to the deep rift magma system before reaching the reservoir, particularly during periods in which the plumbing is being repaired, such as following the 1975 earthquake or the 1924 intrusion. Such magma contributes to the increased magma pressure before eruptions or intrusions but is not itself erupted or intruded at shallow levels. Magma destined for eruption at the summit or on the rift zones is fed either from the liquid core within the main summit reservoir or from the shallower reservoir at about 1-km depth, as illustrated in figure 8.1. Magma destined for intrusion beneath the rift zones is diverted during upward transport below the liquid core, but still within the summit reservoir (fig. 8.1).

Petrologic Contribution to Understanding Magma Storage and Transport

In chapter 4 (fig. 4.13) we presented our interpretation of how the magmas having the chemistry of the three Halema'uma'u eruptions of 1952, 1961, and 1967–68 moved through the rift plumbing. We can now address related questions regarding volumes of the three magma batches, where they entered the rift plumbing, and how long they remained before they were no longer seen in rift eruptions. Volumes of the magma batches seen in both summit eruptions and hybrid east rift eruptions are shown in table 8.1. Volumes of additional magma transferred to the east rift zone between eruptions are also tabulated in table 8.1.

The summit reservoir and its ability to hold magma in different volumes, as shown by the complexity of different internal reservoir deformation centers, is key to the eruption of distinct magma batches. The expected deformation field above expanding magma bodies of differing geometry, such as sills, dikes, or intrusive plugs, has been modeled by Dieterich and Decker (1975). Many deformation cycles documented in Hawaiian Volcano Observatory monthly reports (unpublished) show (1) a migration

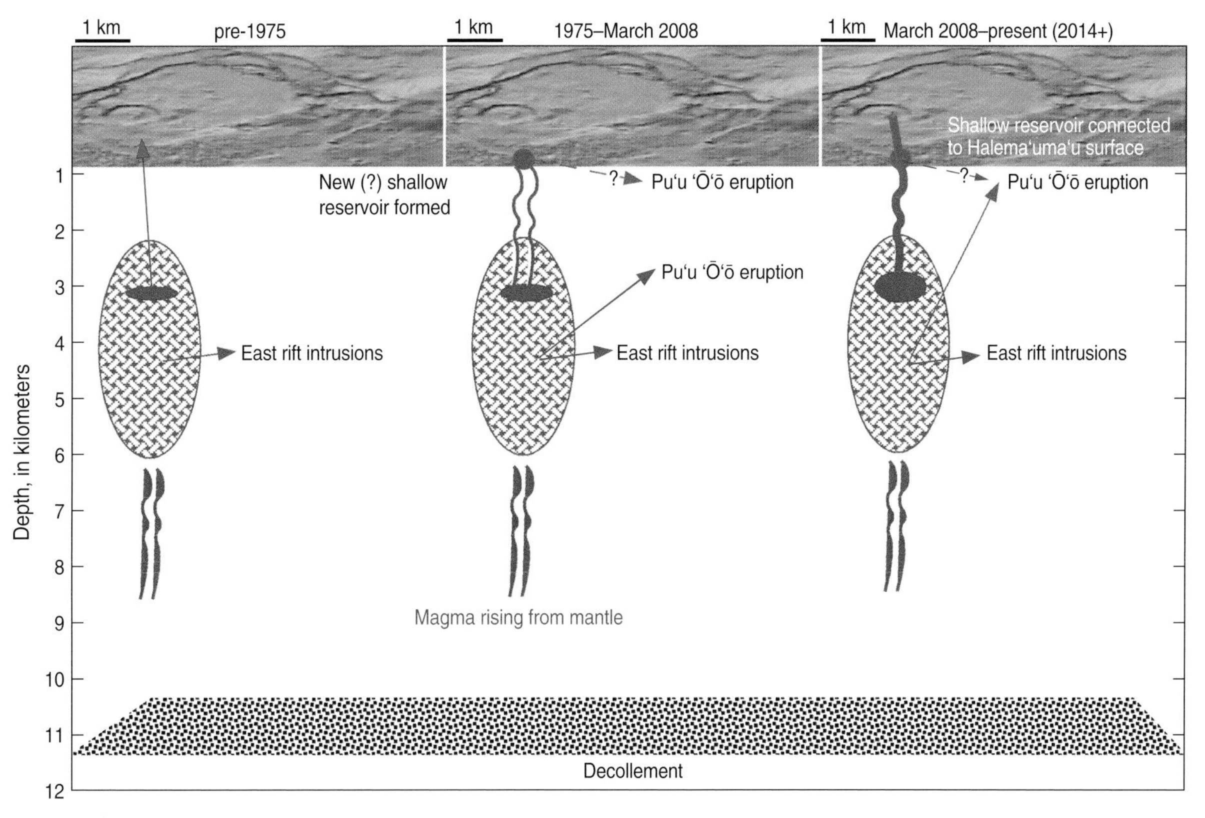

Figure 8.1. Diagram showing three stages in the evolution of Kīlauea's plumbing. Magma from the mantle (wavy red lines) feeds Kīlauea's primary reservoir at 2–6 km depth that supplies all magma for eruption and intrusion. A source depth of 3 km, calculated according to principles laid out by K. Mogi (1958), is shown, averaged from many leveling and tilt surveys. Summit eruptions are fed from a small liquid core within the larger reservoir (see fig. 8.2 and text discussion). Rift eruptions are diverted from the primary vertical conduit at some depth below the liquid core. See text for further discussion. Pre-1975 configuration: A single primary reservoir supplemented by a shallow (<1 km depth) reservoir (not shown) during summit lava lake activity. From 1975 to March 2008: Configuration after appearance of shallow magma reservoir northeast of Halema'uma'u at a depth of less than 1 km. Such a reservoir was also present during the 19th and early 20th century to feed Kīlauea's lava lakes. The interpretation that Pu'u 'Ō'ō is being fed from the reservoir at 1 km depth (Cervelli and Miklius, 2003) has been revised to favor a source within the traditional reservoir at 2-6 km depth (M. Poland, oral commun., 2012). March 2008 to present: Breakdown of the connections between the two reservoirs and the surface, leading to explosive eruption at Halema'uma'u and the existence of a sub-Halema'uma'u lava lake.

of inflation and deflation centers similar to that described by Fiske and Kinoshita (1969) and (2) a greater lateral extension of lines spanning the caldera than that predicted by the spherical Mogi source approximation, implying a larger vertical than horizontal extent of the expanding magma body. We adopt a summit plumbing model of concatenated vertical plugs, following the analysis of Dieterich and Decker (1975), that fits both the migrating centers of inflation and the observed summit ground-deformation patterns (fig. 8.2). The protrusions at the top of the chamber in the figure represent the tops of vertical plugs at the location of the most commonly occupied centers of deflation and inflation.

The volumes of the magma batches erupted at Kīlauea's summit in 1952, 1961, and 1967–68 range from 0.13 to 0.21 km^3 (table 8.1). We use an average area of 1 km^2, based on the extent of the preferred south caldera inflation-deflation centers, and a maximum storage volume of 0.09 km^3 derived from the volumes of magma erupted in Halema'uma'u. These assumptions yield a thickness of the active

Summit magma reservoir

"ears" above inflation/deflation centers

Central liquid core

Figure 8.2. Diagram of Kīlauea's primary magma reservoir. The reservoir is outlined as a magma-permeated zone of concatenated vertical plugs surrounding a much smaller liquid core. The vertical plugs lie above the inflation/deflation centers defined in the buildup to the 1967–68 Halema'uma'u eruption (see fig. 4.2) and are manifested as "ears" sprouting from the reservoir top. The diagram is not to scale, but the larger reservoir is 1–2 km across and extends vertically as much as 4 km, and the liquid core is estimated at tens of meters in diameter.

Table 8.1. Volumes of magma batches entering Kīlauea plumbing.

[dates in m/d/yyyy format; V, volume; do, ditto (same as above)]

a. Magma erupted in homogeneous summit eruption or mixed in hybrid eruptions.

Name	Start date	End date	V (km^3)	Comment	Reference[1]
1952	6/27/1952	11/9/1952	0.0870	Halema'uma'u eruption	4
	3/20/1955	5/26/1965	0.0333	Later part of 1955 eruption; mixing percentage 41.9	9, 1
	1/26/1960	1/29/1960	0.0035	Early part of 1960 eruption; mixing percentage 35.7	9, 2
Total			0.1335		
1954	5/31/1954	6/3/1954	0.0050	Kīlauea Caldera eruption	8
	11/14/1959	11/21/1959	0.0042	1959 eruption episode 1: mixing percentage 45.5	5, 3
	11/22/1959	12/20/1959	0.0027	1959 eruption episodes 2–17: mixing percentage 13.2	5, 3
Total			0.0119		
1961	3/5/1955	3/6/1955	0.0418	Kalalua intrusion—calculated as parent for 1977 eruption	9,
	1/30/1960	2/4/1960	0.0086	Middle part of 1960 eruption; mixing percentage 58.2	5, 2
	2/24/1961	7/17/1961	0.0103	Halema'uma'u eruptions (3)	6
	9/22/1961	9/25/1961	0.0390	East rift eruption: assume 1961 and 1967 in equal volume	6, 15
	12/7/1962	12/9/1962	0.0049	12/1962 east rift eruption	14, 15
	8/21/1963	8/23/1963	0.0022	8/1963 east rift eruption	10, 15
	10/5/1963	10/6/1963	0.0155	10/1963 east rift eruption	11, 15
	3/5/1965	3/15/1965	0.0033	3/1965 east rift eruption	12, 15
	12/24/1965	12/25/1965	0.0036	12/1965 east rift eruption	13, 15
Total			0.1292		
1967–68	2/12/1960	2/18/1960	0.0043	Later part of 1960 eruption: mixing percentage 24.8	5, 2
	9/22/1961	9/25/1961	0.0390	East rift eruption: assume 1961 and 1967 equal volume	6, 15
	12/7/1962	12/9/1962	0.0064	12/1962 east rift eruption	14, 15
	8/21/1963	8/23/1963	0.0016	8/1963 east rift eruption	10, 15
	10/5/1963	10/6/1963	0.0250	10/1963 east rift eruption	11, 15
	3/5/1965	3/15/1965	0.0381	3/1965 east rift eruption	12, 15
	12/24/1965	12/25/1965	0.0177	12/1965 east rift eruption	13, 15
	11/5/1967	7/14/1968	0.0744	Halema'uma'u eruption	7
	8/22/1968	8/28/1968	0.0024	8/1968 east rift eruption	16, 18
	10/7/1968	10/22/1968	0.0234	10/1968 east rift eruption	16, 18
	2/22/1969	2/28/1969	0.0061	2/1969 east rift eruption	17, 18
Subtotal			0.2384		
Total			0.5130	Total amount of unfractionated magma identified[2]	

[1]Helz and Wright, 1992; 2. Wright and Helz, 1996; 3. Wright, 1973; 4. Macdonald, 1952; 5. Richter and others, 1970; 6. Richter and others, 1964; 7. Kinoshita and others, 1969; 8. Macdonald and Eaton, 1954; 9. Macdonald and Eaton, 1964; 10. Peck and Kinoshita, 1976; 11. Moore and Koyanagi, 1969; 12. Wright and others, 1968; 13. Fiske and Koyanagi, 1968; 14. Moore and Krivoy, 1964; 15. Wright and Fiske, 1971; 16. Jackson and others, 1975; 17. Swanson and others, 1976b; 18. Wright and others, 1975.

[2]Add 0.01 km3/year for rift dilation during spreading; lifetime of magma batches ~ 10 years.

b. Additional magma transfer to east rift zone: deflation volumes not associated with eruption.

Start date	End date	V (km^3)	Comment
9/15/1950	12/8/1950	0.0088	Assumed to be magma batch of 1952 chemistry
12/8/1950	12/16/1950	0.0507	do
Subtotal		0.0595	Minimum volume1952 magma intruded into the rift zone.
6/26/1952	10/2/1952	0.0323	Assumed to be magma batch of 1961 chemistry
3/10/1954	3/24/1954	0.0082	do
12/13/1954	1/17/1955	0.0114	do
2/19/1955	12/29/1955	0.1494	do
8/31/1959	11/14/1959	0.0011	do
Subtotal		0.2024	Volume of 1961 magma intruded into the east rift zone
11/15/1959	11/23/1959	0.0213	Assumed to be magma batch of 1967–68 chemistry[3]
12/23/1959	1/17/1960	0.0103	do
1/17/1960	10/21/1960	0.1405	do
9/21/1961	9/30/1961	0.0779	do
10/28/1961	11/4/1961	0.0046	do
12/6/1962	12/9/1962	0.0092	do
5/8/1963	5/12/1963	0.0157	do
6/28/1963	7/2/1963	0.0085	do
8/21/1963	8/22/1963	0.0036	do
10/4/1963	10/10/1963	0.0362	do
11/11/1964	12/1/1964	0.0022	do
3/5/1965	3/9/1965	0.0395	do
12/23/1965	12/29/1965	0.0213	do
10/1/1966	10/7/1966	0.0044	do
8/9/1967	8/18/1967	0.0038	do
Subtotal		0.399	Maximum additional volume of 1967-68 chemistry. May include some 1961 chemistry.
Total		0.6609	

[3]Latest time that magma can enter east rift zone for mixing with the latter part of the 1960 eruption.

storage region of about 115 m, which is sufficient to prevent cooling during the brief time in storage. Mogi models from levelling measurements suggest that this zone of stored high-temperature liquid lies about 3 km below Kīlauea's summit. The percentage of liquid magma within the whole of the seismically and geodetically imaged summit reservoir is quite small. Assumed spherical reservoirs of radii 1 and 2 km yield reservoir volumes of 4.19 and 33.51 km^3, respectively. The estimated percentage of magma to solid rock in storage ranges between 0.3 and 2.1 percent. We consider that magma for the 1952, 1961, and 1967–68 Halema'uma'u eruptions successively occupied a small liquid region like that shown in figure 8.2 and moved from there to eruption in Halema'uma'u as shown in figure 8.1. The fact that each of the three summit eruptions retained a different but uniform chemistry indicates that each magma batch had to be totally removed by eruption before the next magma entered[28].

During each magma batch's time in storage, a variable but high percentage of the magma was partitioned to the east rift zone. The fact that the magmas are seen as components of rift eruptions even before they are seen in Halema'uma'u indicates that they cannot move into the rift zone from the liquid core beneath the caldera, but must move into the rift zone from some greater depth before the liquid core is ready for a new batch of magma to enter from below, as depicted in figure 8.1. The depth at which transfer to the rift occurs can be determined by the CO_2 content of the magma erupted on the rift zone, which averages 0.05 weight percent (Gerlach and others, 2002). That level of CO_2 corresponds to an equilibrium pressure of ~ 1.1 kbar, corresponding to a depth of 4–5 km (Dixon and others, 1995). The transfer depth of 4–5 km places magma in a position to feed the main rift magma conduits at depths of 2–4 km.

Magma occupation of the primary summit reservoir changed in the years following the 1967–68 summit eruption. Beginning with the two 1968 rift eruptions, 10 new olivine-controlled magmas appeared during 1968–75. (Wright and Klein, 2010). Magmas erupted at Mauna Ulu during 1969–74 suggest a more complex hybridization in which olivine-controlled magmas could have been mixed with each other before eruption. The magma batches were smaller in volume than the 1952, 1961, and 1967–68 batches, and thus more than one batch could simultaneously occupy the liquid core depicted in figure 8.2.

The April 1982 summit eruption can be interpreted as a mixture of the two most voluminous magma compositions erupted at Mauna Ulu. Eruptions on the rift zone in 1979 and 1980 were complex mixtures of several olivine-controlled magmas, some of which had undergone limited fractionation. We attribute the greater frequency of chemical changes among magma batches and the resultant more complex mixing to documented increases in magma supply rate. The smaller volume of magma baches allows them to simutaneously occupy the storage area formerly occupied by a single magma batch, thus promoting mixing (see footnote above).

The eruption that began in 1983 has been accompanied by further dramatic increases in magma supply rate, and all vestiges of a sequence of distinct olivine-controlled magmas are gone. A more subtle and continuous variation has been ascribed variously to changes in the chemistry of the source mantle (Marske and others, 2008) or to mixing and flushing of previously existing olivine-controlled magmas by new magma derived from a mantle of constant source chemistry (Thornber, 2001).

Relation Between Eruption and Intrusion Expressed as Eruption Efficiency

Eruption efficiencies (compare Wright and Klein, 2008, their figure 6 and discussion) have been calculated as a ratio of the volume erupted (after correction for vesicularity) to the volume of magma transferred to feed a rift eruption as estimated from the Uwēkahuna tilt deflation magnitudes. Eruption efficiency is a measure of the state of the rift-zone magma plumbing, the confining stress, and the ability of the rift to allow magma to move to the surface. Eruption efficiency estimates magma partitioning between eruptions and intrusions, an alternative to the presentation of eruption/intrusion balances by Dvorak and Dzurisin (1993, figure 10). Eruption efficiencies are plotted in figure 8.3 for the period preceding the 1975 earthquake (fig. 8.3*A*) and during stage IA of the Pu'u 'Ō'ō-Kupaianaha eruption (fig. 8.3*B*). Eruption efficiency is low following a long interval without rift eruption, because magma must fill rift voids created during ongoing flank spreading, and it takes a larger magma pressure and volume to open new conduits to the surface. Eruption efficiency then increases as the rift plumbing becomes hot through repeated use and as conduit widths increase. Reversals to low eruption efficiency shown in figure 8.3*A* are associated with Koa'e activity, possibly because the Koa'e Fault Zone, upon being intruded with rift magma, wedges the east rift apart, forming voids to hold future intrusions without eruption. The Koa'e itself acts as a magma sink, rarely erupts, and so has near-zero eruption efficiency. Lower eruption efficiencies may also follow some suspected deep intrusions on the assumption that some of these intrusions are associated with rift dilation. Magma pressure generated during the suspected deep intrusions in

[28] If two magmas of similar density enter the liquid chamber, the earlier magma will be more degassed and thus more dense than the new arrival, thus promoting mixing of the two magmas.

August and November 1965 (table 4.1; fig. 4.12) may have been one cause of the Koaʻe crisis in December 1965. During Mauna Ulu stage IA from May to December 1969, a brief relative reduction in eruption efficiency followed suspected deep intrusions (fig. 8.3*A*).

Eruption efficiencies for the high-fountaining episodes beginning the Puʻu ʻŌʻō-Kupaianaha eruption (fig. 8.3*B*) scatter widely, but the average is close to 1, as might be expected for a long eruption with little accompanying intrusion. There are significant uncertainties in both the estimates of deflation magnitude and episode volumes. The deflation azimuths from Uwēkahuna given in table 7.1 point to centers between Fiske-Kinoshita centers 1 and 2. Eruption efficiencies for azimuths that point to center 1 are close to 1 at an average depth to the Mogi source of 3.6 km (appendix A, table A1). More northerly azimuths at the same depth require a factor that reduces the eruption efficiency by only 1 percent. To achieve an average eruption efficiency of 1 only by adjusting the volume calculated from tilt magnitude requires depths greater by less than 0.5 km, well within the uncertaintiy in depth determination from Mogi models. Eruption volumes are also uncertain, particularly as flow thickness could only be measured at flow edges (Wolfe and others, 1988, p. 6 and following). Within these uncertainties we cannot reject the null hypothesis that the eruption efficiency during stage IA was 1, meaning that all magma supplied was erupted. This represents a marked departure from earlier periods and attests to the adjustment of the magma transport path to accommodate all magma as eruption rather than intrusion.

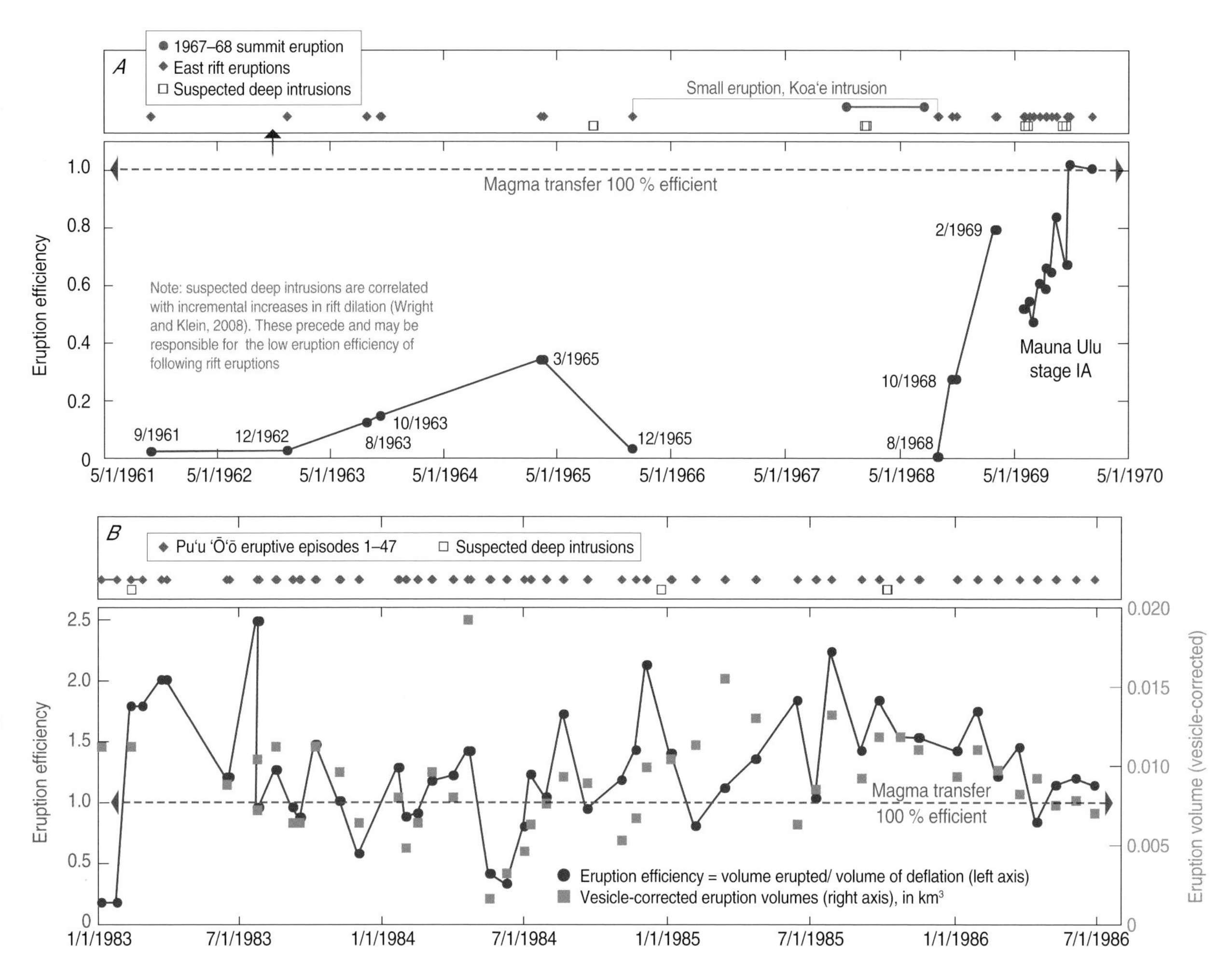

Figure 8.3. Graphs of Kīlauea eruption efficiency plotted against time. See text for further explanation. Dates on figure in m/d/yyyy format. ***A***, 1961–1970. Rift eruptions of the 1960s and Mauna Ulu Stage IA. ***B***, 1983–1986. Stage IA of the Puʻu ʻŌʻō-Kupaianaha eruption. The vesicle-corrected eruption volumes used in the calculation of eruption efficiency are plotted to illustrate their great variability, largely a result of measurement uncertainty.

Traditional and Inflationary Intrusions as Related to Degree of Inflation

It makes sense than an inflated volcano has more magma at the ready to erupt or intrude. A broad view of the relation between frequency of intrusion and the state of inflation as measured by the daily record of the tiltmeter housed in the Uwēkahuna Vault is shown in figure 8.4. The frequency of traditional intrusions associated with tilt drops and a strong south-flank response is greatest at intermediate levels of inflation. Conversely, as the reservoir is filled and the volcano has reached a relatively high state of inflation, inflationary intrusions near the summit with little south-flank anticipation or response become dominant. This observation suggests that the reservoir can sustain high inflation only when intrusions are frequent and small. At these times we envision a full magma reservoir from which magma forces its way or "spills over" into the uppermost (near-summit) parts of the rift zones. Inflationary intrusions are thus more a measure of inflation than of magma supply or the stress state of the rifts.

Most periods of inflationary intrusions are followed by a major traditional intrusion that penetrates deeply into a rift zone. There is a quantitative link between inflation, as measured by high Uwēkahuna tilt values and by summit earthquake counts (Klein, 1982). Inflationary intrusions contribute to high summit earthquake counts. The inflationary intrusions of 1971 immediately preceded the seismic southwest rift zone (SSWR) eruption of 24 September 1971, and the inflationary intrusions of 1974 foretold the SSWR eruption/intrusion of 31 December 1974, as chronicled in chapter 5. The SSWR intrusion of June 1982 and east rift zone eruption of January 1983 were both preceded by episodes of inflationary intrusion. Exceptions exist to the general observation. The inflationary intrusions of January and February 1980 were not immediately followed by a large event, though the SSWR intrusion of February 1981 occurred 1 year later. The inflationary intrusions of January to March 2006 did not halt the Pu'u 'Ō'ō eruption and did not enable a new intrusion, but when combined with the inflationary intrusion of May 2007, they may be causally related to the large Father's day intrusion of 17 June 2007, as chronicled in chapter 7. We therefore come to the important conclusion that the identification of inflationary intrusions, coupled with high summit tilt, indicates an increased likelihood of a major eruption or traditional intrusion.

The intrusion/eruption frequency is thus very sensitive to the inflation state, suggesting that the summit reservoir must be very sensitive to small changes in pressure. For example, tilt changes of about 100 µrad, corresponding to elevation changes of about 5 cm and a magma pressure increase of only 15 gm/cm^2, have a noticeable effect on intrusion frequency. On the other hand, an increase in the magma supply rate has only an indirect effect on the intrusion frequency, but it does promote inflation and increases the volume of magma available to sustain an intrusion.

A consequence of a finite magma supply within a given time period is that large-volume deflations will be followed by less frequent intrusions and a delay in the beginning of re-inflation. Major events that were not immediately followed by sustained inflation were the 1960 eruption and collapse, the 1975 Kalapana earthquake and collapse, and the 1983 Pu'u 'Ō'ō eruption and collapse. These three major collapse events are among the largest in the 1960–2009 record and each may have effectively ripped a large opening in the reservoir that took years to heal until reservoir magma could be retained and major inflation resumed.

Magma Supply to Kīlauea

Magma Supply Rate

The magma supply rate is an important factor in evaluating the long-term history of any volcano. Its estimation is made inherently difficult by the great changes in the viscosity of magma with small reductions of temperature. Magma has to traverse an inhomogenous rock volume to reach the surface, and this may result in temporary delays in transport that may render meaningless all but the longest term estimates of supply rate. Figure 8.5*A* illustrates our best estimates of the rates of both rift dilation associated with south flank spreading (see discussion below) and magma supply over the past 90 years between 1918 and 2008. Figure 8.5*B* illustrates our best estimate of the increase in magma supply rate over the past 50 years. We have included reconstructed supply rates in green to match the reconstructed spreading rates shown in figure 8.6. Figure 8.5 also shows the short-term variability in estimated supply rate as illustrated for the period bracketing the Mauna Ulu eruption.

Magma supply rates before 1950 are estimated from eruption and caldera-filling rates assuming little or no south flank spreading or rift dilation. We infer that the high rates of magma supply and transfer associated with cycles of eruption, draining, and refilling before 1840 are a function of rebound from a massive draining of magma in 1790 that may be represented in Puna by 1790 lava and a possible accompanying intrusion. The 1790 eruption was not a caldera-forming event, but it did drain Kīlauea's entire magmatic system and triggered a very high rate of resupply through lowered pressure above the magmatic system. Filling rates before 1840 are

2–6 times those after 1840 (fig. 2.1). We interpret the decline in magma supply rate after 1840 to be in part a return to a low equilibrium rate and in part to result from the increased activity of Mauna Loa (see discussion below). The typical eruption rates from 1918 to 1924 are virtually the same as the average 19th-century caldera filling rate between 1840 and 1894 shown in chapter 2, figure 2.1.

The 1895–1918 magma-supply rate is difficult to estimate. The amount of magma needed to refill Halema'uma'u during the period from 1899 to 1918 is 0.0225 km^3, assuming filling of the frustrum of an inverted cone (chap. 2), yielding a filling rate of 0.0012 km^3/year, far less than in periods preceding and following. However, the filling shown in figure 2.2 is punctuated by (1) many disappearances of magma before continuous refilling began in 1907, (2) instances of magma withdrawal between 1908 and 1912, and (3) several deflations measured in the Whitney Vault and several drops of the Halema'uma'u lake level accompanied by earthquake swarms between 1912 and 1918 (fig. 2.3*A*). The south flank may have had active spreading during the first two decades of the 20th century, as suggested by the *M*6.7 flank earthquake in September 1908. If each of these categories of events (1–3 above) is interpreted as magma transfer to the rift zone, then we cannot disprove the null hypothesis that the actual magma supply rate was close to the 0.025 km^3/year estimated for the period before 1894 and after 1918 and was not seen as filling of Halema'uma'u lake[29].

We speculate that the low Halema'uma'u filling rate (as contrasted with a magma supply rate) during 1894–1918 (fig. 2.1) was related to the substantial regional caldera uplift of 1918–1919 seen on the Whitney tiltmeter (fig. 2.3*A*) and captured by Wilson's (1935) 1912–22 leveling survey. It is as if the mantle magma supply was choked off at some depth in the upper mantle in the 1890s and was released in 1918 to replenish a large section of the supply conduit, including reservoirs 1, 2, and 3 and below them,

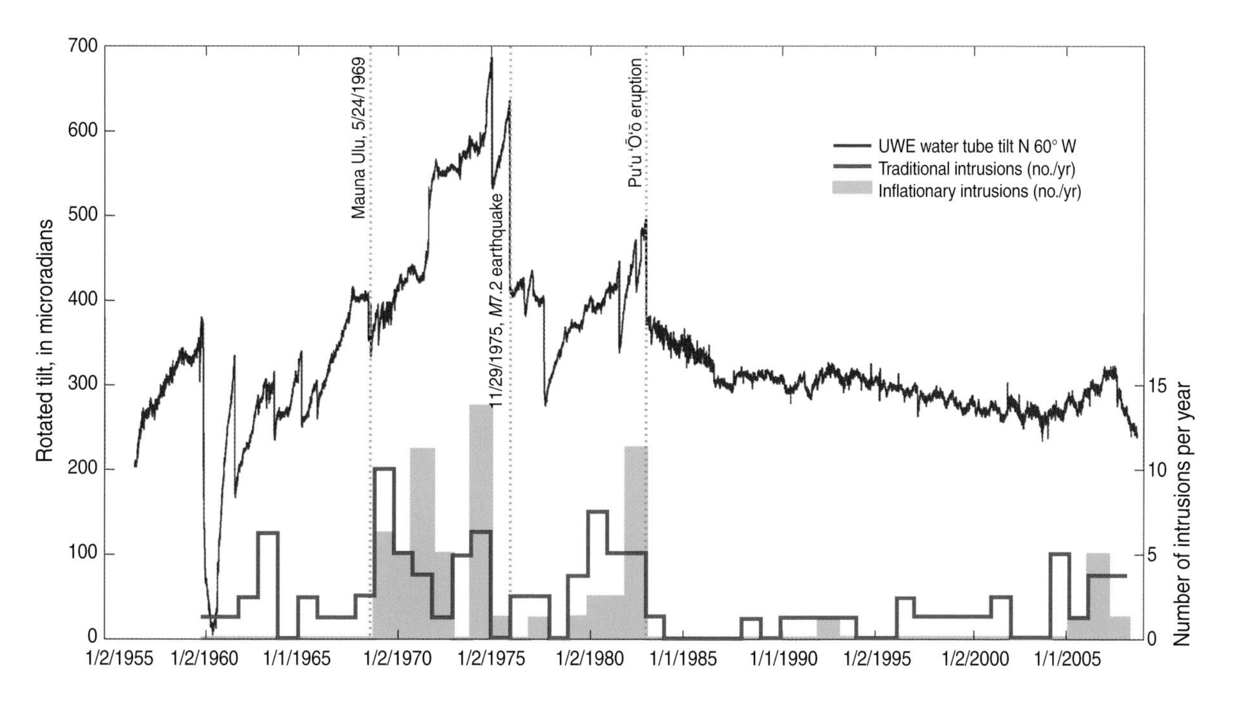

Figure 8.4. Graphs showing frequency of intrusion versus state of inflation at Kīlauea. The Uwēkahuna (UWE) water-tube tilt record is the best continuous measure of Kīlauea's inflation and deflation after 1956. Because Uwēkahuna tilt measures the relative heights of two water pots anchored to the ground, this tilt has good long-term stability. The dashed vertical lines mark three major collapses discussed in this report. The offset caused by shaking during the *M*6.7 Ka'ōiki earthquake on 16 November 1983 has been removed because it did not involve a magmatic change. Tilt is plotted as a single vector rotated to N 60° W., which points away from the typical inflation/deflation center and is a good measure of long-term inflation. Annual numbers of intrusions from event tables in previous chapters are plotted for traditional intrusions (red histogram) and inflationary intrusions (green histogram). Dates on figure in m/d/yyyy format. See text for interpretation.

[29]An alternative approach assumes that the magma supply during this 1894–1918 interval is equal to the bracketing values of 0.0249 km^3/year, then a maximum of 0.6 km^3 of magma entered the Kīlauea plumbing during that 24-year interval, of which 0.49 km^3 represents the volume needed to replenish the deep magma system (see fig. 2.1). Could the remaining 0.11 km^3 be accommodated in the east rift zone? Assuming a rift length of 30 km and height of the active rift zone of 2 km, the remaining 0.11 km^3 results in a rift extension of nearly 2 m over 24 years, or about half of the instantaneous opening observed following the 1975 earthquake (Owen and Bürgmann, 2006) and nearly twice the dilation rate calculated during the Pu'u 'Ō'ō-Kupaianaha eruption (Owen and others, 1995, 2000a). There is no evidence from geodesy or seismicity for this much magma entering the rift and deforming the south flank during 1894–1918.

perhaps in the depth range of 2–30 km. This 1918 release of magma to shallower levels restored the caldera-filling rate and uplifted the caldera within a 4-month period. This charge of new magma provided a ready source for the 1924 east-rift intrusion and formation of the Kapoho dike, which also collapsed the summit back to its pre-1918 level and triggered the 1924 explosive Halema'uma'u eruption. The 1924 evacuation of magma from below the summit encouraged an approximate doubling of the magma supply (periods I to II, fig. 8.5*A*), and it certainly cleared any conduit obstruction that may have existed between 1894 and 1918.

Magma volumes and batches in the post-1950 era are ascribed to changes in magma reservoir 2, except for 1954 and 1959, which involved sources removed from the main plumbing (fig. 4.13). The volume calculations, combined with the observed 10-year time in which successive magma batches are recognized, can be used to estimate the minimum magma-supply rate between the arrival of the 1952 batch in 1950(?) and the using up of the 1967–68 batch in 1969. The total erupted volume is 0.51 km^3 (table 8.1), the total additional volume intruded beneath the east rift zone is 0.66 km^3 (table 8.1), yielding a magma supply rate of about 0.062 km^3/yr over the 19 year interval between 1950 and 1969, less than the magma supply rate of 0.086 km^3/yr estimated from the tilt calculations. This comparison is consistent with the addition to the rift zone of magma that passes through the summit reservoir but is not accessed by eruption to the surface. The aggregate volume of the 1967–68 magma batch exceeds the volumes of the two earlier batches, consistent with the increase in magma supply in the period following the 1967–68 eruption (fig. 8.5).

We conclude the following about Kīlauea's historical magma supply: (1) The magma supply rate from 1840 to 1950 (<0.03 km^3/yr) was far less than what we observe from 1950 and after. (2) In the period between the 1924 collapse and the March 1950 inflation, the volume lost in the collapse was regained at a rate still uncertain. The net magma supply rate before 1950 must have decreased as the lowered pressure caused by the 1924 collapse was equalized, perhaps to a value equal to the 1918–24 filling rate. (3) Summit inflation, driven by increased magma supply, began when the volume lost in the 1924 collapse was recovered, possibly augmented by a shift in the deep mantle plumbing to favor eruption at Kīlauea over eruption at Mauna Loa (see below).

We interpret the 1950–75 period to be one of magma supply steadily increasing to a higher level typical of the rest of the century compared to the first two-thirds of the century (fig. 8.5). Kīlauea accomodated the higher magma supply through (1) more frequent rift zone eruptions and intrusions, including the prolonged 1969–74 Mauna Ulu eruption, (2) a stepwise increase in south flank spreading rate and seismicity, including suspected deep intrusions, (3) increase in the eruption efficiency, defined as the ratio of magma erupted to magma intruded, with continued use of the plumbing, and (4) shifts of the preferred rift zone from the east rift to the southwest rift and back again as the east-rift plumbing was unable to accommodate the magma supply (see chap. 5).

Magma supply rate continued to increase incrementally following the beginning of continuous east rift eruption in 1983. We accept a value of 0.18 km^3/yr in stage I, increasing to 0.2 km^3/yr in stage II following the 1997 eruption and intrusion. Magma supply increased further during stage III following the shift from deflation to inflation at the end of 2003 and the 2.5-fold increase in CO_2 emission in 2004–2005. Poland and others (2012) have suggested that events after 2003 were driven by a surge of deep magma supply that released CO_2 with time delays between (1) the surge and the measured CO_2 increase and (2) between the CO_2 increase and the shallow manifestations of increased magma supply mentioned in chapter 7. In their estimation the Father's Day eruption in 2007 and the connection of the magma transport path beneath Halema'uma'u to the surface in 2008 were driven by a minimum 2-fold increase in magma supply. We incorporate their estimates into figure 8.5, while questioning whether ground deformation and seismic data coincident with the time of CO_2 increase support that magnitude of magma supply increase. Finally, these authors estimate that magma-supply rates gradually returned to pre-2003 levels following the 2008 Halema'uma'u eruption that ends our study.

Magma Storage and Pathways

The primary summit magma reservoir at 2–6 km below the caldera (fig. 8.1) is not the only one active since the founding of HVO in 1912. The sharp deflation in 1924, associated with an explosive Halema'uma'u eruption and an east-rift eruption and intrusion, is correlated with large draining of the lava lake and is consistent with draining magma reservoirs 1 (0.8-km depth), 2 (3.5 km), and 3 (fig. 2.4). Levelling implicates draining of reservoirs at three depths, and the involvement of the deeper reservoir 3 in a summit collapse is unique in the deformation record (see chap. 2). The Whitney tilt record shows an increase in 1918–19 and subsidence in 1924–25 that are consistent with Wilson's (1935) initial conclusion of a 1-foot increase in the altitude of the Volcano House benchmark before the 1921 triangulation and an even greater decrease in altitude across the 1924 collapse. The 1912–22 rise and 1922-27 fall of the

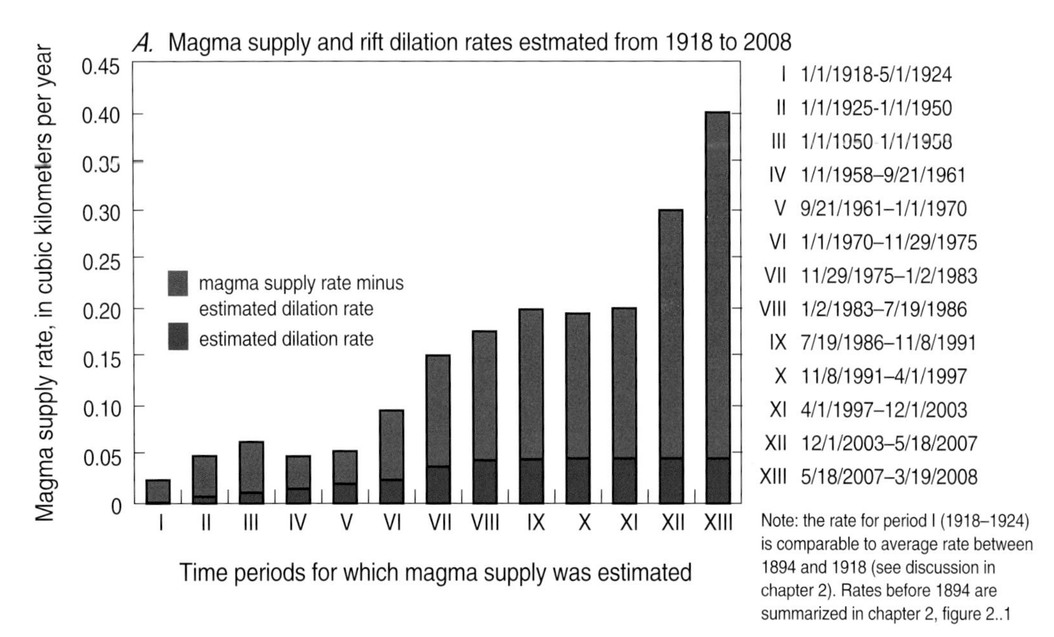

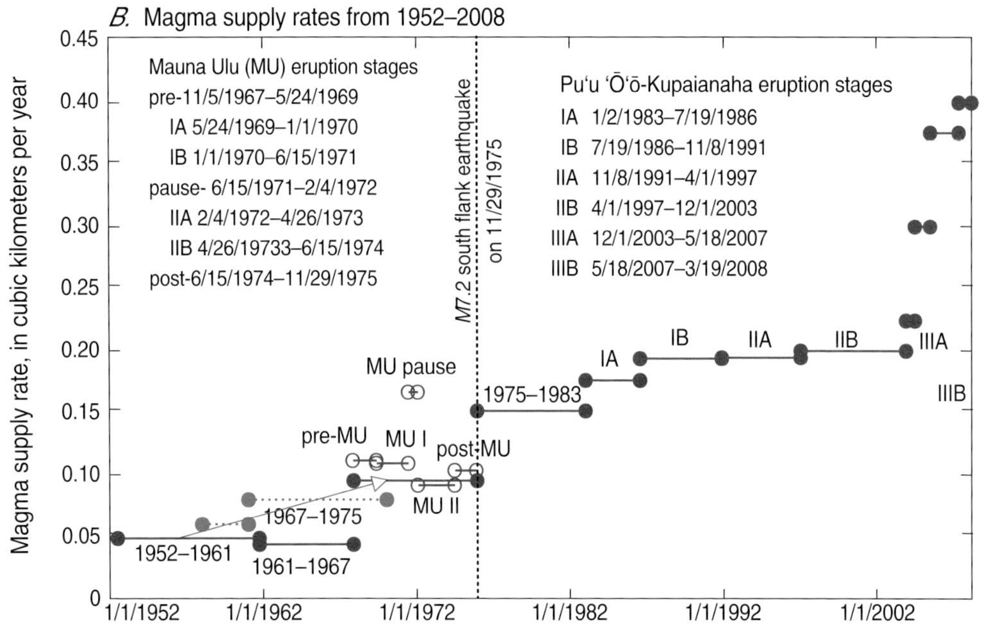

Figure 8.5. Graphs showing long-term rates of magma supply and rift dilation at Kīlauea. Dates on figure in m/d/yyyy format. ***A***, Magma-supply and estimated east-rift dilation rates for the period 1918–2008. For periods before 1950, magma-supply rates are equated to magma filling/eruption rates because the rift dilation (spreading) component is unknown and assumed to be close to zero. The eruption rates from 1918 to 1924 are virtually the same as the average 19th century caldera filling rate between 1840 and 1894 (chap. 2, fig. 2.1). After 1950 magma-supply rates are calculated using the sum of volume-equivalents of deflations at Kīlauea's summit as shown in tables in previous chapters. During long eruptions, the eruption volume is added to the tilt volumes. Dilation rates are assumed between 1950 and 1975. The 1975 Kalapana earthquake had a dilation volume of 0.4 km^3 (Owen and Būrgmann, 2006). Dilation during the eruption that began in 1983 is about 0.0459 km^3/year (Owen and others, 2000a). ***B***, Magma-supply rates from 1952 to 2008. Over the long time span, magma-supply rates increase. Over shorter periods of time, the magma supply calculated as above is quite variable, as illustrated by the various stages of the Mauna Ulu eruption. Individual magma-supply rates have associated errors of about ±20 percent. The green arrow spanning intervals before 1972 indicate increases in magma supply rate that match the increases in south flank spreading rate shown in figure 6.6*A*. A 2.5-fold increase in CO_2 from Kīluaea's summit begins in stage IIIA; decline is gradual after mid-2005. Magma supply increases to 2008, then decreases, following the increase and decrease of CO_2, as indicated by M. Poland and J. Sutton (oral and written commun., 2010). Magma supply rates after the beginning of inflation in 2003 are estimated by Poland and others (2012) and are applied to both 8.5*A* and 8.5*B*, although we remain skeptical as to whether the increase in magma supply was as great as would be suggested by using summit CO_2 emission as a proxy for magma supply rate (Gerlach and others, 2002).

Kea'au benchmark argues that its change is due to inflation and deflation of a deep magmatic source.

The events of the 1953–60 period stand apart from the rest of Kīlauea's recorded history. The deep (40–60 km) seismicity during 1953–60 and the unique chemistry of the 1959 Kīlauea Iki eruption strongly suggest that the magma being supplied was from an alternate deep mantle source above the normal Hawaiian mantle source and that movement of magma from this source to the surface did not follow the typical magma conduit feeding Kīlauea's shallow reservoir—rather, it made a unique path toward Kīlauea Iki (see chap. 4). We thus think of the deep Kīlauea conduit as being multistranded (Wright and Klein, 2006, figure 10 and discussion, p. 63–64).

South Flank Seaward Spreading

Seaward spreading of Kīlauea's south flank is an essential complement to the history of the volcano's magma supply. Spreading is associated with dilation of the rift zones, thus creating additional room for magma to occupy. The south flank has been described as a stress meter whose seismicity responds to changes in rift stresses and magma pressure (Dieterich and others, 2000). Continuous seaward spreading has been documented by numerous triangulation, EDM, and GPS campaigns and recently by continuously recorded GPS stations. It is our assumption that some degree of spreading and rift dilation are essential for eruption (or intrusion) on the rift zones. Construction of a summit shield, such as the 'Ailā'au shield (Clague and others, 1999), probably can occur only when the rift zones are sealed. In this section we address the question of spreading and its relation to eruption and intrusion in terms of the long term history of ground deformation measurements and south flank seismicity.

History of Seaward Movement of Kīlauea's South Flank

The history of Kīlauea's south flank moving southeast away from Mauna Loa is shown in figures 8.6*A–C*, and the differential movement within the lower south flank is shown in figures 8.7*A–C*. The south flank data were obtained by four different methods. The results of the earliest triangulation studies extending from 1896 or 1914 to 1970 are summarized and interpreted by Swanson and others (1976a). We plot their data, but make an adjustment to the long period separating 1914 and 1958 in order to postulate an increase in spreading rate beginning with the reinflation of Kīlauea in March 1950 (fig. 8.6*A*). Later, the higher magma supply may have triggered the offshore south flank earthquake swarm of March and April 1952. We interpret the offshore swarm to have unlocked the south flank, potentially marking the beginning of much less constrained seaward spreading. Subsequent to 1961, horizontal displacements were obtained by EDM (1965–89), campaign GPS (1989–96) and continuously recording GPS (1996–present). Station locations and measured displacements are stored in a computer database (VALVE) maintained by the U.S. Geological Survey's Hawaiian Volcano Observatory (HVO). GPS data were first published in papers by Paul Delaney and others and by authors cited therein (Delaney and others, 1993, 1998). Paul Segall and students of his have collaborated with HVO personnel to continue measurements during the eruption that began in 1983 (see, for example, Owen and others, 1995, 2000a).

It is difficult to compare directly the pre- and post-EDM periods because the EDM network was occupied less frequently than the subsequent GPS network. As GPS stations were installed, EDM stations were replaced or abandoned for a variety of reasons, and the GPS networks were not always colocated with the older EDM networks. For these reasons it is difficult to construct a single south-flank movement curve. Displacement measurements during the Mauna Ulu eruption are incomplete. Data published in the two papers cited above cannot be compared, because the Delaney paper shows detailed measurements of seaward movement of the entire upper south flank, whereas the Swanson paper by contrast shows detailed measurement of movement within the lower south flank (figs. 8.6*B*, 8.7*B*).

Nonetheless it is possible to make some broad comparisons between gross flank motion and internal compression. There is an obvious offset at the time of the 1975 earthquake, with distances lengthening both from Mauna Loa to the south flank (fig. 8.6*A*) and within the south flank (fig. 8.7*A*). Another striking change is the slowing of south-flank spreading rates at the beginning of the Pu'u 'Ō'ō-Kupaianaha eruption (fig. 8.6*A*). At the start of that eruption in 1983 there is a slowing of flank compression within the lower south flank (figs. 8.7*A*,*C*), as documented by Delaney and others (1998), and a possible weak reversal to slow extension after 1997 (fig. 8.7*C*). Since 1983, the south flank motion has been more or less constant at 5–6 cm/yr, with the exception of a small offset accompanying the 1997 intrusion (fig. 8.6*C*). The larger slow-slip events (Montgomery-Brown and others, 2009), discussed in a subsequent section, are apparent in figure 8.6*C* as net south flank motion if two measurements made days apart span the event. Slow-slip events are too small to appear as internal flank extension on figure 8.7*C*, except for the 2007 Father's Day event. Overall, these slow-slip events have only a small effect on the yearly motion.

South-flank motion during the 1969–74 Mauna Ulu eruption can be interpreted volcanologically. The low rate of seaward movement of the upper south

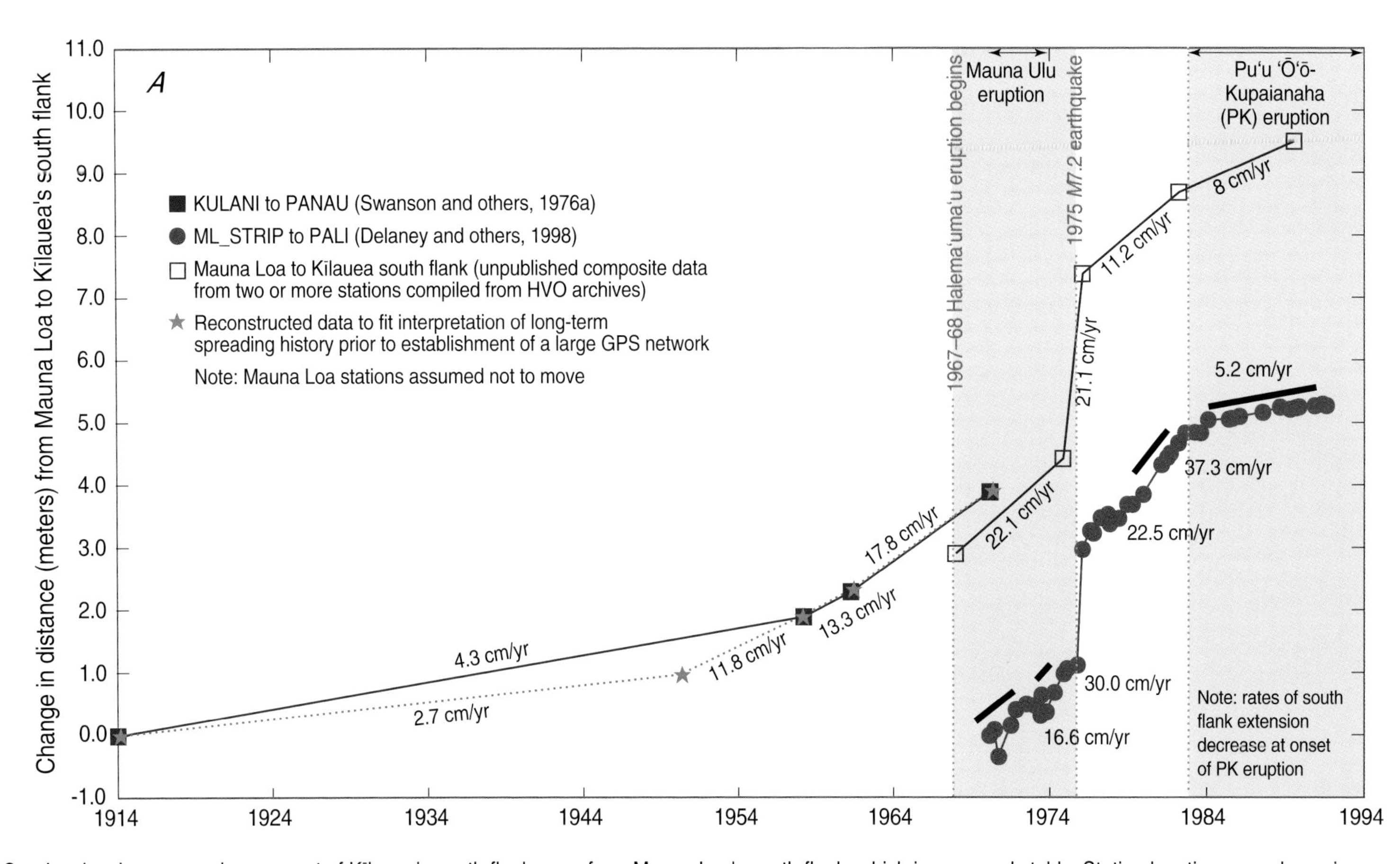

Figure 8.6. Graphs showing seaward movement of Kīlauea's south flank away from Mauna Loa's south flank, which is assumed stable. Station locations are shown in appendix figure H1. ***A***, 1914–1992: Data points shown by blue squares are from triangulation (Swanson and others, 1976a, fig. 5). Data points shown by red filled circles are from electronic distance-measuring (EDM) data from the EDM line connecting the Mauna Loa strip road station ML_STRIP with the Kīlauea station PALI, located on the northern south flank just south of the Koa'e Fault Zone (Delaney and others, 1998, fig. 2b). The latter dataset shows significant extension at the time of the 1975 earthquake, a high rate of extension following the earthquake, and a significant drop in rate after the beginning of the Pu'u 'Ō'ō-Kupaianaha eruption in 1983. Data points shown by hollow purple squares are unpublished data from the Hawaiian Volcano Observatory files. Green stars show a suggested reconstruction of the deformation history consistent with both the observed data and the analysis of spreading presented in this paper. ***B,*** From 1 May 1969 to 1 July 1974: Data on the length of the EDM line ML_STRIP to PALI gathered during part of the Mauna Ulu eruption and shown at an expanded scale, with Kīlauea eruptions and intrusions for the same time period. See text for interpretation. Dates on figure in m/d/yyyy format. ***C,*** From 1982 to 2008: Lengthening of lines connecting Mauna Loa's south flank (assumed to be stable) with stations located near the coast on Kīlauea's southern south flank. Also shown are Kīlauea eruptions and intrusions for the same time period. Measurements are made by EDM (blue squares) and by Global Positioning System (red filled circles and green triangles). Station locations are shown in appendix figure H1. The dotted line is a suggested reconstruction across a large time gap in the data to reflect extension at the beginning of the Pu'u 'Ō'ō-Kupaianaha eruption and a slowing of the extension rate thereafter. Extension rates have remained near-constant during the long eruption, with the exception of small extensions recorded at the times of major east rift intrusions and silent earthquakes.

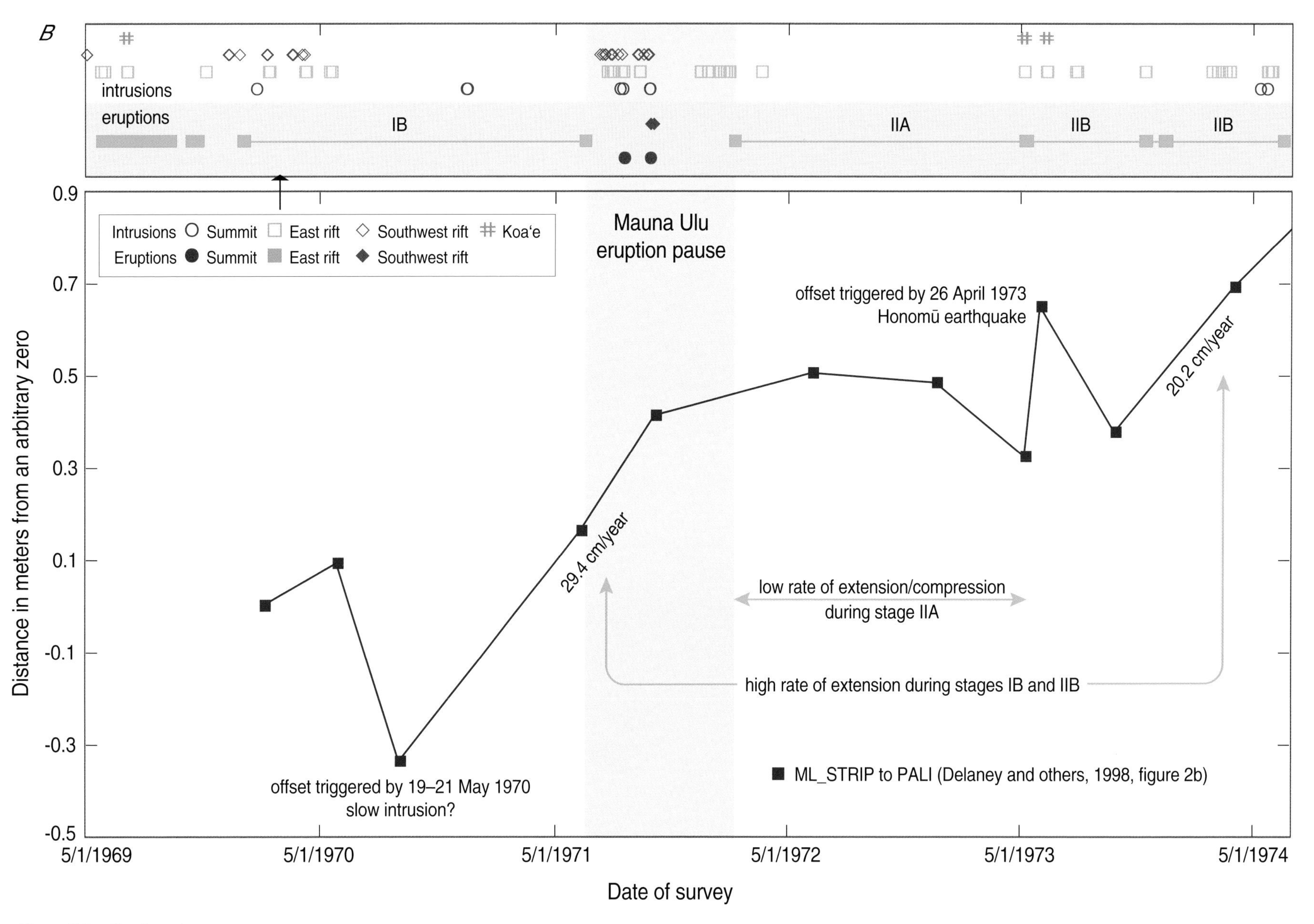

Figure 8.6.—Continued

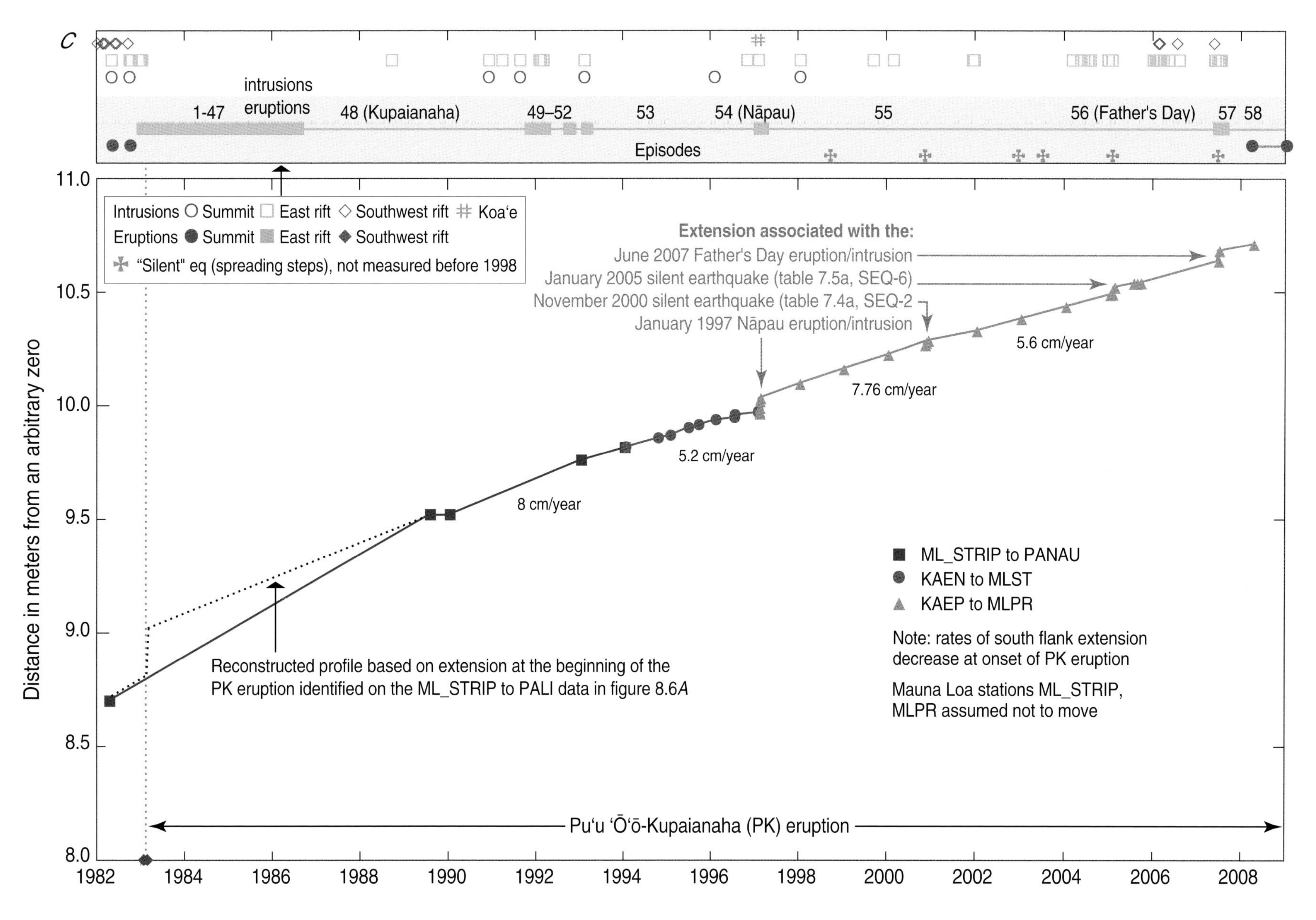

Figure 8.6.—Continued

flank during stage IIA (fig. 8.6*B*) is consistent with our interpretation that this was a period of equilibrium between magma supply and eruption rate as evidenced by summit deflation and low seismicity. The higher rates of flank motion during stages IB and IIB are consistent with our interpretation of increased magma pressure represented by inflation and higher seismicity. Unfortunately the only data available in this period are for the northern south flank, whereas post-1980s spreading rates are measured using south flank stations near the coast.

Line length changes that occur within the south flank are much more difficult to interpret in terms of spreading away from Kīlauea's stable north flank. Internal changes depend both on flank motion and on stresses applied to it. Change of line lengths within the lower south flank (fig. 8.7*A*) show compression between the 1975 earthquake and the beginning of the Pu'u 'Ō'ō-Kupaianaha eruption, shifting to much smaller amounts immediately after the beginning of eruption. We interpret the compression before 1983 as a result of magma added to the rift while the south flank block was buttressed offshore and at its base. We see the extension after 1983 as a release of south flank stress and compression as magma was erupted from the rift. This release of compressive stress at the time of the 1983 eruption was seen by as a fivefold decrease of the stress rate within the south flank (Dieterich and others, 2003). After 1988 the lower south flank varies between extension and compression at much smaller within-flank rates of less than 1 cm/yr (fig. 8.7*C*).

The seismicity of the south flank offers clues to its stress state in addition to its internal compression seen both before and after the 1975 Kalapana earthquake. The Kalapana earthquake profoundly altered stress on the south flank and "softened" it to the stress imposed by magma traversing the east rift. After the earthquake, intrusions outnumbered eruptions, so it was more difficult for magma to experience enough confining stress to reach the surface, whereas before the earthquake, the opposite was true. The high rate of aftershocks in the 17 years following 1975 means the stress released by the mainshock was large, but the relatively high rates of seismicity in the south flank even before the earthquake also means that the stress rate on the flank actually decreased at the time of the earthquake (Klein and others, 2006). This lowering of stress rate in 1975 means that the large number of postearthquake east-rift intrusions without the magma pressure release of long eruptions (like Mauna Ulu 1969–74 and Pu'u 'Ō'ō 1984–2014 ongoing) was offset by a softened and pliable flank such that the internal compression rates before and after the earthquake were about the same (fig. 8.7*A*).

Changes before the Mauna Ulu eruption show compression at a low rate through the 1967–68 summit eruption, a higher rate through the three 1968–69 east rift eruptions that preceded Mauna Ulu, then a return to a low compression rate during stage IA and variable rates during stage IB, when many more measurements were made (fig. 8.7*B*). The small changes during stage IB are difficult to interpret in the absence of measurement of other station pairs.

South Flank and Summit Tilt Response to Eruptions and Intrusions

The south flank responds to magma pressure applied over a range of depths beneath the rift zones. Earthquake swarms beneath Kīlauea's south flank often precede by minutes to days before, and usually follow in the days after, traditional eruptions and intrusions, and also after some inflationary intrusions. Time relations among the beginning of eruptions and earthquake swarms beneath the summit, rift zones, Koa'e, and south flank are tabulated in the opening table of each of chapters 4–7, along with the response of a continuously recording tiltmeter at Kīlauea's summit. Of particular interest is the triggering of south flank earthquake swarms minutes to days ahead of earthquake swarms in the regions where eruption or intrusion occurs. Figure 8.8*A* shows for selected traditional eruptions and intrusions the time elapsed between triggering of an earthquake swarm beneath the south flank and the beginning of a shallow earthquake swarm within the rift. The rift swarm indicates that rock is breaking to initiate formation of a dike before eruption or intrusion. Note that south flank activity preceding intrusions may begin with a few small flank earthquakes hours before the intrusion, and the intrusion will generally trigger a much greater flank response lasting days afterward. Thus the south flank is responding to stress imposed by magma emplaced at a variety of depths within the rift both before and after the intense dike-forming swarm within the rift.

The time delay between flank and rift earthquakes varies. In rare cases, as during the large rift intrusion/eruption in September 1961, the south flank responded seismically only several hours after the beginning of eruption (table 4.1). This may have been because the intrusion was in a part of the rift zone not active for decades, or may have been a function of the limited seismic network at that time when it was difficult to clearly separate earthquakes beneath the rift zone from earthquakes beneath the adjacent south flank. Subsequent time delays between rift and flank earthquakes for intrusions beneath the east rift zone have all been shorter than the 1961 delay. Movement of magma in a shallow (0–4 km) dike can be initiated as a response to the pressure of incoming magma applied to the deep magmatic system. We have no evidence that magma moves upward from the deep rift to feed the shallow intrusion, but only that many

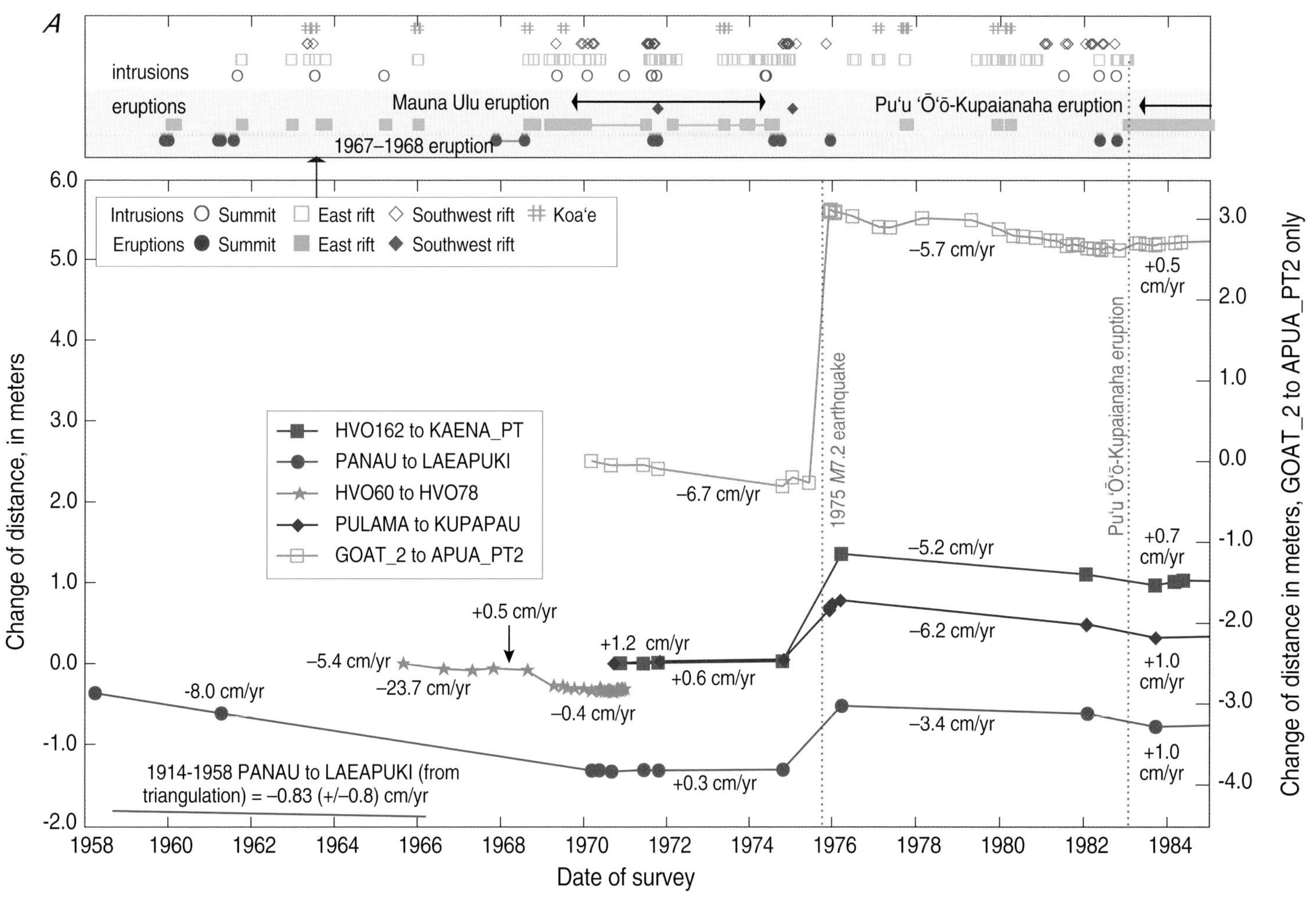

Figure 8.7. Graphs showing changes of line length between stations all located within Kīlauea's south flank. Station locations are shown in appendix figure H1. ***A***, 1958–1985. Line lengths within the near-coast south flank decrease up until the Mauna Ulu eruption, increase slightly till the 1975 earthquake, increase across the earthquake, decrease following the earthquake, and increase slightly following the beginning of the Pu‘u ‘Ō‘ō-Kupaianaha eruption. One line, from the northern south flank GOAT_2 station to the coast APUA_PT2 station, shortens during the time before the 1975 earthquake, in contrast to the other station pairs. See text for explanation. ***B***, 1965–1971. A single pair of stations on Kīlauea's southern south flank was measured repeatedly during part of the Mauna Ulu eruption. It shows shortening before the 1967–68 Halema‘uma‘u eruption, little change during that eruption, a higher shortening rate before the beginning of the Mauna Ulu eruption, and small changes during the eruption. See text for explanation. ***C***, 1982–2008. History of line-length changes within the southern south flank using several different Global Positioning System (GPS) lines. The data show shortening across the beginning of the Pu‘u ‘Ō‘ō-Kupaianaha eruption, partial recovery and further lengthening through Stages I and IIA, slight lengthening during Stages IIB and IIIA, and shortening and partial recovery during and after the Father's Day eruption. Numbers in top panel refer to episodes of the Pu‘u ‘Ō‘ō-Kupaianaha eruption. See text for further explanation.

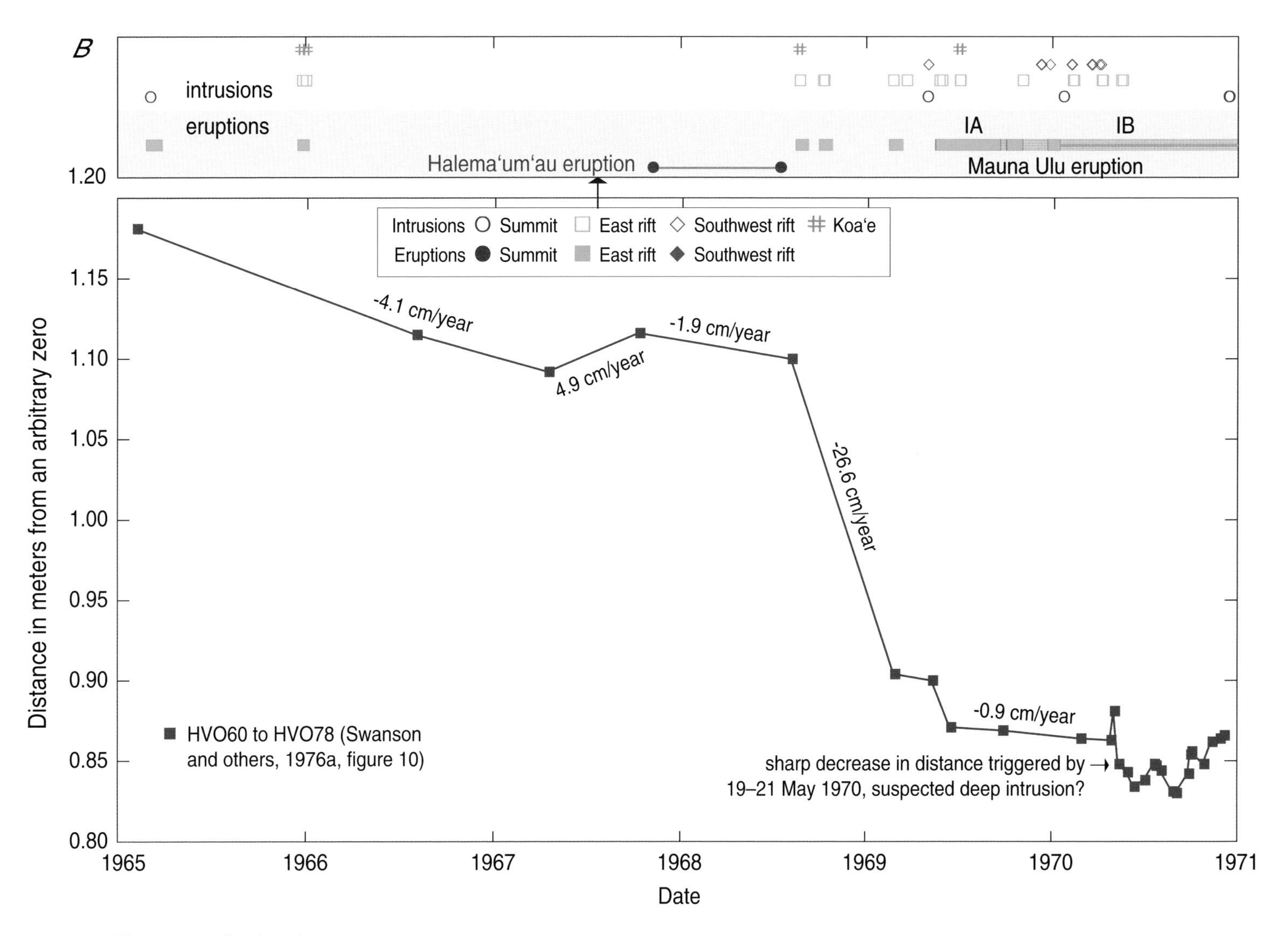

Figure 8.7.—Continued

rift intrusions begin as increased pressure applied to deep rift magma system is manifested in south flank seismicity before the shallow rift dikes expand. The south flank earthquakes (typically 5–9 km deep) are also a direct response to movement of magma into the deeper rift zone. The south flank thus behaves as both an early warning of and later response to increased magma pressure applied beneath the rift zone.

An additional argument suggests that south flank earthquakes that are precursory to rift intrusions respond to magma movement deeper in the rift at 4–10 km, as evidenced by flank seismicity near and downrift of the eruption site and new dike location. One would generally expect rift pressurization only uprift of the eruption site as magma approaches the eruption site. The observation of deeper flank seismicity near and downrift of the eruption site suggests that magma is expanding a deeper part of the rift surrounding the future dike site, and this magma may help to feed the dike from several directions at once.

The seismic southwest rift zone (SSWR) behaves differently to intrusions from the east rift because the flank response may be 1–3 days late. For three large events in September 1971, December 1974, and August 1981 (fig. 8.8*A*) the south flank massively responded hours to days following the initiation of an earthquake swarm beneath the rift, in addition to minimal south flank seismicity preceding the intrusion. It may be that, with less frequent use, the flank adjacent to the SSWR is lightly stressed by the rift to keep up with ongoing flank spreading. Alternatively, the flank adjacent to the SSWR may be "softer" and may require the intrusive stress to be applied for a longer time to achieve south flank slip.

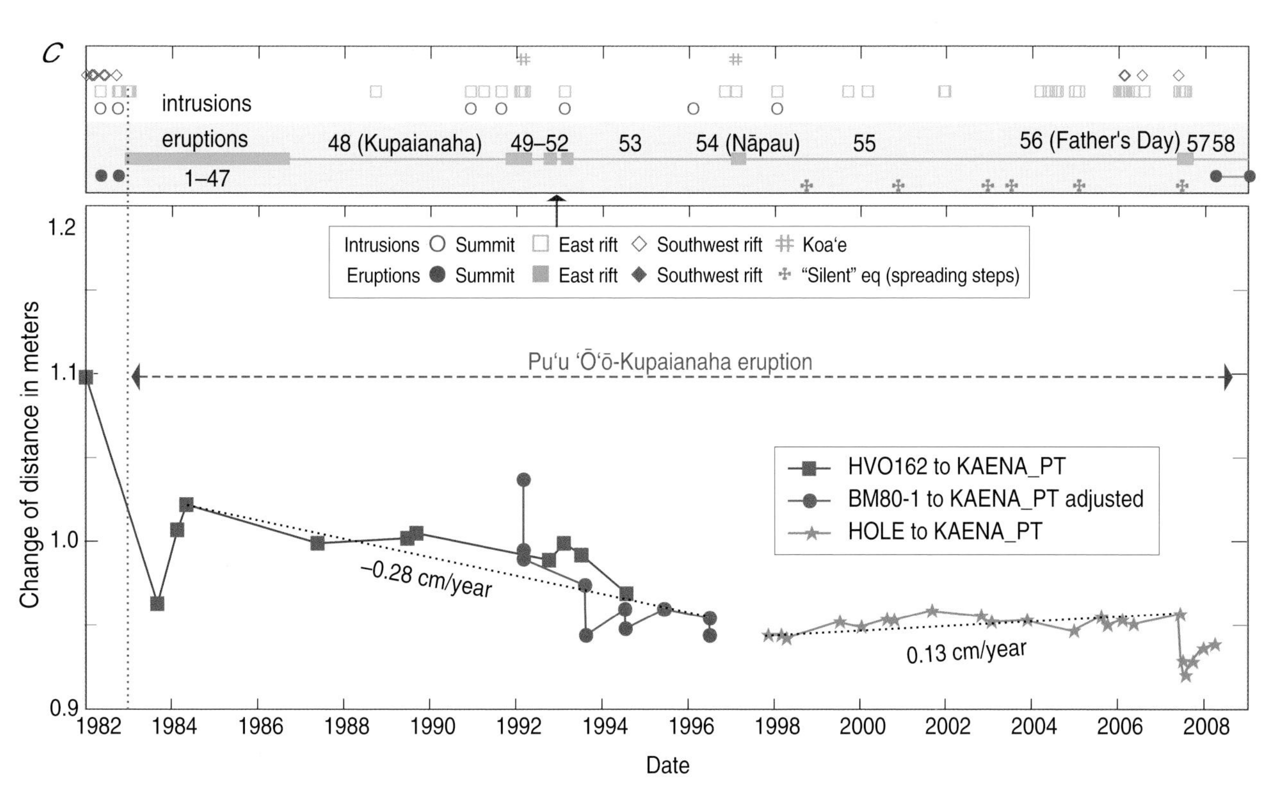

Figure 8.7.—Continued

The continuous summit tilt dataset affords an insight into the nature of resupply of magma from the summit magma reservoir to eruptions on the rift zone (fig. 8.8*B*). For most eruptions, the tilt deflection occurs after the beginning of an earthquake swarm marking eruption or intrusion. This indicates that most rift eruptions are initially fed from magma already stored beneath the rift zone and that the summit reservoir replenishes magma to the rift, a conclusion previously documented in some detail in chapter 6 of the present report and also previously (Klein and others, 1987). A notable exception is the December 1965 eruption, in which the Koa'e Fault Zone underwent significant deformation, and the short eruptions of 1979 and 1980 during the period of recovery from the 1975 earthquake. In these three instances, supply of magma from the summit can be interpreted to have destabilized the Koa'e Fault Zone in 1965 and caused unsustainable eruptions to occur in 1979 and 1980 when magma might have been expected to remain underground as an intrusion.

Figure 8.8. Comparison of onset times of Kīlauea earthquake swarms and the beginning of intrusions or eruptions or deflation at Kīlauea's summit. Dates on figure in m/d/yyyy format. ***A***. Comparison of onset times for rift and south flank earthquake swarms. For most intrusions and eruptions on the east rift, earthquake swarms beneath the south flank precede by minutes to hours earthquake swarms beneath the east rift. For a few large events, such as intrusions in July and December 1974, the south flank anticipation exceeds one day. For intrusions on the seismic southwest rift, earthquake swarms beneath the south flank follow the onset of an earthquake swarm beneath the rift zone with a delay of many hours. ***B***. Comparison of onset times for rift earthquake swarms and summit deflation. For most eruptions and intrusions, the onset of an earthquake swarm beneath the rift precedes the beginning of summit deflation by minutes to hours, as measured by a continuously recording tiltmeter in Uwēkahuna Vault. This is consistent with the idea that rift intrusions sprout from zones where magma has been previously stored. For intrusions on the seismic southwest rift zone, the delay may be as much as several days. An exception is provided by the December 1965 eruption/intrusion into the Koaʻe Fault Zone, in which the onset of rift seismicity followed the beginning of summit deflation by several hours, suggesting that pressure from magma moving from the summit reservoir was instrumental in opening the Koaʻe.

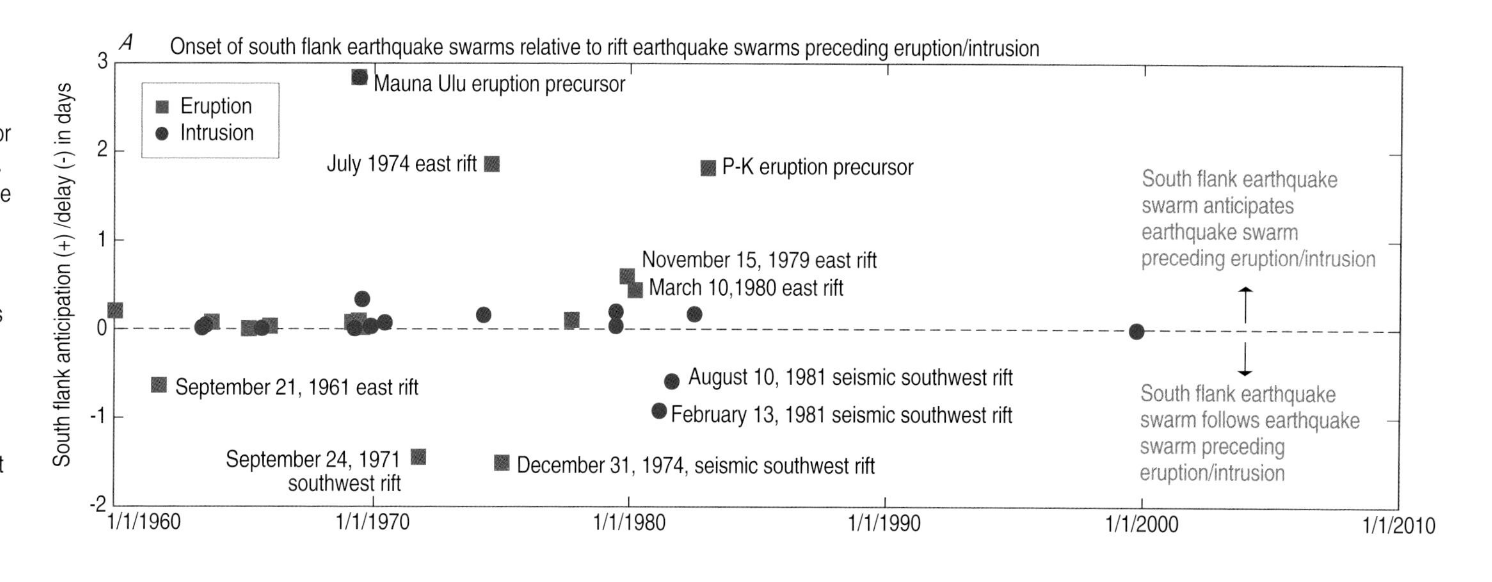

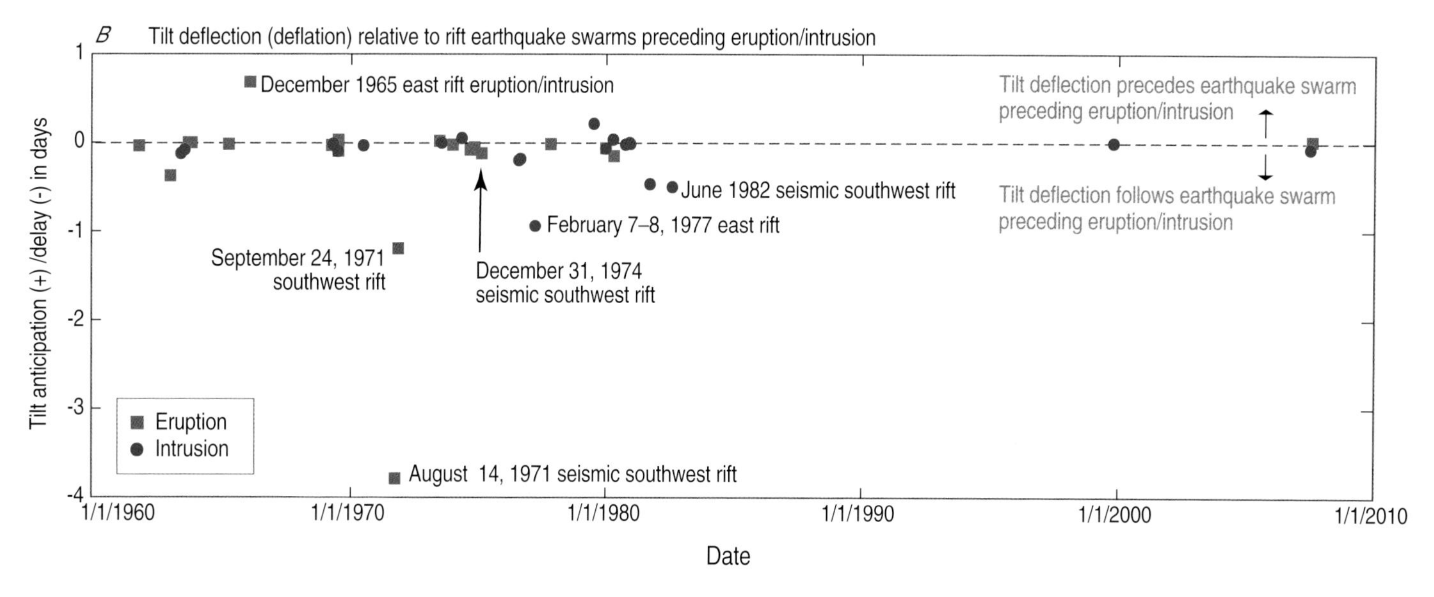

History of Located Earthquake Counts and Seismic Moment Accumulation

The long-term seismic history complements and supports the long-term study of magma supply and south flank spreading. Earthquake counts are difficult to compare before the completion of a comprehensive network and processing facilities in the early 1970s, and even after 1970 their numbers are still subject to variation in the completeness and nonuniformity of small-magnitude earthquakes. In our study we have distinguished short period (SP) and long period (LP) earthquakes, the latter designated in the catalog only after 1972; both are related to stresses from magma flow, but LP earthquakes are more intimately associated with magma conduits. Moment accumulation is dominated by the largest earthquakes, all of which are short-period, and these therefore offer a more reliable comparison of stress.

Seismicity Variation over Intervals of Time

We focus first on short-period seismicity to broadly compare the changes over large periods of time as shown in figures 8.9 (chaps. 4, 5 and 6; 1950–82) and 8.10 (the Pu‘u ‘Ō‘ō-Kupaianaha eruption period of chap. 7; 1982–2008). Earthquake counts include all located earthquakes regardless of magnitude and show the effects both of seismic network improvement and relative seismicity values within periods of stable earthquake processing. Seismic moment, however, is determined by the largest earthquakes and is not affected by increases in small event completeness in this post-1950 period. Moment accumulation in all regions is shown for the period from 1950 to 1982 (fig. 8.9*B*) and for comparison the time period encompassing the eruption that began in 1983 (fig. 8.10*B*). Figure 8.9*A* shows the effects of improved detection of located earthquakes in all regions, particularly in 1969–70, and the 1969–82 values are higher because of this. The central south flank (sf 2.3) dominates both in counts and moment in nearly all time periods. High moment values in figure 8.9*B* are primarily determined by the large-magnitude earthquakes listed. Particularly noticeable is deep magma-supply seismicity (region ms 4.5), which increased beginning in 1961 after the reinflation of Kīlauea in 1950 and continued to be elevated up to the time of the 1975 earthquake.

The period following the 1975 earthquake is dominated by its aftershocks in the south flank regions. We have tried to minimize the effect by beginning this period on 1 January 1976, after the majority of aftershocks had ocurred. Particularly noticeable in this period is the sharp decline in the rate of deep magma-supply earthquakes, probably because stress on this zone was relieved by south-flank slip directly above, which has the same faulting mechanism (Klein and others, 1987). East rift seismicity and, to a lesser extent, seismic southwest rift seismicity increase as a function of the many intrusions occurring in this period.

The period after 1982 and particularly after 1986 shown in figure 8.10 shows greatly decreased seimicity in all regions following episode 1 of the Pu‘u ‘Ō‘ō-Kupaianaha eruption. Seismicity is higher in the period preceding the 1997 east rift intrusion and preceding the Father's Day eruption (stages IIA and IIIA). Overall, the rate of moment release is far below that of the periods preceding the 1975 earthquake.

Continuous Seismicity Variation

Continuous variation in cumulated seismic moment for short-period earthquakes in all regions is shown for the periods between 1950 and 1975 (fig. 8.11) and between 1983 and 2008 (fig. 8.12). We seek in these and succeeding figures to determine changes in strain release over time and correlate it with the magma supply and south flank spreading history. We assume that earthquake moment rates generally increase with magma flow rates, though we cannot verify this quantitatively or statistically. We are also aware that some seismicity deeper than 20 km is related to lithospheric flexure (Klein and Koyanagi, 1989), and we suspect that earthquakes classified as "deep magma supply" may act as a stress trigger when added to the much larger stress imposed by the volcanic load on the oceanic lithosphere. We look at changes of rate (slope) in different regions to better understand how the volcano responds to magma transport and storage. Rate comparisons between regions are visual, and we realize that conclusions drawn from visual comparisons of limited earthquake catalogs must be limited in scope.

We seek to answer the following "chicken and egg" question: On the assumption that magma arriving from more than 20 km depth is represented by moment accumulation in the ms4.5 region, does it trigger a later response in the shallower ms2–3 (5–15 km depth) regions? To answer the question we look at magma-supply seismicity for systematic depth changes with time. Before 1959 it appears that the signal for deep magma supply precedes the signal for shallower magma supply by anywhere from about 1 to nearly 4 years; this inference is based on the interval between the deep *M*6.2 earthquake in April 1951 and the increases in shallower seismicity between 1952 and 1955 (fig. 8.11). Following the seismicity associated with the large subsidence during the 1960 euption, an increase in deep magma supply precedes an increase in shallower magma supply by nearly 6 years (approximately 1962 to 1967). During 1971–75, deep magma supply leads the seismicity

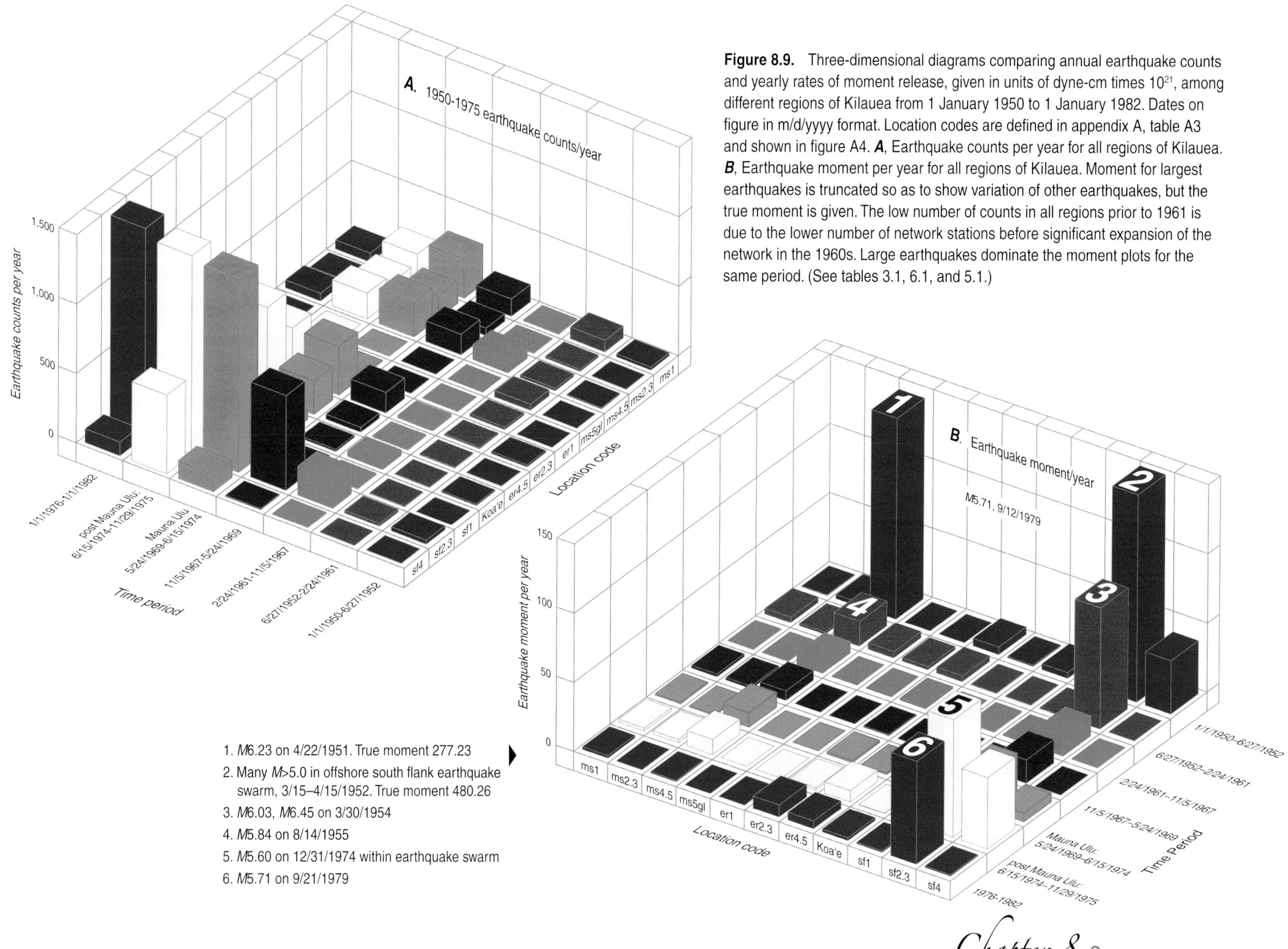

Figure 8.9. Three-dimensional diagrams comparing annual earthquake counts and yearly rates of moment release, given in units of dyne-cm times 10^{21}, among different regions of Kilauea from 1 January 1950 to 1 January 1982. Dates on figure in m/d/yyyy format. Location codes are defined in appendix A, table A3 and shown in figure A4. ***A***, Earthquake counts per year for all regions of Kilauea. ***B***, Earthquake moment per year for all regions of Kilauea. Moment for largest earthquakes is truncated so as to show variation of other earthquakes, but the true moment is given. The low number of counts in all regions prior to 1961 is due to the lower number of network stations before significant expansion of the network in the 1960s. Large earthquakes dominate the moment plots for the same period. (See tables 3.1, 6.1, and 5.1.)

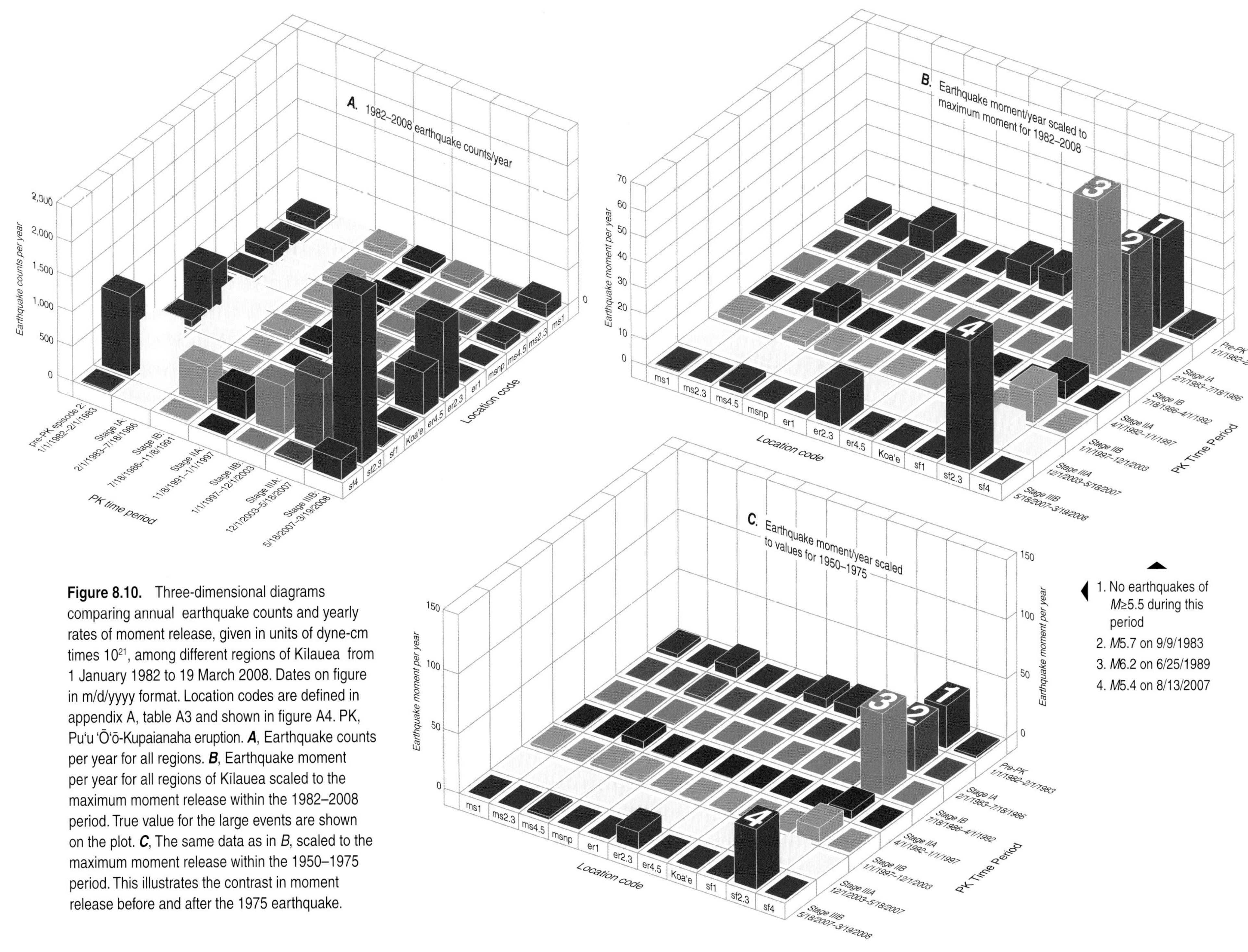

Figure 8.10. Three-dimensional diagrams comparing annual earthquake counts and yearly rates of moment release, given in units of dyne-cm times 10^{21}, among different regions of Kīlauea from 1 January 1982 to 19 March 2008. Dates on figure in m/d/yyyy format. Location codes are defined in appendix A, table A3 and shown in figure A4. PK, Puʻu ʻŌʻō-Kupaianaha eruption. ***A***, Earthquake counts per year for all regions. ***B***, Earthquake moment per year for all regions of Kīlauea scaled to the maximum moment release within the 1982–2008 period. True value for the large events are shown on the plot. ***C***, The same data as in *B*, scaled to the maximum moment release within the 1950–1975 period. This illustrates the contrast in moment release before and after the 1975 earthquake.

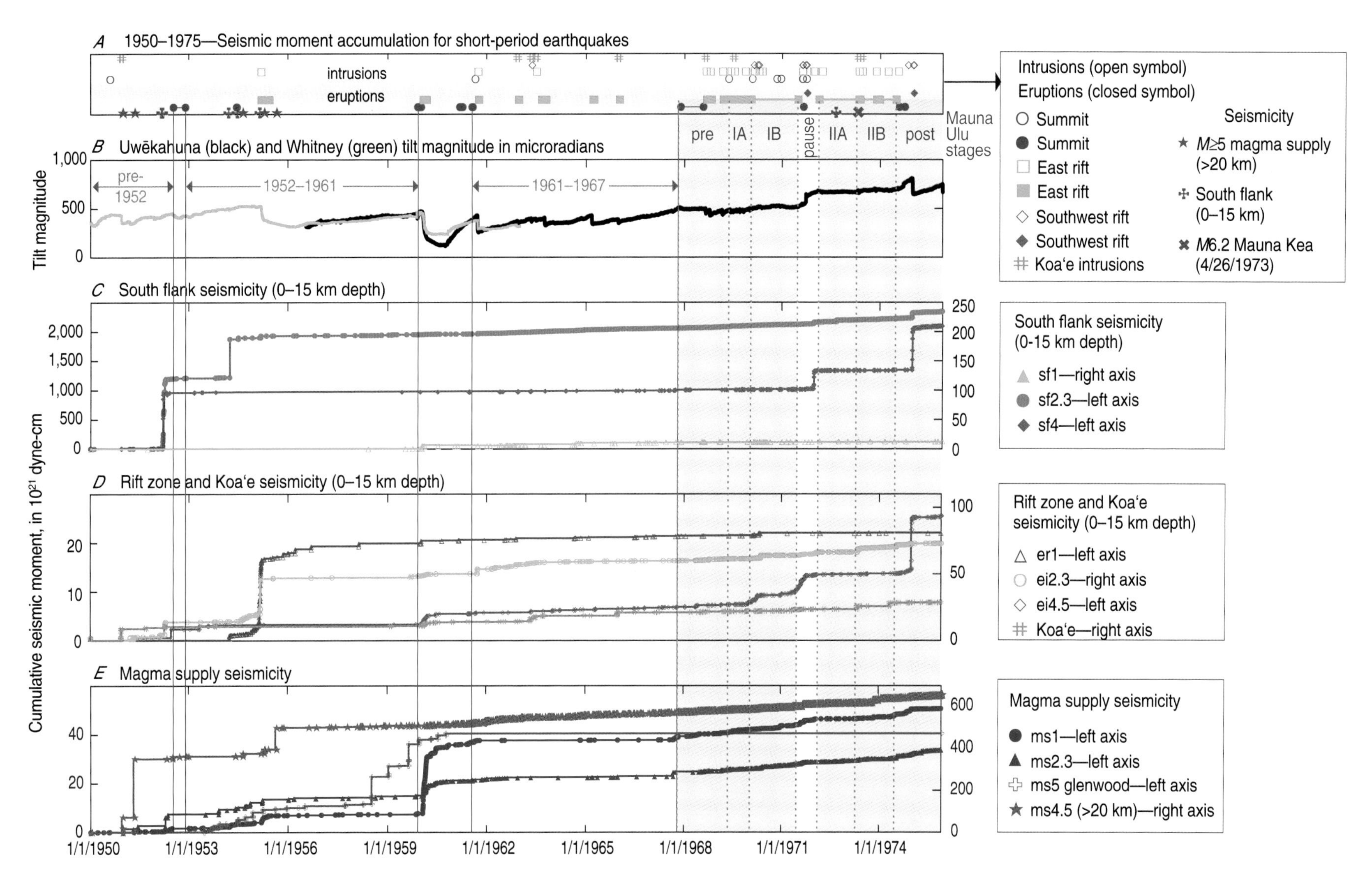

Figure 8.11. Graphs showing earthquake moment accumulation for short-period earthquakes in all regions of Kīlauea from 1 January 1950 to 29 November 1975 (up until the time of the *M*7.2 south flank earthquake). Dates on figure in m/d/yyyy format. Location codes are defined in appendix A, table A3 and shown in figure A4 (for example, er1 and sf2.3). From top to bottom, panel ***A*** shows times of eruption and intrusion; panel ***B*** shows the Uwēkahuna and Whitney tilt data; panels ***C–E*** show, respectively, seismicity beneath the south flank, rift zones and Koa'e, and the magma-supply regions. Vertical lines separate time intervals related to the summit eruptions of 1952, 1961 and 1967–68, as well as the stages of the Mauna Ulu eruption.

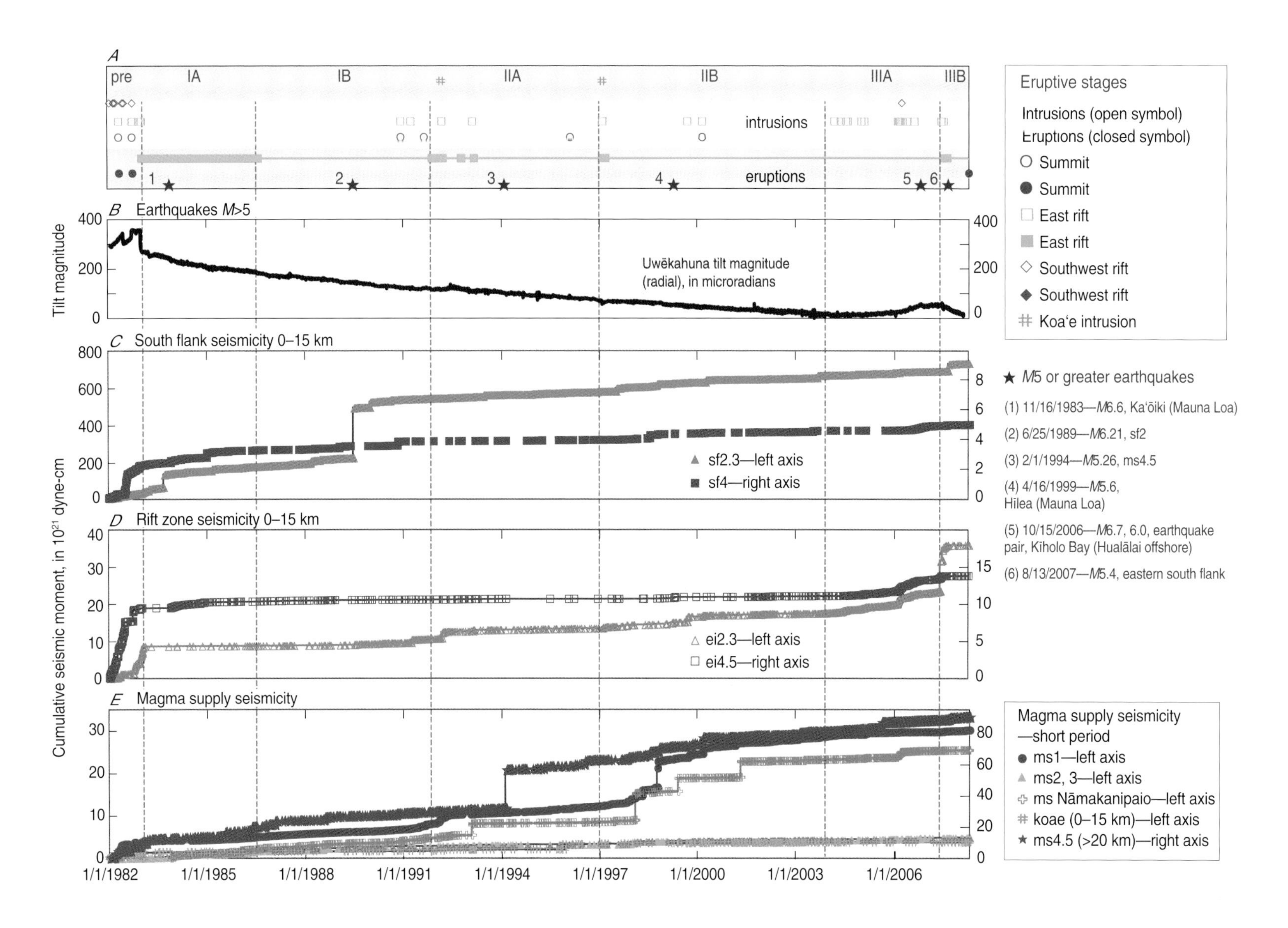

Figure 8.12. Graphs showing earthquake moment accumulation for short-period earthquakes in all regions of Kīlauea from 1 January 1982 to 1 April 2008 (including eruptive stages in the first 26 years of the Pu'u 'Ō'ō-Kupaianaha eruption). Dates on figure in m/d/yyyy format. Location codes are defined in appendix A, table A3 and shown in figure A4 (for example, ms1 and sf2.3). From top to bottom, panel ***A*** shows times of eruption and intrusion; panel ***B*** shows the Uwēkahuna tilt data; panels ***C–E*** show, respectively, seismicity beneath the south flank, rift zones and koae, and the magma-supply regions.

at 5–15 km by less than a month in late 1971 and by nearly a year between late 1973 and late 1974. The relations among deep and shallow long-period seismicity after 1972 (fig. 8.13) are unclear, perhaps because of a much smaller number of earthquakes.

Near Kīlauea's magma supply conduit, deep earthquakes tend to lead shallow earthquakes by years to months. During the eruption that began in 1983, the short-period earthquakes between 5 and 15 km depth (ms2 and ms3) were not active, so we make the comparison of changes in rate of the magma supply deeper than 20 km with magma supply shallower than 5 km (fig. 8.12). For the short-period earthquakes, the deeper magma supply again leads by up to 6 years (1985–1991) and 4 years (1994–1998), the lead time becoming less going forward in time, reaching only 3 months in late 2005-early 2006, and finally showing a nearly simultaneous increase at both depths from 1998 through early 2000. To the extent that the largest earthquake (and therfore the largest moment step) indicates magma supply, there is a 4-year lead of the deep February 1994 earthquake before the shallow September 1998 earthquake. The long-period seismicity shows the relation more clearly, the lead ranging from a difference of about 2 years in 1987–89 to 6 years between 1992 and 1998 (fig. 8.14).

The seismicity results suggest that deep magma supply drives the shallow response, initially with an apparent time delay measured in years. The shallowing of seismicity occurs in a general, diffuse sense, and there is no narrowly defined upward earthquake migration along a specific conduit. With continued use, the magma transport path from mantle to shallow storage becomes more open, resulting in a response within months of the shallow magma system to a change in rate of incoming magma from depth.

Magma Supply and South Flank Spreading—Who's the driver?

The spreading model for Kīlauea postulates a relation between deep magma supply and south flank spreading. Figures 8.15 (1950–75) and 8.16 (1983–2008) compare jumps in the cumulative moment in the deep magma supply (m) and the active central and eastern parts of the south flank (f). Jumps in the rate of moment release result from swarms or large earthquakes. Note that jumps from large-magnitude earthquakes are represented by red up-arrows, with the magnitude and true moment given on the plot.

The close time coincidence in jumps 2, 3, and 4 and the visual similarity of the two moment plots suggest an interpretable relation. In the 1950–75 period (fig. 8.15), the deep magma-supply jumps (m1–m5) lead (by 1.3–0.4 yr) or are very close behind (by 0.6–0.1 yr) the south flank jumps (f1–f5). By contrast, during the 1983–2008 Pu'u 'Ō'ō-Kupaianaha eruption (fig. 8.16), the south flank jumps in earthquake moment (f1–f4) lead (by 5.4 to 0.4 yr, assuming the event associations made in the figures) the jumps in magma-supply earthquake moment (m1–m4). The implications of this shift are discussed below.

The relation of spreading rate to magma supply rate helps us to understand the mechanism of spreading and to answer the question of whether spreading is solely driven by magma supply. The reconstruction of spreading history shown above (figs. 8.6*A* and *C*) shows a low rate of south flank spreading during recovery from the 1924 collapse, and we postulate that the rate increases, beginning with the the pre-1952 inflation and continuing to the 1975 earthquake. There was a 3-m extension during the 1975 earthquake and a spreading rate between the earthquake and the beginning of the Pu'u 'Ō'ō-Kupaianaha eruption that is slightly higher than rates preceding the earthquake (Delaney and others, 1998). During this period the magma supply rate also increased (fig. 8.5). Because the magma supply rate clearly increased following the recovery from the 1924 collapse, and on our assumption of low spreading rate before that time, we conclude that during the period from 1950 to 1983 magma supply drove spreading. This is consistent with the time delay separating deep magma-supply earthquake events from south flank seismicity (fig. 8.15) and also suggests that deep magma supply correlates with, and in 3 of the 5 cases, precedes flank spreading during this period.

During the Pu'u 'Ō'ō-Kupaianaha eruption, motion of Kīlauea's south flank was mostly smooth and slightly declining, with small offsets for large earthquakes and eruptions. Spreading rates were lower at the beginning of eruption (fig. 8.6*A*), while the magma supply rate was higher (fig. 8.5) than in preceding periods. Throughout the eruption magma supply continued to increase, whereas the spreading rate remained constant or slightly declining within ±20 percent of the average rate (fig. 8.6*C*).

In order to explain the relation between magma supply and spreading rates across the beginning of eruption shown in figures 8.6 and 8.5 we propose that the 1975 earthquake reduced friction on the sliding surface (decollement) such that slumping and other gravity-driven forces became the dominant factor governing spreading rate rather than magma supply. Within the Pu'u 'Ō'ō-Kupaianaha eruption, increases in magma supply were accommodated by dilating the rift zone and erupting magma with greater efficiency, rather than by accelerating the spreading rate. This is also consistent with the evidence that jumps in south flank moment release precede jumps in magma supply moment (fig. 8.16).

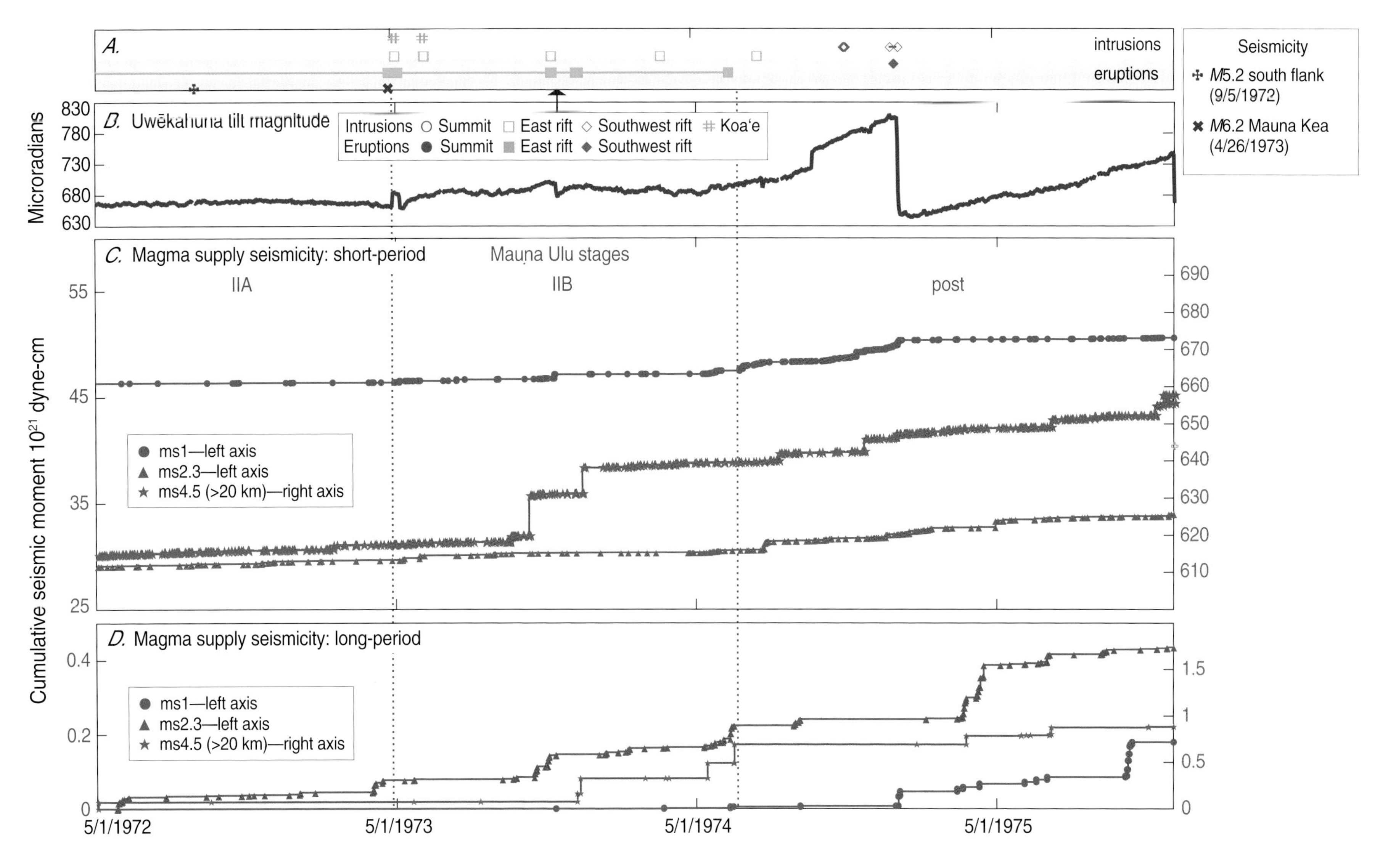

Figure 8.13. Graphs comparing seismic moment accumulation for short- and long-period magma supply earthquakes at Kīlauea for the period from 1 May 1972 to 1 December 1975. Long-period earthquakes were not tabulated before May 1972. Dates on figure in m/d/yyyy format. Location codes are defined in appendix A, table A3 and shown in figure A4 (for example, ms1). From top to bottom, panel ***A*** shows times of eruption and intrusion; panel ***B*** shows the Uwēkahuna tilt data; panels ***C*** and ***D*** show, respectively, short-period and long-period data. The long-period moments are only relative to each other and not absolute because a moment relation has not been defined.

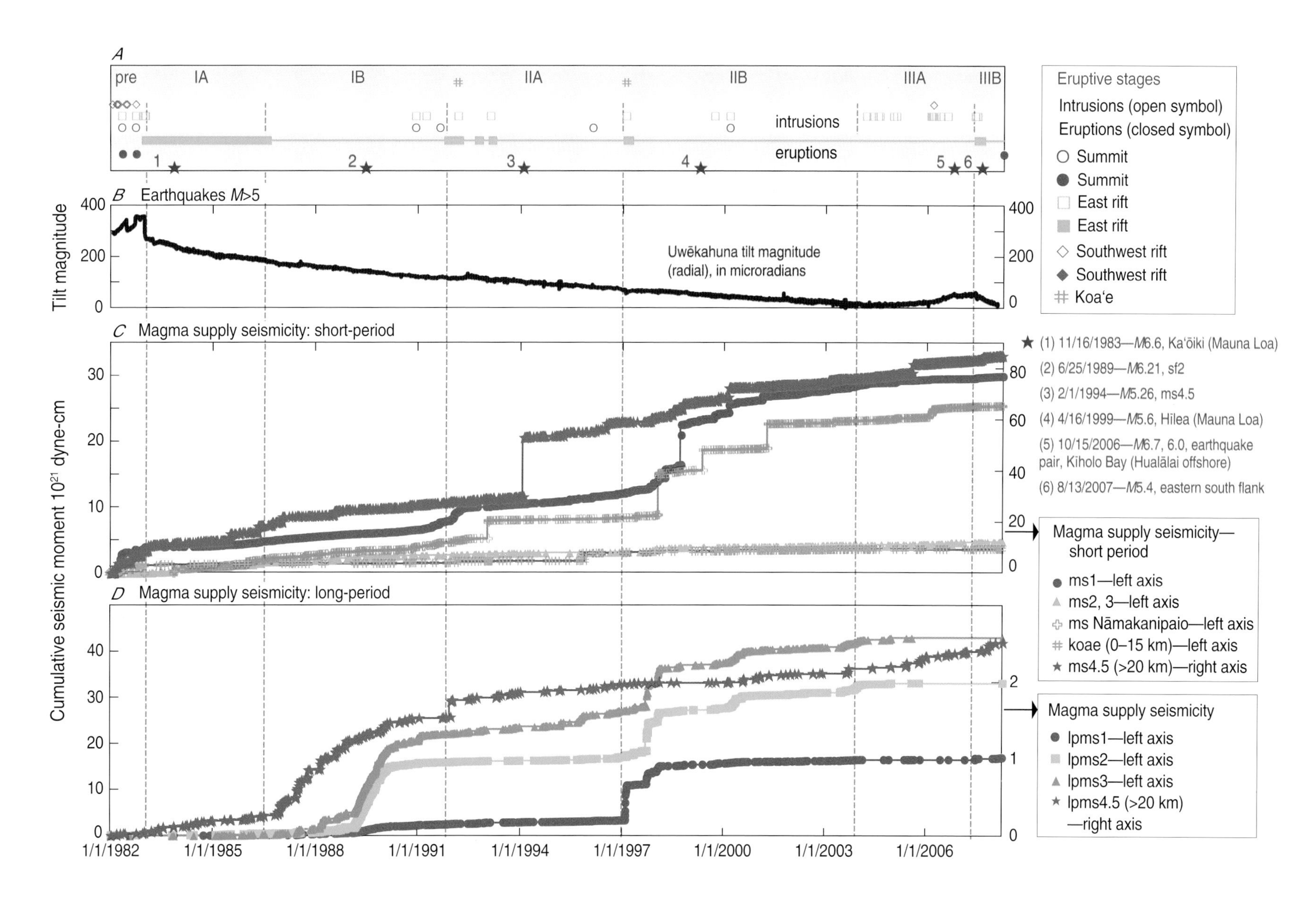

Figure 8.14. Graphs comparing seismic moment accumulation for short- and long-period magma supply earthquakes at Kīlauea for the period from 1 January 1982 to 1 April 2008 (including eruptive stages in the first 26 years of the Puʻu ʻŌʻō-Kupaianaha eruption). Dates on figure in m/d/yyyy format. Location codes are defined in appendix A, table A3 and shown in figure A4 (for example, ms1). From top to bottom, panel ***A*** shows times of eruption and intrusion; panel ***B*** shows the Uwēkahuna tilt data; panels ***C*** and ***D*** show, respectively, short-period and long-period data.

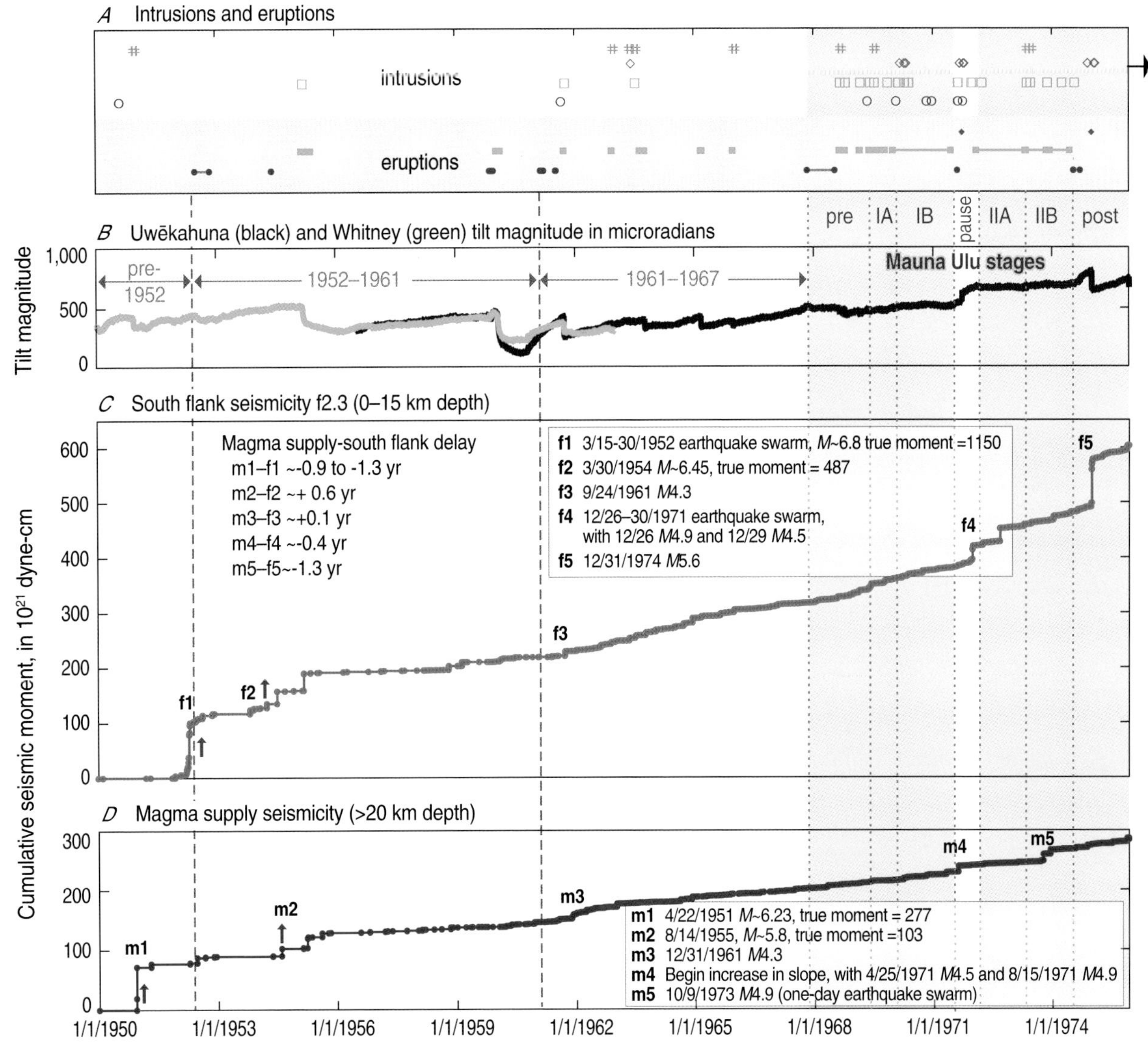

Figure 8.15. Graphs comparing south flank and deep magma-supply moment accumulation at Kīlauea for the period from 1 January 1950 to 29 November 1975. Dates on figure in m/d/yyyy format. Location codes are defined in appendix A, table A3 and shown in figure A4 (for example, ms1). From top to bottom, panel ***A*** shows times of eruption and intrusion; panel ***B*** shows the Uwēkahuna and Whitney tilt data; panels ***C*** and ***D*** show, respectively, south flank and deep magma-supply moment accumulation. Labeled red vertical arrows (for example, f2), indicate times of sharp increase in moment (not shown to scale). m = deep (>20 km) magma supply; f = central and eastern south flank together. Time delays are discussed in the text. Vertical lines separate time intervals related to the summit eruptions of 1952, 1961 and 1967-68, as well as the stages of the Mauna Ulu eruption.

Figure 8.16. Graphs comparing south flank and deep magma-supply moment accumulation at Kīlauea for the period from 1 January 1982 to 1 April 2008 (including the first 26 years of the Pu'u 'Ō'ō-Kupaianaha eruption). From top to bottom, panel ***A*** shows times of eruption and intrusion; panel ***B*** shows the Uwēkahuna tilt data; panels ***C*** and ***D*** show, respectively, south flank and deep magma-supply moment accumulation. Labeled red vertical arrows (for example, f2) indicate times of sharp increase in moment. m = deep (>20 km) magma supply; f = central and eastern south flank together. Time delays are discussed in the text.

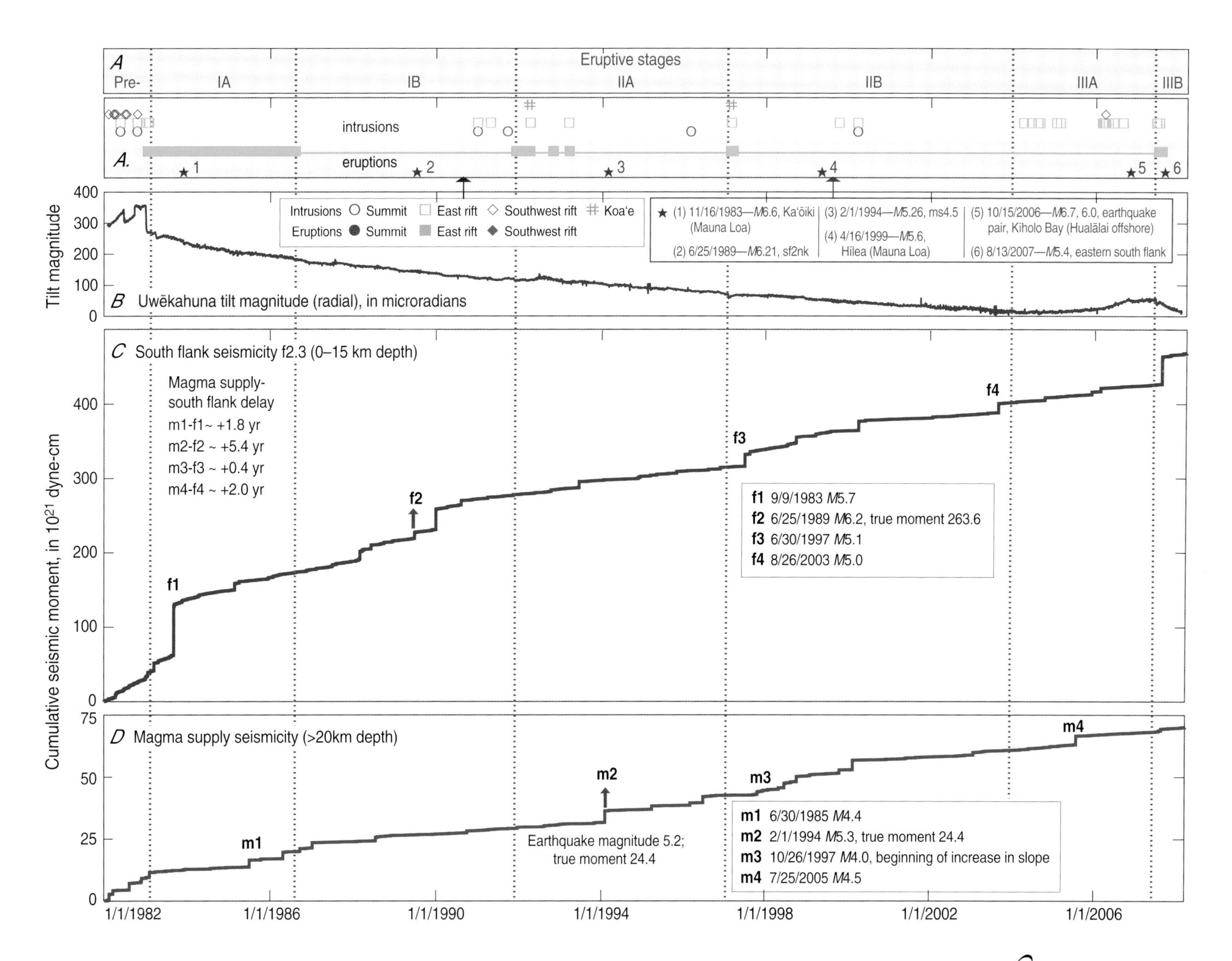
A
Eruptive stages
Pre-
IA
IB
IIA
IIB
IIIA
IIIB
intrusions
A.
eruptions
1
2
3
4
5
6
Intrusions ○ Summit □ East rift ◇ Southwest rift # Koa'e
Eruptions ● Summit ■ East rift ◆ Southwest rift
★ (1) 11/16/1983—M6.6, Ka'ōiki (Mauna Loa)
(2) 6/25/1989—M6.21, sf2nk
(3) 2/1/1994—M5.26, ms4.5
(4) 4/16/1999—M5.6, Hilea (Mauna Loa)
(5) 10/15/2006—M6.7, 6.0, earthquake pair, Kiholo Bay (Hualālai offshore)
(6) 8/13/2007—M5.4, eastern south flank
Tilt magnitude
B Uwēkahuna tilt magnitude (radial), in microradians
C South flank seismicity f2.3 (0–15 km depth)
Magma supply-
south flank delay
m1-f1~ +1.8 yr
m2-f2 ~ +5.4 yr
m3-f3 ~ +0.4 yr
m4-f4 ~ +2.0 yr
f1
f2
f3
f4
f1 9/9/1983 M5.7
f2 6/25/1989 M6.2, true moment 263.6
f3 6/30/1997 M5.1
f4 8/26/2003 M5.0
Cumulative seismic moment, in 10²¹ dyne-cm
D Magma supply seismicity (>20km depth)
m1
m2
m3
m4
Earthquake magnitude 5.2;
true moment 24.4
m1 6/30/1985 M4.4
m2 2/1/1994 M5.3, true moment 24.4
m3 10/26/1997 M4.0, beginning of increase in slope
m4 7/25/2005 M4.5
1/1/1982
1/1/1986
1/1/1990
1/1/1994
1/1/1998
1/1/2002
1/1/2006

Post-1983 History of Continuous Eruption

The Pu'u 'Ō'ō-Kupaianaha eruption embodies many phenomena either not identified or not present during previous periods of activity. Here we discuss these phenomena, which were presented in chapter 7.

Pauses, Surges, Silent Earthquakes, and Suspected Deep Intrusions

Pauses, eruption surges (D-I-D events), suspected deep intrusions, and silent earthquakes all offer clues as to the state of the magma plumbing. Surges, first detected in 1996, can be interpreted to follow temporary blockage of the transport path between the summit reservoir depicted in figure 8.1 and the east rift vent (Cervelli and Miklius, 2003). The timing of inflation at Kīlauea's summit and the eruption site during surges suggests that magma was transmitted from the summit reservoir to the magma source beneath Pu'u 'Ō'ō at rates of 0.54–1.58x10^6 m^3/day (Cervelli and Miklius, 2003, table 1), which translates to an average magma supply rate exceeding 0.20 km^3/year. This may be additional evidence in favor of a surge in magma supply associated with the 1997 eruption. After 2004, the surges lose their distinctive signature, perhaps because blockages were prevented by later increases in magma supply or by smoothing of irregular spots in the magma supply conduit.

In contrast to the surges, we consider that pauses, silent earthquakes, and suspected deep intrusions are all triggered by the deeper magma system beneath the rift zone that was first outlined by Delaney and others (1990). The precursory sequence shown in figure 7.7 of (1) long-period earthquake counts at 5–13 km depth below the summit, followed by (2) a small deflation without an eruption surge, suggests that the pauses were induced by temporary diversion of the deeper summit magma supply to feed intrusion beneath the east rift zone. Pauses lose their distinctive tilt and seismic signatures in 1991 during stage II of the eruption. The later pauses may be simply short interruptions due to temporary blockage of the shallow magma supply path with no precursory signs.

Suspected deep intrusions show a pattern of south flank seismicity similar to that of events documented in recent studies as "slow" or "silent" earthquakes (Cervelli and others, 2002b; Montgomery-Brown and others, 2009; Segall and others, 2006). All suspected deep intrusions are revealed by earthquake swarms beneath the south flank without a main shock, and some are accompanied by a few shallow rift earthquakes. Suspected deep intrusions within the period in which silent earthquakes have been identified are listed in table 8.2. The larger events identifed as silent earthquakes with an earthquake swarm would also be classified by us as suspected deep intrusions. Some of the events with smaller south flank displacements are accompanied by very few earthquakes, insufficient to specify a suspected deep intrusion by seismicity alone. Significantly, there are a few events since the start of continuous geodetic monitoring in 1998 with an earthquake pattern resembling that of a suspected deep intrusion that have not been identified as being associated with a spreading step.

The patterns of south flank earthquakes shown by suspected deep intrusions are shown in figure 8.17. South flank seismicity characteristic of suspected deep intrusions began as early as 1962. The "Kalapana Trail" south flank events mentioned in chapter 4 (fig. 4.12) are possible candidates for suspected deep intrusions because they occur in closely spaced sequences beneath the south flank and are isolated in time from other seismicity. However, because the locations of such small-magnitude events are so uncertain within the seismic network available at that time, we hesitate to label them as such and consider the possibility that they could even represent mislocated rift events. The earliest events that have the pattern of suspected deep intrusions occur in 1965 (figure 8.17*C*).

Suspected deep intrusions likely occur when pressure is applied to the deep magma system beneath the rift zones. The earliest examples may have been triggered following the swarms of magma-supply earthquakes at 30-km depth beneath Kīlauea Caldera that began in 1961 and died out by the end of 1965. These deep swarms are interpreted to accompany increased magma supply and also could trigger suspected deep intrusions. Over the 50 years between 1961 and 2011, we infer that suspected deep intrusions accompanied by a swarm of south flank earthquakes were probably associated with a spreading step. It is possible that a combination of a previous intrusion and the occurrence of suspected deep intrusions may provide a rationale for the absence of seismicity during some east rift eruptions. For example, the intrusion of February 1993 (table 7.3) extended downrift to the site of the January 1997 eruption and intrusion (table 7.4). The force of the 1993 intrusion near the future eruption site, combined with suspected deep intrusions at site SDI 2 on 11–12 September 1993 and 25–26 October 1994 (table 7.3; fig. 7.8) may have dilated the rift zone sufficiently to explain the absence of rift seismicity preceding the January 1997 event.

Relations Between Kīlauea and Mauna Loa

In chapter 1 we noted that the growth of Kīlauea's and Mauna Loa's rift zones has been unequal and complementary—that is, Mauna Loa's long southwest rift zone is paired with Kīlauea's stunted southwest rift zone, and Kīlauea's long east rift zone is paired with Mauna

Table 8.2. Silent earthquakes and suspected deep intrusions[1].

[Rows highlighted in **bold type** represent events identified by us as "suspected deep intrusions" and in the published literature as "silent earthquakes"; dates in m/d/yyyy format. Events identified by us as suspected deep intrusions, but not published as geodetically identified silent earthquakes, are listed in event tables in individual chapters]

Date	P[2]	D[3] (cm)	Loc.[4]	SDI?[5]	No.[6]	Comment	Figure[7]
2/20/1998	0.89	0.12		no	2	Not identified as suspected deep intrusion	-
9/19/1998	**1.0**	**0.72**	**SDI 2**	**yes**	**12**	**South flank 9/19–20/1998**	**G29**
1/10/1999	0.55	0.21			2	Not identified as suspected deep intrusion	-
4/27/1999	0.7	-			0	Not identified as suspected deep intrusion	-
11/21/1999	0.85	0.41			0	Not identified as suspected deep intrusion	-
5/29/2000	**0.75**	**0.17**	**SDI 2**	**yes**	**25**	**South flank 5/29–30/2000**	**7.11**
11/9/2000	1.0	1.43	SDI 2	yes	11	South flank 11/8–9/2000	G36
9/19/2001	0.55	0.06				Not identified as suspected deep intrusion	G39
8/21-22/2002			SDI 2	yes	9	Not identified as a silent earthquake[8]	G44
8/28-29/2002			SDI 2	yes	8	Not identified as a silent earthquake	G44
12/17/2002	0.96	0.12		yes	5	Not identified as suspected deep intrusion	G45
7/3/2003	**0.99**	**0.62**	**SDI 3**	**yes**	**10**	**South flank 7/1–2/2003**	**G48**
12/22/2003			SDI 2	yes	7	Not identified as a silent earthquake	G51
5/6/2004	0.7			no	2	Not identified as suspected deep intrusion	-
1/26/2005	**1.0**	**2.06**	**SDI 2**	**yes**	**55**	**South flank 1/25–27/2005**	**G60**
6/27/2005	0.56	0.22		no		5 south flank earthquakes 6/26–28	-
3/16/2006			SDI 2	yes	13	Not identified as a silent earthquake	G70
4/16/2007	0.55	0.07		no	2	Not identified as suspected deep intrusion	
6/18/2007		4.01	SDI 2	yes			7.13; G79

[1]Events shown in this paper and in table 1 of Montgomery-Brown and others (2009).

[2]Probability of event being a silent earthquake (Montgomery-Brown and others, 2009, table 1, column 7).

[3]Displacements calculated from north and east components of GPS displacements for south flank GPS sation KAEP (Emily Montgomery-Brown, written commun,. 2011).

[4]Referenced to chapter 5, figure 5.2.

[5]yes = south flank swarm corresponds to SDI patterns of chapter 5, figure 5.2.

[6]Number of south flank earthquakes.

[7]Figure references are to chapter 7. Text figures are designated 7.x; appendix G figures are designated Gxx.

[8]Aftershock sequenec following a pair of $M3.6$, 3.0 earthquakes?

Loa's short northeast rift zone. Mauna Loa being older, this implies that its early growth favored the southwest rift zone. Later, the vigor of Mauna Loa's southwest rift development prevented Kīlauea from growing in that direction. Instead Kīlauea was forced to grow to the east, with the result that Mauna Loa's later development was blocked in that direction.

There is a long-term relation between Kīlauea and Mauna Loa eruptive activity, Kīlauea being most active when Mauna Loa is quiet and vice versa (Klein, 1982). Figure 8.18 compares the inferred magma filling rates for Kīlauea with the timing and volume of Mauna Loa eruptions. Eruption volume is the best practical measure of Mauna Loa productivity, whereas long-lived caldera filling and frequent or long-lasting eruptions make cumulative volume the best measure for Kīlauea. The increase in Kīlauea's eruption frequency and magma supply rate, and simultaneous decrease in Mauna Loa's eruption frequency, after 1950 is clearly visible. Unfortunately, the Mauna Loa record only begins in 1843, and little is known of the record before that time. A brief 1832 summit eruption is mentioned in the literature (see, for example, Goodrich, 1833), but its products have been covered by younger lavas. Modern mapping has identified only one eruption whose age could fall between 1790 and 1843 (Frank Trusdell, oral commun., 2011). The absence of significant Mauna Loa volcanism that can be confidently dated between 1790 and 1843 may be one factor controlling the very high filling rates at Kīlauea during the early part of the 19th century, and the reduced activity of Kīlauea after 1840 may correlate with the onset of more frequent activity at Mauna Loa (chap. 2). Thus there are two definite changeover times—from Kīlauea to Mauna Loa in 1840 and back again in 1950[30].

[30] Research by Frank Trusdell and Don Swanson has documented this relationship back in time (see, for example, Trusdell, 2012).

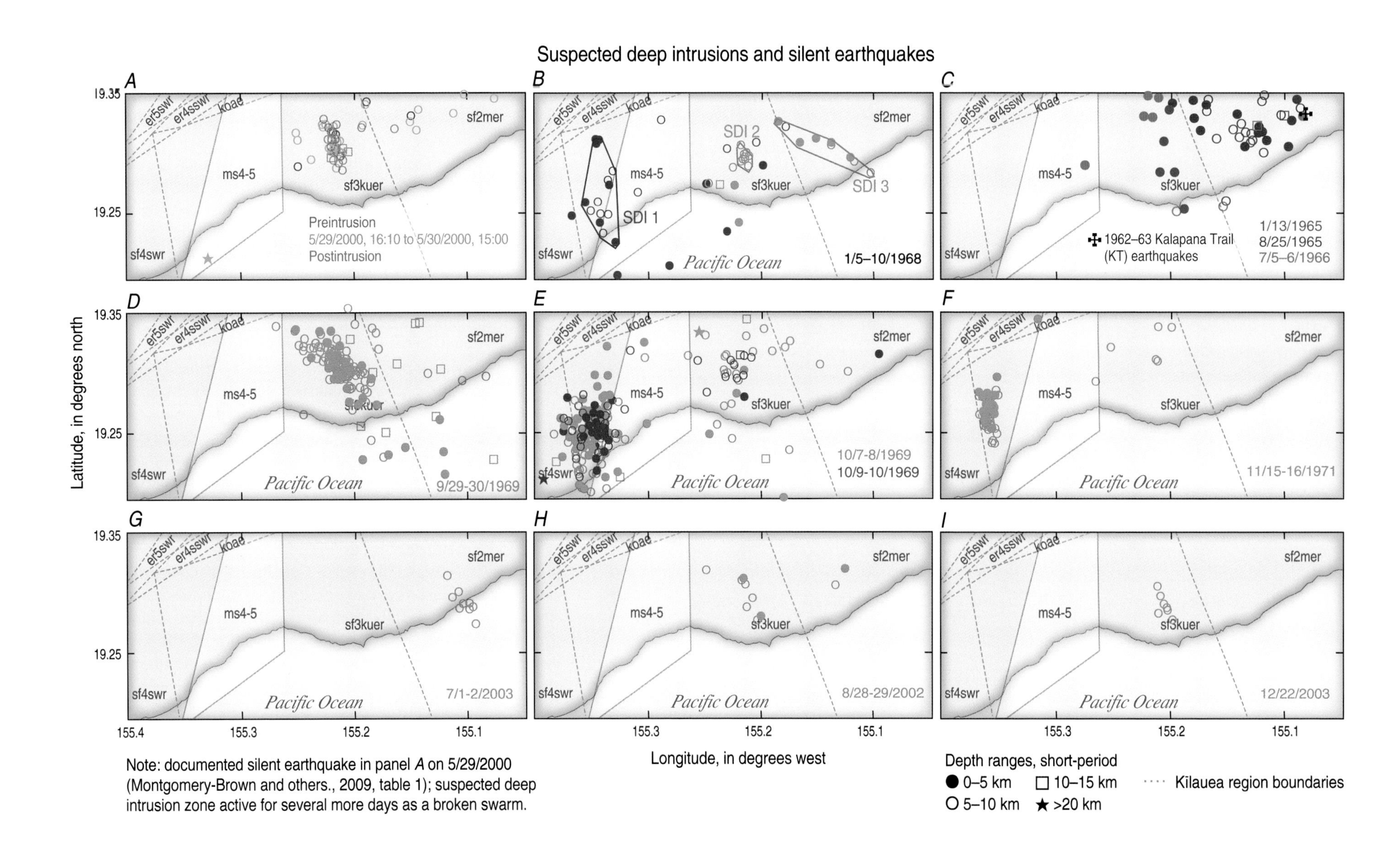
Suspected deep intrusions and silent earthquakes
A
B
C
D
E
F
G
H
I
19.35
19.25
Latitude, in degrees north
155.4
155.3
155.2
155.1
Longitude, in degrees west
er5swr
er4sswr
koae
sf2mer
ms4-5
sf3kuer
sf4swr
Pacific Ocean
Preintrusion
5/29/2000, 16:10 to 5/30/2000, 15:00
Postintrusion
SDI 1
SDI 2
SDI 3
1/5–10/1968
1962–63 Kalapana Trail
(KT) earthquakes
1/13/1965
8/25/1965
7/5–6/1966
9/29–30/1969
10/7–8/1969
10/9–10/1969
11/15–16/1971
7/1–2/2003
8/28–29/2002
12/22/2003
Note: documented silent earthquake in panel A on 5/29/2000 (Montgomery-Brown and others., 2009, table 1); suspected deep intrusion zone active for several more days as a broken swarm.
Depth ranges, short-period
0–5 km
10–15 km
5–10 km
>20 km
Kīlauea region boundaries

Figure 8.17. Maps showing seismicity of silent earthquakes (Montgomery-Brown and others, 2009) and suspected deep intrusions (this report, keyed to figure numbers) at Kīlauea. Dates on figure in m/d/yyyy format. Symbols for depth ranges apply to all figures. ***A***, 29–30 May 2000 (see fig. 7.11). Identified as both a silent earthquake and a suspected deep intrusion. ***B***, 5–10 January 1968 (see fig. 5.2). Suspected deep intrusions are found in three locations labeled, from west to east, SDI 1, SDI 2 and SDI 3. ***C***, 13 January 1965, 25 August 1965, and 5–6 July 1966 (see fig. 4.12). These are the earliest indications of the earthquake pattern associated with suspected deep intrusions. These are located near sites SDI 2 and 3 (see part B for site locations). ***D***, 29–30 September 1969 (see appendix E, fig. E17). An exceptionally strong earthquake swarm beneath site SDI 2. ***E***, 7–10 October 1969. A closely spaced (in time) pair of exceptionally strong earthquake swarms beneath sites SDI 1 and 2 . ***F***, 15–16 November 1971 (see appendix E, fig. E51). An exceptionally strong earthquake swarm beneath site SDI 1. ***G***, 1–2 July 2003 (see appendix G, fig. G48). A suspected deep intrusion that precedes by one day an identified silent earthquake (see table 8.2). ***H***, 28–29 August 2002 (see appendix G, fig. G44). A suspected deep intrusion not identified as a silent earthquake (see table 8.2). ***I***, 22 December 2003 (see appendix G, fig. G51). A suspected deep intrusion not identified as a silent earthquake (see table 8.2).

It is known that Mauna Loa and Kīlauea lavas can be distinguished chemically (see, for example, Wright, 1971), suggesting their derivation from separate mantle sources. Yet over historical time their eruptive activity has favored one volcano over the other, thus suggesting some sort of connection related to magma supply. The relative eruption frequency and chemical differences have been reconciled by postulating partial melting in different parts of the Hawaiian mantle source volume (Wright and Klein, 2006).

There have been instances in historical time during which both volcanoes were in eruption. The continuous activity at Kīlauea between 1907 and 1924 appears to have had no affect on Mauna Loa's eruptive cycles dating from 1843. The Mauna Loa eruption of 1984 occurred during episode 18 of the Kīlauea eruption that began in 1983, with no change in seismicity or ground deformation to indicate that either volcano had influenced the other at a shallow level. Another case in point is the return to activity of Kīlauea in 1952 following Mauna Loa's largest historical eruption in 1950. High seismicity associated with Mauna Loa continued for nearly ten years following the eruption, belying the fact that its activity would be greatly curtailed between then and now (2010) as Kīlauea entered a period of increasing magma supply.

It is by no means obvious what triggers a switch that favors eruption of one volcano over the other, but we suspect that mantle stresses are involved. We have looked for and not found any seismic indication prior to Kīlauea's reinflation in 1950 to indicate that a major switch of eruptive activity would occur, and the cause of a switch from one volcano to the other remains an area for further research (compare Gonnermann and others, 2011).

Evolution of Kīlauea's Plumbing Between 1823 and 2008

As a context for interpreting Kīlauea's behavior after the arrival of European missionaries in 1823, we go back in time to the formation of the ʻAilāʻau shield (whose surface is seen north and east of the present caldera) (Clague and others, 1999) and the present caldera (Clague and others, 1999; Swanson, 2003, 2009; Swanson and others, 2012). The ʻAilāʻau summit shield-building took place over about 50 years, ending at about 1470 C.E.[31]. During that time we presume that Kīlauea's rift zones were unable to accept magma and the south flank was locked, implying the temporary absence of seaward spreading. Soon after, Kīlauea's summit collapsed to an unknown depth to form a caldera during dates estimated at 1470–1510 C.E. A period of explosive eruption ensued, ending in 1790 C.E. with a climactic eruption that killed Hawaiian people marching across Kīlauea's summit (Dibble, 1843; Swanson and Christiansen, 1973; Swanson and Rausch, 2008). A 1790 eruption on Kīlauea's lower east rift zone has been identified by modern mapping (Trusdell and Moore, 2006), indicating that the east rift zone reopened sometime during the period of explosive eruption. This may also have been a time when the deep magma system below the east rift zone and above the decollement was rejuvenated.

When the missionaries arrived in 1823, the floor of Kīlauea Caldera contained an active lava lake over nearly its entire area (Ellis, 1825), which we interpret as recovery from a massive magma draining that probably occurred at the time of the culminating explosive activity of 1790. In successive disappearances of magma from

[31] The Common Era (C.E.) begins with year 1 of the Gregorian calendar.

the caldera during the 19th century we surmise that the entire magma system, encompassing sources 1 and 2, and possibly 3 as shown in chapter 2 (fig. 2.4), was drained and refilled. Earthquake swarms during the 19th century indicate that both the southwest and east rift zones were open to accept intrusions. The diminished rate of earthquake activity after 1840 is ascribed to both recovery from the 1790 draining and the onset of heightened eruptive activity at Mauna Loa. The century ended with Halema'uma'u as a single vent, only sporadically active at its bottom.

Kīlauea has alternated between higher and lower magma activity on a time scale of decades. Two distinct periods of magmatic activity before 1894 and after 1950 are separated by the summit collapse of 1924. Magma gradually returned to Halema'uma'u beginning in 1907, leading to a new lava lake and periodic overflow onto the caldera floor beginning in 1918. The lava lake and overflows were both fed from the shallow source 1. Source 2 was first distinctly evident to modern observation in the collapse of 1922. A broad regional uplift in 1918–19 was matched by an equally large regional subsidence accompanying the summit collapse in 1924. The Halema'uma'u collapse and phreatic eruption of May 1924 ended the lava lake activity, and the last draining of sources 1–3 in 1924 ended with a large intrusion into the lower east rift zone. The events of 1924 stabilized the relation between the rift zones and south flank, such that for several decades up to March 1950, the caldera remained at the post-1924 collapse level and the dominant magma movement was recovery of deep source 3. During the period 1924–50, small eruptions in Halema'uma'u used up the remnants of source 1, and source 2 may have been involved in intrusions of 1937 and 1938. Recovery was complete in March 1950, when inflation began in response to pressure generated within source 2. During the entire period of recovery the south flank remained locked and intrusions occurred with minimal seaward spreading.

Events preceding the 1952 eruption in Halema'uma'u are critical to Kīlauea's subsequent history and demonstrate how changes at different parts of Kīlauea trigger each other. The deep earthquake swarm of December 1950, followed by the single deep earthquake of April 1951, coincided with and perhaps initiated a higher magma supply rate. The higher magma supply also served to trigger the intense offshore south flank swarm of March-April 1952, which we believe led to the unlocking of the south flank[32]. The increase in magma supply rate at Kīlauea is also a consequence of the great decrease in the rate of Mauna Loa activity following its large eruption in June of 1950. By 1952 one can definitely identify the traditional summit reservoir (source 2) as a source for resupply of the lower east rift eruptions of 1955 and 1960, as well as a source for middle and upper east rift eruptions from 1961 onward. Suspected deep intrusions (SDI) occur from at least 1961, at least some of which may have been associated with south flank spreading steps. If SDIs occurred before 1961, they would not have been detectable by the primitive seismic network, so it is unknown when they began. Magma supply and spreading rates increased following the 1967–68 summit eruption and continued to increase throughout the 1969–74 Mauna Ulu eruption.

The inability of Kīlauea during the 1960s to cope with increasing magma supply through release by eruption led to increased intrusive and eruptive activity on the southwest side of the volcano, which in turn led to the 1975 *M*7.2 south flank earthquake. Recovery from the effects of the earthquake occurred over the next 8 years, with episode 1 of the Pu'u 'Ō'ō-Kupaianaha eruption in 1983 being the final event of the 1975 recovery. The consequences of the 1975 earthquake and the associated large component of both seaward spreading and rift dilation led to (1) a rift readily able to initally accept magma through numerous intrusions, (2) the development of an east rift zone plumbing that was able to sustain a decades-long east rift eruption accompanied by continuous deflation of Kīlauea's summit, and (3) the freeing of the south flank to move by gravity alone, evidenced by the decoupling of spreading rate from magma supply rate during the long and still ongoing east rift eruption that began in 1983.

A final crisis in the pre-2010 history ensued following the return to inflation of Kīlauea's summit in 2004 coupled with a great increase in magma supply represented by a large temporary increased output of CO_2 at Kīlauea's summit from mid-2004 to mid-2005. Incomplete accommodation of the still-high magma supply rate in the Pu'u 'Ō'ō-Kupaianaha eruption resulted first in a new small eruption and large east rift intrusion in June 2007, followed a month later by new vents on the east rift. As in the time preceding the 1975 earthquake, these events from additional magma supply were accompanied by summit inflation and increased seismicity in the southwest sector of the volcano. A second area of eruptive activity began in March 2008 as a lava pond visible below the surface of Halema'uma'u. The open vent in Halema'uma'u

[32] The many south flank earthquakes documented in the 19th century (Klein and Wright, 2000) are interpreted to involve stress release without seaward spreading on the decollement, and they probably were triggered by a massive stress redistribution caused by the great 1868 Ka'ū earthquake (Klein and others, 2006). Stress release without a spreading step also occurred during the south flank earthquake of 25 June 1989 (fig. 8.6*C*). During the 1989 earthquake stress relief was interpreted to occur on shallower subhorizontal planes (Arnadottir and others, 1991), with perhaps an additional vertical component (Bryan, 1992). We favor a similar interpretation for the events occurring before March 1950.

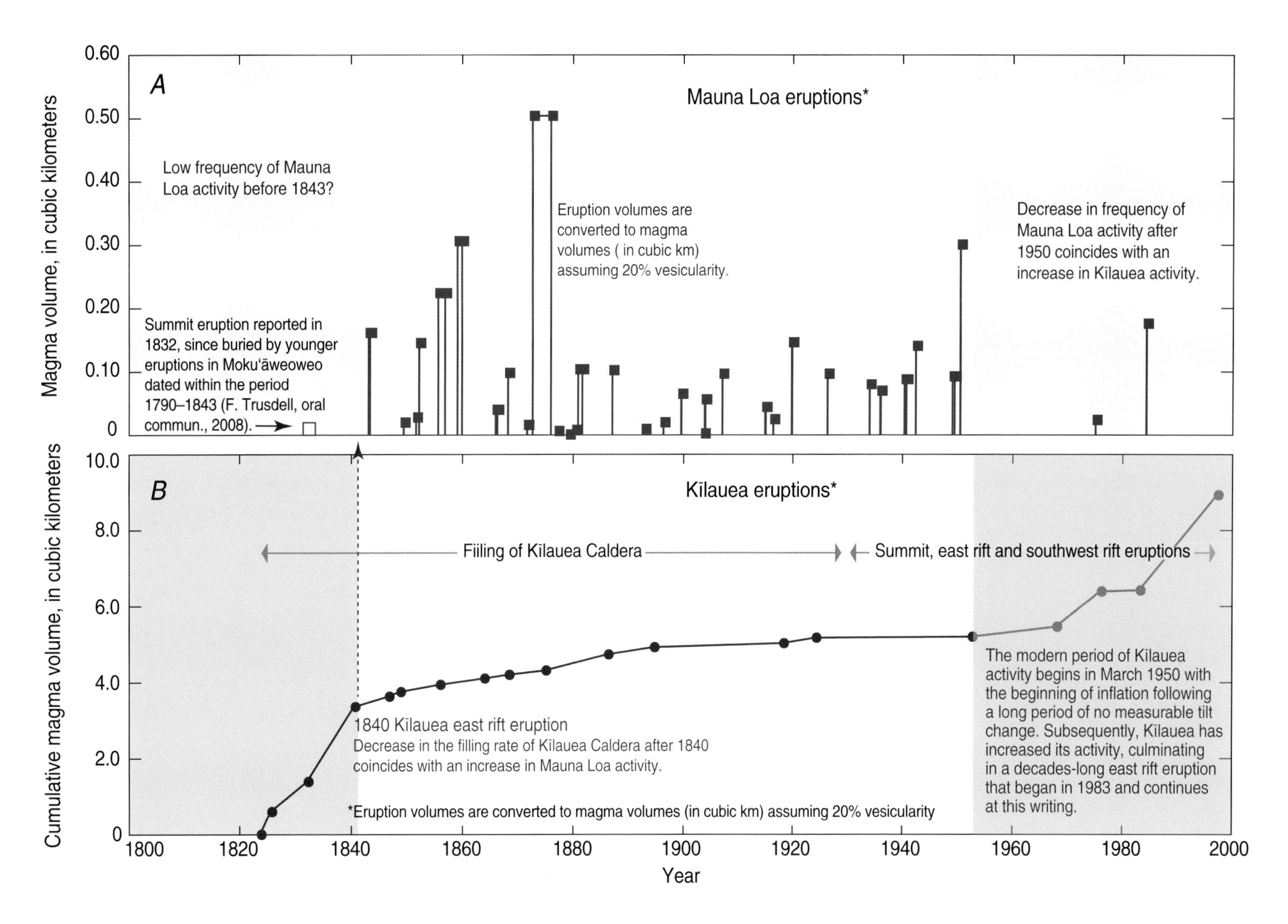

Figure 8.18. Graphs showing rates of filling of Kīlauea Caldera and volumes of later eruptions compared to frequency and volume of Mauna Loa eruptions. Times and volumes of Mauna Loa eruptions (***A***) are compared with filling rates at Kīlauea (***B***). Periods of high Mauna Loa activity correspond to periods of lower Kīlauea activity, as shown by slope of curve of cumulative magma volume, in agreement with previous work (for example, Klein, 1982).

has continued its activity to the present with no waning of the east rift eruption, marking the first time in Kīlauea's recorded history that summit and rift have been simultaneously active.

The combined data for the period 1950–2008 demonstrate a remarkably integrated magmatic system in which the response to the pressure of an increasing magma supply from depth occurs in the following ways: (1) the mantle supplies more magma and the volcano's behavior responds, or (2) the volcano dilates or spreads and encourages the mantle to send more magma, or (3) both happen together because the system is finely balanced. The attempt to erupt magma at Kīlauea's summit is fraught with difficulties related to the unique properties of magma and the interaction of magma with the complex structure of the volcanic edifice. Any resistance to upward movement of magma results in pressure being applied laterally. As long as the rifts are able to accept magma, magma will stay at a lower elevation, and stress is preferentially relieved by rift eruption and intrusion, and by movement of the south flank. In general, after stress is applied during larger rift intrusions, the south flank responds with increased seismicity. However, the response can be reversed when the flank earthquake is very large, as shown by the increased number of rift intrusions in the years following the large south flank earthquake of November 1975.

Eruptions in Halema'uma'u that last longer than a few days offer a different form of stress relief, eventually manifested by the lessening of seismic activity in all parts of the volcano. Yet this stress relief is temporary. The answer to the question posed by one of us many years ago as to why summit eruptions end (Wright and Fiske, 1971, p. 52) can now be viewed as a problem in partitioning of magma between the summit reservoir and the fluid core of the rift zone (compare Wright and Klein, 2008). As long as the rift is available, sustained summit activity is not possible and summit eruptions will end when it becomes easier for magma to move laterally than to move up. The long-term continuity of rift eruptions depends on the development of a stable conduit connecting the fluid core of the rift zone to the surface, such that pressure remains constant and the magma flow is sufficient to resist cooling against the walls. In general, the conduit system is not perfectly open. Cooling and the associated increase of magma viscosity acts to close conduits, causing a back-pressure that results in reinflation of Kīlauea's summit. The ongoing east rift eruption that began in 1983 has built an edifice in much the same way as earlier eruptions such as the one that formed the Heiheiahulu shield and lava field, but the east rift edifice built during this eruption is not high enough to stop the eruption. The most likely circumstance that would end this eruption is a change applied to the mantle source region such as to favor renewed and more frequent activity on Mauna Loa than is seen at present.

During Kīlauea's many cycles of breakdown and recovery, different parts of the plumbing may be more or less active, and the parameters of magma storage and transport may change. These changes are principally because of increased magma activity or the increased occurrence of large, deep, conduit earthquakes. The crisis and recovery cycles offer a sobering reminder for eruption and earthquake prediction, and for hazard evaluation. Crisis and recovery can each be anticipated by looking at both deformation and seismic patterns, but the nature of both crisis and recovery may vary greatly from one instance to the next. Our view is that the volcano's behavior is governed by nonlinear dynamics in which several systems interact. Each crisis changes the ground state of the Kīlauea edifice such that all seismic, deformation, and magma parameters used in monitoring are set to a new and unknown baseline. Continuous monitoring must determine the baselines and departures from it, but changes in the volcano state render long-term eruption and earthquake prediction inherently uncertain.

Supplementary Material

Supplementary material for this chapter appears in appendix H, which is only available in the digital versions of this work—in the DVD that accompanies the printed volume and as a separate file accompanying this volume on the Web at http://pubs.usgs.gov/pp/1806/. Appendix H comprises the following:

Tables H1–H3 contain accumulated earthquake counts and moment for all regions plotted in figures 8.11 and 8.12.

Table H4a gives volumes of the three magma batches erupted at Halema'uma'u in 1952, 1961, and 1967–68 that are identified in eruptions between 1952 and the beginning of the Mauna Ulu eruption in May 1969.

Table H4b gives additional volumes inferred for intrusions over the same period of time. See text for interpretations using these quantities.

Figure H1 shows the location of edm and GPS stations used in constructing figures 8.6 and 8.7.

References

Aki, K., and Koyanagi, R., 1981, Deep volcanic tremor and magma ascent mechanism under Kilauea, Hawaii: Journal of Geophysical Research, v. 86, no. B8, p. 7095–7109.

Alexander, W.D., 1886, Map of the island of Hawaii, Hawaiian Islands (preliminary ed.): Honolulu, Hawaii Government Survey, scale 1:240,000. [Map by W.A. Wall; primary triangulation by W.D. Alexander, C.J. Lyons, and J.S. Emerson; Kau and Puna surveys not completed.]

Almendros, J., Chouet, B., and Dawson, P., 2001, Spatial extent of a hydrothermal system at Kilauea Volcano, Hawaii, determined from array analyses of shallow long-period seismicity, 2. Results: Journal of Geophysical Research, v. 106, no. B7, p. 13581–13597.

Almendros, J., Chouet, B., Dawson, P., and Bond, T., 2002a, Identifying elements of the plumbing system beneath Kilauea Volcano, Hawaii, from the source locations of very-long-period signals: Geophysical Journal International, v. 148, p. 303–312.

Almendros, J., Chouet, B., Dawson, P., and Huber, C., 2002b, Mapping the sources of the seismic wave field at Kilauea Volcano, Hawaii, using data recorded on multiple seismic antennas: Bulletin of the Seismological Society of America, v. 92, no. 6, p. 2333–2351.

Ando, M., 1979, The Hawaii earthquake of November 29, l975; low dip angle faulting due to forceful injection of magma: Journal of Geophysical Research, v. 84, no. B13, p. 7616–7626.

Anonymous, 1868a, Earthquakes and the volcano: Pacific Commercial Advertiser, April 11, p. 3.

Anonymous, 1868b, Latest news!; the lava flowing; great loss of life; destruction of property: Hawaiian Gazette, April 15, p. 2.

Anonymous, 1868c, The volcano: Hawaiian Gazette, May 27, p. 3. [Letter from Sheriff Cone of Hilo, written on the 18th inst.]

Anonymous, 1868d, The volcano, *in* Notes of the week: Pacific Commercial Advertiser, August 8, p. 3.

Anonymous, 1871a, [no title], *in* Local News: Hawaiian Gazette, March 29, p. 3.

Anonymous, 1871b, The volcano active: Hawaiian Gazette, July 5, p. 3.

Anonymous, 1877a, From the crater of Kilauea, *in* Notes of the week: Pacific Commercial Advertiser, May 12, , p. 3.

Anonymous, 1894a, Erratic Kilauea: Hawaiian Annual for 1895, p. 78–85.

Anonymous, 1894b, Here and there: Pacific Commercial Advertiser, July 23, p. 6.

Anonymous, 1894d, Several shocks felt in Kau: Pacific Commercial Advertiser, December 12, p. 1. [Repeated verbatim in Hawaiian Gazette of December 14]

Anonymous, 1894e, Visit to the volcano: The Friend, September, p. 71.

Anonymous, 1896, Condition unchanged; Mokuaweoweo's crater sending out fire: Pacific Commercial Advertiser, April 29, p. 6.

Anonymous, 1902, Halemaumau's lava pit is rapidly filling up with lava while geysers of fire dot the great lake: Pacific Commercial Advertiser, November 17, p. 1.

Anonymous, 1907, Volcano measured; report that crater moved was incorrect: Hawaii Herald, January 10, p. 2. [Repeated in Hilo Tribune, October 15, 1903]

Anonymous, 1918, Quake splits open Kilauea lava floor: Pacific Commercial Advertiser, November 3, p. 1.

Armstrong, W.N., 1894, The lake has sunk again: Hawaiian Gazette, July 20, p. 6. [A communication following that of Thurston; copied from Pacific Commercial Advertiser of July 18.]

Arnadottir, T., Segall, P., and Delaney, P., 1991, A fault model for the 1989 Kilauea south flank earthquake from leveling and seismic data: Geophysical Research Letters, v. 18, no. 12, p. 2217–2220.

Baker, N.A., 1987, A petrologic study of the 1982 summit lavas of Kilauea Volcano, Hawaii: Honolulu, University of Hawaii at Manoa, Honors thesis, 61 p.

Bevens, D., Takahashi, T.J., and Wright, T.L., eds., 1988, The early serial publications of the Hawaiian Volcano Observatory, 3 v.: Hawai'i National Park, Hawai'i Natural History Association, v. 1, 565 p. v. 2, 1273 p.; v. 3, 1224 p.

Borgia, A., 1994, Dynamic basis of volcanic spreading: Journal of Geophysical Research, v. 99, no. B9, p. 17791–17804.

Borgia, A., and Treves, B., 1992, Volcanic plates overriding the oceanic crust; structure and dynamics of Hawaiian volcanoes, *in* Parson, L.M., Murton, B.J., and Browning, P., eds., Ophiolites and their modern oceanic analogues: London, The Geological Society, Special Publication No. 60, p. 277–299.

Bosher, R.F., 1981, Analysis of an earthquake swarm associated with the Christmas, 1965 eruption of Kilauea Volcano, Hawaii: Honolulu, University of Hawaii, Geology and Geophysics, M.S. dissertation, 78 p.

Bosher, R., and Duennebier, F.K., 1985, Seismicity associated with the Christmas 1965 event at Kilauea Volcano: Journal of Geophysical Research, v. 90, no. B6, p. 4529–4536.

Brigham, W.T., 1868, The eruption of the Hawaiian volcanoes, 1868: Boston Society of Natural History Memoirs, v. 1, p. 564–587.

Brigham, W.T., 1887, Kilauea in 1880: American Journal of Science, v. 34, p. 19–27.

Brigham, W.T., 1909, The volcanoes of Kilauea and Mauna Loa on the island of Hawaii: Bernice P. Bishop Museum Memoirs, v. 2, no. 4, 222 p. [Plates, 56 p. reprinted 1974 by Kraus Reprint Co., Millwood, N.Y.]

Brooks, B.A., Foster, J.H., Bevis, M., Frazer, L.N., Wolfe, C.J., and Behn, M., 2006, Periodic slow earthquakes on the flank of Kīlauea volcano, Hawai'i: Earth and Planetary Science Letters, v. 246, no. 3–4, p. 207–216.

Bryan, C.J., 1992, A possible triggering mechanism for large Hawaiian earthquakes derived from analysis of the 26 June 1989 Kilauea south flank sequence: Bulletin of the Seismological Society of America, v. 82, no. 6, Dec., p. 2368–2390.

Buchanan-Banks, J.M., 1987, Structural damage and ground failures from the November 16, 1983, Kaoiki earthquake, island of Hawaii, chap. 44 *of* Decker, R.W., Wright, T.L., and Stauffer, P.H., eds., Volcanism in Hawaii: U.S. Geological Survey Professional Paper 1350, v. 2, p. 1187–1220.

Byron, G., 1826, Voyage of H.M.S. Blonde to the Sandwich Islands, in the years 1824–1825: London, John Murray, 260 p. [p. 169–191 cover the material abstracted.]

Cayol, V., Dieterich, J.H., Okamura, A., and Miklius, A., 2000, High magma storage rates before the 1983 eruption of Kilauea, Hawaii: Science, v. 288, p. 2343–2346.

Cervelli, P.F., and Miklius, A., 2003, The shallow magmatic system of Kīlauea Volcano, *in* Heliker, C.C., Swanson, D.A., and Takahashi, T.J., eds., The Pu'u 'Ō'ō-Kupaianaha eruption of Kīlauea Volcano, Hawai'i—the first 20 years: U.S. Geological Survey Professional Paper 1676, p. 149–163.

Cervelli, P., Segall, P., Amelung, F., Garbeil, H., Meertens, C., Owen, S., Miklius, A., and Lisowski, M., 2002a, The 12 September 1999 Upper East Rift Zone dike intrusion at Kilauea Volcano, Hawaii: Journal of Geophysical Research, v. 107, no. B7, p. ECV 3-1–3-13.

Cervelli, P., Segall, P., Johnson, K., Lisowski, M., and Miklius, A., 2002b, Sudden aseismic fault slip on the south flank of Kilauea Volcano: Nature, v. 415, February 28, p. 1014–1018. [In "Letters to Nature" section]

Chouet, B.A., 1996, Long-period volcano seismicity—its source and use in eruption forecasting: Nature, v. 380, March 28, p. 309–316.

Clague, D.A., 1987, Hawaiian xenolith populations, magma supply rates, and development of magma chambers: Bulletin of Volcanology, v. 49, p. 577–587.

Clague, D.A., and Denlinger, R.P., 1993, The *M*7.9 1868 earthquake; Hawaii's active landslides [abs.]: Eos (American Geophysical Union Transactions), v. 74, no. 43, Fall meeting supplement, p. 635.

Clague, D.A., and Dixon, J.E., 2000, Extrinsic controls on the evolution of Hawaiian ocean island volcanoes: Geochemistry, Geophysics, Geosystems, v. 1, Paper no. 1999GC000023, April 7, 6607 words, 1 table. [Available at http://146.201.254.53/publicationsfinal/articles/1999GC000023/a1999GC000023.html]

Clague, D.A., Hagstrum, J.T., Champion, D.E., and Beeson, M.H., 1999, Kilauea summit overflows; their ages and distribution in the Puna District, Hawai'i: Bulletin of Volcanology, v. 61, p. 363–381.

Clarke, F.L., 1886, The volcano: Pacific Commercial Advertiser, July 12, p. 2.

Coan, T., 1841, Great eruption of the volcano of Kilauea: Missionary Herald, July, p. 283–285. [Reprinted as "Terrible volcanic eruption at the Sandwich Islands": New York, American Repertory of Arts, Sciences, and Manufacturers, 1841, v. 4, p. 139–143. Reprinted as "Great eruption of the volcano Kilauea in 1840": Dwight's American Magazine and Family Newspaper, 1847, v. 3, p. 6–7; 19–20. Also reprinted in Boston Weekly Magazine, 1841, v. 3, no. 48, Aug. 14, p. 377–378. Reprinted, with slight modifications, in Thurston, L.G., 1842, p. 224–230.]

Coan, T., 1851, On the recent condition of Kilauea: American Journal of Science, v. 12, p. 80–82.

Coan, T., 1852, [no title]: The Polynesian, v. 8, no. 47, April 3, p. 1.

Coan, T., 1854, On the present condition of the crater of Kilauea, Hawaii: American Journal of Science, v. 18, p. 96–98.

Coan, T., 1856a, Eruption of Mauna Loa: American Journal of Science, v. 21, p. 139–144.

Coan, T., 1856b, On Kilauea: American Journal of Science, v. 21, p. 100–102.

Coan, T., 1864, Volcano of Kilauea, Hawaii: American Journal of Science, v. 37, p. 415–416.

Coan, T., 1867, Volcanic eruptions in Hawaii: American Journal of Science, v. 43, p. 264–265.

Coan, T., 1868, The great volcanic eruption: Missionary Herald, v. 64, no. 7, [July], p. 219–221.

Coan, T., 1869, Notes on the recent volcanic disturbances of Hawaii: American Journal of Science, v. 47, p. 89–98.

Coan, T., 1870, Volcanic action on Hawaii: American Journal of Science, v. 49, p. 393–394.

Coan, T., 1874, Note on the Hawaiian volcanoes: American Journal of Science, v. 8, p. 467.

Coan, T., 1879, On a recent silent discharge of Kilauea: American Journal of Science, v. 18, p. 227–228.

Coan, T., 1882, Life in Hawaii: New York, Anson D.F. Randolph & Co., 340 p.

Coan, T.M., 1910, Eruptions of Kilauea [in July 1855]: Science, v. 32, p. 716-718.

Cox, D.C., 1980, Source of the tsunami associated with the Kalapana (Hawai‘i) earthquake of November 1975: Hawai‘i Institute of Geophysics, HIG-80-8, December, 46 p.

Crosson, R.S., and Endo, E.T., 1982, Focal mechanisms and locations of earthquakes in the vicinity of the 1975 Kalapana earthquake aftershock zone 1970–1979; implications for tectonics of the south flank of Kilauea Volcano, Island of Hawaii: Tectonics, v. 1, p. 495–542.

Dana, J.D., 1849, Geology; United States Exploring Expedition: New York, George P. Putnam, v. 10, 756 p. [Pages 1–27, 156–226, 278–284, 353–436 contain material pertinent to the geology of the Hawaiian Chain. Relevant sections on Mauna Loa and Kīlauea are reorganized and partly rewritten in American Journal of Science, 2nd ser., v. 9, 1850, p. 347–364; v. 10, 1850, p. 235–244. A short excerpt was published as “On the nature of volcanic eruptions,” in Mather, K.F., and Mason, S.L., eds., A source book in geology, 1400–1900: Cambridge, Harvard University Press, p. 423–429.]

Dana, J.D., 1887a, History of the changes in the Mt. Loa craters: American Journal of Science, v. 33, p. 433–451. [Continuation in v. 34, p. 81–97, 349–364 [includes maps]

Dana, J.D., 1887b, History of the changes in the Mt. Loa craters: American Journal of Science, v. 34, p. 81–97, 349–364. [Continuation from v. 33, p. 433–451 [includes maps]]

Dana, J.D., 1888, History of the changes in the Mt. Loa craters, on Hawaii: American Journal of Science, v. 35, p. 15–34, 213–228, 282–289 [includes maps]; v. 36, p. 14–32, 81–112, 167–175 [includes maps].

Dana, J.D., 1891, Characteristics of volcanoes: New York, Dodd, Mead, and Co., 399 p.

Dawson, P.B., Dietel, C., Chouet, B.A., Honma, K., Ohminato, T., and Okubo, P.G., 1998, Digitally telemetered broadband seismic network at Kilauea Volcano, Hawaii: U.S. Geological Survey Open-File Report 98–108, 126 p.

Dawson, P.B., Chouet, B.A., Okubo, P.G., Villasenor, A., and Benz, H.M., 1999, Three-dimensional velocity structure of the Kilauea caldera, Hawaii: Geophysical Research Letters, v. 26, no. 18, p. 2805–2808.

Dawson, P., Whilldin, D., and Chouet, B., 2004, Application of near real-time radial semblance to locate the shallow magmatic conduit at Kilauea Volcano, Hawaii: Geophysical Research Letters, v. 31, no. 21 (L21606), November 5, p. 1–4; doi:10.1029/2004GL021163.

de Vis-Norton, L.W., 1922, Kilauea Volcano during 1922: Hawaiian Annual for 1923, p. 97–100. [See also “Brief lava flow,” p. 152–153]

de Vis-Norton, L.W., 1923, Kilauea changes in 1923: Hawaiian Annual for 1924, p. 152–153.

Decker, R.W., 1987, Dynamics of Hawaiian volcanoes; an overview, chap. 42 *of* Decker, R.W., Wright, T.L., and Stauffer, P.H., eds., Volcanism in Hawaii: U.S. Geological Survey Professional Paper 1350, v. 2, p. 997–1018.

Decker, R., Okamura, A., Miklius, A., and Poland, M., 2008, Evolution of deformation studies on active Hawaiian volcanoes, U.S. Geological Survey Scientific Investigations Report, 2008–5090.

Delaney, P.T., Denlinger, R.P., Lisowski, M., Miklius, A., Okubo, P.G., Okamura, A.T., and Sako, M.K., 1998, Volcanic spreading at Kilauea, 1976–1996: Journal of Geophysical Research, v. 103, no. B8, p. 18003–18023.

Delaney, P.T., Fiske, R.S., Miklius, A., Okamura, A.T., and Sako, M.K., 1990, Deep magma body beneath the summit and rift zones of Kilauea Volcano, Hawaii: Science, v. 247, March 16, p. 1311–1316.

Delaney, P.T., Wyss, M., Lipman, P.W., and Okamura, A., 1992, Comment on “Precursors to the Kalapana *M* 7.2 earthquake” by Max Wyss, F.W. Klein, and Arch C. Johnston: Journal of Geophysical Research, v. 97, no. B4, p. 4839–4841.

Delaney, P.T., Miklius, A., Arnadottir, T., Okamura, A.T., and Sako, M.K., 1993, Motion of Kilauea Volcano during sustained eruption from the Puu Oo and Kupaianaha vents, 1983–1991: Journal of Geophysical Research, v. 98, no. B10, p. 17,801–17,820.

Denlinger, R.P., and Okubo, P., 1995, Structure of the mobile south flank of Kilauea Volcano, Hawaii: Journal of Geophysical Research, v. 100, p. 24499–24507.

Desmarais, E.K., and Segall, P., 2007, Transient deformation following the 30 January 1997 dike intrusion at Kīlauea volcano, Hawai‘i: Bulletin of Volcanology, v. 69, no. 4, p. 353–363.

Dibble, S., 1839, History and general views of the Sandwich Islands' mission: Lahaina-luna, T.H., Mission Seminary, 464 p. [Simultaneously published in New York by Taylor & Dodd; p. 14–56 of the New York edition cover the material abstracted.]

Dibble, S., 1843, A history of the Sandwich Islands: Lahainaluna, T.H., T.G. Thrum, 451 p. [Reprinted 1909 in Honolulu; p. 1–3, 50–53, 94–95 cover the material abstracted.]

Dieterich, J.H., and Decker, R.W., 1975, Finite element modeling of surface deformation associated with volcanism: Journal of Geophysical Research, v. 80, no. 29, p. 4094–4102.

Dieterich, J., Cayol, V., and Okubo, P., 2000, The use of earthquake rate changes as a stress meter at Kilauea Volcano: Nature, v. 408, November 23, p. 457–460.

Dieterich, J.H., Cayol, V., and Okubo, P., 2003, Stress changes before and during the Puʻu ʻŌʻō-Kūpaianaha eruption, *in* Heliker, C.C., Swanson, D.A., and Takahashi, T.J., eds., The Puʻu ʻŌʻō-Kūpaianaha eruption of Kīlauea Volcano, Hawaiʻi—the first 20 years: U.S. Geological Survey Professional Paper 1676, p. 187–201.

Dixon, J.E., Stolper, E.M., and Holloway, J.R., 1995, An experimental study of water and carbon dioxide solubilities in mid-ocean ridge basaltic liquids, pt. II. Applications to degassing: Journal of Petrology, v. 36, no. 6, p. 1633–1646.

Duffield, W.A., Jackson, D.B., and Swanson, D.A., 1974, The shallow, forceful intrusion of magma and related ground deformation at Kilauea Volcano, May 15–16, 1970, *in* Gonzales-Ferran, O., ed., Proceedings, Symposium on Andean and Antarctic Volcanology Problems, Santiago, Chile, September 9–14, 1974: International Association of Volcanology and Chemistry of the Earth's Interior, Special Series, p. 577–597.

Duffield, W.A., Christiansen, R.L., Koyanagi, R.Y., and Peterson, D.W., 1982, Storage, migration, and eruption of magma at Kilauea Volcano, Hawaiʻi, 1971–1972: Journal of Volcanology and Geothermal Research, v. 13, p. 273–307.

Dvorak, J.J., 1992, Mechanism of explosive eruptions of Kīlauea Volcano, Hawaiʻi: Bulletin of Volcanology, v. 54, p. 638–645.

Dvorak, J., 1994, An earthquake cycle along the south flank of Kilauea Volcano, Hawaii: Journal of Geophysical Research, v. 99, no. B5, p. 9533–9541.

Dvorak, J.J., and Dzurisin, D., 1993, Variations in magma supply rate at Kilauea Volcano, Hawaii: Journal of Geophysical Research, v. 98, no. B12, p. 22225–22268.

Dvorak, J.J., and Okamura, A.T., 1985, Variations in tilt rate and harmonic tremor amplitude during the January-August 1983 east rift erutions of Kilauea Volcano, Hawaii: Journal of Volcanology and Geothermal Research, v. 25, p. 249–258.

Dvorak, J.J., Okamura, A.T., English, T.T., Koyanagi, R.Y., Nakata, J.S., Sako, M.K., Tanigawa, W.T., and Yamashita, K.M., 1986, Mechanical response of the south flank of Kilauea Volcano, Hawaii, to intrusive events along the rift systems: Tectonophysics, v. 124, p. 193–209.

Dvorak, J., Delaney, P.T., Okamura, A.T., and Prescott, W.H., 1989, High-precision geodetic measurements on Hawaiian volcanoes using satellite geodesy [abs.], *in* Abstracts, Continental Magmatism, International Association of Volcanology and Chemistry of the Earth's Interior, General Assembly, Santa Fe, NM, June 25–July 1, 1989: New Mexico Bureau of Mines & Mineral Resources Bulletin 131, p. 78.

Dvorak, J.J., Klein, F.W., and Swanson, D.A., 1994, Relaxation of the south flank after the 7.2-magnitude Kalapana earthquake, Kilauea Volcano, Hawaii: Bulletin of the Seismological Society of America, v. 84, no. 1, p. 133–141.

Dzurisin, D., Anderson, L.A., Eaton, G.P., Koyanagi, R.Y., Lipman, P.W., Lockwood, J.P., Okamura, R.T., Puniwai, G.S., Sako, M.K., and Yamashita, K.M., 1980, Geophysical observations of Kilauea Volcano, Hawaii, 2. Constraints on the magma supply during November 1975–September 1977, *in* McBirney, A.R., ed., Gordon A. Macdonald Memorial Volume (special issue): Journal of Volcanology and Geothermal Research, v. 7, no. 3/4, p. 241–269.

Dzurisin, D., Koyanagi, R.Y., and English, T.T., 1984, Magma supply and storage at Kilauea Volcano, Hawaii, 1956-1983: Journal of Volcanology and Geothermal Research, v. 21, p. 177–206.

Eakins, B.W., Robinson, J.E., Kanamatsu, T., Naka, J., Smith, J.R., Takahashi, E., and Clague, D.A., 2003, Hawaii's volcanoes revealed: U.S. Geological Survey Geologic Investigations Series I-2809, scale 1:85,342.

Eaton, J.P., 1959, A portable water-tube tiltmeter: Bulletin of the Seismological Society of America, v. 49, no. 4, p. 301–316.

Eaton, J.P., 1962, Crustal structure and volcanism in Hawaii, *in* Crust of the Pacific Basin: American Geophysical Union Geophysical Monograph 6, p. 13–29. [National Academy of Science, National Research Council Publication 1035.]

Eaton, J.P., 1986a, History of the development of the HVO seismograph and deformation networks: Hawaiian Volcano Observatory library, unpublished record, 20 p.

Eaton, J.P., 1986b, Narrative account of the instrument developments at HVO from 1953 thru 1961, and a summary of further developments from 1961 thru 1978: Hawaiian Volcano Observatory library, unpublished record, 13 p.

Eaton, J.P., and Fraser, G.D., 1956a, Hawaiian Volcano Observatory summary 3—July-September 1956: U.S. Geological Survey Administrative Report, 5 p.

Eaton, J.P., and Fraser, G.D., 1956b, Hawaiian Volcano Observatory summary 4—October-December 1956: U.S. Geological Survey Administrative Report, 5 p.

Eaton, J.P., and Fraser, G.D., 1957, Hawaiian Volcano Observatory summary 5—January to March 1957: U.S. Geological Survey Administrative Report, 12 p.

Eggers, A.A., and Johnson, D.J., 2006, Residual gravity changes at Kilauea; 1977–2003 [abs.]: Eos , (American Geophysical Union Transactions), v. 87, no. 52, Fall meeting supplement, abs.V51A-1647.

Eissler, H.K., and Kanamori, H., 1986, Depth estimates of large earthquakes on the island of Hawaii: Journal of Geophysical Research, v. 91, no. B2, p. 2063-2076.

Elias, T., and Sutton, A.J., 2002, Sulfur dioxide emission rates from Kīlauea Volcano, Hawai'i, an update; 1998–2001: U.S. Geological Survey Open-File Report 02-460, 10 p.

Elias, T., and Sutton, A.J., 2007, Sulfur dioxide emission rates from Kīlauea Volcano, Hawai'i, an update; 2002–2006: U.S. Geological Survey Open-File Report 2007-1114, p. 37 p.

Elias, T., Sutton, A.J., Stokes, J.B., and Casadevall, T.J., 1998, Sulfur dioxide emission rates of Kīlauea Volcano, Hawai'i, 1979–1997: U.S. Geological Survey Open-File Report 98–462, 40 p.

Ellis, W., 1825, Narrative of a tour through Hawaii, or, Owhyhee: London, H. Fisher, Son, and P. Jackson, 264 p. [Simultaneously published in Boston by Crocker & Brewster. Reprinted 1826, 1827 in London by Fisher and Jackson; reprinted 1917 by the Hawai'ian Gazette Co., Ltd., Honolulu; 1827 London ed. reprinted 1963 as Journal of William Ellis by the Advertiser Publishing Co., Ltd., Honolulu, 342 p.]

Emerson, J.S., 1887, Kilauea after the eruption of March 1886: American Journal of Science, v. 33, p. 87–95 [includes maps].

Finch, R.H., 1924, Seismic sequences of the explosive eruption of Kilauea in May, 1924: Bulletin of the Seismological Society of America, v. 14, p. 217–222.

Finch, R.H., 1925, The earthquakes at Kapoho, island of Hawaii, April 1924: Bulletin of the Seismological Society of America, v. 15, p. 122–127.

Finch, R.H., 1926a, Bulk displacement due to tilting: The Volcano Letter, no. 74, p. 1. [Reprinted in Fiske, R.S., Simkin, T., and Nielsen, E.A., compilers, 1987, The Volcano Letter: Smithsonian Institution Press, 539 p.]

Finch, R.H., 1926b, Twelve years record of earthquakes: The Volcano Letter, no. 54, p. 1. [Reprinted in Fiske, R.S., Simkin, T., and Nielsen, E.A., compilers, 1987, The Volcano Letter: Smithsonian Institution Press, 539 p.]

Finch, R.H., 1944, The November–December 1944 crisis at Kilauea: The Volcano Letter, no. 486, p. 1–3. [Reprinted in Fiske, R.S., Simkin, T., and Nielsen, E.A., compilers, 1987, The Volcano Letter: Smithsonian Institution Press, 539 p.]

Finch, R.H., 1947, The mechanics of the explosive eruption of Kilauea in 1924: Pacific Science, v. 1, p. 237–240.

Finch, R.H., 1950, The December 1950 subsidence at Kilauea: The Volcano Letter, no. 510, p. 1–3. [Reprinted in Fiske, R.S., Simkin, T., and Nielsen, E.A., compilers, 1987, The Volcano Letter: Smithsonian Institution Press, 539 p.]

Finch, R.H., and Macdonald, G.A., 1953, Hawaiian volcanoes during 1950: U.S. Geological Survey Bulletin 996-B, p. 27–89.

Fiske, R.S., and Kinoshita, W.T., 1969, Inflation of Kilauea volcano prior to its 1967–1968 eruption: Science, v. 165, p. 341–349.

Fiske, R.S., and Koyanagi, R.Y., 1968, The December 1965 eruption of Kilauea Volcano, Hawaii: U.S. Geological Survey Professional Paper 607, 21 p.

Fiske, R.S., Simkin, T., and Nielsen, E.A., eds., 1987, The Volcano Letter: Washington, D.C., Smithsonian Institution Press, 539 p. [Compiled and reprinted; originally published, 1925–1955, by the Hawaiian Volcano Observatory]

Fornander, A., 1868, The volcano!: Hawaiian Gazette, April 29, p. 4.

Furumoto, A.S., and Kovach, R.L., 1979, The Kalapana earthquake of November 29, 1975; an intraplate earthquake and its relation to geothermal processes: Physics of the Earth and Planetary Interiors, no. 18, p. 197–208.

Gerlach, T.M., and Graeber, E.J., 1985, Volatile budget of Kilauea Volcano: Nature, v. 313, no. 6000, p. 273–277.

Gerlach, T.M., McGee, K.A., Elias, T., Sutton, A.J., and Doukas, M.P., 2002, Carbon dioxide emission rate of Kīlauea Volcano; implications for primary magma and the summit reservoir: Journal of Geophysical Research, v. 107, no. B9, cit. no. 2189, p. ECV 3-1 to 3-15, doi:10.1029/2001JB000407.

Gillard, D., Rubin, A.M., and Okubo, P., 1996, Highly concentrated seismicity caused by deformation of Kilauea's deep magma system: Nature, v. 384, November 28, p. 343–346.

Gonnermann, H.M., Poland, M., Foster, J.H., Brooks, B., Wolfe, C.J., and Miklius, A., 2011, Mantle-to-surface magma dynamics at Mauna Loa and Kīlauea, Hawai'i [abs.]: Mineralogical Magazine, v. 75, no. 3, p. 932. [Goldschmidt 2011, Prague, Czech Republic, August 14–19, 2011, Abstracts]

Goodrich, J., 1833, Notices of some of the volcanos and volcanic phenomena of Hawaii, (Owyhee) and other islands in that group: American Journal of Science, v. 25, p. 199–203.

Got, J.-L., Monteiller, V., Monteux, J., Hassani, R., and Okubo, P., 2008, Deformation and rupture of the oceanic crust may control growth of Hawaiian volcanoes: Nature, v. 451, no. 7177, 01/24, p. 453–456.

Greenland, P., Rose, W.I., and Stokes, J.B., 1985, An estimate of gas emissions and magmatic gas content from Kilauea Volcano: Geochimica et Cosmochimica Acta, v. 49, p. 125–129.

Guest, J.E., Spudis, P.D., Greeley, R., Taylor, G.J., and Baloga, S.M., 1995, Emplacement of xenolith nodules in the Kaupulehu lava flow, Hualalai volcano, Hawaii: Bulletin of Volcanology, v. 57, p. 179–184.

Hall Wallace, M., and Delaney, P.T., 1995, Deformation of Kilauea Volcano during 1982 and 1983; a transition period: Journal of Geophysical Research, v. 100, no. B5, p. 8201–8219.

Harvey, D., and Wyss, M., 1986, Comparisons of a complex rupture model with the precursor asperities of the 1975 Hawaii $M_s = 7.2$ earthquake: Pure and Applied Geophysics, v. 124, no. 4/5, p. 957–973.

Heliker, C.C., and Mattox, T.N., 2003, The first two decades of the Pu‘u ‘Ō‘ō-Kūpaianaha eruption; chronology and selected bibliography, *in* Heliker, C.C., Swanson, D.A., and Takahashi, T.J., eds., The Pu‘u ‘Ō‘ō-Kūpaianaha eruption of Kīlauea Volcano, Hawai’i—the first 20 years: U.S. Geological Survey Professional Paper 1676, p. 1–27.

Heliker, C.C., and Wright, T.L., 1991, The Pu‘u ‘O‘o-Kupaianaha eruption of Kilauea: Eos (American Geophysical Union Transactions), v. 72, no. 47, p. 521, 530, 531.

Heliker, C.C., Mangan, M.T., Mattox, T.N., Kauahikaua, J.P., and Helz, R.T., 1998, The character of long-term eruptions; inferences from episodes 50–53 of the Pu‘u ‘Ō‘ō-Kūpaianaha eruption of Kīlauea Volcano: Bulletin of Volcanology, v. 59, no. 6, April, p. 381–393.

Heliker, C.C., Swanson, D.A., and Takahashi, T.J., eds., 2003, The Pu‘u ‘Ō‘ō-Kūpaianaha eruption of Kīlauea Volcano, Hawai’i—the first 20 years, U.S. Geological Survey Professional Paper 1676, 206 p.

Helz, R.T., 1987, Diverse olivine types in lava of the 1959 eruption of Kilauea Volcano and their bearing on eruption dynamics, chap. 25 *of* Decker, R.W., Wright, T.L., and Stauffer, P.H., eds., Volcanism in Hawaii: U.S. Geological Survey Professional Paper 1350, v. 1, p. 691–722.

Helz, R.T., 1993, Drilling report and core logs for the 1988 drilling of Kilauea Iki lava lake, Kilauea Volcano, Hawaii, with summary descriptions of the occurrence of foundered crust and fractures in the drill core: U.S. Geological Survey Open-File Report 93–15, 57 p.

Helz, R.T., and Wright, T.L., 1992, Differentiation and magma mixing on Kilauea’s east rift zone; a further look at the eruptions of 1955 and 1960, Part I. The late 1955 lavas: Bulletin of Volcanology, v. 54, p. 361–384.

Hillebrand, W., 1868, The eruption!: Hawaiian Gazette, May 6, p. 2. [excerpted in American Journal of Science, 2nd ser., v. 46, p. 115–121]

Hitchcock, C.H., 1909, Hawaii and its volcanoes: Honolulu, Hawaiian Gazette Co., 314 p. [2nd ed., with supp., 1911, 314 p.]

Holcomb, R.T., 1987, Eruptive history and long-term behavior of Kilauea Volcano, chap. 12 *of* Decker, R.W., Wright , T.L., and Stauffer, P.H., eds., Volcanism in Hawaii: U.S. Geological Survey Professional Paper 1350, v. 1, p. 261–350.

Hon, K., Kauahikaua, J., Denlinger, R., and Mackay, K., 1994, Emplacement and inflation of pahoehoe sheet flows; observations and measurements of active lava flows on Kilauea Volcano, Hawaii: Geological Society of America Bulletin, v. 106, p. 351–370.

Jackson, D.B., Swanson, D.A., Koyanagi, R.Y., and Wright, T.L., 1975, The August and October 1968 east rift eruptions of Kilauea Volcano, Hawaii: U.S. Geological Survey Professional Paper 890, 33 p.

Jackson, M.D., Endo, E.T., Delaney, P.T., Arnodottir, T., and Rubin, A.M., 1992, Ground ruptures of the 1974 and 1983 Kaoiki earthquakes, Mauna Loa Volcano, Hawaii: Journal of Geophysical Research, v. 97, no. B6, p. 8775–8796.

Jaggar, T.A., Jr., 1912, [Kilauea in 1909–1912], *in* [First Special] Report of the Hawaiian Volcano Observatory: Boston, Society of Arts of the Massachusetts Institute of Technology, p. 1–74. [Compiled and reprinted in Bevens, D., Takahashi, T.J., and Wright, T.L., eds., 1988, The early serial publications of the Hawaiian Volcano Observatory: Hawai‘i National Park, Hawai‘i Natural History Association, v. 1, p. 1–80]

Jaggar, T.A., Jr., 1925, Plus and minus volcanicity: Journal of the Washington Academy of Sciences, v. 15, p. 416–417. [Reprinted in Bulletin Volcanologique, ser. 2, p. 327–328]

Jaggar, T.A., Jr., 1931, Events preceding the great eruption 1924: The Volcano Letter, no. 329, p. 1–4. [Reprinted in Fiske, R.S., Simkin, T., and Nielsen, E.A., compilers, 1987, The Volcano Letter: Smithsonian Institution Press, 539 p.]

Jaggar, T.A., 1947, Origin and Development of Craters: Geological Society of America, Memoir 21, 508 p.

Jaggar, T.A., Jr., and Finch, R.H., 1924, The explosive eruption of Kilauea in Hawaii, 1924: American Journal of Science, v. 8, p. 353–374.

Jaggar, T.A., Jr., and Finch, R.H., 1928, Tilting and level changes at Pacific volcanoes, *in* The National Research Council of Japan, ed., Proceedings, Pan-Pacific Science Congress, 3rd, Tokyo, Japan, October 30-November 11, 1926: Pan-Pacific Science Congress, 3rd, Proceedings, p. 672–686.

Jarves, J.J., 1840, Gleanings from the editor's notebook—Hawaii. No. 6: The Polynesian, v. 1, no. 12, Aug. 29, p. 45–46. [Included in the original is an engraving of the three sand hills (littoral cones) thrown up at Nanawale.]

Jarves, J.J., 1844, Scenes and scenery in the Sandwich Islands: Boston, James Munroe & Co., 341 p. [p. 224–258 cover the material abstracted.]

Johnson, D.J., 1992, Dynamics of magma storage in the summit reservoir of Kilauea Volcano, Hawaii: Journal of Geophysical Research, v. 97, no. B2, p. 1807–1820.

Johnson, D.J., Sigmundsson, F., and Delaney, P.T., 2000, Comment on "Volume of magma accumulation or withdrawal estimated from surface uplift or subsidence, with application to the 1960 collapse of Kilauea Volcano" by P.T. Delaney and D.F. McTigue: Bulletin of Volcanology, v. 61, p. 491–493.

Kauahikaua, J.P., 1993, Geophysical characteristics of the hydrothermal systems of Kilauea Volcano, Hawaii: Geothermics, v. 22, no. 4, p. 271–299.

Kauahikaua, J.P., and Miklius, A., 2003, Long-term trends in microgravity at Kīlauea's summit during the Pu‘u ‘Ō‘ō-Kūpaianaha eruption, *in* Heliker, C.C., Swanson, D.A., and Takahashi, T.J., eds., The Pu‘u ‘Ō‘ō-Kūpaianaha eruption of Kīlauea Volcano, Hawai‘i—the first 20 years: U.S. Geological Survey Professional Paper 1676, p. 165–171.

Kauahikaua, J., Mangan, M., Heliker, C., and Mattox, T., 1996, A quantitative look at the demise of a basaltic vent; the death of Kupaianaha, Kilauea Volcano, Hawai‘i: Bulletin of Volcanology, v. 57, p. 641–648.

Kauahikaua, J., Hildenbrand, T., and Webring, M., 2000, Deep magmatic structures of Hawaiian volcanoes, imaged by three-dimensional gravity models: Geology, v. 28, no. 10, p. 883–886.

Kinoshita, W.T., 1967, May 1963 earthquakes and deformation in the Koae fault zone, Kilauea Volcano, Hawaii: U.S. Geological Survey Professional Paper 575-C, p. C173–C176.

Kinoshita, W.T., Okamura, A.T., and Koyanagi, R.Y., 1965 (?), Hawaiian Volcano Observatory summary 35—July, August, and September 1964: U.S. Geological Survey Administrative Report, 31 p.

Kinoshita, W.T., Koyanagi, R.Y., Wright, T.L., and Fiske, R.S., 1969, Kilauea Volcano; the 1967–1968 summit eruption: Science, v. 166, p. 459–468.

Kinoshita, W.T., Swanson, D.A., and Jackson, D.B., 1974, The measurement of crustal deformation related to volcanic activity at Kilauea Volcano, Hawaii, in Civetta, L., Gasparini, P., Luongo, G., and Rapolla, A., eds., Physical volcanology: Amsterdam, Elsevier Scientific Publishing Co., Developments in solid Earth geophysics 6, chap. 4, p. 87–115.

Klein, F.W., 1982, Patterns of historical eruptions at Hawaiian volcanoes: Journal of Volcanology and Geothermal Research, v. 12, p. 1–35.

Klein, F.W., 2007, The 15 October 2006 *M*6.7 Kiholo earthquake under the Island of Hawai‘i as caused by Pacific Plate flexure [abs.]: Seismological Research Letters, v. 78, no. 2, March/April, p. 297. [Seismological Society of America 2007 annual meeting, Kona, Hawaii, April 11–13, 2007]

Klein, F.W., and Koyanagi, R.Y., 1980, Hawaiian Volcano Observatory seismic network history 1950–1979: U.S. Geological Survey Open-File Report 80–302, 84 p.

Klein, F.W., and Koyanagi, R.Y., 1989, The seismicity and tectonics of Hawaii, *in* Winterer, E.L., Hussong, D.M., and Decker, R.W., eds., The eastern Pacific Ocean and Hawaii: Boulder, Colo., Geological Society of America, The Geology of North America, v. N, chap. 12, The Hawaiian-Emperor Chain, p. 238–252; pls. 1A, 1B, 2A, 2B, 3A, and 5 (folded maps) in separate box include Hawaii.

Klein, F.W., and Wright, T.L., 2000, Catalog of Hawaiian earthquakes, 1823–1959: U.S. Geological Survey Professional Paper 1623, 90 p. [A CD-ROM containing (1) all catalog and reading files and (2) all bibliographic sources is included with the paper.]

Klein, F.W., and Wright, T.L., 2008, Exponential decline of aftershocks of the *M*7.9 1868 great Kau earthquake, Hawaii, through the twentieth century: Journal of Geophysical Research, v. 113, B9, B09310.

Klein, F.W., Koyanagi, R.Y., Nakata, J.S., and Tanigawa, W.R., 1987, The seismicity of Kilauea's magma system, chap. 43 *of* Decker, R.W., Wright, T.L., and Stauffer, P.H., eds., Volcanism in Hawaii: U.S. Geological Survey Professional Paper 1350, v. 2, p. 1019–1185.

Klein, F.W., Frankel, A.D., Mueller, C.S., Wesson, R.L., and Okubo, P.G., 2001, Seismic hazard in Hawaii; high rate of large earthquakes and probabilistic ground-motion maps: Bulletin of the Seismological Society of America, v. 91, no. 3, p. 479–498.

Klein, F.W., Wright, T., and Nakata, J., 2006, Aftershock decay, productivity, and stress rates in Hawaii; indicators of temperature and stress from magma sources: Journal of Geophysical Research, v. 111, B7, B07307. [pdf in hand]

Koyanagi, R.Y., Krivoy, H.L., and Okamura, A.T., 1963, Hawaiian Volcano Observatory summary 25—January, February, and March 1962: U.S. Geological Survey Administrative Report, 35 p.

Koyanagi, R.Y., Okamura, A.T., Krivoy, H.L., and Yamamoto, A., 1964a, Hawaiian Volcano Observatory summary 29—January, February, and March 1963: U.S. Geological Survey Administrative Report, 36 p.

Koyanagi, R.Y., Okamura, A.T., Kinoshita, W.T., Moore, J.G., and Powers, H.A., 1964b, Hawaiian Volcano Observatory summary 31—July, August, and September 1963: U.S. Geological Survey Administrative Report, 40 p.

Koyanagi, R.Y., Okamura, A.T., and Powers, H.A., 1965a, Hawaiian Volcano Observatory summary 33—January, February, and March 1964: U.S. Geological Survey Administrative Report, 44 p.

Koyanagi, R.Y., Okamura, A.T., Kinoshita, W.T., and Powers, H.A., 1965b, Hawaiian Volcano Observatory summary 36—October, November, and December 1964: U.S. Geological Survey Administrative Report, 31 p.

Koyanagi, R.Y., Okamura, A.T., and Kinoshita, W.T., 1966, Hawaiian Volcano Observatory summary 39—July, August, and September 1965: U.S. Geological Survey Administrative Report, 33 p.

Koyanagi, R.Y., Okamura, A.T., and Powers, H.A., 1969, Hawaiian Volcano Observatory summary 40—October, November, and December 1965: U.S. Geological Survey Administrative Report, 30 p..

Koyanagi, R.Y., Unger, J.D., Endo, E.T., and Okamura, A.T., 1974, Shallow earthquakes associated with inflation episodes at the summit of Kilauea Volcano, Hawaii, *in* Gonzales-Ferran, O. ed., Proceedings, Symposium on Andean and Antarctic Volcanology Problems, Santiago, Chile, September 9–14, 1974: International Association of Volcanology and Chemistry of the Earth's Interior, Special Series, p. 621–631.

Krivoy, H.L., Koyanagi, R.Y., and Okamura, A.T., 1963, Hawaiian Volcano Observatory summary 24—October, November, and December 1961: U.S. Geological Survey Administrative Report, 32 p.

Krivoy, H.L., Koyanagi, R.Y., Okamura, A.T., and Kojima, G., 1964, Hawaiian Volcano Observatory summary 28—October, November, and December 1962: U.S. Geological Survey Administrative Report, 51 p.

Krivoy, H.L., Kinoshita, W.T., Okamura, A.T., and Koyanagi, R.Y., 1965, Hawaiian Volcano Observatory summary 30—April, May, and June 1963: U.S. Geological Survey Administrative Report, 32 p.

Lipman, P.W., Lockwood, J.P., Okamura, R.T., Swanson, D.A., and Yamashita, K.M., 1985, Ground deformation associated with the 1975 magnitude-7.2 earthquake and resulting changes in activity of Kilauea Volcano, Hawaii: U.S. Geological Survey Professional Paper 1276, 45 p.

Lipman, P.W., Sisson, T.W., Ui, T., Naka, J., and Smith, J.R., 2001, Ancestral submarine growth of Kīlauea volcano and instability of its south flank, *in* Hawaiian volcanoes; deep underwater perspectives: Washington, D.C., American Geophysical Union, Geophysical Monograph 128, p. 161–191.

Lockwood, J.P., Tilling, R.I., Holcomb, R.T., Klein, F., Okamura, A.T., and Peterson, D.W., 1999, Magma migration and resupply during the 1974 summit eruptions of Kīlauea Volcano, Hawai'i: U.S. Geological Survey Professional Paper 1613, 37 p.

Lyman, C.S., 1851, On the recent condition of Kilauea: American Journal of Science, v. 12, p. 75–80.

Ma, K.-F., Kanamori, H., and Satake, K., 1999, Mechanism of the 1975 Kalapana, Hawaii earthquake inferred from tsunami data: Journal of Geophysical Research, v. 104, no. B6, p. 13153–13167.

Maby, J.H., 1886, Volcanic changes: Pacific Commercial Advertiser, March 15, p. 2. [Letter written by J.H. Maby, manager of the Volcano House, to S.G. Wilder, dated March 8, 1886; reprinted in Honolulu Daily Bulletin of March 15, 1886]

Maby, J.H., 1891, [On Kilauea in 1891]: Pacific Commercial Advertiser, March 12, p. 3.

Macdonald, G.A., 1951, The Kilauea earthquake of April 22, 1951, and its aftershocks: The Volcano Letter, no. 512, p. 1–3. [See Fiske and others, 1987]

Macdonald, G.A., 1952, The 1952 eruption of Kilauea: The Volcano Letter, no. 518, p. 1–10. [Reprinted in Fiske, R.S., Simkin, T., and Nielsen, E.A., compilers, 1987, The Volcano Letter: Smithsonian Institution Press, 539 p.]

Macdonald, G.A., 1954, Activity of Hawaiian volcanoes during the years 1940–1950: Bulletin Volcanologique, v. 15, p. 119–179.

Macdonald, G.A., 1955, Hawaiian volcanoes during 1952: U.S. Geological Survey Bulletin 1021–B, p. 15–107. [Includes plates.]

Macdonald, G.A., 1959, The activity of Hawaiian volcanoes during the years 1951–1956: Bulletin Volcanologique, ser. 2, v. 22, p. 3–70, 20 plates.

Macdonald, G.A., and Abbott, A.T., 1970, Volcanoes in the sea: Honolulu, University of Hawai‘i Press, 441 p. [2nd ed., Macdonald, Abbott, and Peterson (1983)]

Macdonald, G.A., and Eaton, J.P., 1954, The eruption of Kilauea Volcano in May, 1954: The Volcano Letter, no. 524, p. 1–9. [See Fiske and others, 1987.]

Macdonald, G.A., and Eaton, J.P., 1955, Hawaiian volcanoes during 1953: U.S. Geological Survey Bulletin 1021–D, p. 127–166.

Macdonald, G.A., and Eaton, J.P., 1956, Hawaiian Volcano Observatory summary 2—April–June 1956: U.S. Geological Survey Administrative Report, 5 p.

Macdonald, G.A., and Eaton, J.P., 1957, Hawaiian volcanoes during 1954: U.S. Geological Survey Bulletin 1061–B, p. 17–72 (1 plate).

Macdonald, G.A., and Eaton, J.P., 1964, Hawaiian volcanoes during 1955: U.S. Geological Survey Bulletin 1171, p. 1–170; plates 1–5 in pocket.

Macdonald, G.A., and Wentworth, C.K., 1954, Hawaiian volcanoes during 1951: U.S. Geological Survey Bulletin 996–D, p. 141–216.

Maley, R.P., 1986, Strong-motion results from the November 16 Hawaii earthquake: U.S. Geological Survey Circular 971, p. 13.

Mangan, M.T., Heliker, C.C., Mattox, T.N., Kauahikaua, J.P., and Helz, R.T., 1995, Episode 49 of the Pu‘u ‘O‘o-Kupaianaha eruption of Kilauea Volcano—breakdown of a steady-state eruptive era: Bulletin of Volcanology, v. 57, p. 127–135.

Marsh, B.D., 1981, On the crystallinity, probability of occurrence, and rheology of lava and magma: Contributions to Mineralogy and Petrology, v. 78, p. 85–98.

Marske, J.P., Garcia, M.O., Pietruszka, A.J., Rhodes, J.M., and Norman, M.D., 2008, Geochemical variations during Kīlauea's Pu‘u ‘Ō‘ō eruption reveal a fine-scale mixture of mantle heterogeneities within the Hawaiian plume: Journal of Petrology, v. 49, no. 7, p. 1297–1318.

Mastin, L.G., 1997, Evidence for water influx from a caldera lake during the explosive hydromagmatic eruption of 1790, Kilauea Volcano, Hawaii: Journal of Geophysical Research, v. 102, no. B9, p. 20093–20109.

Mitchell, H.C., 1930, Triangulation in Hawaii: U.S. Coast and Geodetic Survey Special Publication no. 156, 219 p. [Includes maps. Updated in 1969 (U.S. Department of Commerce, 1969)]

Mogi, K., 1958, Relations between the eruptions of various volcanoes and the deformations of the ground surfaces around them: Bulletin of the Earthquake Research Institute, v. 36, p. 111–123.

Montgomery-Brown, E.K., Segall, P., and Miklius, A., 2009, Kilauea slow-slip events—identification, source inversions, and relation to seismicity: Journal of Geophysical Research, v. 114, B00A03.

Montgomery-Brown, E.K., Sinnett, D.K., Poland, M., Segall, P., Orr, T., Zebker, H., and Miklius, A., 2010, Geodetic evidence for en echelon dike emplacement and concurrent slow slip during the June 2007 intrusion and eruption at Kilauea volcano, Hawaii: Journal of Geophysical Research, v. 115, B07405.

Moore, J.G., and Koyanagi, R.Y., 1969, The October 1963 eruption of Kilauea Volcano, Hawaii: U.S. Geological Survey Professional Paper 614–C, p. C1–C13.

Moore, J.G., and Krivoy, H.L., 1964, The 1962 flank eruption of Kilauea Volcano and structure of the east rift zone: Journal of Geophysical Research, v. 69, no. 10, p. 2033–2045.

Moore, J.G., and Mark, R.K., 1992, Morphology of the Island of Hawaii: GSA Today, v. 2, no. 12, p. 257–259, 262.

Moore, R.B., Helz, R.T., Dzurisin, D., Eaton, G.P., Koyanagi, R.Y., Lipman, P.W., Lockwood, J.P., and Puniwai, G.S., 1980, The 1977 eruption of Kilauea Volcano, Hawaii, *in* McBirney, A.R., ed., Gordon A. Macdonald Memorial Volume (special issue): Journal of Volcanology and Geothermal Research, v. 7, no. 3/4, p. 189–210.

Morgan, J.K., and McGovern, P.J., 2005, Discrete element simulations of gravitational volcanic deformation, 2. Mechanical analysis: Journal of Geophysical Research, v. 110, no. B5, B05403, doi:10.1029/2004JB003253.

Morgan, J.K., Moore, G.F., Hills, D.J., and Leslie, S., 2000, Overthrusting and sediment accretion along Kilauea's mobile south flank, Hawaii; evidence for volcanic spreading from marine seismic reflection data: Geology, v. 28, no. 7, p. 667–670.

Morgan, J.K., Moore, G.F., and Clague, D.A., 2003, Slope failure and volcanic spreading along the submarine south flank of Kilauea volcano, Hawaii: Journal of Geophysical Research, v. 108, no. B9, 2415, doi:10.1029/2003JB002411, p. EPM 1-1 – 1-23.

Murata, K.J., and Richter, D.H., 1966, Chemistry of the lavas of the 1959–60 eruption of Kilauea volcano, Hawaii: U.S. Geological Professional Paper 537-A, p. A1–A26.

Nakata, J., comp., 2007a, Hawaiian Volcano Observatory 1959 Quarterly Administrative Reports: U.S. Geological Survey Open-File Report 2007–1319, 209 p. [Available at http://pubs.usgs.gov/of/2007/1319.]

Nakata, J., comp., 2007b, Hawaiian Volcano Observatory 1960 Quarterly Administrative Reports: U.S. Geological Survey Open-File Report 2007–1320, 112 p. [2nd to 4th quarters of 1960 never prepared]

Nielsen, N.N., Furumoto, A.S., Lum, W., and Morrill, B.J., 1977, Honomu, Hawaii earthquake of April 26, 1973, *in* Proceedings, World Conference on Earthquake Engineering, 6th, New Delhi, Jan. 10–14, 1977: World Conference on Earthquake Engineering, 6th, Proceedings, v. 1, no. 6, p. 227–232.

Okamura, A.T., Koyanagi, R.Y., and Krivoy, H.L., 1963, Hawaiian Volcano Observatory summary 26—April, May, and June 1962: U.S. Geological Survey Administrative Report, 44 p.

Okamura, A.T., Kojima, G., and Yamamoto, A., 1964, Hawaiian Volcano Observatory summary 27—July, August, and September 1962: U.S. Geological Survey Administrative Report, 49 p.

Okamura, A.T., Koyanagi, R.Y., Kinoshita, W.T., Moore, J.G., Peck, D.L., and Powers, H.A., [1964?], Hawaiian Volcano Observatory summary 32—October, November, and December 1963: U.S. Geological Survey Administrative Report, 35 p.

Okamura, A.T., Koyanagi, R.Y., and Powers, H.A., 1966, Hawaiian Volcano Observatory summary 37—January, February, and March 1965: U.S. Geological Survey Administrative Report, 42 p.

Okubo, P., and Nakata, J.S., 2003, Tectonic pulses during Kīlauea's current long-term eruption, *in* Heliker, C.C., Swanson, D.A., and Takahashi, T.J., eds., The Pu'u 'Ō'ō-Kūpaianaha eruption of Kīlauea Volcano, Hawai'i—the first 20 years: U.S. Geological Survey Professional Paper 1676, p. 173–186.

Orr, T.R., Patrick, M.R., Wooten, K.M., Swanson, D.A., Elias, T., Sutton, J., Wilson, D.C., and Poland, M.P., 2008, Explosions, tephra, and lava; a chronology of the 2008 summit eruption of Kilauea Volcano, Hawai'i [abs.]: Eos (American Geophysical Union Transactions), Fall meeting supplement, abs. V11B-2018.

Owen, S.E., and Bürgmann, R., 2006, An increment of volcano collapse; kinematics of the 1975 Kalapana, Hawaii, earthquake, *in* Poland, M.A., and Newman, A., eds., The changing shapes of active volcanoes—recent results and advances in volcano geodesy: Journal of Volcanology and Geothermal Research, 150, 1–3, 163–185.

Owen, S., Segall, P., Freymuller, J., Miklius, A., Denlinger, R., Arnadottir, T., Sako, M., and Bürgmann, R., 1995, Rapid deformation of the south flank of Kilauea volcano, Hawaii: Science, v. 267, p. 1328–1332.

Owen, S., Segall, P., Lisowski, M., Miklius, A., Denlinger, R., and Sako, M., 2000a, Rapid deformation of Kilauea Volcano; global positioning system measurements between 1990 and 1996: Journal of Geophysical Research, v. 105, no. B8, p. 18983–18998.

Owen, S., Segall, P., Lisowski, M., Miklius, A., Murray, M., Bevis, M., and Foster, J., 2000b, January 30, 1997 eruptive event on Kilauea Volcano, Hawaii, as monitored by continuous GPS: Geophysical Research Letters, v. 27, no. 17, p. 2757–2760.

Parfitt, E.A., and Wilson, L., 1994, The 1983–86 Pu'u 'O'o eruption of Kilauea Volcano, Hawaii: a study of dike geometry and eruption mechanisms for a long-lived eruption: Journal of Volcanology and Geothermal Research, v. 59, p. 179–205.

Peck, D.L., and Kinoshita, W.T., 1976, The eruption of August 1963 and the solidification of Alae lava lake, Hawaii: U.S. Geological Survey Professional Paper 935-A, p. A1–A33.

Phillips, K.A., Chadwell, C.D., and Hildebrand, J.A., 2008, Vertical deformation measurements on the submerged south flank of Kilauea volcano, Hawai'i reveal seafloor motion associated with volcanic collapse: Journal of Geophysical Research, v. 113, B5, B05106.

Pietruszka, A.J., and Garcia, M.O., 1999, A rapid fluctuation in the mantle source and melting history of Kilauea Volcano inferred from the geochemistry of its historical summit lavas (1790–1982): Journal of Petrology, v. 40, no. 8, p. 1321–1342.

land, M.P., and Sutton, A.J., 2008, Kilauea summit activity during 2007–2008; a failed eruption and an eruption that should have failed [abs.]: Eos (American Geophysical Union Transactions), Fall meeting supplement, abs. V51D-2073.

Poland, M., Miklius, A., Sutton, A.J., and Orr, T., 2007, The consequences of increased magma supply to Kilauea Volcano, Hawaii [abs.]: Eos (American Geophysical Union Transactions) v. 88, no. 52, Fall meeting supplement, abs. V52B-01

Poland, M., Miklius, A., Orr, T., Sutton, J., Thornber, C., and Wilson, D., 2008, New episodes of volcanism at Kilauea Volcano, Hawaii: Eos (American Geophysical Union Transactions), v. 89, no. 5, January 29, p. 37–38.

Poland, M.P., Sutton, A.J., and Gerlach, T.M., 2009, Magma degassing triggered by static decompression at Kīlauea Volcano, Hawai'i: Geophysical Research Letters, v. 36, no. L16306, 5 p.

Poland, M.P., Miklius, A., Sutton, A.J., and Thornber, C.R., 2012, A mantle-driven surge in magma supply to Kīlauea Volcano during 2003-2007: Nature Geoscience, v. 5, no. 4, Apr., doi:10.1038/ngeo1426, p. 295-300.

Powers, H.A., 1946, Reanalyzing tilt records at the Hawaiian Volcano Observatory: The Volcano Letter, no. 492, p. 1–6. [Reprinted in Fiske, R.S., Simkin, T., and Nielsen, E.A., compilers, 1987, The Volcano Letter: Smithsonian Institution Press, 539 p.]

Powers, H.A., 1947, Annual tilt pattern at the Hawaiian Volcano Observatory: The Volcano Letter, no. 495, p. 1–5. [Reprinted in Fiske, R.S., Simkin, T., and Nielsen, E.A., compilers, 1987, The Volcano Letter: Smithsonian Institution Press, 539 p.]

Powers, H.A., Okamura, A.T., Koyanagi, R.Y., and Kinoshita, W.T., 1966, Hawaiian Volcano Observatory summary 38—April, May, and June 1965: U.S. Geological Survey Administrative Report, 30 p.

Richter, D.H., and Moore, J.G., 1966, Petrology of the Kilauea Iki lava lake, Hawaii: U.S. Geological Professional Paper 537-B, p. B1–B26.

Richter, D.H., Ault, W.U., Eaton, J.P., and Moore, J.G., 1964, The 1961 eruption of Kilauea Volcano, Hawaii: U.S. Geological Survey Professional Paper 474-D, p. D1–D34.

Richter, D.H., Eaton, J.P., Murata, K.J., Ault, W.U., and Krivoy, H.L., 1970, Chronological narrative of the 1959–60 eruption of Kilauea volcano, Hawaii: U.S. Geological Survey Professional Paper 537-E, p. E1–E73.

Rivalta, E., and Segall, P., 2008, Magma compressibility and the missing source for some dike intrusions: Geophysical Research Letters, v. 35, no. 4, paper no. L04306, February 28, 5 p.

Rowland, S.K., and Munro, D.C., 1993, The 1919–1920 eruption of Mauna Iki, Kilauea; chronology, geologic mapping, and magma transport mechanisms: Bulletin of Volcanology, v. 55, p. 190–203.

Rubin, A.M., and Gillard, D., 1998, Dike-induced earthquakes; theoretical considerations: Journal of Geophysical Reseach, v. 103, no. B5, p. 10017–10030.

Rubin, A.M., Gillard, D., and Got, J.-L., 1998, A reinterpretation of seismicity associated with the January 1983 dike intrusion at Kilauea Volcano, Hawaii: Journal of Geophysical Research, v. 103, no. B5, p. 10003–10015.

Ryan, M.P., 1987, Neutral buoyancy and the mechanical evolution of magmatic systems, *in* Mysen, B.O., ed., Magmatic processes; physiochemical principles: University Park, Pa., The Geochemical Society, Special Publication No. 1, p. 259–287.

Ryan, M.P., 1988, The mechanics and three-dimensional internal structure of active magmatic systems; Kilauea Volcano, Hawaii: Journal of Geophysical Research, v. 93, no. B5, p. 4213–4248.

Ryan, M.P., Blevins, J.Y.K., Okamura, A.T., and Koyanagi, R.Y., 1983, Magma reservoir subsidence mechanics; theoretical summary and application to Kilauea Volcano, Hawaii: Journal of Geophysical Research, v. 88, no. B5, p. 4147–4181.

Segall, P., Desmarais, E.K., Shelly, D., Miklius, A., and Cervelli, P., 2006, Earthquakes triggered by silent slip events on Kīlauea volcano, Hawaii: Nature, 442, 7098, 71–74. [correction in Nature, 2006, v. 444, p. 235.]

Smith, J.R., Malahoff, A., and Shor, A.N., 1999, Submarine geology of the Hilina slump and morpho-structural evolution of Kilauea Volcano, Hawaii, *in* Elsworth, D., Carracedo, J.C., and Day, S.J., eds., Deformation and flank instability of oceanic island volcanoes; a comparison of Hawai'i with Atlantic island volcanoes: Journal of Volcanology and Geothermal Research, v. 94, no. 1–4, p. 59–88.

Soule, S.A., Cashman, K.V., and Kauahikaua, J.P., 2004, Examining flow emplacement through the surface morphology of three rapidly emplaced, solidified lava flows, Kilauea Volcano, Hawai'i: Bulletin of Volcanology, v. 66, no. 1, January, p. 1–14.

Stearns, H.T., 1925, The explosive phase of Kilauea Volcano, Hawaii, in 1924: Bulletin Volcanologique, ser. 2, v. 5–6, p. 193–208.

Stearns, H.T., 1926, The Keaiwa or 1823 lava flow from Kilauea Volcano, Hawaii: Journal of Geology, v. 34, p. 336–351.

Sutton, A.J., Elias, T., and Kauahikaua, J.P., 2003, Lava-effusion rates for the Pu‘u ‘Ō‘ō-Kūpaianaha eruption derived from SO_2 emissions and very low frequency (VLF) measurements, *in* Heliker, C.C., Swanson, D.A., and Takahashi, T.J., eds., The Pu‘u ‘Ō‘ō-Kūpaianaha eruption of Kīlauea Volcano, Hawai‘i—the first 20 years: U.S. Geological Survey Professional Paper 1676, p. 137–148.

Sutton, A.J., Elias, T., Gerlach, T.M., Lee, R.C., Miklius, A., Poland, M.P., Werner, C.A., and Wilson, D.C., 2009, Volatile budget of Kilauea Volcano receives stimulus from the magma bank [abs.]: Eos (American Geophysical Union Transactions), Fall meeting supplement.

Swanson, D.A., 2003, Kilauea's caldera and the Keanakakoi ash [abs.], *in* Cities on Volcanoes 3 meeting, Hilo, Hawai'i, July 14–18, 2003, Abstracts volume, p. 125.

Swanson, D., 2009, What caused Kilauea's caldera to form? [abs.]: Eos (American Geophysical Union Transactions, v. 90, no. 52, Fall meeting supplement, abs. V43F-2324.

Swanson, D.A., and Christiansen, R.L., 1973, Tragic base surge in 1790 at Kilauea Volcano: Geology, v. 1, p. 83–86.

Swanson, D.A., and Rausch, J., 2008, Human footprints in relation to the 1790 eruption of Kilauea [abs.]: Eos (American Geophysical Union Transactions), Fall meeting supplement, abs. V11B-2022.

Swanson, D.A., Duffield, W.A., and Fiske, R.S., 1976a, Displacement of the south flank of Kīlauea Volcano; the result of forceful intrusion of magma into the rift zones: U.S. Geological Survey Professional Paper 963, 39 p.

Swanson, D.A., Jackson, D.B., Koyanagi, R.Y., and Wright, T.L., 1976b, The February 1969 east rift eruption of Kilauea Volcano, Hawaii: U.S. Geological Survey Professional Paper 891, p. 1–30.

Swanson, D.A., Duffield, W.A., Jackson, D.B., and Peterson, D.W., 1979, Chronological narrative of the 1969–71 Mauna Ulu eruption of Kilauea Volcano, Hawaii: U.S. Geological Survey Professional Paper 1056, 55 p.

Swanson, D., Kenedi, K., Hanley, D., McGeehin, J., Schuster, L., Wulzen, W., Fiske, D., and Rose, T., 1999, Explosions from Kilauea in the 15th to 18th centuries: evidence from field observations and Hawaiian history [abs.], in Big Island Science Conference, 15th, Hilo, HI, Apr. 15-17, 1999, Proceedings: Hilo, HI, University of Hawaii at Hilo, v. 15, p. 1. [Sponsored by Sigma XI, UH Hilo Chapter and University of Hawaii at Hilo]

Swanson, D.A., Rose, T.R., Fiske, R.S., and McGeehin, J.P., 2012, Keanakāko‘i Tephra produced by 300 years of explosive eruptions following collapse of Kīlauea caldera in about 1500 C.E.: Journal of Volcanology and Geothermal Research, v. 215–216, February 15, p. 8–25, doi.org/10.1016/j.jvolgeores.2011.11.009.

Thornber, C.R., 2001, Olivine-liquid relations of lava erupted by Kīlauea Volcano from 1994 to 1998; implications for shallow magmatic processes associated with the ongoing east-rift-zone eruption, *in* Phase equilibria in basaltic systems; a tribute to Peter L. Roeder, part 1: The Canadian Mineralogist, v. 39, pt. 2, April, p. 239–266.

Thornber, C.R., Heliker, C.C., Reynolds, J.R., Kauahikaua, J., Okubo, P., Lisowski, M., Sutton, A.J., and Clague, D., 1996, The eruptive surge of February 1, 1996; a highlight of Kilauea's ongoing East Rift Zone eruption [abs.]: Eos (American Geophysical Union Transactions), v. 77, no. 46, Fall meeting supplement, p. F798.

Thornber, C.R., Heliker, C., Sherrod, D.R., Kauahikaua, J.P., Miklius, A., Okubo, P.G., Trusdell, F.A., Budahn, J.R., Ridley, W.I., and Meeker, G.P., 2003, Kilauea east rift zone magmatism; an episode 54 perspective: Journal of Petrology, v. 44, no. 9, p. 1525–1559.

Thurston, L.A., 1894, The lake has sunk again: Hawaiian Gazette, 07/20, p. 6. [See also p. 4.]

Tikku, A.A., Poland, M., Roeker, S., and Okubo, P., 2006, Continuous gravity measurements from Kilauea volcano, Hawai‘i, 2007–2008 [abs.]: Eos (American Geophysical Union Transactions), v. 89, no. 53, Fall meeting supplement, abs. V11B-2019.

Tilling, R.I., 1976, The 7.2 magnitude earthquake, November 1975, island of Hawaii: Earthquake Information Bulletin, v. 8, no. 6, p. 5–13.

Tilling, R.I., and Dvorak, J.J., 1993, Anatomy of a basaltic volcano: Nature, v. 363, no. 6425, p. 125–133.

Tilling, R.I., Christiansen, R.L., Duffield, W.A., Endo, E.T., Holcomb, R.T., Koyanagi, R.Y., Peterson, D.W., and Unger, J.D., 1987, The 1972–1974 Mauna Ulu eruption, Kilauea Volcano: an example of quasi-steady-state magma transfer, chap. 16 *of* Decker, R.W., Wright, T.L., and Stauffer, P.H., eds., Volcanism in Hawaii: U.S. Geological Survey Professional Paper 1350, v. 1, p. 405–469.

Trusdell, F.A., 1991, The 1840 eruption of Kilauea Volcano; petrologic and volcanologic constraints on rift zone processes: Honolulu, University of Hawai‘i, M.S. thesis, 109 p., map in pocket.

Trusdell, F., 2012, Does activity at Kilauea influence eruptions at Mauna Loa? [abs.]: American Geophysical Union, 2012 Chapman Conference Abstracts, p. 82. [AGU Chapman Conference on Hawaiian Volcanoes: From Source to Surface, Waikoloa, HI, Aug. 20-24, 2012]

Trusdell, F.A., and Moore, R.B., 2006, Geologic map of the middle east rift geothermal subzone, Kīlauea Volcano, Hawai‘i: U.S. Geological Survey Geologic Investigations Series I-2614, scale 1:24,000.

Walter, T.R., and Amelung, F., 2006, Volcano-earthquake interaction at Mauna Loa Volcano, Hawaii: Journal of Geophysical Research, v. 111, 17 p.

Whitney, H.M., 1868, The eruption of 1868!; full details of the earthquakes, mud-flow, shower of ashes and lava stream, as seen by eye-witnesses: Pacific Commercial Advertiser, May 9, p. 1, 4.

Williamson, C.G., 1868a, [On the earthquake swarm and eruption of 1868]: Hawaiian Gazette, April 29, p. 2.

Williamson, C.G., 1868b, [On the earthquake swarm and eruption of 1868]: Hawaiian Gazette, April 29, p. 3.

Wilkes, C., 1845, Narrative of the U.S. Exploring Expedition during the years 1838–1842: Philadelphia, Lee and Blanchard, v. 4, 539 p. [p. 87–231 cover the material abstracted.]

Wilson, D., Elias, T., Orr, T., Patrick, M., Sutton, J., and Swanson, D., 2008, Small explosion from new vent at Kilauea’s summit: Eos (American Geophysical Union Transactions), v. 89, no. 22, May 27, p. 203.

Wilson, R.M., 1935, Ground surface movements at Kilauea Volcano, Hawaii: University of Hawaii Research Publication 10, 56 p.

Wingate, E.G., 1933, Puna triangulation: The Volcano letter, no. 400, p. 1–2. [Reprinted in Fiske, R.S., Simkin, T., and Nielsen, E.A., compilers, 1987, The Volcano Letter: Smithsonian Institution Press, 539 p.]

Wolfe, E.W., ed., 1988, The Puu Oo eruption of Kilauea Volcano, Hawaii—episodes 1 through 20, January 3, 1983, through June 8, 1984: U.S. Geological Survey Professional Paper 1463, 251 p.; 5 pls. (folded maps), scale 1:50,000.

Wolfe, E.W., Garcia, M.O., Jackson, D.B., Koyanagi, R.Y., Neal, C.A., and Okamura, A.T., 1987, The Puu Oo eruption of Kilauea Volcano, episodes 1–20, January 3, 1983, to June 8, 1984, chap. 17 *of* Decker, R.W., Wright, T.L., and Stauffer, P.H., eds., Volcanism in Hawaii: U.S. Geological Survey Professional Paper 1350, v. 1, p. 471–508.

Wolfe, E.W., Neal, C.A., Banks, N.G., and Duggan, T.J., 1988, Geologic observations and chronology of eruptive events, chap. 1 *of* Wolfe, E.W., ed., The Puu Oo Eruption of Kilauea Volcano, Hawaii; Episodes 1 through 20, January 3, 1983, through June 8, 1984: U.S. Geological Survey Professional Paper 1463, p. 1–98; 5 pls. (folded maps), scale 1:50,000).

Wood, H.O., 1912, Earthquake rings a bell and starts machine to working; scientists at Kilauea installing instruments at Observatory—records to be made of every earth quiver: Pacific Commercial Advertiser, July 6, p. 9.

Wood, H.O., 1917, On cyclical variations in eruption at Kilauea, *in* Second Special Report of the Hawaiian Volcano Observatory: Cambridge, Massachusetts Institute of Technology, 59 p. [Compiled and reprinted in Bevens, D., Takahashi, T.J., and Wright, T.L., 1988, The early serial publications of the Hawaiian Volcano Observatory: Hawaii National Park, Hawaii Natural History Association, v. 1, p. 81–143.]

Wright, T.L., 1971, Chemistry of Kilauea and Mauna Loa lava in space and time: U.S. Geological Survey Professional Paper 735, 40 p.

Wright, T.L., 1973, Magma mixing as illustrated by the 1959 eruption, Kilauea Volcano, Hawaii: Geological Society of America Bulletin, v. 84, p. 849–858.

Wright, T.L., and Doherty, P.C., 1970, A linear programming and least squares computer method for solving petrologic mixing problems: Geological Society of America Bulletin, v. 81, p. 1995-2008.

Wright, T.L., and Fiske, R.S., 1971, Origin of the differentiated and hybrid lavas of Kilauea Volcano, Hawaii: Journal of Petrology, v. 12, no. 1, p. 1–65.

Wright, T.L., and Helz, R.T., 1996, Differentiation and magma mixing on Kilauea’s east rift zone; a further look at the eruptions of 1955 and 1960, Part II. The 1960 lavas: Bulletin of Volcanology, v. 57, p. 602–630.

Wright, T.L., and Klein, F.W., 2006, Deep magma transport at Kilauea volcano, Hawaii, *in* Edwards, B.R., and Russell, J.K., eds., Mantle to magma—lithospheric and volcanic processes in Western North America: Lithos, 87, 1–2, p. 50–79.

Wright, T.L., and Klein, F.W., 2008, Dynamics of magma supply to Kīlauea volcano, Hawai'i; integrating seismic, geodetic and eruption data, *in* Annen, C., and Zellmer, G.F., eds., Dynamics of crustal magma transfer, storage and differentiation: Geological Society of London Special Publication 304, p. 83–116.

Wright, T.L., and Klein, F.W., 2010, Magma transport and storage at Kīlauea volcano, HI [abs.]: American Geophysical Union, Fall Meeting Abstracts, abs. V43C-2401. [AGU fall meeting, San Francisco, Calif, December 13–17, 2010, abstracts available online at http://www.agu.org/meetings/fm10/program/]

Wright, T.L., and Takahashi, T.J., 1998, Hawai'i bibliographic database: Bulletin of Volcanology, v. 59, no. 4, p. 276–280.

Wright, T.L., and Tilling, R.I., 1980, Chemical variation in Kilauea eruptions 1971–1974: American Journal of Science, v. 280-A, p. 777–793.

Wright, T.L., Kinoshita, W.T., and Peck, D.L., 1968, March 1965 eruption of Kilauea Volcano and the formation of Makaopuhi lava lake: Journal of Geophysical Research, v. 73, no. 10, p. 3181–3205.

Wright, T.L., Swanson, D.A., and Duffield, W.A., 1975, Chemical compositions of Kilauea east-rift lava, 1968–1971: Journal of Petrology, v. 16, no. 1, p. 110–133.

Wright, T.L., Takahashi, T.J., and Griggs, J.D., 1992, Hawai'i volcano watch; a pictorial history, 1779–1955: Honolulu, University of Hawai'i Press, 162 p.

Wyss, M., and Koyanagi, R.Y., 1992, Isoseismal maps, macroseismic epicenters, and estimated magnitudes of historical earthquakes in the Hawaiian Islands: U.S. Geological Survey Bulletin 2006, 93 p.; addendum, 1 p.

Wyss, M., Klein, F.W., and Johnston, A.C., 1981, Precursors to the Kalapana $M = 7.2$ earthquake: Journal of Geophysical Research, v. 86, no. B5, p. 3881–3900.

Wyss, M., Koyanagi, R.Y., and Cox, D.C., 1992, The Lyman Hawaiian earthquake diary, 1833–1917: U.S. Geological Survey Bulletin 2072, 34 p.

Wyss, M., Klein, F., Nagamine, K., and Wiemer, S., 2001, Anomalously high *b*-values in the south flank of Kilauea Volcano, Hawaii: evidence for the distribution of magma below Kilauea's east rift zone: Journal of Volcanology and Geothermal Research, v. 106, p. 23–37.

Zablocki, C.J., 1976, Mapping thermal anomalies on an active volcano by the self-potential method, Kilauea, Hawaii: Geothermal Resource Abstracts, no. 2, p. 1299–1309.

Zablocki, C., and Koyanagi, R.Y., 1979, An anomalous structure in the lower east rift zone of Kilauea Volcano, Hawaii, inferred from geophysical data [abs.], *in* Decker, R.W., Drake, C., Eaton, G., and Helsley, C. eds., Hawaii Symposium on Intraplate Volcanism and Submarine Volcanism, Hilo, HI, July 16–22, Abstract volume: Hawaii National Park, HI, U.S. Geological Survey, Hawaiian Volcano Observatory, p. 177.

Zuniga, F.R., Wyss, M., and Scherbaum, F., 1988, A moment-magnitude relation for Hawaii: Bulletin of the Seismological Society of America, v. 78, no. 1, p. 370-373.

Appendix A

Measurement and Application of Ground Tilt and Classification of Earthquakes and Earthquake Swarms at Kīlauea

Discusses the establishment of the long-term viability of daily tilt measured in the Whitney Vault, provides definitions of the seismic regions of Kīlauea Volcano and of earthquake swarms, and presents our methodology for calculating magma supply rates.

Webcams on the edge of the Kīlauea Caldera, monitored by HVO.

Tilting of the Ground

Tilt in volcanology is the change in slope of the ground measured between two instants in time. It is important to specify the beginning and ending times of measurement to evaluate a tilt change, and in a time series tilt is measured relative to the initial value of the slope. Tilt is an angle measured in microradians[1] or seconds of arc. Tilting of the ground at Kīlauea's summit is one method by which one can infer the amount of inflation or deflation of the volcano and locate the depth of the magma reservoir from which eruptions are fed. Tilt is the only frequent measure of ground deformation that extends from the present back in time to the founding of the Hawaiian Volcano Observatory (HVO) in 1912.

Tilt Measurement

Tilt is a vector measured as two perpendicular components, north-south (u) and east-west (v), measured from an arbitrary starting point (u_0, v_0). Tilt values are expressed as a vector with a magnitude and an azimuth. Tilt magnitudes extending back in time to an arbitrary starting point are calculated as $[(u-u_0)^2+(v-v_0)^2]^{1/2}$. Tilt magnitudes depend on the choice of origin (u_0, v_0), but changes in tilt magnitude are relatively free of origin choice if (1) the origin is far from the current tilt value, (2) the tilt change is much smaller than the total size of the tilt magnitude, and (3) the azimuth does not change much during the measurement interval. Tilt magnitudes for daily measurements made in the two vaults at Kīlauea are shown on the time-series plots in this report as calculated from arbitrary values assigned at the time of installation of the tiltmeter. Such plots show the cycles of inflation and deflation of Kīlauea's summit over the lifetime of the tiltmeter. The tilt azimuth formulas are constructed such that pairs of north-south and east-west vectors, respectively, [1, 1], [–1, 1], [–1, –1], [1, –1] yield azimuths of 45, 135, 225, and 315 degrees. Tilt vectors are conventionally plotted as arrows pointing in the downward (deflation) direction at the calculated azimuth.

[1] 1 microradian (μr) = 0.2063 sec; 1 sec = 4.8468 μr

Nature of the HVO Tilt Networks

The original site for tilt measurement was the Whitney Vault, located adjacent to the present-day Volcano House Hotel on the northeast rim of Kīlauea Caldera. Tilt was measured daily using a Bosch-Omori seismometer, and components of the tilt vector were recorded in the Weekly and Monthly Bulletins of the Hawaiian Volcano Observatory (HVO) (Bevens and others, 1988) and later in The Volcano Letter (Fiske and others, 1987) to give a continuous time history (see figs. in chaps. 2–4). A vault built in 1948 a few hundred meters back of Uwēkahuna bluff near the Volcano Observatory was refurbished in 1957. Jerry Eaton installed a water-tube tiltmeter of his design (Eaton, 1959) in both the Uwēkahuna and Whitney Vaults (fig. A2 and appendix I), and these instruments were also read daily in parallel from 1957 until the abandonment of the Whitney Vault at the end of 1962 (fig. A1). The three datasets were never published, but they were obtained from the HVO data archives. Plotted together with simple linear scaling, the three datasets show remarkable agreement (fig. A1). This gives us confidence to use the long-term (1912–62) Whitney seismic tilt as a reliable indicator of ground movements at Kīlauea's summit. Using both tiltmeters, we choose the 1960–61 and 1966–67 time periods for our example calculations (appendix I). The 1960–61 cycle adds credence to the use of Whitney tilt back in time as a surrogate for volume of summit magma accumulation. During the period between 1961 and 1967, tilt and level surveys were measured frequently and several large inflation-deflation cycles were captured.

Eaton also installed a field network of arrays of concrete piers, arranged in triangles approximately 50 meters on a side (Eaton, 1959), on which water-tube tiltmeters could be placed and read. This network was occupied periodically, at intervals of several months and always after a major event such as an eruption, intrusion, or very large earthquake. Station locations are also shown in appendix I. The water-tube tilt network had two limitations common to any periodic surveys to measure vertical and horizontal movements of the ground; these are that (1) one could always measure the state of the ground following an event, but one had to be very lucky to occupy a network just preceding such an event, and (2) the surveys took several days to complete, during which the ground was moving, and not necessarily in the same way at each station.

Source of Tilt Changes

At the time of the founding of the Hawaiian Volcano Observatory (HVO) there were only sketchy ideas about the source of magma for Kīlauea's eruptions. The Japanese seismologist Kiyoo Mogi modeled leveling and triangulation surveys made both before and after the large collapse at Kīlauea's summit in 1924 (Mogi, 1958). From this study and his study of Japanese volcanoes he concluded that ground deformation measured at various volcanoes could be modeled as responding to a spherical or point source of magma located several kilometers

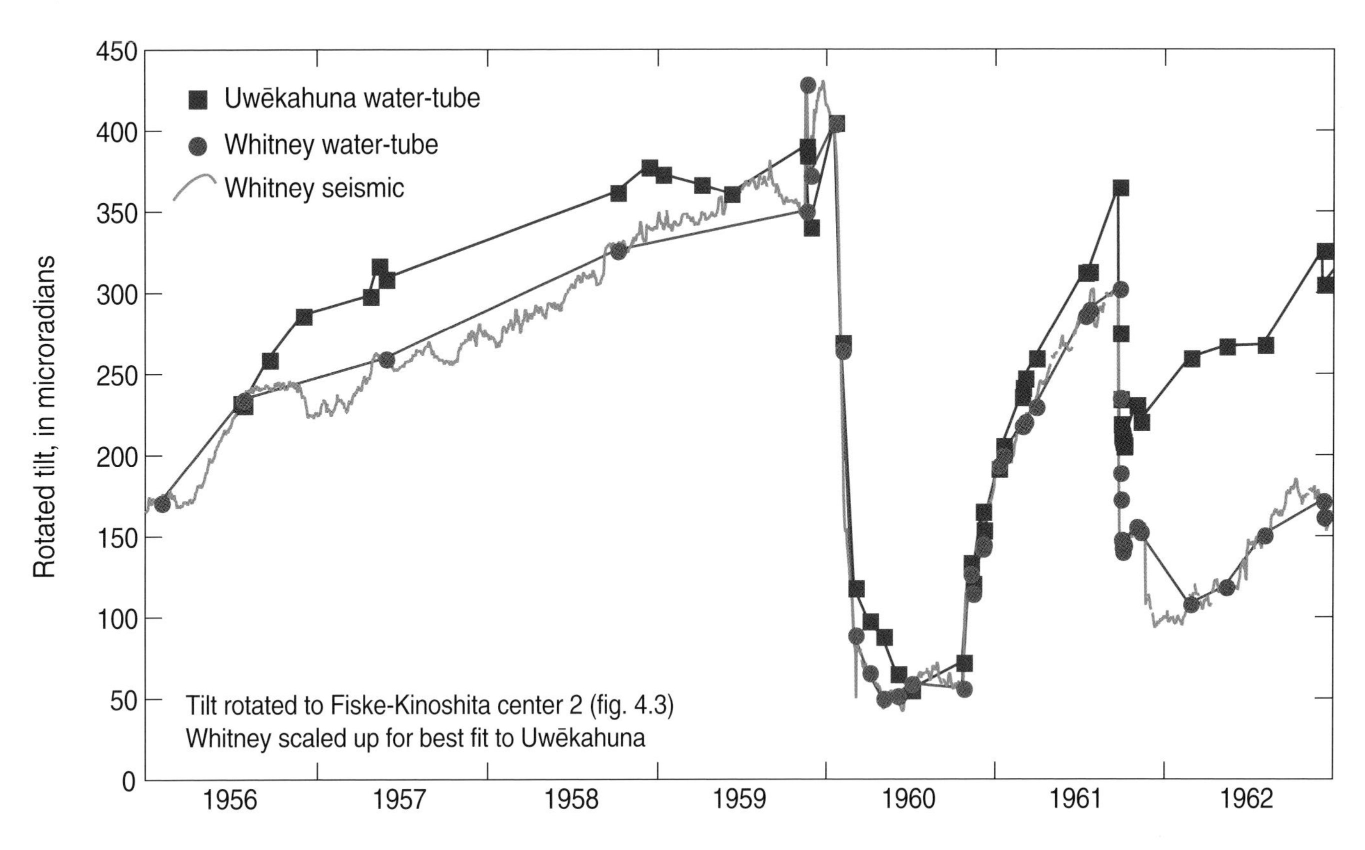

Figure A1. Graph comparing three tilt records at Kīlauea for the overlap time after the installation of the Uwēkahuna Vault in 1956 and before the abandonment of the Whitney Vault in 1963. Comparison is of measurements made by short-base water tube tiltmeters installed in Uwēkahuna (blue points) and Whitney (red points) Vaults with the tilt measured by the Bosch-Omori seismometer installed in The Whitney Vault (green line). The time of overlap is between mid-1956 and the end of 1962. Tilt data are rotated to Fiske-Kinoshita center 2 (fig. A2), and the Whitney water-tube data are scaled to fit the Uwēkahuna water-tube data. The excellent agreement indicates that the tiltmeters track each other and provide a firm basis for interpreting the Whitney seismic tilt, the only available tilt measurement between 1912 and 1956. The time period covered includes the major eruptions and deflations in January 1960 and September 1961. The approximate coincidence of the three tilt functions demonstrates they all measure the same inflation/deflation bulge. The divergence of Uwēkahuna and Whitney tilt after August 1961 means the deflation and inflation location shifted relative to the inflation in the 1950s and the 1960 collapse and recovery. The tilt scale on the figure refers to the Uwēkahuna water tube tilt. The Uwēkahuna tilt vector τ is measured frequently, and we seek tilt rotated to the azimuth –67 degrees pointing from the Fiske-Kinoshita (1969) deflation center 2. The plotted Uwēkahuna water tube tilt function is $T_U = [\cos(-67°)^*\tau_N + \sin(-67°)^*\tau_E]+500$. The offset of 500 microradians is the arbitrary baseline chosen when measurements started and is not determined by any data fitting. The Whitney seismic tilt vector τ was measured daily as the offset of the pendulum of the north and east components of the Bosch-Omori seismometer, and we seek the component rotated to the azimuth 45 degrees pointing from the deflation center 2 in figure A2. The plotted Whitney seismic tilt function is $T_{WS} = 8.2\ [\cos(45°)^*\tau_N + \sin(45°)^*\tau_E]-150$. The linear scaling factor (8.2) and the offset (–150) are determined empirically to fit the tilt functions from the two vaults. The Whitney water tube tilt vector τ was measured frequently, and we plot the tilt rotated to the azimuth 45 degrees pointing from the inflation center 2 in figure A2. The plotted Whitney water tube tilt function is $T_{WW} = 8.2\ [\cos(45°)^*\tau_N + \sin(45°)^*\tau_E]—165$. The scaling factor (8.2) was determined by the seismometer tilt fit, and the offset (–165) adjusts for the arbitrary baseline.

beneath the volcano's summit. Jerry Eaton expanded on Mogi's work to clearly recognize the existence of an approximately spherical magma reservoir located about 4 km beneath Kīlauea's summit (Eaton, 1962). Addition of magma to this reservoir caused the ground above to expand upward and outward (inflation), and the reverse happened (deflation) when magma left the reservoir to feed eruptions and intrusions. Thus the tiltmeters located around Kīlauea's summit faithfully recorded cycles of inflation and deflation.

As a context for interpreting the time-history of tilt changes at Kīlauea's summit we rely on the careful leveling surveys made in the period preceding the 1967–68 eruption in Halema'uma'u (Fiske and Kinoshita, 1969, fig. 5). They documented an apparent migration of centers of inflation from north to south and then to the west, with some movement back and forth. The locations of these centers, and the two vaults where tilt was measured daily, are shown in figure A2 . Mogi and Eaton's idea of a single point source was modified to identify multiple centers of inflation and deflation within a broader, and only approximately spherical, source region covering a depth range of 2–6 km beneath Kīlauea's summit. Most modeled centers occur between 2 and 4 km depth. The consequences of this source complexity is that tilt vectors from the summit network almost never intersect at a point because each station responds more to the nearest source(s) and less to the more distant sources. One additional complexity is the existence of deformation at several places along the east rift zone, documented in several published studies (see, for example, Wright and Klein, 2006, fig. 2 and references cited therein). The Whitney Vault appears to respond to deformation within the east rift zone as well as to summit inflation/deflation cycles, because the azimuths of Whitney tilt often point to the east of the Fiske-Kinoshita centers.

In this report we are mainly concerned with tilt changes during the short periods of time associated with intrusions, eruptions, and earthquake swarms. These values are independent of any assumptions regarding starting values and, subject to the assumption of similar sources, allow accurate comparison of tilt changes for events occurring at different times in Kīlauea's history. Complete formulations of the tilt equations, including the conversion from tilt magnitude to magma volume, are given in appendix I. During the days associated with a volcanic event, tilt vectors may change direction during deflation and reinflation to indicate draining and filling of different parts of the reservoir complex as represented by the Fiske-Kinoshita centers shown in figure A2. Many events show a hysteresis involved in deflation and reinflation, in which deflation vectors may migrate clockwise (from the Uwēkahuna Vault), indicating draining of the northernmost parts of the magmatic system first, and reinflation also occurs first to the north. However, extended periods of deflation or inflation may show vectors that point outside of the range of Fiske-Kinoshita centers. For example, in the early stages of deflation the north-south and east-west tilt components change together. Later the north-south or east-west component may continue in the same direction while the other component remains unchanging or may begin to reverse direction. By simple vector addition one can interpret this behavior as the beginning of resupply to one source while another source continues to drain.

Whitney Tilt Volume Calculations

In an effort to determine magma center volume changes from a single tiltmeter, we examine the simpler period of 20 January to 1 April 1960, during which Kīlauea's summit underwent a massive deflation associated with the east rift eruption of 13 January to 19 February 1960. Unfortunately there is no level survey for this interval, but the deflation is included in a longer interval from January 1958 to May 1960 for which a level survey exists (appendix I, table I1, event 1). We can use this 1960 deflation as an example of estimating source depth from two tiltmeters and volume change from a single tiltmeter. The tilts are 275 and 199 microradians, and the source distances are 3.2 and 3.8 km for the short-base water tubes at Uwēkahuna and Whitney, respectively (fig. A3). Equation 6 in appendix I thus yields a source depth of 3.0 km. Comparison of measured tilts and a Mogi source at 3.0 km depth (appendix I) confirm the 3.0-km depth and a volume of 0.107 km^3 as a good fit to the tilt data, excluding Ke'āmoku as an amplified site. The tradeoff between volume and source depth of the three curves of appendix I, figure I4, means that for variation of +/-0.5 km from its average source depth of 3.0 km, the magma volumes vary by +/-20 percent (table A1). Thus either tiltmeter, when used alone, gives usable volume estimates. The factor of 0.00045 used by Dvorak and Dzurisin (1993) corresponds to a depth between 3 and 3.4 km at the centers in the south caldera most frequently seen in tilt vectors for inflation and deflation (table A1). From table A1 the ratio of Whitney to Uwēkahuna tilt for a depth of 3 km is 1.38. We apply this ratio to the Dvorak factor for Uwēkahuna of 0.00045 km^3/microradian to yield 0.000621 km^3/microradian for the conversion of Whitney tilt magnitude to volume. We use this factor for the volume calculations during periods before 1960 when the only tilt magnitudes were calculated in seconds of arc at the Whitney Vault.

The theoretical volume factor for the Whitney tiltmeter and the 1960 deflation source is D_{WT}=0.00049 km^3/microradian, which is somewhat larger than the Uwēkahuna factor because Whitney is farther from the source. The empirical factor for Whitney is D_{WE}=0.00054

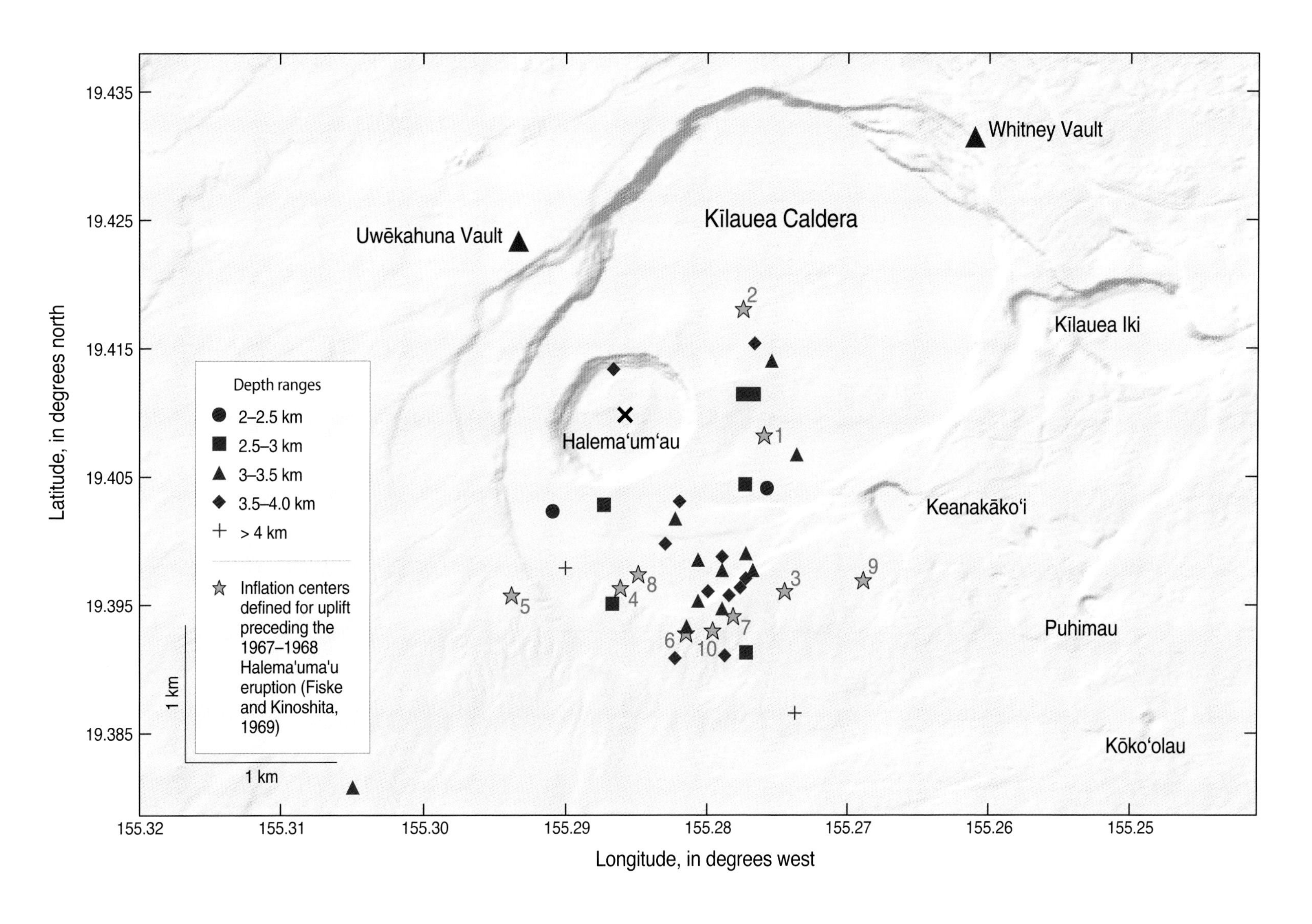

Figure A2. Map showing locations of inflation-deflation centers at Kīlauea. Tiltmeter locations are shown as red triangles. The green stars are centers of inflation from leveling surveys conducted during the inflation preceding the 1967–68 Halema'uma'u eruption (Fiske and Kinoshita, 1969, figure 5). The blue symbols (keyed to depth) are inflation and deflation centers determined in this study and listed in appendix I table I1.

Table A1. Volume factors (km³/μrad) for continuous tiltmeters for various summit Mogi sources at Kīlauea[1].

Caldera centers	South centers[2]	South centers 6, 7, 10[3]	Central center 1	SE center 9	NE center 2
		A. Uwēkahuna			
Distance (km)	3.2	3.5	2.7	4	1.77
Depth (km)					
2.5	0.000317	0.000386	0.000229	0.000536	0.000140
3	0.000388	0.000455	0.000303	0.000597	0.000222
3.5	0.000491	0.000556	0.000410	0.000695	0.000344
4	0.000632	0.000695	0.000556	0.000831	0.000519
		B. Whitney			
Distance (km)	3.8	4.2	2.9	3.5	2.16
Depth (km)					
2.5	0.000471	0.000611	0.000261	0.000387	0.000168
3	0.000536	0.000668	0.000334	0.000456	0.000245
3.5	0.000636	0.000764	0.000439	0.000557	0.000358
4	0.000773	0.000898	0.000583	0.000696	0.000516

[1]Values include a +10-percent empirical correction. See text.

[2]Calculated for the average location of unnumbered centers of inflation and deflation plotted in figure A2.

[3]Numbered centers shown in figures A2 and A3

Table A2. Whitney tilt rainfall correlation and magnitude error.

[Dates in m/dd/yyyy format]

Dates	Az.	M (μrad)	std[1]	Δt[2]	in/day[3]	std[4]	Comment
9/13/1925–2/13/1926	207	0.65	5.41	153	0.14	0.27	Flat tilt
7/9/1932–9/6/1932	259	2.97	3.59	59	0.09	0.11	do[5]
5/8/1938–9/2/1938	255	3.32	3.77	117	0.17	0.18	do
11/17/1929–1/16/1930	192	14.18		60	0.41	1.01	Sharp deflation
2/2/1932–3/25/1932	184	42.86		52	0.39	0.53	do
12/25/1936–2/21/1937	188	66.97		58	0.86	1.30	do
8/19/1920–12/7/1920	349	1.48	3.13	111	0.21	0.34	No volcanic event
8/25/1921–10/20/1921	84	2.92	2.58	56	0.24	0.38	No volcanic event

[1]Standard deviation of magnitude for the time intervals of little tilt change. Not calculated for periods of large tilt change

[2]Time interval in days.

[3]The standard deviation is calculated for the days within the period for which rainfall was recorded.

[4]Standard deviation in inches of rain per day.

[5]Do, ditto (same as above).

km³/microradian. The theoretical volume factors for Whitney also require a 10-percent correction. Using equation 5 and typical deformation centers (appendix I; table A1), we can calculate a table of tiltmeter volume factors for various Mogi locations and depths and apply the 10-percent empirical correction (see table A1). The Dvorak factor for Uwēkahuna that we use (0.00045 km³/microradian) yields a depth of ~3.8 km for the central caldera source in table A1.

Uncertainty of Whitney Tilt Magnitudes

Whitney tilt data were considered in the early literature to be affected by seasonal variations to explain the apparent tilt cycles in the tilt record between 1925 and 1950. It is easy to dismiss the lack of precise correlation of the Whitney tilt with events at Halema‘uma‘u as due to errors in the use of a recently installed instrument, and one primarily designed to measure earthquakes. Indeed, the instrument used as a tiltmeter was said to be affected by factors related to both diurnal and seasonal changes (Jaggar and Finch, 1928; Powers, 1946; Powers, 1947). In fact, no seasonal changes could be verified (Powers, 1947). We tested the effects of heavy rain and calculated the standard deviation of tilt changes in periods of little expected volcanic tilt change (table A2). For three large apparent deflations in the 1925–39 period uncorrelated with volcanic events we show a strong correlation with high average rainfall (table A2). Periods of little tilt change are correlated with low average rainfall (table A2). We suspect that much of the cyclic change in the 1925–50 period is due to high rainfall and to recovery from it.

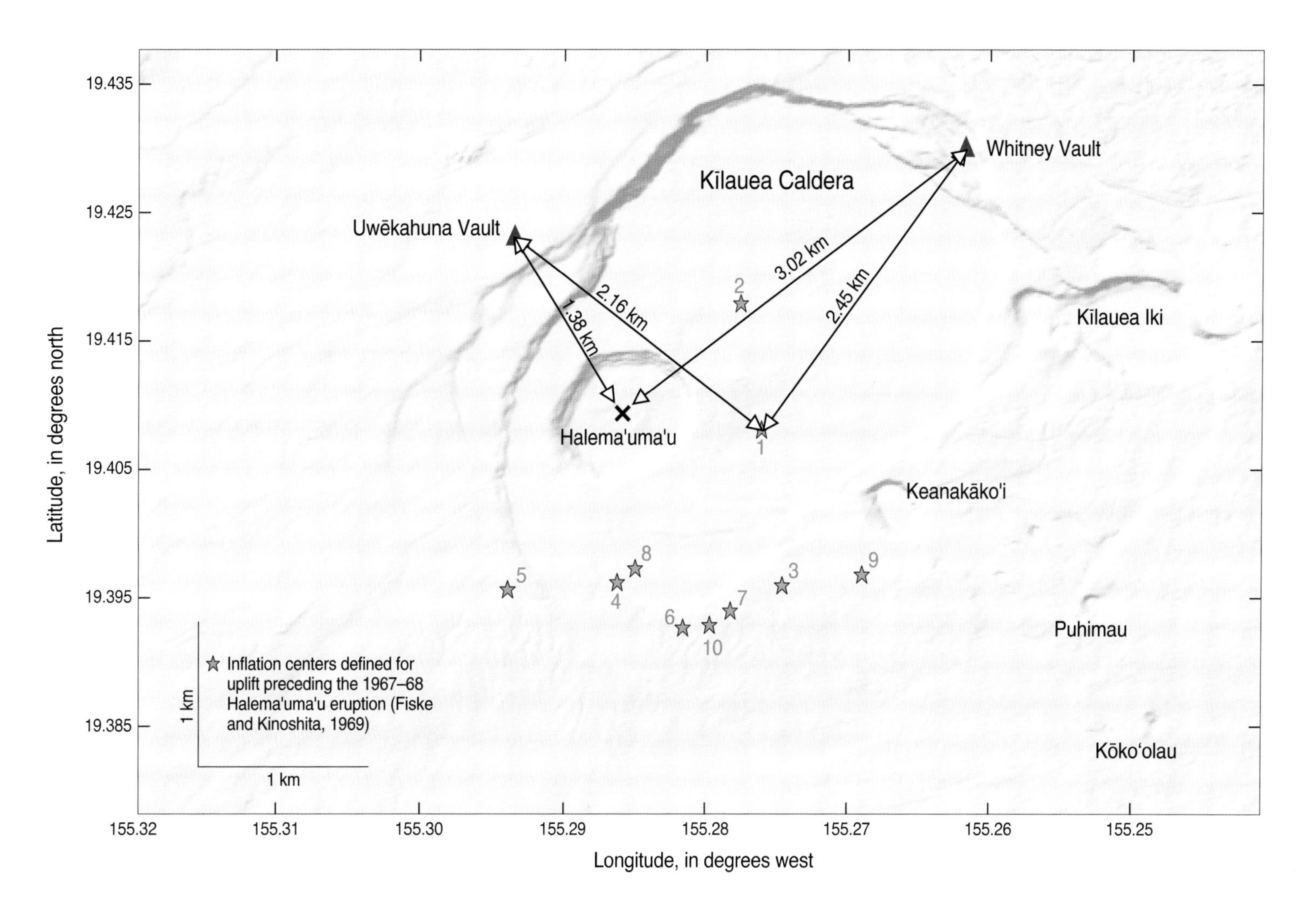

Figure A3. Map comparing geographic relations of the Whitney and Uwēkahuna Vaults. The Whitney Vault is at a much greater horizontal distance from Halema'uma'u, so that removal of magma from a shallow (<1 km) source, including the rise and fall of lava in Halema'uma'u lava lake is not measured at the Whitney Vault, whereas such events are easily detected at Uwēkahuna. The slant distances to the shallow reservoir at 2-6 km depth are more nearly comparable, such that both vaults record magma addition and removal from that reservoir, including addition of magma below inflation centers identified during the buildup to the 1967–68 Halema'uma'u eruption. These centers are also the locus of inflation/deflation for both earlier and later intrusions and eruptions. The Whitney Vault is close to Kīlauea Iki Crater and closer to the upper east rift zone compared to the Uwēkahuna Vault and therefore can record changes associated with shallow magma storage at both locations that could not be noticed at Uwēkahuna.

Table A3. Presentation and interpretation of seismicity.

A. Earthquake Classification by Tectonic Region.

Code	Depth (km)[1]	Location (see text figure 1.3)
ms1	0–5	Magma supply path beneath Kīlauea's summit, including some earthquakes north of Kīlauea Caldera
ms2	5–10	Magma supply path beneath Kīlauea's summit, including some earthquakes north of Kīlauea Caldera
ms3	10–20	Magma supply path beneath Kīlauea's summit, including some earthquakes north of Kīlauea Caldera
ms4	20–35	Magma supply path from the Hawaiian hotspot to beneath Kīlauea's summit
ms5	>35	Magma supply path from the Hawaiian hotspot to beneath Kīlauea's summit
ms5gln	>35	Unique earthquake swarms north of Kīlauea Caldera and east rift zone
ei1ler	0–15	Lower east rift zone—continuation of east rift zone east from Heiheiahulu; continues offshore for 50 km
ei2mer	0–15	Middle east rift zone—east-trending east rift zone defined by surface fractures, shields and cinder cones and pit craters; active vents extend from Mauna Ulu to Heiheiahulu
ei3uer	0–15	Upper east rift zone—southeast-trending rift zone defined by surface fractures, cones and pit craters. Turns east at intersection with Koaʻe fault system; active vents extend from Puhimau to Mauna Ulu
ei4sswr	0–15	Seismic southwest rift zone—locus of earthquake swarms trending south from Halemaʻumaʻu, then southwest to join southwest rift zone in the vicinity of the Kamakaiʻa Hills; locus of intrusions
ei5swr	0–15	Southwest rift zone—defined by surface fractures, cones and pit craters; locus of eruptions and intrusions
koae	0–15	Koaʻe Fault Zone—the region between the southwest and east rift zones; bounded on the north by the northernmost set of active east-west fractures; bounded on the south by the Kalanaokuaiki north-facing fault scarp
sf1ler	0–15	Far eastern south flank—south of lower east rift zone
sf2mer	0–15	Eastern south flank—south of middle east rift zone; site of large-magnitude south flank earthquakes
sf3kuer	0–15	Central south flank—south of Koaʻe Fault Zone and west of upper east rift zone
sf4swr	0–15	Western south flank—east of southwest rift zone and southwest of Koaʻe Fault Zone

[1]Depths are coded on time-series plots. Earthquakes at depths less than 20 km are mostly within the Kīlauea edifice above the decollement. Most events are shallower than 12 km. The few events whose depths are between 15 and 20 km may be mislocated. Earthquakes deeper than 20 km beneath the middle and lower east rift zones and eastern and far eastern south flank fall off the magma supply path and are related to other mantle processes, including response to Pacific Plate motion. These infrequent events are plotted but not interpreted in this report.

B. Definition of Earthquake Swarms

Earthquake swarms
1. Periods of eruption and (or) shallow intrusion in regions ms1, ei1-5 and koae. Depth range 0–5 km.
Earthquake swarms comprise a contiguous period during which events in a region occur at a frequency of greater than 1 event every 6 hours.
Earthquake swarms commonly consist of at least 10 events.
2. Elevated subcaldera activity in regions ms2 and ms3 or rift activity in regions ei2-5. Depth range 5–20 km.
Earthquake swarms comprise a contiguous period during which events in a region occur at a frequency of greater than 1 event every 6 hours.
Earthquake swarms commonly consist of at least 5 events.
3. Elevated deep activity along magma supply path in regions ms4-5, koae, ei3-5, and sf 2-4. Depth range >20 km.
Earthquake swarms comprise a contiguous period during which events in a region occur at a frequency of greater than 1 event every 6 hours.
Earthquake swarms commonly consist of at least 5 events.
4. Elevated south flank activity. Depth range 0–15 km.
Earthquake swarms comprise a contiguous period during which events in a region occur at a frequency of greater than 1 event every 6 hours.
Earthquake swarms commonly consist of at least 5 events.

We also calculated the standard deviation of calculated magnitude for periods of low tilt change (table A2, column 4) and found two standard deviations to be less than 10 microradians, in agreement with an uncertainty estimate made during transcription of written records to a computer file (A. Okamura, oral commun., 2011).

Earthquake Swarms

Earthquakes are assigned to regions of the volcano, which are named and described in table A3*A* and shown in map view on figure A4. Earthquake swarms precede nearly all eruptions and are critical to the identification of intrusions. Earthquake swarms in all regions are defined as a sequence of at least 5–10 earthquakes within a single region (fig. A4) with a time difference from one to the next of less than 6 hours. Additional description of plotting parameters is given in table A3*B*.

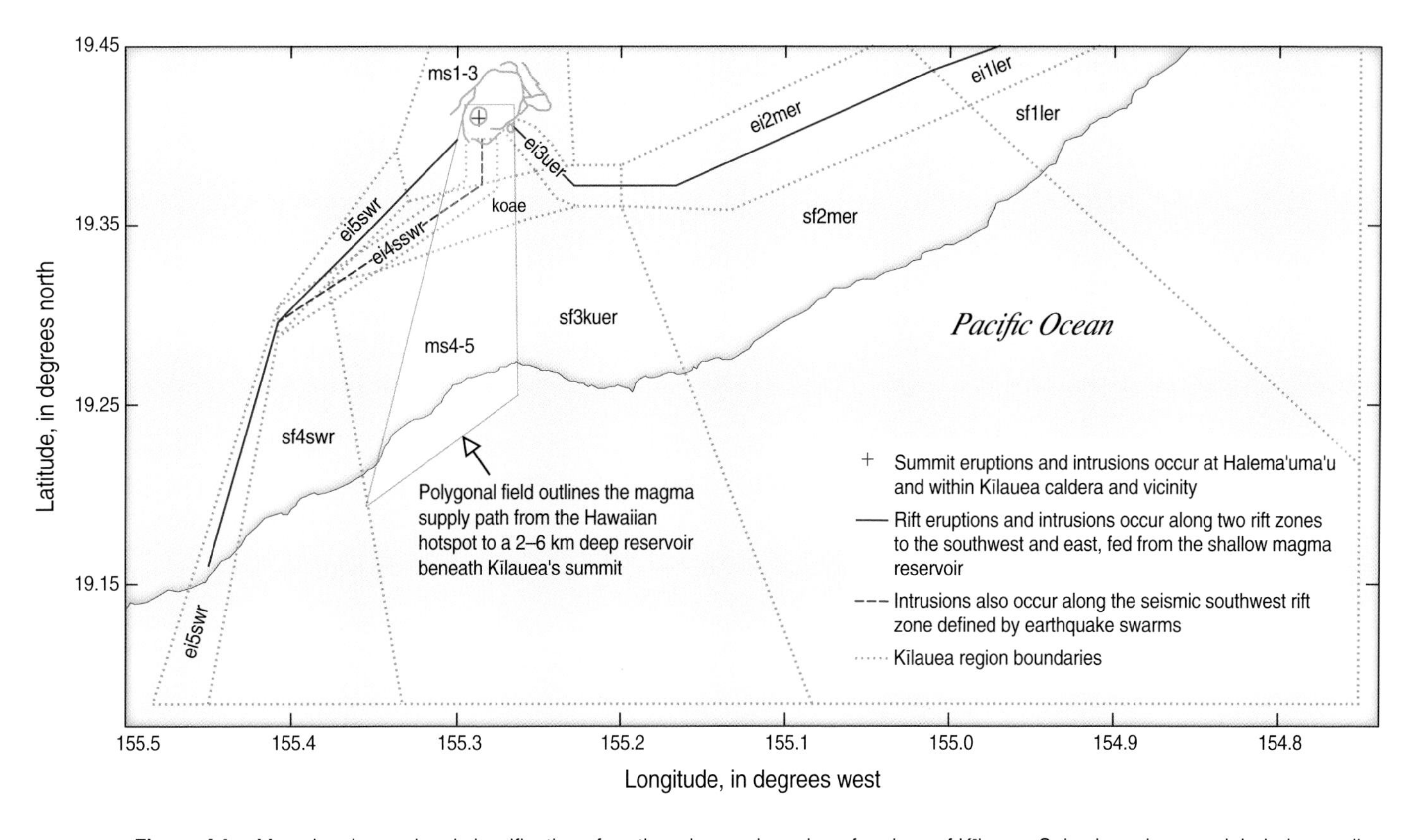

Figure A4. Map showing regional classification of earthquakes and naming of regions of Kilauea. Seismic regions are labeled according to tectonic regions described in table A3. The region abbreviations are used in numerous figures in this report: "ms" stands for m(agma) s(upply), and numbers from 1 to 5 correspond to different depth ranges along the magma supply path from the mantle defined by Wright and Klein (2006). Magma supply regions 1–3 lie below Kīlauea's summit at depths shallower than 15 km. Magma supply regions 4 (20–35 km) and 5 (>35 km) extend from north of Kīlauea's summit south and southwest beneath the Koa'e Fault Zone and both onshore and offshore parts of Kīlauea's central and western south flank The magma supply path boundary is shown in the background of all plots that show earthquake locations. "ei" stands for e(ruption) i(ntrusion), and regions 1–5 correspond to earthquakes beneath the rift zones extending in a counterclockwise direction from the easternmost "lower east rift" through the westernmost "southwest rift." "koae" separately designates the Koa'e Fault Zone, which connects the two rift zones. "sf" stands for s(outh)f(lank), and south flank regions 1–4 extend counterclockwise adjacent to the rift segments and the Koa'e Fault Zone. Rift, south flank, and Koa'e earthquakes occur at depths shallower than 15 km, and, allowing for some depth error, their hypocenters lie above the decollement at 10–12 km.

Calculation of Magma Supply and Eruption Efficiency

A principal focus of this study is to calculate both spreading rates and magma supply rates over time. Magma supply and spreading rates in the modern era have been summarized elsewhere (Wright and Klein, 2008) and are reevaluated in this report. We rely on rates at which magma is added to the surface to represent a minimum supply rate. Volumes of vesiculated lava are reduced by 20 percent to yield an equivalent magma volume before calculating the rate as km^3/yr.

19th Century: Calculation of Magma Supply from Caldera Filling Rates

For events in the 19th century we estimate magma supply rate from the volumes of filling and draining of Kīlauea Caldera, to which are added a hypothesized estimate of deep endogenous growth related to rift dilation during intrusion. Magma filling is a gradual process, for which rate calculation is possible, but drainings are generally sudden events, and only a volume estimate can be made. Just like the tilt record and inferred inflation level measured in later years of observation by HVO, the level of a lava lake is a proxy for the inflation state and can be visualized as a gradual up and sudden down sawtooth function. We calculate draining volumes (chapter 2, table 2.2) as combinations of cylinder, cone frustum, or half an oblate spheroid, depending on the description given in the literature cited. Filling (chapter 2, table 2.5) is calculated using the same formulations, but with lava volumes converted to magma equivalents as above. Filling rates are calculated for intervals bounded by more definitive data on lava added to the caldera and (or) prior to major drainings. The volume calculated includes the amount of magma needed to fill any draining volumes before the end of the period added to the volume of new magma equivalents added to the caldera floor.

1894–1950: Calculation of Magma Supply from Filling of Halema'uma'u Crater and Tilt

In this period we apply the same methodology as above for filling and draining of Halema'uma'u and factor in volume changes from deflations accompanying the 1922 east rift eruption and the 1924 draining of Halema'uma'u.

Post-1950: Calculation of Magma Supply from Tilt

From the beginning of reinflation of Kīlauea in 1950, it is possible to consider volumes of inflation and deflation to augment eruption rates in calculating magma supply. Spreading rates are crudely estimated for times preceding the installation of a continuously recording geodetic network. We use summit tilt magnitude as a surrogate for magma volume moving in or out of the summit reservoir. Summit inflation is equated to magma addition to the summit reservoir. Summit deflation is equated to magma transfer to the rift zone(s). We assume that, because of buoyancy, magma never retreats down the vertical supply path, which could be an alternate interpretation for summit deflation. We calculate volumes using the same tilt-to-volume factor as previously published (Dvorak and Dzurisin, 1993). At times when the Whitney Vault houses the only tiltmeter we increase the volume factor by about 1.6, based on comparison of Uwēkahuna and Whitney tilt during the period of overlap (fig. A1). Our calculations ignore the possibility of magma compression during storage or decompression during transfer to the rift zone (Johnson, 1992; Johnson and others, 2000; Rivalta and Segall, 2008)[2]. Our estimates of volume thus may be slightly underestimated for summit inflation and slightly overestimated for summit deflation. We treat tilt data differently in the magma supply calculations for periods of (1) noneruption, (2) summit and continuous rift eruptions, and (3) short rift eruptions accompanied by summit deflation. During noneruptive times an increase of tilt magnitude when the azimuth indicates inflation is interpreted as addition of magma to Kīlauea's summit reservoir and is treated as a positive magma volume quantity. A short period of deflation reflects transfer of magma to the rift zones and is also treated as a positive quantity, but magma counting is based on the deflation volume and the succeeding period of recovery is not independently counted. During eruption at Kīlauea's summit, an inflation azimuth reflects addition of magma to Kīlauea's shallow reservoir, whereas a deflation azimuth reflects transfer of magma to the rift zones. The tilt magnitudes associated with both are counted as positive quantities in the magma supply calculations, and the rate of magma supplied from depth over the period of eruption always exceeds the rate of eruption. At the beginning of summit eruptions the initial eruption rate may temporarily exceed the supply rate, resulting in a small deflation reflecting a temporary drawdown of the summit reservoir. Rapid inflation of the summit during summit eruptions

[2]The effects of compression occur only at the end stages of long inflations, thus have little effect on the overall tilt change. Decompression occurs virtually immediately at the onset of deflation preceding and accompanying magma transfer to the rift zone. Thus the overall tilt change associated with deflation is likewise barely affected by the decompression.

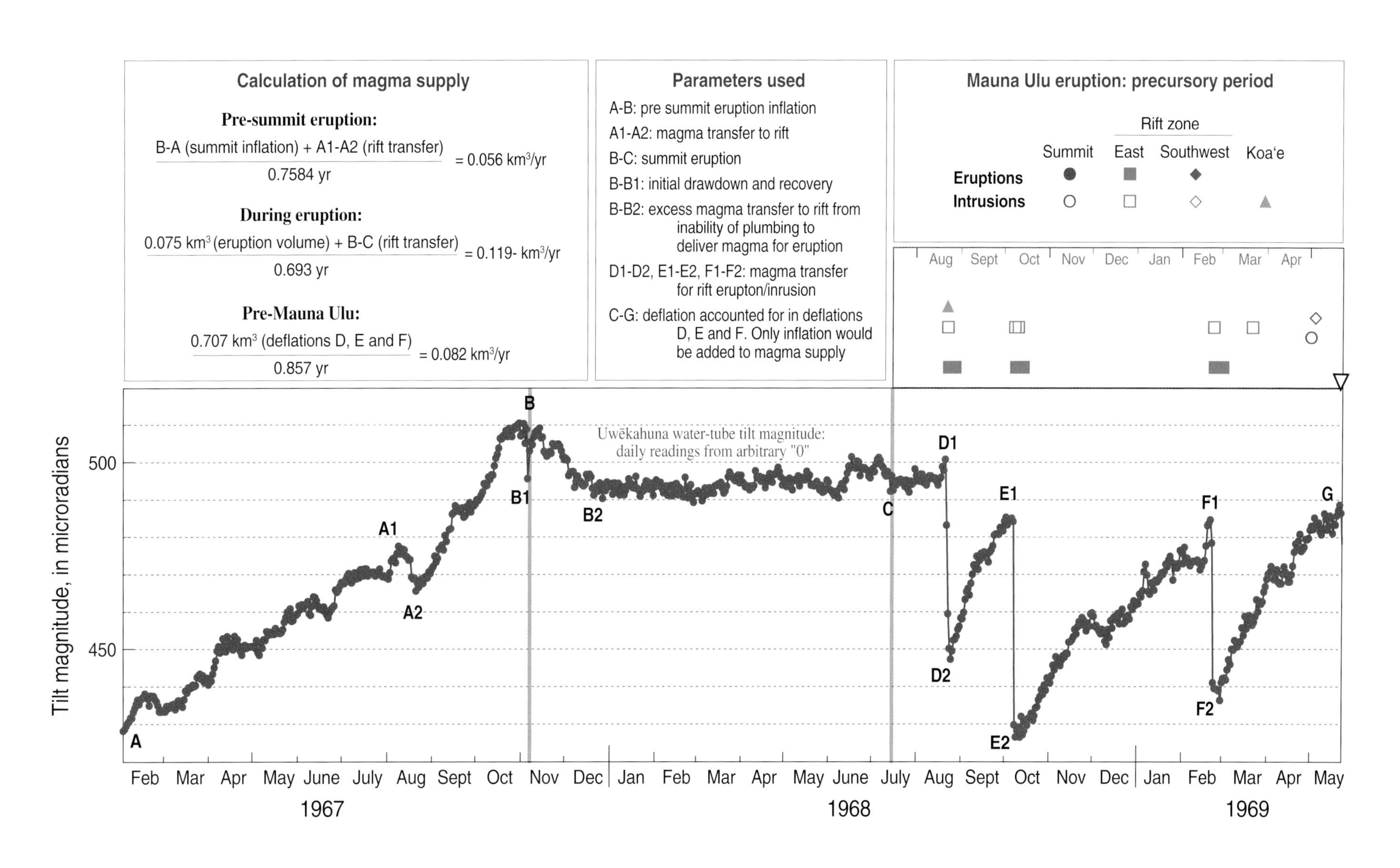

Figure A5. Calculation of magma supply—an example. Graphical depiction of the assumptions that go into our magma-supply calculations. See text for further explanation.

indicates shallow intrusion between the summit reservoir and the ground surface. For both cases the only quantity included in the calculations of magma supply rate is the net change over the entire eruptive period. To summarize, we calculate magma supply as follows (fig. A5):

Magma supply rate = $\Sigma V_1 + V_2 + V_3$ (km^3)/ Δ time (years).

V_1 = Lava volume of continuous eruption corrected to 80 percent of the published value to account for assumed 20 percent vesicles.

V_2 = Volume of summit inflation over the time period.

V_3 = Sum of all volumes of summit deflation over the time interval, used as a proxy for total volume of magma transfer to the rift zone(s).

Conversion of tilt to volume is calculated using the factor of 0.00045 km^3/microradian (Dvorak and Dzurisin, 1993).

Δ time (years) = (ending date–beginning date)/365.25.

Eruption Efficiency

Rift eruptions are accompanied by a rapid deflation. In most cases the volume calculated from the deflationary tilt magnitude exceeds the volume of rift eruption and is used in the magma supply calculations as the amount of magma transferred to the rift zone. The ratio of erupted volume to tilt volume is recorded as "eruption efficiency," varying between 0 (no eruption, all magma left underground as intrusion) and 1 (eruption volume equals tilt volume with negligible magma left as intrusion). This is an alternative way of looking at partitioning of magma between summit and rift to that published previously (Dvorak and Dzurisin, 1993, figure 10), and the results help validate the tilt-to-volume factor for intruded magma. For cases where the eruption efficiency exceeds 1, the tilt volumes are used in the magma supply calculations[3].

[3] We assume that the uncertainties of lava volume estimates are larger than the uncertainties associated with assuming that the deflation source lies at a constant distance and depth relative to the summit tiltmeter.

Summary

Looking at the entire Kīlauea history we face the same dilemma that John Dvorak faced in arriving at a tilt-to-volume factor, namely how to account for a range of latitude, longitude, and depth for modeled centers of inflation and deflation (Dvorak and Dzurisin, 1993). We find that changing the depth and (or) location within the limits of ±0.5 km shown by the modeling does not compromise any of the conclusions reached regarding the variation of eruption efficiency with time. Thus we come to the same conclusion reached by Dvorak, namely that assumption of a constant for converting tilt magnitude to volume is appropriate, and therefore we use the Dvorak value.

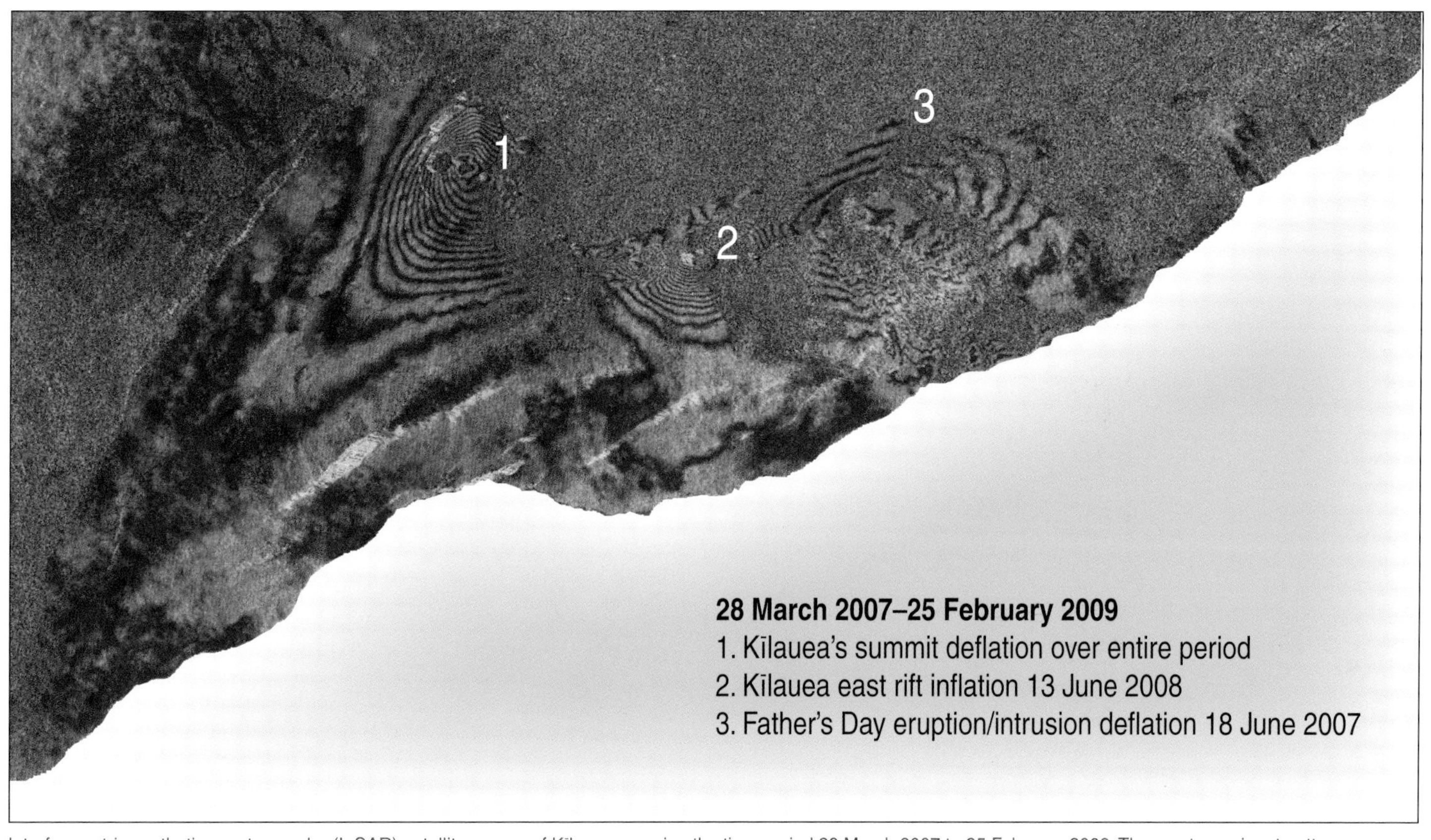

Interferometric synthetic aperture radar (InSAR) satellite survey of Kilauea covering the time period 28 March 2007 to 25 February 2009. The most prominent pattern seen is deformation associated with the 18 June 2007 Father's Day eruption. In the colored interference bands, red inside blue indicates deflation (increased distance from the satellite to the ground), red outside blue indicates inflation (decreased distance). Hawaiian Volcano Observatory image courtesy of Michael Poland.

Menlo Park Publishing Service Center, California
Manuscript approved for publication July 17, 2014
Edited by Peter H. Stauffer
Layout and design by Jeanne S. DiLeo

BACK COVER:
View of Kilauea's summit lava lake from the rim of Halema'uma'u Crater at dusk. The lava lake is in the informally named "Overlook" crater (from its location immediately below the former Halema'uma'u visitor overlook), set within the larger Halema'uma'u Crater. The lava surface is about 50 m (160 ft) below the rim of Overlook crater. At the right (southeast) margin of the lake, a persistent spattering source ejects spatter more than halfway up the crater wall. (USGS photograph by Matthew Patrick, taken 1 Feburary 2014.)